Y0-BYK-905

Aldehydes—
Photometric Analysis

THE ANALYSIS OF ORGANIC MATERIALS

An International Series of Monographs

edited by R. BELCHER and D. M. W. ANDERSON

1. S. VEIBEL: The Determination of Hydroxyl Groups. 1972.
2. M. R. F. ASHWORTH: The Determination of Sulphur-containing Groups. *Volume 1.* The analysis of sulphones, sulphoxides, sulphonyl halides, thiocyanates, isothiocyanates; and isocyanates. 1972. *Volume 2.* The determination of thiol groups. 1976.
3. F. KASLER: Quantitative Analysis by NMR Spectroscopy. 1973.
4. T. R. CROMPTON: Chemical Analysis of Organometallic Compounds *Volume 1.* Elements of groups I–III. 1973. *Volume 2.* Elements of groups IVA-B. 1974. *Volume 3.* Elements of group IVB. 1974. *Volume 4.* Elements of group V. 1975. *Volume 5,* Elements of groups VIA, VIB, VIIA, VIIB, Aluminium and Zinc. 1977.
6. K. MÜLLER: Functional Group Determination of Olefinic and Acetylenic Unsaturation. *Translated from the German by* M. R. F. ASHWORTH. 1975.
7. L. C. THOMAS: Identification of Functional Groups in Organophosphorus Compounds. 1975.
8. N. K. MATHUR and C. K. NARANG: Determination of Organic Compounds with N-Bromosuccinimide and Allied Reagents. 1975.
9. E. SAWICKI and C. R. SAWICKI: Aldehydes—Photometric Analysis. *Volumes 1, 2,* 1975; *Volumes 3, 4,* 1976; *Volume 5,* 1978.
10. T. S. MA and A. D. LADAS: Functional Group Analysis *via* Gas Chromatography. 1976.
11. J. BARTOS and M. PESEZ: Colorimetric and Fluorimetric Analysis of Steroids. 1976.
12. H. E. MALONE: The Analysis of Rocket Propellants. 1976.
13. R. BELCHER. (Ed.): Instrumental Organic Elemental Analysis. 1977.

Aldehydes— Photometric Analysis

Volume 5

Formaldehyde Precursors

EUGENE SAWICKI
and
CAROLE R. SAWICKI

Raleigh,
North Carolina,
U.S.A.

1978
Academic Press
London · New York · San Francisco

A Subsidiary of Harcourt Brace Jovanovich, Publishers

ACADEMIC PRESS INC. (LONDON) LTD.
24/28 Oval Road,
London NW1

United States Edition published by
ACADEMIC PRESS INC.
111 Fifth Avenue
New York, New York 10003

Library of Congress Catalog Card Number 75-11373
ISBN: 0-12-620505-1

Printed in Great Britain by
Page Bros (Norwich) Ltd, Norwich

PREFACE

This fifth volume of the series is concerned with the huge variety of formaldehyde precursors in the environment and in biological tissues and fluids. Photometric methods of analysis of these precursors through their derived formaldehyde are discussed. Representative procedures of analysis are given for many of these compounds. In addition some indirect methods of analyses are described for enzymes and other compounds where a secondary reactant is the formaldehyde precursor and analysis is for the test substance in terms of the formaldehyde derived from the reactant.

In this volume we are also concerned with the formation of formaldehyde from the aspects of (i) possible analytical utility of the reaction in the future, (ii) the structure and physiological and environmental prevalence of the precursors, (iii) the reactions involved in the formation of formaldehyde, (iv) the type of environment or living tissue wherein these reactions take place, (v) the physiologically useful or toxic phenomenon resulting directly or indirectly from the formation of formaldehyde in the environment or living tissue, and (vi) the types of structures or bonds formed in the reaction of metabolically-derived formaldehyde with the chemicals and biopolymers in living tissue.

Examples of some of the biological processes in which formaldehyde formation and formaldehyde reactions may play a role are mutagenicity, carcinogenicity, schizophrenia, clastogenicity, teratogenicity, atmospheric lachrymation and aging. There are a fairly large number of carcinogens from which formaldehyde can be derived analytically and/or metabolically. These include 1,2-aminoalkanols, azomethanes, azoxymethanes, cycasin, α-chloromethyl ethers, methylnitrosamines, epoxides, ethylenimine, chlorobutadiene, methylhydrazines, methyltriazenes, methyltetrazenes, N-methyl aromatic amines, N-methyl azo dyes, β-propiolactone, safrole, vinyl chloride, vinylidene chloride, etc.

We do not think that the important roles of physiologically derived formaldehyde in biological processes have as yet been adequately investigated or explained, probably due to the lack of adequate analytical techniques. We do know that formaldehyde can be metabolically derived from many drugs, carcinogens and natural products. We do know that formaldehyde can be used constructively in living tissue and even destructively. There are indications that formaldehyde could form hallucinogenic drugs in the brain; there are indications that it could affect the vital reactions of DNA. Thus, much needs to be done in improving analytical methods for the analysis of formaldehyde and its precursors in the environment and in living tissue and in investigating the many ramifications of the formation, reaction and biological effects of formaldehyde in the environment and in living tissue.

Raleigh, N.C.

EUGENE SAWICKI
CAROLE R. SAWICKI

CONTENTS

Preface v

62. FORMALDEHYDE PRECURSORS 1
- I. Introduction 1
- II. $R_2C=CH_2$ Compounds 21
  - A. Introduction 21
  - B. 1-Alkenes 23
    - 1. Photochemical smog 23
    - 2. Oxidative MBTH test 28
    - 3. Spectrophotofluorimetric assay 29
  - C. Polar $R_2C=CH_2$ compounds 30
    - 1. Colorimetric methods 30
    - 2. Fluorimetric methods 31
  - D. Miscellaneous hydrocarbons 32
- III. Alcohols 32
  - A. Miscellaneous alcohols 32
  - B. 2-Nitro-l-alkanols 33
- IV. Aldosterone 47
- V. Amines 47
  - A. 1,2-Aminoalkanols 48
  - B. Ethylenediamines 48
  - C. Hexosamines 48
  - D. Glycine 48
  - E. Aminomethyl ketones 50
  - F. Aminopyrine 50
- VI. Azomethane Derivatives 51
  - A. Azoalkanes 53
  - B. Azoxymethanes 53
  - C. Cycasin 54
- VII. Bilirubin 55
- VIII. α-Chloromethyl Ethers 56
- IX. Lignin Products 58
- X. Cellulose and Derivatives 58
  - A. Carboxymethylcellulose 58
  - B. Cellulose 59
- XI. Chloroacetates 60
- XII. 3-(4-Chlorophenyl)-l-methylurea 60
- XIII. Cholesterol 61
- XIV. 2-Deoxyribose 61
- XV. l,l-Dichloroethylene 62

XVI. Dimethylnitrosamine 63
A. Introduction 63
B. Source and environmental concentrations 63
C. Carcinogenic activity 67
D. Formaldehyde formation and biological activity 69
1. Introduction 69
2. Methylation of DNA 70
3. Toxicity 70
4. Mutagenesis 73
5. Mutagenesis and carcinogenesis 74
E. Human health effects 76
F. Enzymatic reactions and formaldehyde formation 77
G. Other nitrosamine reactions 78
XVII. Epoxides 79
XVIII. Ethanolamine and Ethylenimine 82
XIX. Ethylene 83
XX. Ethylenediamines, N-Phenothiazinyl 83
XXI. Ethylene Glycol 84
XXII. Eugenol 86
XXIII. Formals 86
XXIV. Formates 87
XXV. Formyl and Isonitrile Groups 89
XXVI. Glucaric Acid 89
XXVII. Glucose and Derivatives 90
A. Glucose 90
B. Glucose phenylosazone 90
XXVIII. Glycerol Derivatives 91
A. Glycerides 91
B. α-Glycerophosphates 93
C. Glycerophosphatides 93
XXIX. Glycidyl Stearate 94
XXX. Glycine 94
XXXI. Glycolic Acids 99
A. Introduction 99
B. Glycolic acid 99
1. 2,7-Dihydroxynaphthalene 99
2. Chromotropic acid 100
3. β-Naphthol 101
C. Phenoxyacetic acids 101
D. Alkoxyacetic acids 102
XXXII. 1,2-Glycols 103
A. Introduction 103
B. Atmospheric α-glycols 107
C. Miscellaneous reagents 107
D. Chromatographic localization 108
E. Sialoglycoproteins 111
XXXIII. Glycoproteins and Polysaccharides 111
XXXIV. Glyoxylic Acids 113
XXXV. Heptitols, Hexitols, Tetritols and Other Sugar Alcohols 114
XXXVI. Hexahydro-1,3,5-trinitro-s-triazine 118

XXXVII. Hexamethylenetetramine 119
XXXVIII. Hexamethylmelamine 120
XXXIX. HFSH and HCG 121
XL. Hydrazines, Triazenes and Tetrazenes 121
A. Introduction 121
B. Methylhydrazines 121
C. Alar 124
D. Triazenes 124
E. Hydrazones of formaldehyde 126
XLI. Hydrocarbons, Unsaturated 128
XLII. p-Hydroxybenzyl Alcohols 129
XLIII. Hydroxylysine 131
XLIV. Hydroxymethylketones or Acetols 131
XLV. 4-Hydroxystyrenes 132
XLVI. Lactose and Maltose 134
XLVII. Mannitol 135
XLVIII. Methanol and Methyl Esters 136
A. Methanol—chemical oxidation 136
B. Methanol—enzymatic oxidation 138
C. Methyl esters 138
D. Indirect reactions 139
XLIX. Methionine Sulphoxide 140
L. Methoxone 141
LI. Methoxyflurane 142
LII. Methoxymethyl Methanesulphonate 142
LIII. N-Methyl Compounds 142
A. Introduction 142
B. Aliphatic amines 155
C. Aromatic amines 163
D. Heterocyclic amines 166
E. Azo dye amines 168
F. Heterocyclic 172
G. Nitrosamines 179
H. Hydroxylamines 186
LIV. O-Methyl Compounds 187
LV. S-Methyl Compounds 192
LVI. Methylene, Hydrolyzable 193
A. Introduction 193
1. Compound types 193
2. Formaldehyde standards 193
3. Physiological activity 195
4. Atmospheric formaldehyde 196
5. Methods 198
6. Formation of formaldehyde precursors 204
B. $-O-CH_2-O-$ compounds 212
C. $-O-CH_2-N\langle$ compounds 217
D. $-O-CH_2-S-$ compounds 222
E. $\rangle N-CH_2-N\langle$ compounds 223

F. $>N-CH_2-S-$ compounds 227
G. $-S-CH_2-S-$ compounds 228
H. $-O-CH_2-\overset{|}{\underset{|}{C}}-$ compounds 230
I. Methylene halides 231
J. Formaldehyde and the genotoxic methylene group 232
LVII. 5-Methyltetrahydrofolic Acid 234
LVIII. Nitrites, Alkyl 242
LIX. Olefins 243
LX. Oxalic acid 244
LXI. Oxetanes 246
LXII. Ozonides 247
LXIII. Pilocarpine 248
LXIV. Polymethylene Oxides 249
LXV. Polyols 251
LXVI. β-Propiolactone 251
LXVII. Sophorose and Sophoritol 253
LXVIII. Thiol-bound Formaldehyde 254
LXIX. Triglycerides 255
A. Introduction 255
B. Chromotropic acid 256
C. MBTH 260
D. 2,4-Pentanedione 260
1. Colorimetric manual 260
2. Colorimetric automated 264
3. Fluorimetric manual 266
4. Fluorimetric semiautomated 267
5. Fluorimetric automated 267
LXX. Uric Acid 268
LXXI. Vinyl Chloride 269
A. Introduction 269
B. Atmospheric and metabolic reactions 269
1. Atmospheric oxidation 269
2. Metabolism 271
C. Physiological activity 273
1. Toxicity 273
2. Clastogenicity 273
3. Mutagenicity 273
4. Carcinogenicity 273
5. Teratogenicity 274
D. Analysis 274
LXXII. Vinyl Compounds 274
A. Vinyl butyl ether and acrolein 275
B. 2-Vinylpyridine and 5-vinyl-2-picoline 275
LXXIII. Xylonic, Cellobionic and Gluconic Acids 275
LXXIV. Basic Precursor Structures 276
LXXV. Reagents and Reactions in Formaldehyde Formation 277
LXXVI. Chromogen-Forming Reagents 283
LXXVII. Fluorogen-Forming Reagents 283

To the good guys,
wherever you are.
Though you're losing,
this one's for you.

62. FORMALDEHYDE PRECURSORS

I. INTRODUCTION

Among the largest families of widely dissimilar compounds containing the same hidden functional group are the formaldehyde precursors. These compounds can be characterized or determined through the formation of a chromogen or fluorogen from the formaldehyde obtained from the test substance. Examples of this type of functional group analysis are presented in Tables 1 to 4. This type of analysis has been briefly reviewed from two different viewpoints.[521,522]

The formaldehyde precursors are probably one of the most prevalent groups of organic chemicals in the human environment. Consequently, in attempting to understand and control the environment in terms of these compounds a vast variety of mixtures have to be collected and sometimes separated before analysis. Analysis of the various types of precursors are discussed in the following sections.

Since formaldehyde is reported to be carcinogenic[523] and mutagenic to *Drosophila* with[524] and without[525, 526] hydrogen peroxide, to *Neurospora cassida* with hydrogen peroxide,[527] and to *E. coli*,[528] the formaldehyde precursors can be even more important physiologically especially if formaldehyde can be derived from them metabolically at the target tissue polymer.

There is some evidence of the importance of formaldehyde addition products (essentially potential formaldehyde precursors) in mutagenic reactions. Thus, foods treated with formaldehyde are transformed into effective mutagens for *Drosophila*, e.g. casein treated with formaldehyde and subsequently washed is such a mutagen.[529]

In this respect both free amino acids and amino acids of protein molecules are involved in the reaction of formaldehyde with nucleic acid components.[530] Similarly chromosomal proteins can be joined to DNA by treatment with formaldehyde.[531] The rate of reaction of formaldehyde with nucleotides increases in the presence of amino acids and lysine rich proteins.[530] The reaction is accompanied by a degradation of DNA. The formation of potential formaldehyde precursors under these conditions may be of interest for the study of the cytostatic and mutagenic action of formaldehyde and its precursors.

Table 1. Formation of formaldehyde from precursors

Precursors	Reagent	Reaction[a]	Ref.
2-Acetylamino-2-deoxy-D-glucose diethyl dithio-acetal	Periodate	(O)	1, 2
N-Acetylglucosamines	Periodate	(O)	3, 4
N-Acetylhexosamines	Periodate	(O)	5
N-Acetylneuraminic acid	Periodate	(O)	6
Acrylamide	Permanganate–periodate	(O)	7, 8
Acrylic acid	Permanganate–periodate	(O)	7, 8
Acrylonitrile	Permanganate–periodate[b]	(O)	9
Adenosine	Periodate	(O)	4
Adonitol	Periodate	(O)	4
Alar (N-Dimethylaminosuccinamic acid)	Alkali → selenium dioxide	H → (O)	10
Albumin[c]	Periodate	(O)	11
Aldohexoses	Acid	D → (O)	12
Aldoses	Periodate	(O)	5, 13
Aldosterone		(O)	14
Alkanolamines	Periodate	(O)	15
1-Alkenes[d]	Ozone	(O)	16–18
1-Alkenes[e]	Air	Photooxidation	19–23
1-Alkenes	Permanganate–periodate	(O)	7, 9, 24
0-Alkylglycerols	Periodate	(O)	25
Allyl acetate	Permanganate–periodate	(O)	9
Allyl alcohol	Permanganate–periodate	(O)	7, 9, 26, 27
Allylamine	Permanganate–periodate	(O)	7, 9
Allylbenzene	Permanganate–periodate	(O)	7
Allyl bromide	Permanganate–periodate	(O)	7
Allyl chloride	Permanganate–periodate	(O)	7
Allyl ethyl ether	Permanganate–periodate	(O)	9
1-Allylthiourea	Permanganate–periodate	(O)	9

2-Amino-1-alkanols	Periodate	(O)	28–30
2-Amino-2-deoxy-D-glucose diethyl dithioacetal	Periodate	(O)	1, 2
2-Amino-1-ethanol	Periodate	(O)	26, 30, 31
2-Amino-2-hydroxymethyl-1,3-propanediol	Periodate	(O)	32
2-Amino-2-deoxy-D-glucose	Periodate	(O)	3, 11, 30, 33
2-Amino-2-deoxy-D-galactose	Periodate	(O)	3, 11
Aminopyrine	Air + enzyme	(O)	34–36
Aminopyrine	N-Demethylase	(O)	37, 38
Angelic acid	Periodate	(O)	7
Anhydro-D-xylo-benzimidazole	Periodate	(O)	39, 40
Anisyl alcohol	H_2SO_4	(O)	41, 42
Apiole	H_2SO_4	(O)	43
Arabinose	Periodate	(O)	4, 44
Arabitol	Periodate	(O)	4, 44, 45
Aramite[f]	Alkali → Periodate	H → (O)	46, 47
Asarinin	H_2SO_4	H	48
Ascorbic acid	Periodate	(O)	44
Azomethane	Aq. H_2SO_4	Taut → H	49
Azoxymethane	Air + enzyme	(O)	50
Benzphetamine	N-Demethylase	(O)	37
Betaine	Bacterial cells[g]	(O)[h]	51, 52
Bilirubin	ArN_2^+	Displacement	53–56
Bilirubin[i]	MBTH + $FeCl_3$	Diaz. + (O)	57
1,2-Bis(4-hydroxy-3-methoxyphenyl)-1,3-propanediol	Strong acid	DP	58
Bis-chloromethyl ether	Water	H	59, 60
2-Bromo-2-nitropropan-1,3-diol	Acid	H	61
3-Bromopropene	Permanganate + periodate	(O)	9
2,3-Butanedione	Sulfuric acid (hot)	(O)	62
1,2,4-Butanetriol	Periodate	(O)	4
3-Butenenitrile	Permanganate–periodate	(O)	9
1-Buten-3-ol	Permanganate–periodate	(O)	7, 8

Table 1—*continued*

Precursors	Reagent	Reaction[a]	Ref.
3-Buten-1-ol	Permanganate–periodate	(O)	7, 8
Calciferol	RuO_2 + periodate	(O)	63, 64
Camphene	RuO_2 + periodate	(O)	63, 64
Carbohydrates	Periodate	(O)	3–5, 33, 65, 66
Carbon dioxide	Mg + HCl	(H)	67
Carboxymethylcellulose	H_2SO_4	H → (O)	68–71
Carboxymethyloxysuccinate	Acid	H	72
Cellobiose	Periodate	(O)	73
Cellulose	$NaBH_4$ → acid → periodate	(H) → H → (O)	74
Cellulose acetate formal	Aqueous H_2SO_4	H	75
Cellulose formal	Aqueous H_2SO_4	H	76
Chloracetic acid	H_2SO_4	H → DP	77, 78
Chloramphenicol	H_2SO_4	H → DC	79
Chloramphenicol	Acid → periodate	H → (O)	80
2-Chlorethanol	Periodate	H → (O)	81
1-Chloro-2,3-epoxypropane	Aq. H_2SO_4 → periodate	H → (O)	82
Chlorogenic acid	Permanganate–(periodate)	(O)	4, 9
Chloromethyl methyl ether	Water (warm)	H	59, 60
Chlorophenoxyacetic acids[j]	H_2SO_4 [150° (67) or 165°(68)]	H → DP	83, 84
3-(4-Chlorophenyl)-1-methylurea	N-Demethylase	(O)	85
Chlorphenesin	Periodate	(O)	86
Codeine	N-Demethylase	(O)	37, 38
Corticosteroids[k]	Permanganate–periodate	(O)	7, 8
Corticosteroids	Periodate	(O)	4, 26, 87–94
Corticosteroids	Bismuthate	(O)	95, 96
Cortisol	Periodate	(O)	97
Cycasine[l]	Acid	H	98–100
Cyclohexylazomethane	10% H_2SO_4	Taut → H	49
Cytidine	Periodate	(O)	4

1-Decene	Permanganate–periodate	(O)	7, 101
2-Deoxy-D-arabinohexose diethyl dithioacetal	Periodate	(O)	1, 2
2-Deoxy-D-glucose	Periodate	(O)	4, 8
2-Deoxy-D-ribose	Periodate	(O)	4, 8
2,4-Dichlorophenoxyacetates	H_2SO_4 (hot)	H → DP	77, 83, 84, 102–107
Diethanolamine	Periodate	(O)	15, 30, 108
Diethoxymethane[m]	Aqueous acid	H	109
Diethyl and Dimethyl ether	Air + $h\nu$	(O)	110
2,2-(Diethylthio)ethanol	Periodate	(O)	1, 2
Dihydrodehydrodiconiferyl alcohol	Strong acid	DP	58
Dihydrosphingosine	Periodate	(O)	111
Dihydrostreptomycin	Periodate	(O)	26, 113–115
Dihydroxyacetone	Periodate	(O)	4, 8, 44, 116
Dimethoate or Rogor[n]	HCl → ninhydrin	H → (O)	117
N,N-Dimethylalkylamine oxides[o]	$SO_2 + H_2O$ (hot)	Adduct formn → H	118
Dimethyl- and methylalkyl amines	2,4-Dinitrochlorobenzene → chromate	Formn of $ArNR_2$ → (O)	119
N,N-Dimethyl-*p*-aminobenzaldehyde	$FeCl_3$ + MBTH	(O)	120
N,N-Dimethyl-4-aminobenzylidene aniline	$FeCl_3$ + MBTH	(O)	121
N,N-Dimethylaniline	O_2(enzymatic) → enz. hydrol.	(O) → H	122, 123
N,N-Dimethylaniline	CrO_3	(O)	124
N,N-Dimethylaniline N-oxide	Enzyme	H	123
N,N-Dimethylanilines, *p*-substituted	$FeCl_3$ + MBTH	(O)	120
1,2-Dimethylhydrazine	O_2 + enzyme	(O)	125
1,1-Dimethylhydrazine	H_2O (100° for 20 hr)[75] or $HgSO_4$[121]	(O) → Taut → H	126, 127
1,2-Dimethylhydrazine	$HgSO_4$ + Aq. H_2SO_4	Taut → H	49
Dimethylnitrosamine	Enz.	(O)	50, 128, 129, 131
Dimethylsulfoxide	Acetic anhydride → H_2SO_4	Acetylation → H	130
1,1-Diphenylethylene	Permanganate–periodate	(O)	7
1,5-Diphenyl-3-pentadienone	Permanganate–periodate	(O)	7

Table 1—*continued*

Precursors	Reagent	Reaction[a]	Ref.
1,8-Diphenyl-1,3,5,7-octatetraene	Permanganate–periodate	(O)	7
1,9-Diphenyl-1,3,6,8-nonatetraen-5-one	Permanganate–periodate	(O)	7
1,4-Diphenyl-2,3-butadiene	Permanganate–periodate	(O)	7
Disaccharides	Periodate	(O)	132
EDTA	Cobalt(III) or manganese(III)	(O)	133–136
Elemicin	H_2SO_4	(O)	43
Epichlorohydrin	Acid → periodate	H → (O)	137
2,3-Epoxy-1-propanol	Periodate	(O)	138
Erythritol	Periodate	(O)	4, 7, 8, 139
Erythrulose	Ketotetrose aldolase	DP	140, 141
Ethanolamine	Periodate	(O)	15, 28–30, 44, 87, 142, 143
Ethion[p]	Perbenzoate → alkali	(O) → H	144
Ethyl acrylate	Permanganate + periodate	(O)	145
Ethyl N,N-bis-hydroxymethyl carbamate	Acid	H	146
Ethylene	Periodate + permanganate	(O)	9, 147, 148
Ethylene	Air + *hv*	(O)	20, 149
Ethylene	Periodate	(O)	150
Ethylenediamine	Periodate	(O)	15
Ethylenediamines	Persulfate	(O)	151
Ethylene glycol	Periodate	(O)	1, 112, 137, 152–158
Ethylene oxide	Acid → periodate	H → (O)	150, 159–164
Ethylenimine	Acid → periodate	H → (O)	165
Ethyl N-hydroxymethylcarbamate	Acid	H	146
Ethylmorphine	Enzymatic demethylation	(O)	37, 38, 166
Eugenol	Ozone	(O)	167, 168
2-(2-Fluorenylamino)ethanol	Periodate	(O)	30
Formaldehyde dimethylacetal	Alkali	H	169
Formaldehyde-tanned collagen	H_2SO_4–H_2O (1:2)	H	170

Formaldehyde 2,4-dinitrophenylhydrazone	H_2SO_4	H	41
Formaldehyde hydrazone	H_2SO_4	H	41
Formic acid	Mg + HCl	(H)	171–178
Formyl group	NaOH → Mg + HCl	H → (H)	179
Fructose	Periodate	(O)	4, 44, 47
Furfural	Permanganate + periodate	(O)	7
β-2-Furylacrylophenone	Permanganate + periodate	(O)	7
Galactonic acid	Periodate	(O)	2
Galactose	Periodate	(O)	4, 44
α-D-Galactose diethyl dithioacetal	Periodate	(O)	1, 2
Gelsemine	Permanganate + periodate	(O)	9
Glucaric acid	Periodate → acid	(O) → H	180, 181
Gluconic acid	Periodate	(O)	4, 116
Glucosamine	Periodate	(O)	30, 44
Glucose	Periodate	(O)	3, 4, 12, 44, 116, 158, 182, 183
Glucose	Heat	DP	184
Glucose	H_2SO_4 (hot)	(O)	12, 185
Glucose	Coupled oxidn of CH_3OH[q]	(O)	186–188
Glucose phenylosazone	Periodate	(O)	189
Glucosone	Periodate	(O)	190
Glucuronic acid	Periodate	(O)	4
Glyceraldehyde	Periodate	(O)	4, 182
Glyceric acid	Periodate	(O)	4, 44, 191
Glyceride glycerol	Periodate	(O)	192
Glycerides	Periodate	(O)	193–231
Glycerol	Periodate	(O)	4, 8, 139, 142, 153, 158, 232–238
α-Glycerophosphates	Periodate	(O)	158, 239
Glycerophosphatides	Heat → periodate	H → (O)	240
Glycidyl stearate	Periodate (in AcOH–$CHCl_3$)	(O)	241, 242
Glycine	Ninhydrin	(DC + DA + (O))	183, 243–247

Table 1—*continued*

Precursors	Reagent	Reaction[a]	Ref.
Glycine	Nitrite → H_2SO_4	DA → DP	248–250
Glycine	Chloramine T	DA + DC + (O)	251–253
Glycolaldehyde	Periodate	(O)	158
Glycolaldehyde	H_2SO_4 (heat)	DP	254
Glycolic acid	H_2SO_4 (heat)	DP	41, 72, 81, 182, 255–270
Glycolic nitrile	Alkali (→ H_2SO_4)	DP	271
Glycolipids	Periodate	(O)	272
1,2-Glycols	Periodate[r]	(O)	2, 3, 8, 26, 44, 66, 182, 273–276
1,2-Glycols	O_2 + Co(II)	(O)	277–278
N-Glycolylneuraminic acid	Aq. H_2SO_4 → H_2SO_4	H → (O)	267, 279, 280
Glycoproteins	Periodate	(O)	281
Glyoxal	H_2SO_4 (heat)	(O)	83
Glyoxylic acid	H_2SO_4 (heat)	DC	78, 255
Glyoxylic acid	Mg → H_2SO_4	(H) → DP	266
Glyoxylic acid[s]	Phenylhydrazine → ferricyanide	PHFm → Diaz. → DC	282
Guaiacylglycerol-β-(2-anisyl)ether	Strong acid	DP	58
Guaran	Periodate → borohydride → periodate	(O) → (H) → (O)	283–285
Guthion[t]	HCl	H	286
1-Heptene	Periodate	(O)	7
Hexahydro-1,3,5-trinitro-s-triazine	H_2SO_4	H	287
Hexamethylenetetramine	Acid	H	41, 78, 288–293
1-Hexene	Periodate	(O)	78
5-Hexen-1-ol	Ozone	(O)	294
5-Hexen-2-one	Ozone	(O)	294
Hexitols	Periodate	(O)	4, 7, 8, 26, 295
Hexosamines	Periodate	(O)	5

Hexuronic acids	Periodate	(O)	5
Hydrocarbons, unsaturated	Nitric oxide + air	(O)	296
Hydrochlorothiazide	Acid or alkali	H	297, 298
Hydrocortisone	Periodate	(O)	4, 8
Hydroxyacetone	Periodate	(O)	4
5-Hydroxy-3-indolylacetic acid			299
17-Hydroxy-17-ketolsteroids	H_2O_2	(O)	300
Hydroxylactone	Permanganate + periodate	(O)	9
Hydroxylactone[u]	Ozone	(O)	301
5-Hydroxylysine[v]	Periodate	(O)	30, 87, 302–313
1-Hydroxymethyl-5,5-dimethylhydantoin	Acid	H	61
N-Hydroxymethylurea	Acid	H	314, 315
4-Hydroxy-2-oxobutyrate	Hydroxyoxobutyrate aldolase (ED 4.1.2.1)	DP[w]	316
Hydroxyproline[x]	Periodate	(O)	317, 318
Imidan[y]	Acid	H	319
3-Indoleacetic acid	H_2SO_4 (150° for 30 min)	DP	83
Inositol	Periodic acid	(O)	320
3-Iodopropene	Permanganate + periodate	(O)	7
Isoatisine	Permanganate + periodate	(O)	9
Isoeugenol	Permanganate + periodate	(O)	7
Isoeugenol methyl ether	H_2SO_4	(O)	43
Isomaltitol	Periodate	(O)	321
Isonitrile group	NaOH → Mg + HCl	H → (H)	179
Isosafrole	Periodate	(O)	41
Isotenulin	H_2SO_4	(O)	322
Itaconic acid	Permanganate + periodate	(O)	4
20,21α-Ketols	Periodate	(O)	97
Ketones, dialkyl	Air and NO	Photooxidation	20
Ketones, methyl	Air and NO	Photooxidation	20
Lactose	Periodate	(O)	3, 323
Lactulose	Periodate	(O)	324

Table 1—*continued*

Precursors	Reagent	Reaction[a]	Ref.
Laminarin	Periodate	(O)	325, 326
Levulose	Periodate	(O)	116
Lignin	Strong acid	DP	58
Limonene	Ozone	(O)	294
Maleic acid hydrazide	Permanganate + periodate	(O)	7
Maltitol	Periodate	(O)	327
Maltose	Periodate	(O)	3, 27
Mannitol	Periodate	(O)[z]	4, 8, 328–331
Mesitylene	NO and air	Photooxidation	20
Mesityl oxide (or 4-methyl-3-pentene-2-one)	NO and air	Photooxidation	20
Metasaccharinic acids	Periodate	(O)	129
Methacrylamide	Permanganate + periodate	(O)	9
Methane	Nitric acid	(O)	332
Methanol	Permanganate	(O)	172, 333–366
Methanol	H_2O_2	(O)	367
Methanol	Persulfate	(O)	368
Methionine hydantoin sulfoxide	$Ac_2O \rightarrow H_2SO_4$	Acetylation → H	128, 369
Methionine sulfoxide[aa]	$Ac_2O \rightarrow H_2SO_4$	Acetylation → H	130, 370
Methoxone	Acid permanganate or H_2SO_4	(O)	371
Methoxy compounds[bb]	H_2SO_4 → permanganate	H → (O)	372, 373
4-Methoxybenzoate	O_2 + enzyme	(O)	374
4-(*p*-Methoxyphenyl)-3-buten-2-one	Permanganate + periodate	(O)	7
Methylal	Acid	H	375
N-Methyl-N-alkylarylamines	Chromate	(O)	124
N-Methyl-N-alkylarylamines	$Pb(OAc)_4$	(O)	376
2-Methylaminoethanol	Periodate	(O)	30, 108
N-Methylaniline	O_2 + enzyme	(O)	377, 378
Methylazobutane	Aqueous H_2SO_4	Taut → H	49
Methylazoxymethanol acetate	Alkali	H	379

1-Methyl-2-butylhydrazine	$HgSO_4$ → Aqueous H_2SO_4	(O) → Taut → H	49
1-Methyl-2-cyclohexylhydrazine	$HgSO_4$ → Aq. H_2SO_4	(O) → Taut → H	49
1-Methyl-1-cyclohexylhydrazine	$HgSO_4$ → Aq. H_2SO_4	(O) → Taut → H	127
Methyldialkylamines	Nitrite	(O)	380
N,N-Methylene-bis-acrylamide	H_2SO_4	H	41
Methylene chloride	NaOH	H	381
Methylenedioxo + analogous compounds	Acid	H	382, 383
Methylenedioxyphenyl compounds[cc]	H_2SO_4	H	41. 42, 48, 373, 384–397
Methylenedioxyphenyl compounds[cc]	HI	H	398
Methylenedithio compounds[dd]	H_2SO_4	H	399
Methyl esters	Trypsin → permanganate	H → (O)	400–402
Methyl methacrylate	NaOH → permanganate	H → (O)	403
N-Methylmorpoline N-oxide	SO_2 → water (hot)[ee]	Adduct formn → H	118
1-Methyl-1-p-nitrophenyl-2-methoxyazonium tetrafluoroborate	1-Quinoline oxide → alkali	Diaz. → H	404
Methylpropylhydrazine	$HgSO_4$ → Aq. H_2SO_4	(O) → Taut → H	49
α-Methylstyrene	Permanganate + periodate	(O)	7
Methyl vinyl ketone	Permanganate + periodate	(O)	7
Morphine N-oxide	SO_2 → water (hot)	Adduct formn → H	118
Myanesin[ff]	Periodate	(O)	405
Myosalvarson[gg]	H_2SO_4 (hot)	H	406
Myrcene[hh]	Permanganate + periodate	(O)	101
Myristicin	H_2SO_4	(O)	43
N-1-Naphthylethylenediamine	Periodate	(O)	30
Narceine[ff]	H_2SO_4 (hot)	H	407
Narcotine (Noscapine)[ff]	H_2SO_4 (hot)	H	408
Nialate[ff]	HCl	H	286
2-Nitro-1-alkanols[ff, ii]	NaOH	H	26, 409
Nitromethanol[ff]	NaOH	Taut → H	410
Oligosaccharides	$NaBH_4$ → periodate	(H) → (O)	326
Orosomucoid[jj]	Permanganate	(O)	11

Table 1—*continued*

Precursors	Reagent	Reaction[a]	Ref.
Oxalic acid[kk]	Mg + dilute acid → H_2SO_4	(H) → DP	266, 411
Oxalic acid[ll]	Zn + HCl → H_2SO_4	(H) → DP	412
Oxalic acid[ff, mm]	Zn + H_2SO_4 → H_2SO_4	(H) → DP	413, 414
2-Oxapentoate	Enzyme	H	415
Oxetanes	hv	DP	416–418
Papaverine[ff]	H_2SO_4 (hot)	(O)	419
Paraformaldehyde	Heat	H	375, 420, 421
Paraformaldehyde	H_2SO_4 (hot)	H	83
Penicillin V	H_2SO_4 (hot)	H → DP	422
Pentose	Periodate	(O)	423
Phenoxyacetic acids	H_2SO_4 (hot)	H → DP	83[ff], 84[nn], 424[ff], 425[ff]
Phenylazomethane	Aqueous H_2SO_4	Taut → H	49
Phenylephrine	Halogen + alkali	(O)[oo]	426
1-Phenyl-3,3-dimethyltriazenes	Microsomal fraction of rat liver	(O)	50, 427[ff]
Pentoses (furanose configuration)	Periodate	(O)	423[ff]
Phenethanol	Vanadate	(O)	428
Phorate (thimet)	Perbenzoic acid → NaOH	(O) → H	144, 429–431
Phospatidylglycerol	Periodate (1 %)	(O)[pp]	272
Pilocarpine	Benzoyl peroxide	(O)	432
Piperine	H_2SO_4	H	109, 433–435
Piperonyl butoxide	H_2SO_4	H	436[qq], 437[rr]
Piperonylic acids	H_2SO_4	H	41, 78, 438[ss]
Podophyllotoxin	H_2SO_4	H	439
Polyols[tt]	Periodate	(O)	4, 11, 26, 44, 45, 116, 273, 440, 441
α-Polyoxymethylene	Heat	H	421
1,2-Propanediol phosphate	Acid → periodate	H → (O)	442
4-Propenylveratrole	Permanganate + periodate	(O)	7

Propylene oxide	Periodate	(O)	443
Protoporphyrin monomethyl ester	H_3PO_4 + permanganate	H → (O)[ff]	400
Pseudouridine phosphates	Periodate (alkaline)	(O)	444, 445
Quinine	Permanganate + periodate	(O)	9, 26
Raceophenidol	NaOH → periodate	H → (O)	446
RDX	H_2SO_4	H	447
Riboflavin	Periodate	(O)[ff]	448
Ribose	Periodate	(O)	2, 8
Sabinene	Permanganate + periodate	(O)	101
Safrole[uu]	H_2SO_4	H	43, 449
Safrole	Permanganate + periodate	(O)	7, 8
Serine	Periodate	(O)	11, 26, 28, 30, 33, 44, 87, 142, 183, 234, 318, 450–456
Sesamin	H_2SO_4	H	457
Sialic acids	Periodate	(O)[vv]	33
Sialoglycoprotein	Periodate	(O)	458
Sophoritol	Periodate	(O)	459
Sophorose	Periodate	(O)	459
Sorbitol	Periodate	(O)	4, 8, 26, 331, 460, 461
Steroids[ww]	Bismuthate	(O)	462–463
Steroids[ww]	Periodate	(O)	4, 7, 8, 26, 88[vv], 90[vv], 92–94, 464–479
Streptomycin	Periodate	(O)	113, 115
Styrenes	NO and air	Photooxidation	20
Styrene	Periodate	(O)	9
Styrolene[ii]	Permanganate + periodate	(O)	26
Sugar sulfates	Periodate	(O)	480
Sulpyrine	H_2SO_4	(O)	481
Tartronate[kk]	Aqueous H_2SO_4 → H_2SO_4	DC → DP	482, 483
Terpenes[ff]	Permanganate + periodate	(O)	101
1-Tetradecene	Permanganate + periodate	(O)	7

Table 1—*continued*

Precursors	Reagent	Reaction[a]	Ref.
N,N-α,α-Tetramethylheptylamine oxide[xx]	SO_2 → water (hot)	Adduct formn → H	118
Tetramethylhydrazine[ff]	$HgSO_4$	(O)	127
N,N-β,β-Tetramethylphenethylamine oxide[xx]	SO_2 → water (hot)	Adduct form → H	118
Tetramethyltetrazene[ff]	$HgSO_4$	(O)	118
Tetritols	Periodate	(O)	295
Thiazolidine-4-carboxylic acid	Iodine	H → (O)	484
Thiol-bound formaldehyde	$HgCl_2$	H	383
Toluene[ff]	Nitrogen dioxide	Photooxidation	19
p-Tosyl arginine methyl ester[xx]	Trypsin → permanganate	H → (O)	401
Triethylamine	NO and air	Photooxidation	20
Triglycerides	NaOH → Periodate	H → (O)	485
Triglycerides	KOH → Periodate	H → (O)	486–497
Triglycerides	KOH + $Ba(OH)_2$ → Periodate	H → (O)	498
Triglycerides	NaOEt → Periodate	H → (O)	499
Triglycerides	NaOME → Periodate	H → (O)	500–504
Triglycerides	HCl → Periodate	H → (O)	505
Triglycerides	KOEt → Periodate	H → (O)	506
Trimethyl and dimethylamines	H_2O_2	(O)	507
Trimethylamine oxide	Enzyme	H	508–513
Trimethylamine oxide	SO_2 → hot water	Adduct formn → H	515
Trimethylamine oxide	NO_2^-	(O)	380
Trimethylamine oxide	Acid (heat)	(O)	507–514
Trimethylhydrazine	$HgSO_4$ → Aq. H_2SO_4	(O) → Taut → H	127
2,4,4-Trimethyl-1-pentene	Permanganate–periodate	(O)	7
sym-Trioxane	Aqueous or conc. H_2SO_4	H	41, 516
sym-Trithiane	H_2SO_4	H	41
Trithion[yy]	Acid	H	144
Tropine N-oxide	SO_2 → water (hot)	Adduct formn → H[xx]	118

10-Undecanoic acid[zz]	Permanganate + periodate	(O)	147
Urobilinogen	ArN_2^+	(O)	517
Vinyl acetate	Permanganate + periodate	(O)	145
Vinyl butyl ether	Permanganate + periodate	(O)	518
Vinyl chloride	Periodate	(O)	519
5-Vinyl-2-picoline	Permanganate + periodate	(O)	520
2-Vinylpyridine	Permanganate + periodate	(O)	520
m-Xylene	NO + air	Photooxidation	20
Xylitol	Periodate	(O)	4, 8
d-Xylose [aaa]	Periodate	(O)	44, 405

[a] CA = chromotropic acid, D = dehydration, DA = deamination, DC = decarboxylation, Diaz = diazotization, DP = disproportionation, H = hydrolysis, (H) = reduction, MBTH = 3-methyl-2-benzothiazolinone hydrazone, (O) = oxidation, PHFm = phenylhydrazone formation, and Taut = tautomerism.

[b] 50% yield of CH_2O by oxidizing at pH 10 to form glycol followed by oxidation at pH 7 to give CH_2O.

[c] Contains 2-amino-2-deoxy-D-glucose and mannose, both giving formaldehyde on oxidation.

[d] 40 to 105% yield of formaldehyde from twelve 1-alkenes. After separation of hydrocarbons by gas chromatography, ozonolysis followed by reaction with chromotropic acid used to identify the 1-alkenes.

[e] 0·3 to 0·7 mole of aldehyde per mole of starting hydrocarbon.

[f] 2-(p-t-butylphenoxy)isopropyl 2-chloroethylsulfite.

[g] A chromobacter cholinophagum.

[h] Three methyl groups of betaine converted to formaldehyde. Similar oxidation of dimethylglycine and sarcosine.

[i] CH_2O is byproduct in determination of bilirubin with MBTH.

[j] Phenoxyacetic acid and 6 chloro derivatives studied.

[k] See steroids.

[l] A carcinogen with structure $CH_3NO{=}N{-}CH_2{-}O$-glucose.

[m] Recommended as standard in formaldehyde and precursor methods.

[n] O,O-Dimethyl S-(N-methylcarbamoylmethyl)phosphorodithoate, $(MeO)_2P({=}S){-}S{-}CH_2{-}CO{-}NHMe$.

[o] Eight N-oxides studied.

[p] $(EtO)_2{-}P({=}S){-}S{-}CH_2{-}S{-}P({=}S)(OEt)_2$. Better yield of CH_2O by preliminary oxidation to sulfone.

Table 1—*continued*

[q] Glucose oxidase–catalase system.
[r] Other oxidants which have been used include chromic acid, sodium persulfate, manganese acetates, sodium perbismuthate, and lead tetraacetate.[274]
[s] To be discussed more fully under glyoxylic acid.
[t] (1,2,3-benzotriazin-4(3H)-one-3-yl)$-CH_2-S-P(=S)(OMe)_2$. Oxygen analog also reacts.
[u] 24% yield of formaldehyde.
[v] Free and also protein-bound.[306–311]
[w] Pyruvate is a byproduct.
[x] 0·6 mole CH_2O produced per mole compound in one hr under the conditions of the serine determination.[318,319]
[y] $(MeO)_2-P(=S)-S-CH_2-N$(phthalimido)
[z] One mole of mannitol gives 2 moles of CH_2O and 4 of formic acid.[328]
[aa] And many other derivatives such as the acetyl, benzoyl, carbobenzoxy and 2,4-dinitrophenyl derivatives of methionine sulfoxide, etc.
[bb] Anisaldehyde, Methyl cellosolve, cocaine, codeine, methylcellulose, pectin, vanillin, etc.[372]
[cc] Environmental aspects of methylenedioxyphenyl and related derivatives have been reviewed.[384]
[dd] In polycaprolactam fibers. CA is reagent for analysis.
[ee] CH_2O determination with 2,4-pentanedione at λ 412.
[ff] Reagent is CA.
[gg] Also aminopyrine, bilamide, neoarsphenamine, sulfarsphenamine and sulfoxone sodium.
[hh] Also camphene, 1-decene, geraniol, linalool, linalyl acetate, nerol, cis-ocimene, 1-octadecene, β-pinene and sabinene.
[ii] Reagent is ethyl acetoacetate.[26]
[jj] Reagent is 2,4-pentanedione. One hundred times more sensitive than colorimetric method.
[kk] Reagent is 2,7-dihydroxynaphthalene.
[ll] Reagent is phenylhydrazine.
[mm] 98%[413] and 86%[414] recovery.

[nn] J-Acid and phenyl J-acid as reagents.
[oo] Periodate would probably be a better oxidizing agent.
[pp] Detection of liberated CH_2O on TLC with *p*-rosaniline-SO_2 reagent.
[qq] Determined in flour, grains and oil base materials. Reagent is tannic acid.
[rr] Determined in wheat, pinto beans, Alaska peas, hulled rice, oats and barley. Reagent is CA.
[ss] Sprayed with 10% CA in 62·5% H_2SO_4 after TLC.
[tt] Reagents for CH_2O are ethyl acetoacetate,[26] 2,4-pentanedione,[11, 273] MBTH[2] and CA for remainder.
[uu] Detection in food and drugs with gallic acid.
[vv] Detection on paper chromatograms with 2,4-pentanedione.
[ww] See corticosteroids. CA used.[462–479]
[xx] 50 to 60% yield of CH_2O which is determined with 2,4-pentanedione.
[yy] *p*-$(EtO)_2—P(O)—S—CH_2—S—C_6H_4Cl$.
[zz] 63% yield of CH_2O determined with CA.
[aaa] Also 2,3-dimethylglucose, galactose and mannitol.[405]

In addition, formaldehyde and/or its precursors could be potent cofactors for the polynuclear hydrocarbon carcinogens. This appears plausible with the suggested role of 6-hydroxymethylbenzo[a]pyrene as a proximate carcinogen of benzo[a]pyrene (BaP) and 6-methylBaP.[532] BaP is metabolized by fortified rat-liver homogenates to 3-HO–BaP, 6-HO–BaP, BaP–1,6-quinone, BaP–3,6-quinone and one metabolite which is indistinguishable from 6-$HOCH_2$–BaP by either TLC or UV absorption spectra. Authentic

Table 2. Aldehyde yields from photooxidation of hydrocarbon–nitrogen oxide mixtures[19]

Moles of aldehyde per mole of hydrocarbon[a]			
Hydrocarbon	Formaldehyde	Acrolein	Aliphatic Aldehydes
Ethylene	0·3–0·4		0·3–0·4
Propylene	0·4		0·6–0·8
1-Butene	0·4		0·65
Isobutene	0·5–0·7		0·5–0·7
Trans-2-butene	0·35		1·25–1·55
1,3-Butadiene	0·3–0·5	0·2–0·3	0·5–0·8
1-Pentene	0·5		1·0
2-Methyl-2-butene	0·3–0·5		0·8–1·2
Cis-2-hexene			0·9–1·0
Toluene	0·05		0·1
Xylenes	0·15–0·2		0·2–0·3
1,3,5-Trimethylbenzene	0·15–0·2		0·3–0·4

[a] Based on hydrocarbon initially present as reactant.

6-$HOCH_2$–BaP is a potent carcinogen when administered by subcutaneous injection to rats. Thus, a lethal synthesis could involve the methylation of BaP followed by hydroxylation of the methyl group, or hydroxymethylation of BaP could take place as has been shown in the hydroxymethylation of a benzene ring by a C-1 donor bound to a macromolecule.[533] On the basis of this data Flesher and Sydnor have postulated[532] that methylation or hydroxymethylation is one of the first steps in metabolic activation of carcinogenic polycyclic hydrocarbons. The ultimate carcinogen could then be formed by reaction of 6-$HOCH_2$–BaP with sulfate in the presence of a sulfotransferase. The derived sulfate would be a powerful alkylating agent with a good leaving group which would generate the highly reactive carbonium ion, 6-BaP–CH_2^+.

It would appear that formaldehyde precursors could play an important role in hydroxymethylation. These precursors could probably be determined with appropriate tests either in tissue, in solution or in foods.

Table 3. Some formaldehydogenic steroids[90]

21	CH_2OH	CH_2OH	CH_2OH	CH_2OH
	|	|	|	|
20	CHOH	C=O	CHOH	CO
	|	|	|	|
	CH	CH	COH	COH
	/ \	/ \	/ \	/ \
	17			

17α,20β,21-trihydroxy-pregn-4-en-3-one	20-dihydro-compound "S"
11β,17α,20β, 21-tetrahydroxy-pregn-4-en-3-one	compound "E" (Reichstein)
11α,17α,20β,21-tetrahydroxy-pregn-4-en-3-one	compound *epi* "E"
17α,20β,21-trihydroxy-pregn-4-ene-3,11-dione	compound "U" (Reichstein)
17α,20α,21-trihydroxy-pregn-4-ene-3,11-dione	compound *epi* "U"
pregn-4-ene-3,20-dione	progesterone
21-hydroxy-pregn-4-ene-3,20-dione	cortexone
11β,21-dihydroxy-pregn-4-ene-3,20-dione	corticosterone
11α,21-dihydroxy-pregn-4-ene-3,20-dione	*epi*-corticosterone
14α,21-dihydroxy-pregn-4-ene-3,20-dione	14α-hydroxy-cortexone
16α,21-dihydroxy-pregn-4-ene-3,20-dione	16α-hydroxy-cortexone
17α,21-dihydroxy-pregn-4-ene-3,20-dione	compound "S" (Reichstein)
18,21-dihydroxy-pregn-4-ene-3,20-dione	18-hydroxy-cortexone
19,21-dihydroxy-pregn-4-ene-3,20-dione	19-hydroxy-cortexone
11β,21-dihydroxy-18-al-pregn-4-ene,3,20-dione	aldosterone
6β,11β,21-trihydroxy-pregn-4-ene-3,20-dione	6β-hydroxy-corticosterone
7α,14α,21-trihydroxy-pregn-4-ene-3,30-dione	7α,14α-dihydroxy-cortexone
11β,17α,21-trihydroxy-pregn-4-ene-3,20-dione	cortisol
11α,17α,21-trihydroxy-pregn-4-ene-3,20-dione	*epi*-cortisol
17α,19,21-trihydroxy-pregn-4-ene-3,20-dione	17α,19-dihydroxy-cortexone
16α,17α,21-trihydroxy-pregn-4-ene-3,20-dione	16α-hydroxy-compound "S"
2β,11β,17α,21-tetrahydroxy-pregn-4-ene-3,20-dione	2β-hydroxy-cortisol
2α,11β,17α,21-tetrahydroxy-pregn-4-ene-3,20-dione	2α-hydroxy-cortisol
11β,16α,17α,21-tetrahydroxy-pregn-4-ene-3,20-dione	16α-hydroxy-cortisol
21-hydroxy-pregn-4-ene-3,11,20-trione	11-dehydrocorticosterone
17α,21-dihydroxy-pregn-4-ene-3,11,20-trione	cortisone
6β,17α,21-trihydroxy-pregn-4-ene-3,11,20-trione	6β-hydroxy-cortisone
16α,17α,21-trihydroxy-pregn-4-ene-3,11,20-trione	16α-hydroxy-cortisone

Table 4. Some basic formaldehyde precursor structures

Structure	Example
1. $-O-CH_2-O$	
$ROCH_2OR$	Diethoxymethane
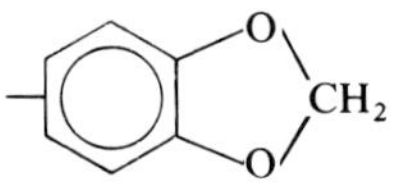 [3,4-methylenedioxyphenyl group: benzene ring fused with $-O-CH_2-O-$]	Piperonylic acid
2. $RO-CH_2SR$	Acetoxymethyl methylsulfide
3. $RO-CH_2-NR_2$	
4. $RS-CH_2-SR$	Phorate, s-trithiane
5. $RS-CH_2-NR_2$	Thiazolidine-4-carboxylic acid
6. $R_2N-CH_2NR_2$	Allantoic acid
7. $RSO_2-CH_2-SO_2R$	Phorate sulfone
8. CH_3OAr	Anisole
9. $CH_3N\langle$	
CH_3RN-Ar	N,N-Dimethylaniline
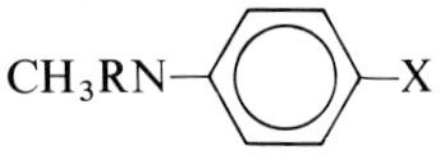 $CH_3RN-C_6H_4-X$	*p*-Dimethylaminobenzaldehyde
$CH_3NHet.$	Caffeine
10. $CH_3-N{=}N-R$	Azomethane
11. $CH_3-N^+(O^-)$ Sat.	Tropine N-oxide
12. $CH_3-N^+(O^-)R_2$	Methyldiethylamine oxide
13. $(CH_3)_2N^+(O^-)-R$	Dimethylethylamine oxide
14. $(CH_3)_2N-NX_2$	$X_2 = H_2$ or O
15. $CH_3RN-NRCH_3$	R = H, CH_3
16. $CH_3NH-NHR$	R = H, alkyl
17. 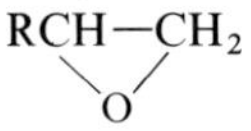 $RCH-CH_2$ (epoxide, bridged by O)	R = H, alkyl, aryl
18. CH_3-SO-R	R = alkyl, aryl
19. $CH_2{=}CH-X$	X = alkyl, aryl, $CONH_2$, COOH, CH_2NH_2, CH_2OH
20. $ROCH_2-C(=O)-X$	R, X = H, OH, alkyl, aryl

Table 4—*continued*

Structure	Example
21. $H_2NCH_2{-}C({=}O){-}X$	X = OH, alkyl, aryl
22. $HOCH_2{-}CHOH{-}R$	Ethylene glycol, mannitol
23. $HOCH_2CH(NHR){-}X$	Adrenaline
24. $RNH{-}CH_2{-}CH(OH){-}X$	Diethanolamine
25. $RNH{-}CH_2{-}CH(NHR)X$	Ethylenediamine
26. $HOCH_2{-}CH(NO_2)R$	2-Nitroethanol
27. $ROCH_2CH_2X$, X = halogen	2-Chloroethanol
28. $ClCH_2{-}O{-}R$	Bis-chloromethyl ether
29. $HOOC{-}X$	X = H, OH, CH_2Cl, CH_2OH, CHO, COOH
30. $R{-}CO{-}CO{-}R$	R = H, CH_3

In urban communities formaldehyde values can average around 10 $\mu g/m^3$ air with a range of about 0·5 to 60 $\mu g/m^3$. At heavily trafficked highways and in the atmosphere of shoe, textile and furniture industries concentrations ranging from 1000 to 46 000 $\mu g/m^3$ can be found. Levels as high as 50 mg/liter have been reported in some industrial waters. Some of the material reported as formaldehyde is probably easily hydrolyzable formaldehyde precursors. Suitable end-capped formaldehyde polymers are probably ubiquitous and could possibly even be found in interstellar space.

II. $R_2C{=}CH_2$ COMPOUNDS

A. Introduction

These types of molecules are present throughout the environment. They can be oxidized analytically or metabolically to formaldehyde and aldehydes or ketones. Polluted air contains large amounts and varieties of these types of molecules, e.g. ethylene, propylene, 1,3-butadiene, styrene, 1-hexene, 1-hexadecene, etc. Some of these compounds under favorable conditions could play a role in carcinogenesis. Thus, 1-hexadecene could be a precarcinogen or even a carcinogen since it can be biotransformed by hepatic microsomes to the carcinogenic 1,2-epoxyhexadecane.[534] The formation of the epoxide was observed only when the olefin was incubated in the presence of the epoxide hydrolase inhibitor, 1,2-epoxydecane. This oxide is the most potent inhibitor and the poorest substrate for epoxide hydrolase,[535] an enzyme system which detoxifies arene and alkene oxides. 1,2-Epoxyhexadecane causes skin carcinoma in the mouse.[536] These studies indicate that there are probably many more olefins in the environment that could form carcinogenic oxides on metabolism or upon oxidation in the environment.

The $R_2C{=}CH_2$ functional group is present in a large number of compounds, Tables 1 and 2. Formaldehyde is usually formed by permanganate–periodate oxidation of the molecule to a 1,2-glycol and then to formaldehyde. 1-Alkenes, polyunsaturated 1-alkenes and $C{=}CH_2$ compounds containing miscellaneous hetero atoms have been analyzed by means of the derived formaldehyde.

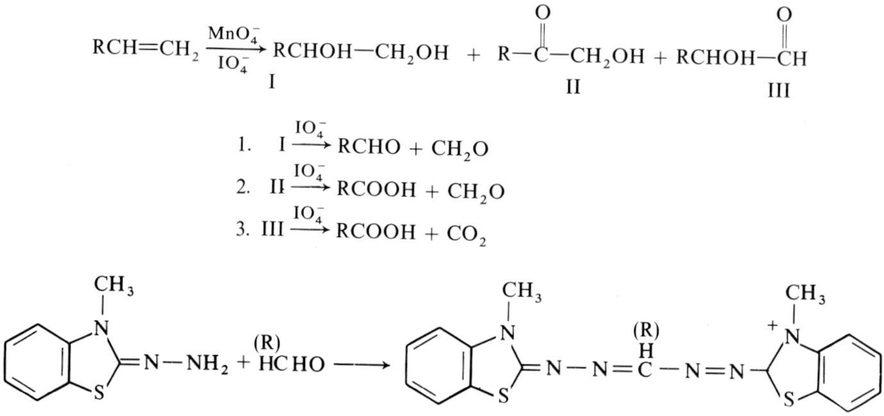

$$\mathrm{RCH{=}CH_2} \xrightarrow[\mathrm{IO_4^-}]{\mathrm{MnO_4^-}} \underset{\mathrm{I}}{\mathrm{RCHOH{-}CH_2OH}} + \underset{\mathrm{II}}{\mathrm{R{-}\overset{O}{\overset{\|}{C}}{-}CH_2OH}} + \underset{\mathrm{III}}{\mathrm{RCHOH{-}\overset{O}{\overset{\|}{C}}H}}$$

1. $\mathrm{I} \xrightarrow{\mathrm{IO_4^-}} \mathrm{RCHO} + \mathrm{CH_2O}$
2. $\mathrm{II} \xrightarrow{\mathrm{IO_4^-}} \mathrm{RCOOH} + \mathrm{CH_2O}$
3. $\mathrm{III} \xrightarrow{\mathrm{IO_4^-}} \mathrm{RCOOH} + \mathrm{CO_2}$

FIG. 1. Reaction sequence in the determination of olefinic compounds with the oxidative MBTH method.[7]

For the water-insoluble compounds of this type an aqueous acetic acid test solution is used as solvent in their analysis.[7] Permanganate–periodate is used to oxidize these and other $C{=}CH_2$ compounds to formaldehyde.[7, 9, 24, 101] The postulated equations for the oxidation and determination of olefins with MBTH are presented in Fig. 1.[7] The permanganate used in the oxidation is regenerated continuously by the periodate.[9, 147] Intermediary steps in the oxidation have been postulated.[537] The structure of the test substance, the pH, temperature and solvent composition will determine the types of products formed. The formation of I is favored with an aqueous test solution and when R equals COOH, $CONH_2$, CH_2OH and CH_2NH_2. Formation of II seems to be favored with an acetic acid test solution. Further oxidation of III just results in the formation of a carboxylic acid and carbon dioxide. The aldehydes formed in the reactions react readily with MBTH to give the blue formazan cation. The intermediary steps in the formation of the dye have been discussed.[538]

B. 1-Alkenes

1. *Photochemical smog* first appeared in Los Angeles in the 1940s and has grown into a hazardous problem that has increased and spread to other American cities and other countries. The main effects are eye irritation, reduced visibility, plant damage and damage to property. These effects are caused by aldehydes and other pollutants resulting from the sunlight-initiated photo-oxidation of exhaust gases from internal combustion engines, under special meteorological and geographical conditions.[539]

The role of hydrocarbons and nitrogen oxides in photochemical smog formation is usually investigated in smog chambers wherein the chemicals, the lighting and other conditions can be controlled and where analyses are more readily obtained than in the outside air. Samples of auto exhaust with varying hydrocarbon and NO_x levels or appropriate mixtures of pure chemicals are irradiated in a smog chamber under conditions that result in levels of smog manifestations similar to those observed in the atmosphere.[540]

The mechanisms of photochemical air pollution have been described.[541, 542] 1-Alkenes irradiated with NO_x or O_3 were found to form formaldehyde as one of the products. Thus, single, binary, and tertiary mixtures of ethylene, propylene and cis-2-butene irradiated in the presence of nitrogen oxides gave formaldehyde as one of the products.[543] The major product in the ozonolysis of ethylene in air are water, carbon monoxide, carbon dioxide, formic acid and formaldehyde.[544] The ultraviolet irradiation of a mixture of ethylene and nitrogen dioxide in a chamber shows an interesting growth and decay of reactants and products, Fig. 2.[545] After 240 min irradiation approximately 82% of the ethylene had reacted. The formaldehyde concentration became constant with increasing irradiation time after 160 min irradiation. At 190 minutes ozone concentration reached a maximum of 1·23 ppm.

Many other smog chamber studies have shown that the 1-alkenes and the methylbenzenes are formaldehyde precursors. The formaldehyde yields show extremely well-defined dependencies on the concentrations of hydrocarbons and NO_x (nitric oxide and nitrogen dioxide). A typical example involving auto exhaust is shown in Fig. 3 where formaldehyde dosage appears to be a linear function of the hydrocarbon concentration, with the slope depending slightly on NO_x.

When low concentrations of exhaust gases from automobiles are irradiated with ultraviolet light, (a) the unsaturated hydrocarbons are oxidized and disappear from the system; (b) the secondary pollutants: aldehydes, ketones, nitrates and peroxyacetyl nitrate appear; (c) nitric oxide is oxidized to nitrogen dioxide; and (d) ozone is formed when all the nitric oxide is consumed. An example of this phenomenon can be seen in the change in concentration of the

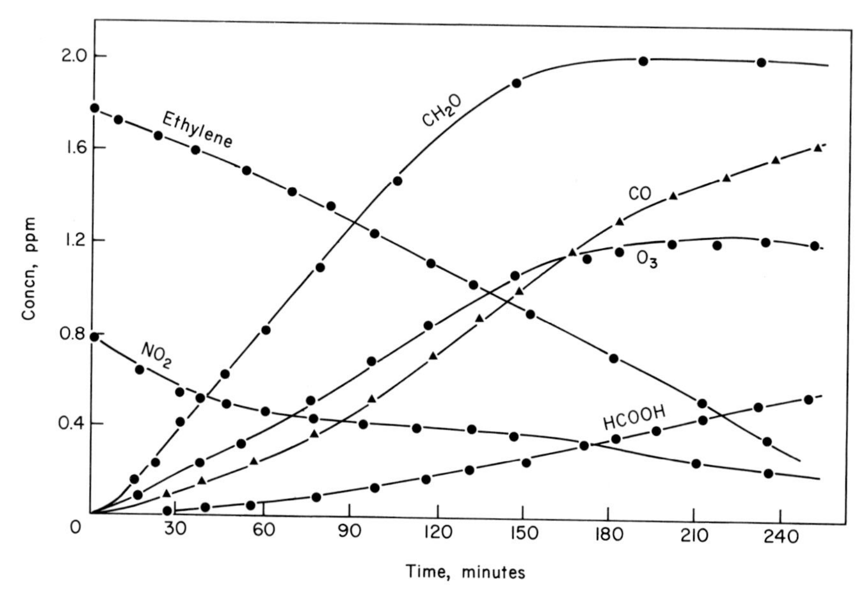

FIG. 2. Irradiation of a mixture of ethylene and NO_2.[545]

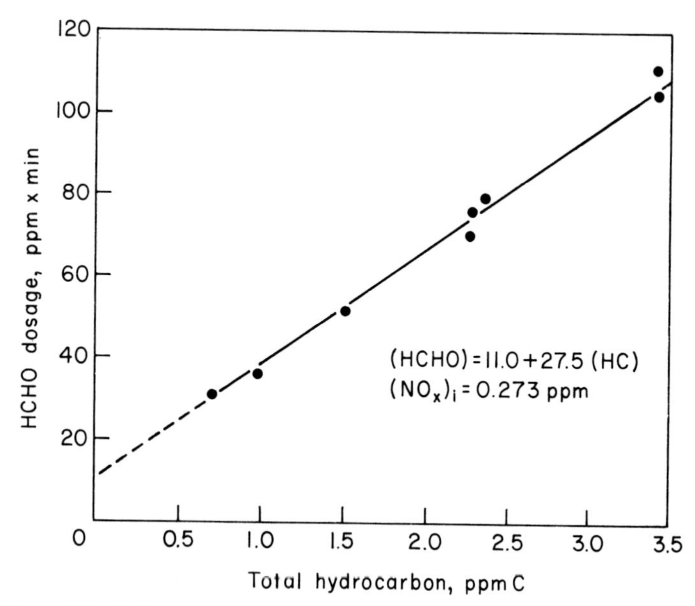

FIG. 3. Formaldehyde-dosage reactivity of exhaust as a function of total hydrocarbon at $NO_x = 0{\cdot}273$ ppm.

various products with time on the irradiation with ultraviolet light of a mixture of isobutene and nitric oxide (plus some nitrogen dioxide) in air, Fig. 4.[22, 546]

Because of the indirect role of hydrocarbons as reactant species in the photochemical smog process, the establishment of criteria and standards for such compounds is rather complex and not well defined. Primarily because of

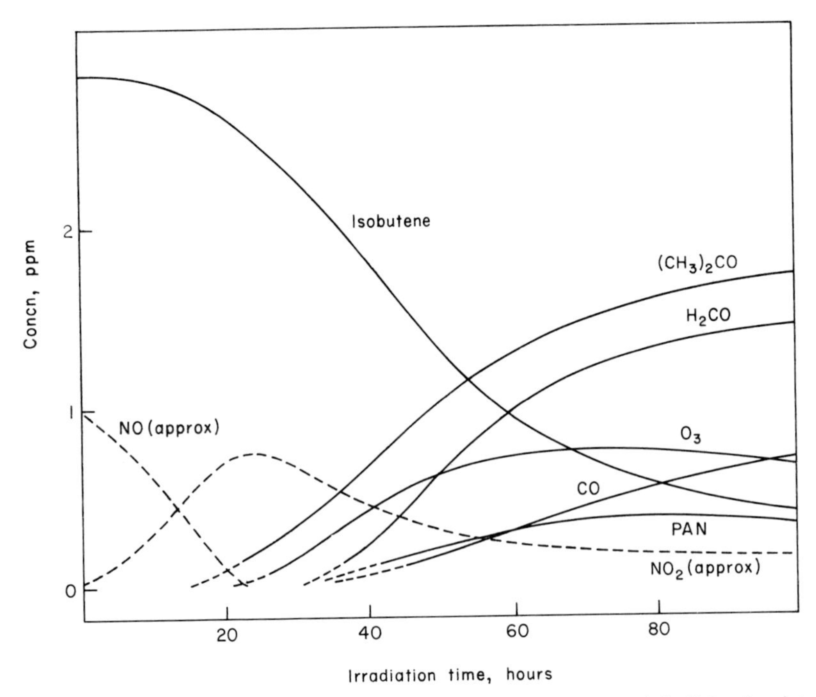

FIG. 4. Products and their variation with time for the irradiation with ultraviolet light of a mixture of isobutene and nitric oxide (plus some nitrogen dioxide) in air.[22, 546]

this ill-defined role, attempts have been and are being made to regulate the concentrations of the hydrocarbons in accordance to their reactivity as defined by their capability to produce the products and manifestations of photochemical smog. In one simplistic approach it was suggested that in order to maintain the oxidant level in the atmosphere below a desirable 0·1 ppm (one hour average) the readily measured non-methane hydrocarbons in the atmosphere be maintained below 0·3 ppm. To understand the smog problem a little more realistically the reactivity of a large number of solvent materials have been examined.[20]

Since photochemical smog is manifested in a variety of ways, it is very difficult to develop reactivity scales which can satisfy the chemical, subjective, physiological and/or the material manifestations of the problem, e.g. eye irritation, plant damage and visibility reduction. Reactivity can be expressed in terms of rate measurements and product yields. Rate measurements can be in terms of NO_2 formation and depletion, solvent consumption and ozone or aldehyde formation, and product yields in terms of NO_2, ozone, aldehyde, peroxyacetyl nitrate, oxidized hydrocarbon derivatives, etc.

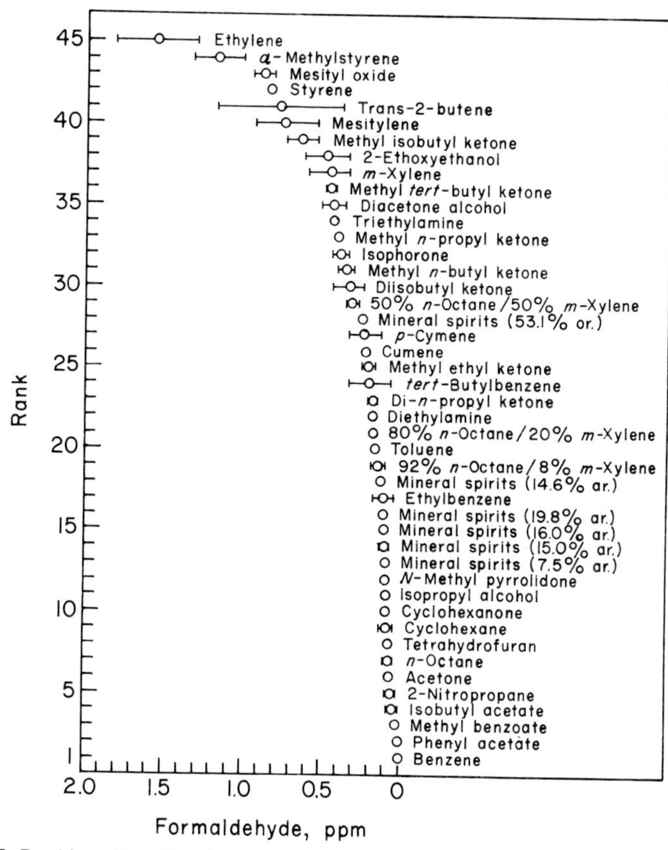

FIG. 5. Ranking of irradiated compounds based on their formaldehyde production.[20]

Four rankings were investigated, e.g. based on NO_2 t-max, oxidant production, eye response data and formaldehyde production. None of these correlated with each other. The ranking based on formaldehyde production is shown in Fig. 5 and Table 5.[20]

Of particular importance in the afternoon of a smoggy sunny day are the reactions of ozone which builds up to high levels with time and then reacts with

Table 5. Summary of smog chamber experiments[20]

Solvent: 4 ppm; NO: 2 ppm

Solvent	Aldehyde		
	Total, ppm	HCHO, ppm	Standard Error of Mean, ppm
1. Acetone	0·14	0·08	0·0
2. Benzene	0·02	0·01	0·0
3. Trans-2-Butene	6·60	0·80	0·23
4. Butylbenzene (tert.)	0·65	0·23	0·07
5. Cumene	0·59	0·24	0·01
6. Cyclohexane	0·30	0·10	0·02
7. Cyclohexanone	0·30	0·10	0·0
8. *p*-Cymene	0·59	0·25	0·05
9. Diacetone Alcohol	0·92	0·46	0·04
10. Diethylamine	2·60	0·20	0·005
11. Di-iso-butyl Ketone	1·15	0·35	0·05
12. Di-n-propyl Ketone	1·27	0·20	0·01
13. 2-Ethoxyethanol	1·42	0·50	0·07
14. Ethylbenzene	0·46	0·13	0·03
15. Ethylene	2·38	1·58	0·13
16. iso-Butyl Acetate	0·11	0·04	0·006
17. Isophorone	0·90	0·40	0·02
18. iso-Propyl Alcohol	0·28	0·11	0·0
19. Mesitylene	1·80	0·77	0·10
20. Mesityl Oxide	2·20	0·90	0·03
21. Methyl Benzoate	0·11	0·03	0·0
22. Methyl Butyl Ketone (tert.)	0·80	0·48	0·01
23. Methyl iso-Butyl Ketone	0·85	0·67	0·05
24. Methyl n-Butyl Ketone	1·10	0·37	0·02
25. Methyl Ethyl Ketone	0·25	0·23	0·02
26. Methyl n-Propyl Ketone	0·90	0·42	0·01
27. N-Methyl Pyrrolidone	0·12	0·11	0·01
28. α-Methylstyrene	1·86	1·19	0·08
29. Mineral Spirits (7·5% ar.)	0·45	0·11	0·01
30. Mineral Spirits (14·6% ar.)	0·52	0·14	0·01
31. Mineral Spirits (15·0% ar.)	0·34	0·12	0·02
32. Mineral Spirits (16·0% ar.)	0·41	0·12	0·01
33. Mineral Spirits (19·8% ar.)	0·45	0·12	0·003
34. Mineral Spirits (53·1% ar.)	0·76	0·27	0·01
35. 2-Nitropropane	0·13	0·07	0·01

Table 5—*continued*

Solvent	Aldehyde		
	Total, ppm	HCHO, ppm	Standard Error of Mean, ppm
36. n-Octane	0·32	0·08	0·01
37. 92% n-Octane/8% m-Xylene	0·65	0·16	0·02
38. 80% n-Octane/20% m-Xylene	0·76	0·19	0·01
39. 50% n-Octane/50% m-Xylene	0·91	0·33	0·02
40. Phenyl Acetate	0·02	0·01	0·0
41. Styrene	2·98	0·85	0·005
42. Tetrahydrofuran	1·25	0·09	0·005
43. Toluene	0·42	0·18	0·01
44. Triethylamine	5·31	0·45	0·01
45. *m*-Xylene	1·14	0·48	0·07

olefins to form aldehydes and alkylperoxy radicals,[21] e.g.

$$O_3 + CH_3CH{=}CH_2 \longrightarrow CH_3\dot{C}HO\dot{O} + CH_2O$$

In these reactions chemiluminescence also results.[18] Dependent upon the structure of the olefin the spectral emission arises from electronically excited aldehydes, such as formaldehyde, glyoxal or pyruvaldehyde.

2. *Oxidative MBTH test.* Hydrocarbons as large as docosene can be assayed by the oxidative MBTH method.[7] The parameters of this method with 1-hexene as test substance were investigated. Beer's law was obeyed from 2 to 34 μg; the relative standard deviation for 12 runs was ±1·51%.

Polyunsaturated 1-alkenes also react in the procedure, although in this case a wider variety of oxidation products are formed in addition to the formaldehyde, as shown for cis-ocimene, Fig. 6.[101]

The aliphatic fractions or benzene (or cyclohexane or acetone) extracts of urban airborne particulates or automobile exhaust tars[547, 548] give positive results in the oxidative MBTH procedure for olefins as described in the next paragraph.[7] The results indicate that olefinic compounds are present in the aliphatic fractions of particulate samples in appreciable amounts.

Oxidative MBTH procedure for olefins.[7] To 0·5 ml of the acetic acid test solution add 0·1 ml of aqueous 0·03 M potassium permanganate. Let stand for 1 min. Add 1 ml of aqueous 0·02 M sodium periodate. Let stand for 10 min. Mixing the solution after each step, add 1 ml of aqueous 1·5 M sodium arsenite followed by 1 ml of 2 N sulfuric acid and 1 ml of aqueous 0·8% MBTH. Heat the mixture and the usual blank on a water bath for 3 min. Cool under the tap to room temperature. Add 2 ml of aqueous 0·5% iron(III) chloride. Read the absorbance at 629 nm at 20–30 min after the last addition.

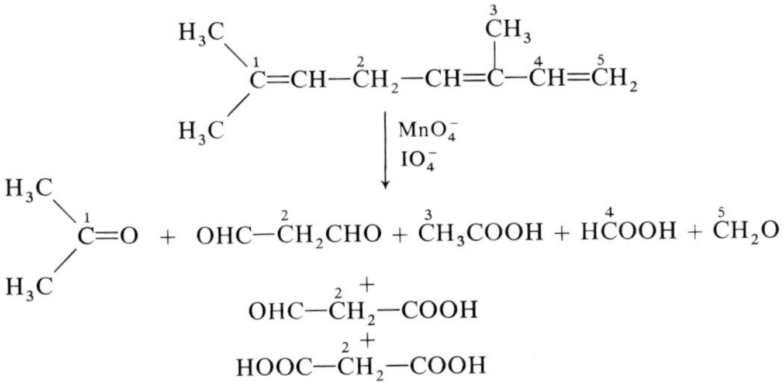

$$(H_3C)_2\overset{1}{C}{=}CH{-}\overset{2}{C}H_2{-}CH{=}\overset{\overset{3}{C}H_3}{C}{-}\overset{4}{C}H{=}\overset{5}{C}H_2 \xrightarrow{MnO_4^-,\ IO_4^-}$$

$$(H_3C)_2\overset{1}{C}{=}O + OHC{-}\overset{2}{C}H_2CHO + \overset{3}{C}H_3COOH + H\overset{4}{C}OOH + \overset{5}{C}H_2O$$

$$+\ OHC{-}\overset{2}{C}H_2{-}COOH$$

$$+\ HOOC{-}\overset{2}{C}H_2{-}COOH$$

FIG. 6. Products formed in the oxidation of cis-ocimene.[101]

The procedure for the water-soluble olefinic compounds was the same as that for the olefins except that the test solvent was water and the final reading was made 25 min after the addition of the ferric chloride.

3. *Spectrophotofluorimetric assay.* A fluorescent procedure for 1-alkenes is also available.[8] The method has been standardized with 1-hexene. The relative standard deviation was ±4·86. Hexene could be determined over a linear range of 0·8 to 17 μg. The fluorescence intensity is stable for longer than 60 min. The procedure is fairly simple.

J-Acid procedures for 1-alkenes or water-soluble olefins.[8] To 0·5 ml of an acetic acid test solution (1-alkenes) or an aqueous test solution (water-soluble olefins) add 0·1 ml of aqueous 0·03 M potassium permanganate followed by a 1-minute wait and 0·5 ml of 0·02 M sodium periodate solution. After 10 minutes add 0·5 ml of 1·5 M sodium arsenite solution followed by 0·5 ml of 2 N sulfuric acid and 5 ml of 0·01% J-acid in sulfuric acid. Allow to react for 15 minutes before diluting to 10 ml with concentrated sulfuric acid. Cool under the tap and read at *F*462/518.

Somewhat similar oxidation methods have been described for the analysis of 1-alkenes in industrial atmospheres.[549] Compounds such as 1,3-butadiene and isoprene were analyzed with chromotropic acid as the reagent. Formaldehyde and other 1-alkenes would be interferences.

C. Polar $R_2C{=}CH_2$ compounds

1. *Colorimetric methods.* Many of these compounds are readily soluble in water so that their aqueous test solutions can be readily analyzed by the oxidative MBTH method. The parameters of this method were investigated with acrylic acid.[7] The relative standard deviation for 13 determinations was $\pm 3{\cdot}2\%$. Beer's law was obeyed from 2·2 ($A = 0{\cdot}1$) to at least 36 μg of acrylic acid. The blank run against water gave an absorbance reading of 0·10 from wavelength 625 to 670 nm.

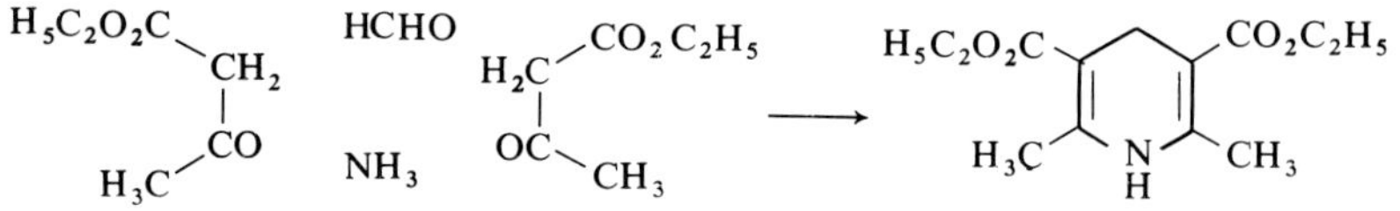

$H_5C_2O_2C{-}CH_2{-}CO{-}CH_3$ + HCHO + NH_3 + $H_2C(CO_2C_2H_5){-}OC{-}CH_3$ ⟶ $H_5C_2O_2C$, $CO_2C_2H_5$, H_3C, N–H, CH_3 (dihydropyridine)

FIG. 7. Reaction in the ethyl acetoacetate determination of formaldehyde.[27]

Where the compound is insoluble in water, and acetic acid or aqueous acetic acid is used as the test solvent, the molar absorptivity is decreased. Thus, when water is used as the test solvent instead of acetic acid the molar absorptivities obtained with acrylic acid, acrylamide, allylamine, and allyl alcohol are approximately 12½, 5, 3½, and 2½ times as high, respectively. Examples of olefinic compounds containing a 1-ene and a polar group giving high millimolar absorptivities are acrylamide—85, allylamine—71, allyl alcohol—52, acrylic acid—50, 1-buten-3-ol—50, chlorogenic acid—49, and 3-buten-1-ol—48.

The acid fraction and the aqueous ammonia extract of urban airborne particulates reacted readily in the oxidative MBTH procedure for olefins.[7] Any 1,2-glycols in these samples would also react in the method.

An alternative method for polar ethylenic compounds with the structure $RCH{=}CH_2$ has been described.[27] Oxidation is with periodate—permanganate with ethyl acetoacetate as the condensation reagent, Fig. 7. Compounds such as allyl alcohol and quinine can be determined with allyl alcohol giving an mε of about 6·5 at λ 378 nm. The following procedure is used.

Determination of polar ethylenic compounds.[27] *Reagents.* Aqueous 0·05 N sodium metaperiodate, aqueous 0·005 N potassium permanganate, 0·1 N aqueous potassium arsenite, 1% ethyl acetoacetate in 20% aqueous ammonium acetate, freshly prepared.

Procedure. To 1 ml of the pyridine test solution add 0·5 ml of the periodate solution and 1 ml of the permanganate solution. Mix and after 20 minutes of standing add 1 ml of the arsenite solution and 1 ml of the ethyl acetoacetate. Heat for 20 min at 60°, then cool to room temperature in ice water, and read the absorbance at 378 nm.

Similar types of compounds, e.g. acrylic and methacrylic acids, have been analyzed in industrial atmospheres by a similar oxidation procedure with chromotropic acid as the reagent.[549]

A fairly sensitive method for the determination of allyl alcohol, acrylic, and methacrylic acids in highly polluted atmospheres has been described.[550] Air is sampled into 2% ammonium acetate at a rate of 20 liter/hr. The mixture is oxidized with permanganate–periodate and the resulting formaldehyde is determined by the chromotropic acid procedure. The method is not selective since allyl alcohol, acrylic acid, methacrylic acid, formaldehyde and other methylenic compounds would be determined in the procedure. This is shown in a recently described procedure where methacrylic acid and formaldehyde jointly present in the air can be determined separately as follows.[551] The formaldehyde is determined before and after oxidation. The amount of formaldehyde in the air (before oxidation) is subtracted from the total formaldehyde (after oxidation) to give the amount of methacrylic acid that is present, assuming no other oxidizable methylenic compounds are present.

The ozone reaction with olefins to form aldehydes has been used indirectly also. Thus, atmospheric ozone can be determined by its reaction with eugenol to form formaldehyde which is then determined by the pararosaniline–sulfur dioxide method,[168] e.g.

$$\text{HO}-\text{C}_6\text{H}_3(\text{OCH}_3)-\text{CH}_2\text{CH}=\text{CH}_2 \xrightarrow{O_3} \text{HO}-\text{C}_6\text{H}_3(\text{OCH}_3)-\text{CH}_2\text{CHO} + \text{CH}_2\text{O}$$

2. *Fluorimetric methods.* Polar olefins can also be determined by oxidative fluorimetric methods using J-acid, dimedone or 2,4-pentanedione.[8] Compounds, such as allyl alcohol, allyl amine, acrylamide, acrylic acid and 3-buten-1-ol have been analyzed by these sensitive procedures. The oxidation of one of these compounds is shown in Fig. 8.[101]

Many of these polar olefins are listed in Table 1, e.g. eugenol, quinine, safrole, terpenes and vinyl compounds. Their analysis could be accomplished with the oxidative MBTH or J-acid procedures or with the procedures described in the next paragraph.

2,4-*Pentanedione and dimedone procedures for water-soluble olefins.*[8] To 0·5 ml of aqueous test solution add 0·1 ml of 0·03 M potassium permanganate solution. After 1 min add 1 ml of 0·02 M aqueous sodium periodate solution. Let stand 10 min. Add in consecutive order 1 ml of 1·5 M sodium arsenite solution, 1 ml of 2 N sulfuric acid and 2 ml of the reagent.

In the procedure using 2,4-pentanedione the reagent contains 15 g

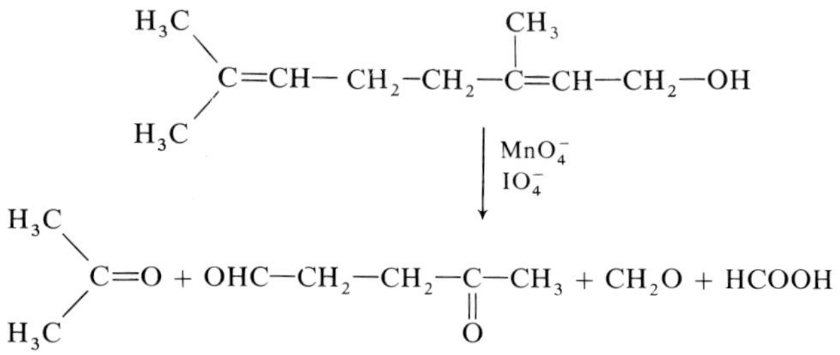

$$(H_3C)_2C{=}CH{-}CH_2{-}CH_2{-}C(CH_3){=}CH{-}CH_2{-}OH \xrightarrow{MnO_4^-,\ IO_4^-} (H_3C)_2C{=}O + OHC{-}CH_2{-}CH_2{-}C({=}O){-}CH_3 + CH_2O + HCOOH$$

FIG. 8. Products formed in the oxidation of geraniol.[101]

ammonium acetate, 0·2 ml 2,4-pentanedione and 0·3 ml of distilled acetic acid diluted to 100 ml with water. Once this reagent is added, heat the mixture 10 min at 57°, cool and read at *F*410/510.

In the procedure using dimedone the reagent contains 25 g ammonium acetate, 0·2 ml 2,4-pentanedione, and 0·3 ml of distilled acetic acid diluted to 100 ml with water. Once the 2 ml of this reagent is added, heat the mixture at 100° for 10 min using a cold finger for condensation of the vapors. Cool and read at *F*395/460.

D. Miscellaneous hydrocarbons

Owing to the tremendous amounts of natural and man-made pollutants spewed into the air, the atmosphere is a rich soup of a vast variety of chemicals, many of which are interacting with each other. With the help of oxidizing agents and sunlight some of the formaldehyde and other aldehydes are regularly synthesized and just as regularly converted into other chemicals or transferred to the soils and waters of the earth.

Laboratory experimentation has shown that many types of hydrocarbons are converted to formaldehyde and other aldehydes through photooxidation, Table 2.[19]

III. ALCOHOLS

A. Miscellaneous alcohols

Some of the monohydroxy alcoholic compounds which can be analyzed through the chromogen or fluorogen formed from the formaldehyde derived from the alcohol include methanol and methyl esters (discussed in a later section), unsaturated alcohols, benzyl alcohols, chloroethanols, 2-nitro-1-alkanols, aminoalkanols (discussed in another section), and glycolic acids (discussed in a later section).

Appropriately unsaturated primary alcohols can be oxidized to formaldehyde, as shown for geraniol, an important constituent of many essential oils, Fig. 8. The diversity of photometric methods available for the analysis of these types of compounds is shown in Table 6 for allyl alcohol where reagents such as chromotropic acid, ethyl acetoacetate, MBTH, dimedone, J-acid and 2,4-pentanedione have been used in the analysis of this compound. Many other examples are given in the Tables.

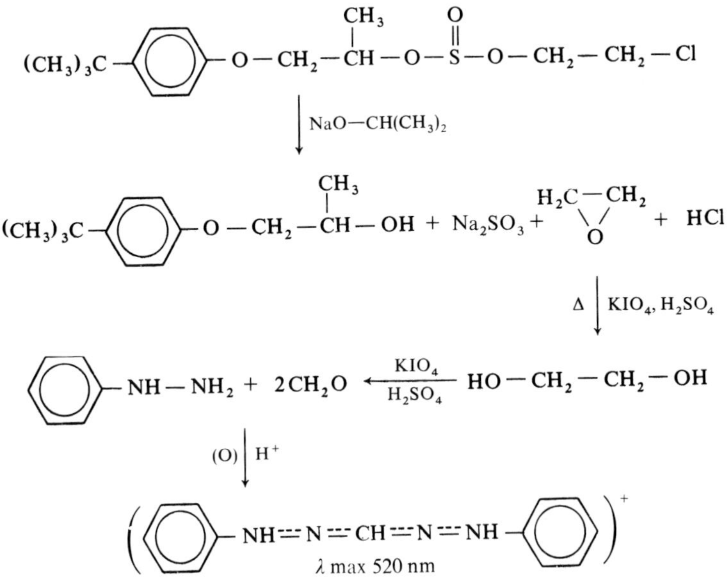

$$(CH_3)_3C-C_6H_4-O-CH_2-CH(CH_3)-O-S(=O)-O-CH_2-CH_2-Cl$$

$$\downarrow NaO-CH(CH_3)_2$$

$$(CH_3)_3C-C_6H_4-O-CH_2-CH(CH_3)-OH + Na_2SO_3 + H_2C(-O-)CH_2 + HCl$$

$$\downarrow \Delta \; KIO_4, H_2SO_4$$

$$C_6H_5-NH-NH_2 + 2CH_2O \xleftarrow[H_2SO_4]{KIO_4} HO-CH_2-CH_2-OH$$

$$\downarrow (O) \; H^+$$

$$(C_6H_5-NH{\cdots}N{\cdots}CH{\cdots}N{\cdots}NH-C_6H_5)^+$$

λ max 520 nm

FIG. 9. Reaction sequence in the determination of aramite through condensation of its derived formaldehyde with phenylhydrazine.

Benzyl alcohols can also be analyzed through the formation of formaldehyde through oxidative hydrolysis.

2-Chloroethanol and its precursors can be hydrolyzed and then oxidized to formaldehyde, as shown for aramite in Fig. 9. In this case phenylhydrazine is used as the reagent and the formazan cation is measured.

B. 2-Nitro-1-alkanols

In alkaline solution these compounds essentially disproportionate to give nitroalkanes and formaldehyde, e.g.

$$RCH(NO_2)-CH_2OH \longrightarrow RCH_2NO_2 + CH_2O$$

Table 6. Determination of formaldehyde precursors

Precursor	Reagent[a]	*F* exc/em or λ_{max} ($m\varepsilon$)	Other data[b]	Ref.
N-Acetylglucosamine	MBTH	624 (9·0)	Other λ 659s (9·0)	2
N-Acetylneuraminic acid	2,4-Pentanedione	412	Amino sugars + sugars react	3
Acrylamide	MBTH	629 (85)	Other λ 666 (83·0)	7
Acrylamide	Dimedone	395/460		8
Acrylamide	J-Acid	460/518		8
Acrylamide	2,4-Pentanedione	410/510		8
Acrylic acid	MBTH	629 (50)	Other λ 666 (48·0)	7
Acrylic acid	Dimedone	395/460	%S = ±3·3 Fluor. stable 3 min.	8
Acrylic acid	J-Acid	460/518	%S = ±4·0 Fluor. stable >60 min.	8
Acrylic acid	2,4-Pentanedione	410/510	%S = ±1·5 Fluor. stable >60 min.	8
Adenosine	MBTH	656 (27)	Other λ 622 (26·0)	2
Adonitol	MBTH	630 (99)	Other λ 664s (94·0)	2
1-Alkenes	CA	570	40 to 105% yield of CH_2O	16
Allyl compounds	CA	570	42 to 100% yield of CH_2O	9
Allyl alcohol	Ethyl acetoacetate	366/470	Det. limit 2 µg	26
Allyl alcohol	MBTH	629 (52)	Other λ 666, $m\varepsilon$ 50	7
Allyl alcohol	Dimedone	395/460		8
Allyl alcohol	J-Acid	460/518		8
Allyl alcohol	2,4-Pentanedione	410/510		8
Allyl amine	MBTH	629 (71)	Other λ 666, $m\varepsilon$ 67	7
Allyl benzene	J-Acid	460/520		8
N-Allyl-N-methylaniline	MBTH	629 (46)	Other λ 666, $m\varepsilon$ 44	7
2-Aminoacetophenone	MBTH	629 (17)	Other λ 666s (16·0)	30
2-Amino-1-alkanols[c]	CA	570	CC before oxidation	29
2-Amino-1-alkanols	MBTH	629	Also λ 666s	30
2-Aminoethanol	MBTH	629 (109)	Also λ 666s, $m\varepsilon$ 101	30
2-Aminoethanol	Ethyl acetoacetate	366/470	Det. limit 1 µg	26

2-Amino-2-hydroxymethyl-1,3-propanediol	CA	570	2 moles CH_2O from 1 mole amine	32
2-Amino-2-deoxy-D-glucose	MBTH	629 (17)	Also λ 666s, mε 15	30
Anisyl alcohol	CA	578 (8·8)	Sens. could be improved	41
Arabinose	CA	570	~94% yield of CH_2O	44
Arabinose	MBTH	620 (44)	Other λ 650s, mε 43	2
Arabitol	MBTH	627 (110)	Other λ 670s, mε 100	2
Arabitol	Dimedone	392/460		8
Arabitol	J-Acid	460/515		8
Arabitol	2,4-Pentanedione	405/510		8
Arabitol	CA	570	After ion exchange chrom.	44, 45
Aramite	Phenylhydrazine	520		46
Aramite	2,4-Pentanedione	412	Color stable 2 hrs	47
Ascorbic acid	CA	570	With periodate + H_2SO_4 distn → CH_2O	44
Azomethane	CA	560	Should work by this proc.	49
Bis-chloromethyl ether	MBTH	625 (97·0)		59
2-Bromo-1-butene	MBTH	629 (7·0)	and λ 666 (7·0)	7
2,3-Butanedione	CA	570 (~15·0)	Reaction at 100° for 1 hr	62
1,2,4-Butanetriol	MBTH	625 (77)	and λ 654 (77)	2
3-Butenenitrile	CA	570 (7·8)	50% yield of CH_2O	9
1-Buten-3-ol	MBTH	629 (50)	and λ 666 (45)	7
1-Buten-3-ol	Dimedone	395/460		8
1-Buten-3-ol	J-Acid	460/518		8
1-Buten-3-ol	2,4-Pentanedione	410/510		8
3-Buten-1-ol	MBTH	629 (48)	and λ 666 (45)	7
3-Buten-1-ol	Dimedone	395/460		8
3-Buten-1-ol	J-Acid	416/518		8
3-Buten-1-ol	2,4-Pentanedione	410/510		8
Calciferol	MBTH	660 (31)		63, 64
Camphene	MBTH	660 (17)		63, 64
Carbohydrates	CA	570 (15·7)		1, 12, 65
Carbohydrates	MBTH	620 (50)		2, 66

Table 6—*continued*

Precursor	Reagent[a]	*F* exc/em or λ_{max} ($m\varepsilon$)	Other data[b]	Ref.
Carbohydrates	2,4-Pentanedione	YG fluor.	Location on PC or TLC	33
Carboxymethylcellulose	2,7-Naphthalenediol	530	Intermediate is glycolic acid	68, 69
Cellulose acetate formal	*p*-Rosaniline-SO_2	550	Schiff's reagent method	75
Cellulose formal	*p*-Rosaniline-SO_2	590	Schiff's reagent method	76
Chloracetic acid	J-Acid	462/512	on PC or TLC	78
Chloramphenicol	CA	570		80
2-Chlorethanol	2,4-Pentanedione	412	CH_2O + ethylene glycol interference	81
Chlorogenic acid	MBTH	629 (49)	and λ 666 (45·0)	2, 7
1-Chloro-2,3-epoxypropane	CA	580	Det. limit −0·5 μg/liter air	82
Chloromethyl methyl ether	MBTH	625 (58)	Read at 14 to 30 min.	59
2-Chlorophenoxyacetic acid[d]	J-Acid	580 (22·2)	Color stable 3 hrs	84
2-Chlorophenoxyacetic acid[d]	Phenyl J-acid	635 (39·1)	Color stable 6 hrs	84
Chlorphenesin	CA	570 (15·6)	Det. in presence of its 1-carbamate	86
Cinchonidine sulfate	MBTH	629 (22)	And λ 666 (26·0)	7
Cinnamaldehyde	MBTH	629 (10)		7
Cinnamic acid	MBTH	629 (12)		7
Corticosteroids	MBTH	630 (41 to 57)	5 studied	2, 7
Corticosteroids[e]	2,4-Pentanedione	YG fluor.	Location on PC,[88] TLC[89,90]	87–90
Corticosteroids	CA	~570	Conway cell[91] or after dist.[93–94]	91–94
Corticosteroids	Ethyl acetoacetate	366/470	4 steroids investigated	26
Corticosteroids	J-Acid	460/520	6 steroids investigated	8
Cyclohexylazomethane	CA	560 (17·5)	100% yield of CH_2O	49
Cysteine	2,4-Pentanedione	410/510	CH_2S may be intermediate	11
Cytidine	MBTH	606 (12)	and λ 633 (11·0)	2
1-Decene	MBTH	629 (36)	and λ 666 (34·0)	7
2-Deoxy-D-glucose	Dimedone	392/460		8
2-Deoxy-D-glucose	J-Acid	460/515		8
2-Deoxy-D-glucose	2,4-Pentanedione	405/510		8

2-Deoxy-D-ribose	Dimedone	392/460	Fluor. stable 5 min	8
2-Deoxy-D-ribose	J-Acid	460/515	Fluor. stable >60 min	8
2-Deoxy-D-ribose	2,4-Pentanedione	405/510	Fluor. stable >60 min	8
2-Deoxy-D-ribose	MBTH	630 (53)	and λ 671 s (49·0)	2
Dextrose	MBTH	660 (22)		2
Diallyl ether	CA	570 (14·1)		9
Dibenzalacetone	MBTH	629 (18)	and λ 666 (18·0)	7
2,4-Dichlorophenoxyacetates	CA	565–580	mε 13·3 (46·0)	83, 102–104, 106, 107
2,4-Dichlorophenoxyacetates	J-Acid	580 (23·2)	Ident. limit—4 µg	84
2,4-Dichlorophenoxyacetates	Phenyl J-acid	635 (39·2)	Ident. limit—2 µg	84
Diethanolamine	MBTH	629 (125)	Also λ 666s (116)	30
Diethanolamine	CA	570 (60·4)	Detm of 20 ppb in 25 liters air	108
Dihydrostreptomycin	CA	570 (16)		113–115
Dihydrostreptomycin	Ethyl acetoacetate	336/470		26
Dihydroxyacetone	CA	570 (15·5)	~97% yield of CH_2O	44
Dihydroxyacetone	MBTH	622 (52)	Also λ 657 (49)	2
Dihydroxyacetone	Dimedone	392/460		8
Dihydroxyacetone	J-Acid	460/515		8
Dihydroxyacetone	2,4-Pentanedione	405/510		8
Dimethoate	CA	570		117
N,N-Dimethylalkylamine oxides	2,4-Pentanedione	412 (~3·9)	10^{-3} M HSO_3^- prevents color formn.	118
N,N-Dimethyl-p-aminobenzaldehyde[f]	MBTH	635 (70)	Also λ 670 (55)	120
N-(N,N-Dimethyl-4-aminobenzylidene)aniline[f]	MBTH	625 (71)	Also λ 670 (69)[g]	121
1,2-Dimethylhydrazine	CA	570 (15·7)		49
1,1-Dimethylhydrazine	CA	570 (15·7)		127
1,1-Diphenylethylene	MBTH	629 (23)	Also λ 666 (22)	7
N,N′-Diphenylethylenediamine	MBTH	629 (73)	Also λ 666 (69·0)	30
1,5-Diphenyl-3-pentadienone	MBTH	629 (22)	Also λ 666 (22·0)	7
1,8-Diphenyl-1,3,5,7-octatetraene	MBTH	629 (15)	Also λ 666 (15·0)	7

Table 6—*continued*

Precursor	Reagent[a]	F exc/em or λ_{max} ($m\varepsilon$)	Other data[b]	Ref.
Dulcitol	Dimedone	392/460		8
Dulcitol	J-Acid	460/515		8
Dulcitol	2,4-Pentanedione	405/510		8
Dulcitol	MBTH	626 (99)	Also λ 657s (93·0)	2
Erythritol	MBTH	632 (110)	Also λ 665s (104)	2,7
Erythritol	Dimedone	395/460	Very sensitive	8
Erythritol	J-Acid	460/518	Very sensitive	8
Erythritol	2,4-Pentanedione	410/510	Very sensitive	8
Erythritol	CA	570		139
Ethanolamine	2,4-Pentanedione	YG fluor.	Identification on P.C.	87
Ethanolamine	MBTH	629 (109)	Also λ 666s (101)	30
Ethanolamine	CA	570 (15·7)	Phospholipid hydrolyzate	44, 142
Ethion	CA	570		144
Ethylene	CA	570	Mole of CH_2O per mole of ethylene	9, 147, 150
Ethylene glycol	CA	570		152
Ethylene glycol	Phenylhydrazine	520	Anion determined	153
Ethylene glycol	p-Rosaniline	555		154, 155[h]
Ethylene glycol	Ethyl acetoacetate	366/470		26
Ethylene glycol	MBTH	628 (73)	Also λ 654 s (69·0)	2
Ethylene glycol	Dimedone	392/460		8
Ethylene glycol	J-Acid	460/515		8
Ethylene glycol	2,4-Pentanedione	412	Recoveries of 100 %	156
Ethylene glycol	2,4-Pentanedione	405/510		17
Ethylene oxide	CA	570	2 moles CH_2O per mole ts	159–163
Ethylene oxide	2,4-Pentanedione	412		161, 164
Ethylmorphine	2,4-Pentanedione	412		166
Eugenol	MBTH	629 (22)	Also λ 666 (20·0)	7

2-(2-Fluorenylamino)ethanol	MBTH	625 (78)	Also λ 666 (73)	30
Formaldehyde dimethylacetal	Phenylhydrazine		Measured as formazan cation	100
Formald.-tanned collagen	CA	570		101
Formald. 2,4-DNPH[i]	CA	578 (19·1)	External heat needed	41
Formald. 2,4-DNPH	J-Acid	467 (24)	External heat needed	41
Formald. 2,4-DNPH	J-Acid	580 (12.6)	External heat needed	41
Formald. 2,4-DNPH	Phenyl J-acid	655 (19·5)	External heat needed	41
Formaldehyde hydrazone	CA	578 (14·7)	External heat needed	41
Formaldehyde hydrazone	J-Acid	467 (25·5)	External heat needed	41
Formaldehyde hydrazone	J-Acid	590 (23·9)	External heat needed	41
Formaldehyde hydrazone	Phenyl J-acid	655 (34·5)	External heat needed	41
Formic acid	CA	570 (~5)	~30% yield of CH_2O	171–176, 178
Fructose	MBTH	628 (68)	Also λ 658 s (63)	2
Fructose	Dimedone	392/460		8
Fructose	J-Acid	460/515		8
Fructose	2,4-Pentanedione	405/510		8
Fructose	CA	570 (14)		44
Furfural	MBTH	629 (11)	Also λ 666 (11)	7
β-2-Furylacrylophenone	MBTH	629 (18)	Also λ 666 (17)	7
Galactonic acid	MBTH	629 (77)	Also λ 666s (74)	2
Galactose	MBTH	618 (55)	Also λ 652 (54)	2
Galactose	CA	570 (14·9)	Distillation proc.	44
Gelsemine	CA	570 (13·3)	Neutral oxidation	9
Gluconic acid	MBTH	629 (98)	Also λ 666s (93)	2
Gluconolactone	MBTH	626 (74)	Also λ 659s (70)	2
Glucosamine	MBTH	629 (17)	Also λ 666s (15)	30
Glucosamine	CA	570	~33% yield CH_2O in 45 min.	44
Glucose	CA	575 (3·0)[j]	Hexoses react[k]	1, 2, 12, 44, 182–187
Glucose	MBTH	620	Beer's law 1–20 μg ml ts.	188
Glucuronic acid	MBTH	618 (24)	Also λ 659 (24)	2

Table 6—*continued*

Precursor	Reagent[a]	*F* exc/em or λ_{max} ($m\varepsilon$)	Other data[b]	Ref.
Glyceraldehyde	CA	570		182
Glyceric acid	MBTH	621 (91)	Also λ 666s (87)	2
Glyceric acid	CA	570 (17)		44, 191
Glycerides	CA	570		193–215
Glycerides	2,4-Pentanedione	405/505	Semi-automated proc.	216–218
Glycerides	2,4-Pentanedione	412	Detm limit 10 μg	219–221
Glycerides	Phenylhydrazine	520 or 530		222–225
Glycerides	MBTH	630		227
Glycerides	2,4-Dinitrophenylhydrazine	350	Measured in hexane	228
Glycerol	CA	570	Glucose a minor interference	139, 142, 232–237
Glycerol	Dimedone	392/460		8
Glycerol	J-Acid	460/515		2, 8
Glycerol	MBTH	625 (103)	Also λ 665s (94)	2
Glycerol	2,4-Pentanedione	405/510	0·05–5-μg assayed	8, 238
Glycerol	2,4-Pentanedione	412	2–10 μg assayed	241
Glycerol	Phenylhydrazine	520	Formazan anion assayed	153, 317
Glycine	CA	580	CH_2O + precursors interfere	243, 244, 248–250
Glycolaldehyde	Pyrogallol	600	CH_2O + precursors interfere	254
Glycolic acid	CA	570 (18·5)[l]	CH_2O + precursors interfere	41, 260–265
Glycolic acid	2,7-Dihydroxynaphthalene	530 (23·2)[m]	More sens. than CA(267)	265–269
Glycolic acid	2,7-Dihydroxynaphthalene	530	CH_2O + glyoxylic acid do not interfere	270
Glycolic acid	J-Acid	612 (7·5)	Sens. could be increased	41
Glycolipids	*p*-Rosaniline	Blue	Detect lipids on TLC	272

1,2-Glycols	MBTH	~630 (100)	Other glycols would react	2, 66
1,2-Glycols	J-Acid	460/515		2, 8
1,2-Glycols	Dimedone	392/460		8
1,2-Glycols	2,4-Pentanedione	405/510		8
1,2-Glycols	2,4-Pentanedione	412	Arabitol, dulcitol + sorbitol detn	273
1,2-Glycols	*p*-Rosaniline		Staining of tissues	274
1,2-Glycols	Ethyl acetoacetate	366/470		26
N-Glycolylneuraminic acid	2,7-Dihydroxynaphthalene	535	After hydrol. +CC	267, 279, 280
Glyoxal	CA	570 (12·5)		83
Glyoxylic acid	J-Acid	375, 462/512	Characterizn on PC + TLC	78
Glyoxylic acid	2,7-Dihydroxynaphthalene	530		266
Glyoxylic acid	Phenylhydrazine	537 (49·3)	See discussion	282
Guthion	CA	570		286
1-Heptene	MBTH	629 (34)	Also λ 666 (32)	7
Hexamethylenetetramine	CA	578 (28·8)	Also λ 480 (15·6)	41, 288
Hexamethylenetetramine	J-Acid	468 (54)	Also λ 376 (38·5)	41
Hexamethylenetetramine	J-Acid	375, 462/512	Characterized on chromatogram	78
Hexamethylenetetramine	J-Acid	612 (71·3)		41
Hexamethylenetetramine	Phenylhydrazine			292
Hexamethylenetetramine	Phenyl J-acid	660 (122·5)		41
1-Hexene	MBTH	629 (30)	Also λ 666 (29)	7
1-Hexene	J-Acid	460/520	Fluor. stable >60 min.	8
Hexitols	Ethyl acetoacetate	366/470		26
Hexitols	MBTH	629 (~100)		7
Hexitols	Dimedone	392/460		8
Hexitols	J-Acid	460/515		8
Hexitols	2,4-Pentanedione	405/510		8
Hydrochlorothiazide	*p*-Rosaniline		Ring opening → CH_2O	297
Hydrocortisone	MBTH	630 (55)	Also λ 666 (53)	2
Hydrocortisone	J-Acid	460/520		8

Table 6—*continued*

Precursor	Reagent[a]	*F* exc/em or λ_{max} ($m\varepsilon$)	Other data[b]	Ref.
Hydroxyacetone	MBTH	626 (26)	Also λ 665 (25)	2
17-Hydroxy-17-ketolsteroids	HBT	580 (47·9)[n]		300
Hydroxylactone	CA	570 (10·5)		9
Hydroxylysine	MBTH	629 (57)	Also λ 666s (52)	30
Hydroxylysine	2,4-Pentanedione		Ident. on PC	87
Hydroxylysine	Phenylhydrazine	530	From protein hydrolyzate	312
Imidan[o]	CA	570	GC anal. preferred	318
3-Indoleacetic acid	CA	570 (12·2)		83
Inositol	2,7-Naphthalenediol	535		320
3-Iodopropene	MBTH	629 (27)	Also λ 666 (26)	7
Isoatisine	CA	570 (13)		9
Isoeugenol	MBTH	629 (14)	Also λ 666 (13)	7
Isosafrole	CA	578 (7·7)		41
Isotenulin	CA	411[p]		322
Itaconic acid	CA	570 (14·5)		9
Maleic acid hydrazide	MBTH	629 (23)	Also λ 666 (22)	7
Mannitol	MBTH	627 (95)	Also λ 666s (91)	2
Mannitol	Dimedone	392/460		8
Mannitol	J-Acid	460/515		8
Mannitol	2,4-Pentanedione	405/510		8
Mannitol	CA	570	Used as primary std (329)	328–330
Methacrylamide	CA	570 (12·1)		9
Methanol	CA	570	Much more sensitive than fuchsin (356)	332–356, 364, 365
Methanol	2,7-Naphthalenediol			357
Methanol	*p*-Rosaniline			358, 359
Methanol	2,4-Pentanedione	410/510		360, 366
Methanol	2,4-Pentanedione	412	Poor sensitivity	360

Methanol	Phenylhydrazine	520		361, 362, 368
Methionine hydantoin sulf-oxide	CA	580 (17·5)	Hexamethylenetetramine as std	130
Methionine sulfoxide[q]	CA	580 (10·5)	Hexamethylenetetramine as std	130
4-(p-Methoxyphenyl)-3-butene-2-one	MBTH	629 (13)		7
2-Methylaminoethanol	MBTH	629 (96)	Also λ 666s (90)	30
2-Methylaminoethanol	CA	570 (31·2)	Theoretical mε = ~35	108
Methylazobutane	CA	560 (12·0)	Also 31 % yield butyraldehyde	49
1-Methyl-2-butylhydrazine[r]	CA	560 (9·1)	Also 48 % yield butyraldehyde	49
1-Methyl-2-cyclohexylhydrazine	CA	560 (14·9)		49
N,N′-Methylene-bis-acrylamide	CA	578 (16·9)	Also λ 480 (6·4)	41
N,N′-Methylene-bis-acrylamide	J-Acid	467 (19·8)	Also λ 376 (13·2)	41
N,N′-Methylene-bis-acrylamide	J-Acid	585 (9·9)		41
N,N′-Methylene-bis-acrylamide	Phenyl J-acid	660 (39·3)		41
Methylene chloride	CA	574	Chlorinated alkanes—no interfer.	381
Methylenedioxyphenyl group[s]	CA	570	Detect on TLC (48, 390)	41, 385–391
Methylenedioxyphenyl group	Gallic acid		Solvent = H_3PO_4 + CH_3COOH	392–394
Methylenedioxyphenyl group	Tannic acid		Detect on TLC (396)	395–397
Methyl methacrylate	*p*-Rosaniline		Det. limit—35 µg/ml	403
Methylpropylhydrazine	CA	560 (6·8)	Byproduct 61 % propionaldehyde	49
α-Methylstyrene	MBTH	629 (28)	Also λ 666 (27)	7
Methylvinyl ketone	MBTH	629 (13)	Also λ 666 (13)	7
N-1-Naphthylethylenediamine	MBTH	590 (13)	Also λ 575s (12)	30
1-Nonadecene	MBTH	629 (16)	Also λ 666 (15)	7
1-Octene	MBTH	629 (34)	Also λ 666 (30)	7
Paraformaldehyde	CA	580 (12·0)		83
Phenylazomethane	CA	570 (15·8)	90 % yield of CH_2O	49
Phorate (thimet)	CA	570 (13·4)[t]	Oxygen + oxygenated analogs react	429–431
Piperine[u]	CA	580 (16·5)	Piperettine also reacts	109, 433, 434
Piperine[v]	Gallic acid	660 (23·0)	Color stable >4 hrs	435
Piperonylic acid[w]	CA	578 (16·8)	Also λ 480 (8·9)	41
Piperonylic acid[w]	Phenyl J-acid	660 (52·5)		41

Table 6—*continued*

Precursor	Reagent[a]	*F* exc/em or $\lambda_{max}(m\varepsilon)$	Other data[b]	Ref.
Piperonylic acid[w]	J-Acid	465 (21·5)	Also λ 375 (15)	41
Piperonylic acid[w]	J-Acid	612 (34·8)		41
Piperonylidene acetone	MBTH	629 (27)	Also λ 666 (25)	7
Podophyllotoxin	CA	570	Beer's law 5–40 μg	439
4-Propenylveratrole	MBTH	629 (24)	Also λ 666 (23)	7
Propylene glycol	MBTH	613 (80)	Also λ 654s (72)	2
Quinine	CA	570 (12·4)	71 % yield of CH_2O	9
Quinine	Ethyl acetoacetate	366/470		26
Raffinose	MBTH	612 (23)	Also λ 658s (21)	2
RDX[x]	2,7-Dihydroxynaphthalene	540		447
Ribose	MBTH	619 (43)	Also λ 654 (42)	2
Ribose	J-Acid	460/515		8
Ribose	CA	570 (18)		44
Safrole	MBTH	629 (46)	Also λ 666 (44)	7
Safrole	J-Acid	460/520		8
Serine	CA	570 (16)		28, 44, 142, 183, 234, 319, 450–456
Serine	MBTH	629 (100)	Also λ 666s (94)	30
Serine	2,4-Pentanedione	410/510	Automated (49)	11, 33[y], 87[y]
Serine	Ethyl acetoacetate	366/470		26
Sorbitol	CA	570	Interferences removed by degradation + CC	460, 461
Sorbitol	MBTH	631 (106)	Also λ 666s (101)	2
Sorbitol	Ethyl acetoacetate	366/470		26
Sorbitol	Dimedone	392/460		8
Sorbitol	J-Acid	460/515		8

Sorbitol	2,4-Pentanedione	405/510		8
Styrene	CA	570 (11·2)		9
1-Tetradecene	MBTH	629 (33)	Also λ 666 (32)	7
Thiazolidine-4-carboxylic acid	MBTH	670 (58)	Color stable 30 min.	484
Trimethylhydrazine	CA	570 (31)	200% yield of CH_2O	127
2,4,4-Trimethyl-1-pentene	MBTH	629 (16)	Also λ 666 (15)	7
sym-Trioxane	CA	578 (44)	Also λ 480 (25·7)	41
sym-Trioxane	Phenyl J-Acid	660 (141·9)		41
sym-Trioxane	J-Acid	467 (60·6)	Also λ 376 (39·4)	41
sym-Trioxane	J-Acid	612 (98)		41
sym-Trithiane	CA	578 (51·6)	Also λ 480 (26·8)	41
sym-Trithiane	Phenyl J-acid	655 (5·0)		41
sym-Trithiane	J-Acid	468 (22)	Also λ 376 (16·0)	41
sym-Trithiane	J-Acid	605 (23·8)		41
Vinyl chloride	CA	574	$CH_3OH + C_2H_4$ interfere	519
Xylitol	MBTH	628 (101)	Also λ 661s (97)	2
Xylitol	Dimedone	392/460		8
Xylitol	J-Acid	460/515		8
Xylitol	2,4-Pentanedione	405/510		8
d-Xylose	CA	570 (16)		44

[a] CA—chromotropic acid = 1,8-dihydroxy-3,6-naphthalenedisulfonate, J-Acid = 6-amino-1-naphthol-3-sulfonate, MBTH = 3-methyl-2-benzothiazolinonehydrazone, and phenyl J-acid = 6-anilino-1-naphthol-3-sulfonate.

[b] CC = column chromatography, PC = paper chromatography, Sens. = sensitivity, %S = relative standard deviation, TLC = thin layer chromatography, and ts = test solution.

[c] Phenylalaninol, leucinol, isoleucinol, valinol, tyrosinol, propanolamine, aspartidol, and ethanolamine.

[d] Five other chlorophenoxyacetates studied. Negative results with analogous propionic and butyric acids and with the chloracetic acids. Could also react with CA.[83]

[e] Thirteen formaldehydogenic steroids with 20,21-ketolic, 17,20,21-trihydroxy and 17,21-dihydroxy-20-keto structures gave positive results.[89] In addition twenty-seven formaldehydogenic Δ^4-3-oxo-C_{21}-steroids gave positive results after TLC separation.[90]

[f] Probable formaldehyde precursors. Many more *p*-substituted dimethylanilines listed in the paper gave positive results.

[g] Also band at λ 580, mε 84 probably due to MBTH azo dye, derived from aniline formed during the reaction.

[h] Visual estimation after TLC → (O) → reagent.

[i] DNPH = dinitrophenylhydrazone.

Table 6—*continued*

[j] At λ_{max}367, mε 6·8.[185]
[k] Pentoses do not react; disaccharides do.
[l] 80% alcoholic test solution decreases mε to 6·4;[265] mε = 13·7[81] and 13·4[41] also reported.
[m] 80% alcoholic test solution decreases mε to 3·1.[265]
[n] For triamcinolone.
[o] Could also determine barthrin, ethion, ethyl guthion, methyl trithion, carbophenothion, Bismetan, dazomet (Mylone), Dowco 184, Norda Ketal Synergist, Geigy 28029 (phenkapton), Safroxan, and sesamex in the same fashion.
[p] Not due to CH_2O. Final solution reported to have red purple color believed to be due to CH_2O chromogen.
[q] And many other derivatives with mε from 7·5 to 15·8.
[r] Negative results with 1-methyl-2-phenyl- and 1-methyl-2-benzylhydrazines.
[s] Alkaloids, synergists, etc.
[t] Ref. 429; mε 9·2.[430]
[u] Piperonal, piperonyl butoxide and safrole also determined after TLC, spraying with CA, and direct densitometry.[433]
[v] Piperic acid, piperettine, chavicine, compounds with the methylenedioxy group, glycolic, tartaric, glyceric and glyoxylic acids will respond to this test.
[w] Piperonal and β-piperonylacrylic acid also reacted.
[x] Hexahydro-1,3,5-trinitro-s-triazine.
[y] Detection on chromatograms.

These compounds can be analyzed by any of the aldehyde reagents. The ethyl acetoacetate procedure is described. 2,6-Dimethyl-3,5-dicarbethoxy-1,4-dihydropyridine, IV, is the fluorogen formed in the reaction.

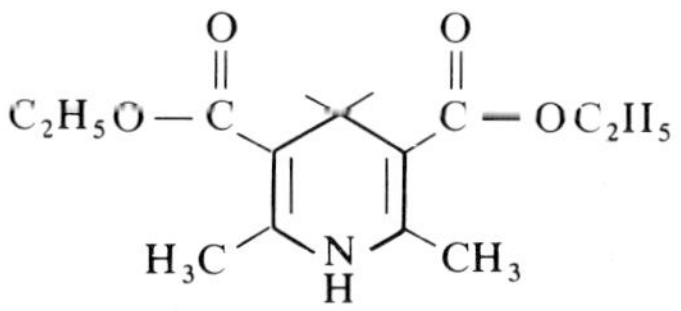

$C_2H_5O-C(=O)$ ... $C(=O)-OC_2H_5$

H_3C N H CH_3

IV

Ethyl acetoacetate determination of 2-nitro-1-alkanols.[26] To 1 ml of aqueous test solution add 1 ml of 0·5 N NaOH. After 5 min. add 1 ml 1 N HCl and 1 ml of reagent (2 % ethylacetoacetate in 20 % aqueous ammonium acetate). Heat at 60°C for 20 min. and then cool to room temperature in an ice-water bath. Read at *F*366/470.

In the determination of formaldehyde in cosmetic products with the 2,4-pentanedione fluorescence reaction interferences were found to be hydroxymethyl-5,5-dimethylhydantoin, Germall 115 (a substituted imidazolidinyl urea compound), and 2-bromo-2-nitropropan-1,3-diol (Bronopol).[61] A quantitative determination of Bronopol and the hydantoin indicated that formaldehyde was completely released from the hydantoin but not from Bronopol. If the chromotropic acid test is used to determine formaldehyde in shampoos, positive results are obtained with some of the perfume ingredients which are formaldehyde precursors.

IV. ALDOSTERONE

This steroid can be estimated through its derived formaldehyde.[14] The method is based on the oxidation of the ketol side chain of aldosterone to formaldehyde and other products. The formaldehyde is condensed with ammonia and 2,4-pentanedione to form a highly fluorescent 3,5-diacetyl-2,6-dimethyl-1,4-dihydropyridine.

V. AMINES

Some of the amines which can be oxidized to formaldehyde include 1,2-substituted aminoalkanols, 1,2-diaminoalkyl derivatives, hexosamines, glycine and methylamine, the latter two to be much more fully discussed in sections under glycine and methylamines.

A. 1,2-Aminoalkanols

Compounds of this type determined by the oxidative MBTH procedure include 2-aminoethanol, phenylephrine, serine, 2-(2-fluorenylamino)-ethanol, adrenalin, arterenol, etc. In the determination of serine Beer's law was obeyed from 0·53 ($A = 0{\cdot}1$) to 5·5 μg. Since the absorbance changed with time, it had to be read at 15 min. In spite of this factor the relative standard deviation for 12 determinations was ±1·39.[30] Serine gave a millimolar absorptivity of 100 at λ_{max} 629 nm.

B. Ethylenediamines

In these compounds the Ns can be monosubstituted with alkyl or aryl groups. Ethylenediamine gives a millimolar absorptivity of 42 at λ_{max} 629 nm.

Oxidative MBTH determination of compounds containing 2-aminoethanol and ethylenediamine structures.[30] To 0·5 ml aq. sample soln., add 1 ml 0·02 M aq. sodium metaperiodate. Mix and allow the mixture to stand 10 min. Add 1 ml 1·5 M aq. sodium metaarsenite followed by 1 ml 1 M aq. H_2SO_4 and 1 ml 0·8% aq. MBTH·HCl soln. Heat 3 min on a boiling water bath and then cool to room temp. in running tap water. Add 2 ml 0·5% aq. iron(III) chloride soln and allow the mixture to stand 15 min. Read the absorbance in 1-cm cells at 629 nm.

The manganese(III)[134,135] and cobalt(III)[133,136] ethylenediaminetetraacetic acid complexes decompose on heating to give formaldehyde. The manganese complex oxidizes free excess ethylenediaminetetraacetic acid in preference to the bound material. As for the cobalt complex it is probable that even in the presence of a ligand such as $[(NH_2CH_2{-}CH_2)_2N{-}CH_2]_2$, which preferentially stabilizes the cobalt(III) state to a very large extent, some oxidation can occur under extreme conditions.

C. Hexosamines

Hexosamines can also be determined by analogous oxidative methods. Thus, D-glucosamine gives mε 17 at λ 629 nm.[30] The sensitivity could be considerably improved by optimizing the variables.

D. Glycine

Glycine can be readily converted to formaldehyde with ninhydrin through decarboxylation and oxidative deamination (see following procedure) or through treatment with chloramine. The ninhydrin method can also be used to

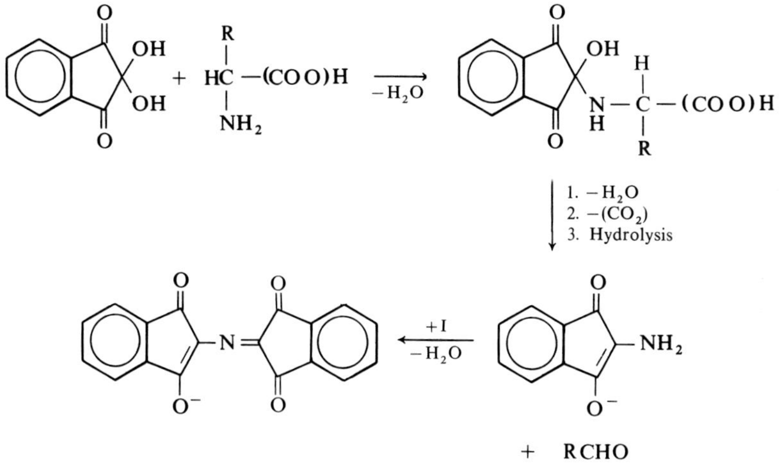

FIG. 10. Byproduct formation of aldehydes in the ninhydrin determination of α-amino acids.(552–554)

form formaldehyde from methylamine or various types of aldehydes from a variety of aliphatic amines and amino acids, as shown in Fig. 10.(552–554)

Glycine can be determined by the following ninhydrin procedure.

Chromotropic acid determination of glycine.(243) Using an all-glass distillation apparatus distill rapidly a mixture of 5 ml of a 1 to 50 dilution of urine, 2 ml of pH 5·5 phosphate buffer (3·5 g of K_3PO_4 added to 100 ml of a 20% solution of KH_2PO_4), 1 ml of 1% ninhydrin and a glass bead. After about 7 ml of liquid has been collected, cool the receiving flask to room temperature, add 2 ml of water and continue the distillation to dryness. Total time for distillation should not exceed 15 min. Make distillate up to 10 ml. Pipet 5 ml into a test tube and add 4 ml of concentrated sulfuric acid while cooling and shaking in an ice bath. Allow to cool to room temperature and add 0·1 ml 5% chromotropic acid. Shake test tube, stopper and heat at 100°C for 39 min. Cool to room temperature and read absorbance at 575 nm.

Glycine can also be determined through oxidative deamination with nitrous acid followed by disproportionation in strong acid, e.g.

$$H_2NCH_2COOH \xrightarrow{HONO} HOCH_2COOH \xrightarrow{H_2SO_4} CH_2O + CO + H_2O$$

The formaldehyde can then be determined with chromotropic acid as in the following procedure. Some of the other more sensitive or selective reagents for formaldehyde could be used here in place of chromotropic acid.

Alternative chromotropic acid determination of glycine.[249] Add 5 ml test solution containing about 500 μg of glycine to 2 ml of fresh 7% sodium nitrite. Then add 2 ml of 1:35 sulfuric acid and mix. Heat at 100° for 45 min. Cool and dilute to 10 ml with water. Add 0·2 ml of an aliquot to 20 mg of solid stannous chloride. Add 5 ml concentrated sulfuric acid followed by 0·1 ml of 5% aqueous chromotropic acid. Heat at 100° for 20 min and store in the dark for 45 min. Read at 580 nm.

See also Glycine Section.

E. Aminomethyl ketones

Aminomethyl ketones can also be oxidized to formaldehyde which can then be determined with MBTH.[30]

F. Aminopyrine

Many of these enzymatic reactions will be discussed much more fully in the methylamines section. N-Methylated derivatives such as aminopyrine, benzphetamine, codeine and ethylmorphine, can be used as substrates in the automated determination of hepatic N-demethylase activity.[37] Ethyl morphine was clearly the most readily demethylated substance and codeine and aminopyrine were also useful substrates for this enzyme. Since the administration of many drugs causes the induction of liver microsomal enzymes, liver N-demethylase activity is often used as a marker for such activity in drug metabolism and animal toxicity studies. The assay for this enzyme is based on its ability to demethylate many substrates[555] with the liberation of formaldehyde which is subsequently assayed with 2,4-pentanedione.[38, 556, 557]

The substrates could, of course, be assayed through this type of reaction.

Potent inducers of aminopyrine-N-demethylase include phenobarbital,[34] 2,3,7,8-tetrachlorodibenzo-*p*-dioxin,[34] ethylenediaminetetraacetic acid,[38, 558] probably NADH in the presence of NADPH,[559, 560] and 3-methylcholanthrene.[34] Essentially, phenobarbital treatment of the postmitochondrial supernatant fraction of a rat liver homogenate produced a 5-fold increase in the rate of N-demethylation of aminopyrine (as shown by formaldehyde production) and a 4-fold increase in NADPH-cytochrome c reductase activity. 2,3,7,8-Tetrachlorodibenzo-*p*-dioxin and 3-methylcholanthrene each produced about a 60% stimulation of aminopyrine-N-demethylase activity and about a 30% increase in the NADPH–cytochrome c reductase activity.

There appears to be a close relationship between lipid peroxidation and the activities of drug metabolizing enzymes.[558] The inhibition of lipid peroxidation by the addition of EDTA in the incubation mixture produces a marked increase in the activities of ethylmorphine, aminopyrine, and codeine demethylases. Oxidative demethylase activities were assayed with the formaldehyde precursors—ethylmorphine, aminopyrine and codeine—as substrates and the amount of formaldehyde formed was measured by reaction with 2,4 pentanedione.[556] N Demethylase activities of ethylmorphine and aminopyrine are increased twofold by addition of EDTA to the incubation mixture.

In contrast to observations in the rat and guinea pig, early human intrauterine development microsomal fractions of the whole embryo as well as of embryonal liver show increasing activity of N-demethylases, as shown by the steadily increasing amounts of formaldehyde obtained from substrate aminopyrine with aging of the embryo.[557]

Inhibition of the N-demethylases has been reported. Demethylase activities of the formaldehyde precursors—aminopyrine, ethylmorphine and codeine—are markedly decreased by the stimulation of microsomal lipid peroxidation, as for example with carbon monoxide or ascorbic acid.[38] The decrease in the activities of the oxidative demethylases is probably due to the disintegration of the microsomal membranes through lipid peroxidation. Addition of malonaldehyde, a product of lipid peroxidation, had no effect on the activity of aminopyrine N-demethylase. The inhalation of low doses of carbon disulfide (about 20 ppm) inhibits oxidative N-demethylation in man[561] and rat.[562] In line with this is the significant reduction in the hexobarbital sleeping time after an 8-hr inhalation of carbon disulfide concentrations as low as 20 ppm which is the current threshold limit value in industry. Conversely, one would expect an enhancement of hexobarbital sleeping time with pro-oxidants, such as ozone and nitrogen dioxide.

Thus, by using a formaldehyde precursor and determining the generated formaldehyde with an appropriate formaldehyde test (usually with 2,4-pentanedione) and by following the lipid peroxidation with appropriate aldehyde tests (usually with TBA), much information can be gathered on the physiological functions and the liver microsomal enzyme systems.

These various effects of N-demethylation in living tissue need much more thorough investigation since the generation of formaldehyde adjacent to highly reactive functional groups in biopolymers could have a drastic effect eventually on the physiological functioning of the living tissue.

VI. Azomethane Derivatives

Druckrey and Preussman have investigated in a series of invaluable studies the genotoxicity of a large number of alkylating agents. Three large groups of procarcinogens have been comparatively tested for organotropic and transplacental carcinogenicity, and for teratogenicity in BD— rats.[50] These three groups of procarcinogens, most of which are discussed in other parts of this section, include dialkylnitrosamines; 1-aryl-3,3- dialkyltriazenes; and 1,2-dialkylhydrazines and azo- and azoxy-alkanes. Some of the metabolic reactions, e.g. the lethal syntheses, postulated for these compounds are shown in Fig. 11. All of these compounds are procarcinogens (i.e. they need metabolic conversion to the ultimate carcinogen before a carcinogenic reaction can be

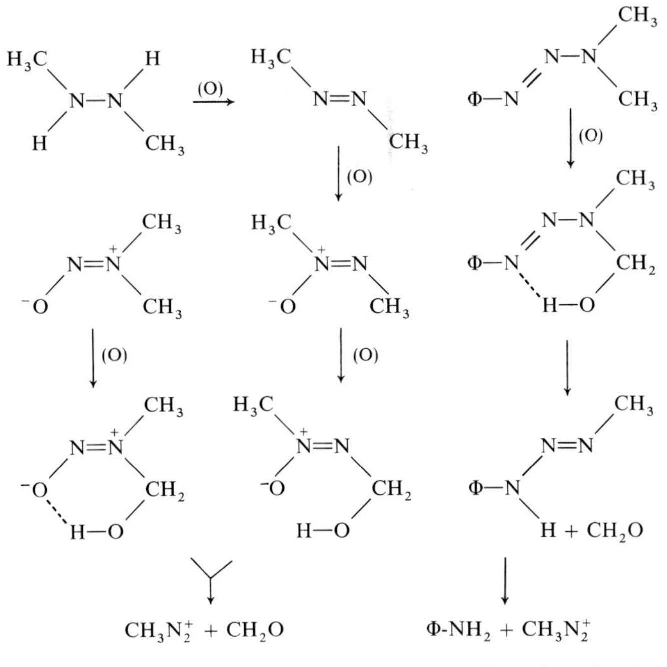

$CH_3N_2^+ + CH_2O$ $\Phi\text{-}NH_2 + CH_3N_2^+$

FIG. 11. Byproduct formation of formaldehyde in the metabolism of 1,2-dimethylhydrazine, azomethane and 1-phenyl-3,3-dimethyltriazene.[50]

obtained) as shown by the fact that subcutaneous injection in rats gave no local sarcomas but only tumors in remote organs. Preussman *et al.*[427,563] demonstrated that oxidative dealkylation by incubation of the compound with the microsomal fraction of rat liver in the presence of a NADPH-regenerating system and molecular oxygen yielded the corresponding aldehyde with all three groups of substances. The resulting monoalkyl compounds decompose to form an alkylating compound, e.g. alkyldiazonium or diazohydroxide or the alkyl carbonium ion. This alkylating compound may be the proximate or even ultimate carcinogen, as indicated by the identification of alkylated RNA and DNA in both *in vitro* and *in vivo* experiments. As to be expected the hydrazo, azo and azoxy alkanes with identical alkyl groups show the same organotropic carcinogenic effect.[50]

Aldehyde precursors, such as dimethylnitrosamine, diethylnitrosamine, dimethylhydrazine and azoxymethane are reported to be non-teratogenic to BD-rats,[50] while acetaldehyde-precursors such as 1,2-diethylhydrazine and azoxyethane are potent teratogens even at the tenth day after coitus. Methylazoxymethanol is teratogenic when given between the fourteenth and sixteenth day at a dose of 20 mg/kg.[50] The activation of dimethylhydrazine and azoxymethane to the "proximate" carcinogen probably requires enzymic C-hydroxylation in the fetus, which is not possible. The hydroxylases for the ethyl derivatives are different and specific so that de-ethylation occurs at a significantly earlier stage of fetal development than demethylation. The role, if any, of the formaldehyde (and acetaldehyde) formed in these dealkylations during the genotoxic phenomena has not as yet been studied.

A. Azoalkanes

Alkyl and aryl azomethanes can be readily tautomerized to the hydrazone from which formaldehyde can be derived, e.g.

$$R{-}N{=}N{-}CH_3 \longrightarrow R{-}NH{-}N{=}CH_2 \xrightarrow{H_2O} R{-}NH{-}NH_2 + CH_2O$$

This reaction will be discussed more fully in the hydrazines section.

For the determination of the azomethanes the chromotropic acid method described in the next paragraph can be used. If necessary, some of the more sensitive colorimetric or fluorimetric formaldehyde reagents can be substituted for chromotropic acid.

CA determination of azomethanes.[49] To 0·1 to 2 μmole of the azoalkane in water in a 50 ml distilling flask add 4 ml of 10% aqueous sulfuric acid and water to a volume of 25 ml. To 10 ml of the distillate add 2 ml of 5% chromotropic acid and then dilute to 25 ml with concentrated sulfuric acid with cooling. Heat 30 min on the water bath; cool to room temperature and read at λ 560 nm.

B. Azoxymethanes

The enzymatic α-hydroxylation of the carcinogen azoxymethane results in the formation of methylazoxymethanol which also appears to be the proximate carcinogen of cycasin. Methylazoxymethanol is unstable and readily decomposes to formaldehyde and an alkylating group.

The initiation of cancer in living tissue by a carcinogen seems to involve a chemical "memory", sometimes one of long duration. The "memory" seemingly consists of a new chemical structure stable over long periods of time. Druckrey has given data showing that brain tumors can be developed in rats 500 to 800 days after a single transplacental dose of carcinogen.[50] Thus, neurogenic cancer may be caused during intrauterine life when the sensitivity of the nervous system is so exceedingly high. Formation of alkylating agents (and aldehydes?) appears to play an important role in this process. Of some interest is Druckrey's report that azo- and azoxy-alkanes, as well as dialkyltriazenes undergo de-ethylation (acetaldehyde formation) at a significantly earlier stage of fetal development than demethylation (formaldehyde formation).

The cancer memory may even extend to later generations. In some species the carcinogenic effects of chemicals administered to pregnant animals may extend not only to offspring exposed in utero, but to later generations as well. Thus, with 7,12-dimethylbenz[a]anthracene in mice[564] and nitrosomethylurethan in rats[565] a significantly increased tumor incidence was observed in untreated second generation animals when both parents, or the female parent alone, had been exposed transplacentally to a carcinogen. Tanaka postulates that this effect is mediated by damage to the female germ cells, the multiplication of which is confined to fetal life. Obviously, this concept has important implications for the genesis of cancer and short and long term inheritance. It constitutes an important new area of research that needs accelerated study. Interestingly enough both carcinogens utilized in these investigations could be formaldehyde precursors in living tissue under appropriate conditions.

C. Cycasin

Cycasin (methylazoxymethyl-β-D-glucopyranoside), V, and its aglycone methylazoxymethanol, VI, are alkylating agents (and formaldehyde precursors) which are found in the seeds, roots and leaves of cycad plants.

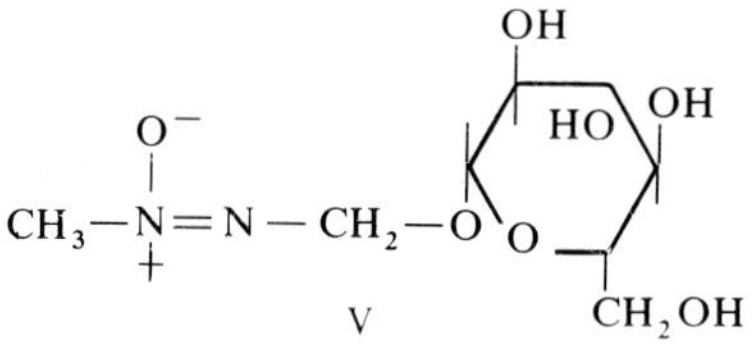

$$CH_3-\overset{\overset{O^-}{|}}{\underset{+}{N}}=N-CH_2-O-C_6H_{10}O_5 \text{ (glucopyranosyl: OH, HO, OH, } CH_2OH) \quad \text{V}$$

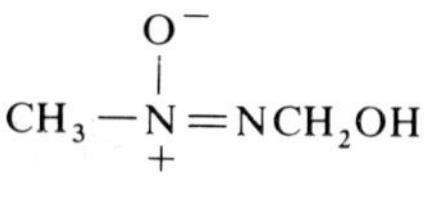

$$CH_3-\overset{\overset{O^-}{|}}{\underset{+}{N}}=NCH_2OH \quad \text{VI}$$

There is a high incidence of human and neurological disease in areas of the world where cycads are utilized as foods and medicines. The background facts on cycasin have been reviewed.[566–570]

Cycasin is a procarcinogen that produces tumors in adult animals only when given orally.[571] In the intestines the glucose moiety is removed by microbial glucosidases, and the proximate carcinogen spreads out into the tissues, maternal and fetal, where it then spontaneously decomposes to the methyldiazonium salt and formaldehyde. Methylazoxymethanol appears to be the proximate carcinogen of cycasin, since it causes neoplasms after subcutaneous and intraperitoneal injection as well as feeding while cycasin causes tumors in mature animals only if fed.[572]

Cycasin is a transplacental carcinogen in rats[573] while methylazoxymethanol is teratogenic in the golden hamster[574] and mutagenic in Drosophila[575] and *Salmonella typhimurium*.[576]

Cycasin present in the kernels of *C. circinalis* can be extracted in 80% methanol, chromatographed on Whatman No. 1 paper with n-butanol–acetic acid–water (4:1:1), and assayed for its combined formaldehyde with chromotropic acid.[100] The values obtained were comparable to the chromotropic acid method of Matsumoto and Strong.[99] Thin-layer chromatographic procedures can also be used for the separation of cycasin and methylazomethanol from complex mixtures.[577] The developing solvent can be n-butanol–acetic acid–water (4:1:1), benzene–carbon tetrachloride (2:1), or ethyl acetate–hexane.[570] Chromotropic acid, 2,4-pentanedione or MBTH can be used to assay for these formaldehyde precursors.

Another formaldehyde precursor with methylazoxymethanol as the aglycone is macrozamin (β-primeverosyloxyazoxymethane) which was isolated from the seeds of Australian cycad, *Macrozamia spiralis*.[578] Macrozamin occurs in the seeds of a number of Australian cycads.[579]

Another somewhat similar formaldehyde precursor is methylazoxymethanol acetate. This compound is also a potent carcinogen; a single dose of 35 mg/kg given to rats produced intestinal and liver tumors after 6–7 months.[580] Nucleic acids are alkylated in the fetal rat brain by this compound.[581] Formaldehyde and the methyl diazonium and methyl cations would be the reactive products formed in the *in vivo* breakdown of methylazomethanol acetate. The acetate is readily hydrolyzed to formaldehyde, especially under alkaline conditions.

In vivo aldehyde formation in tissues just before and during the carcinogenic reaction needs a more thorough study, especially just before, during, and after the stages of proximate or ultimate carcinogen formation.

VII. BILIRUBIN

The reaction between bilirubin and diazotized sulfanilic acid is used to determine the concentrations of bilirubin and its conjugates in sera. Essentially bilirubin reacts with diazotized aromatic amines and is cleaved at the central methylene bridge to form two isomeric azo pigments and formaldehyde, Fig. 12.[55]

A somewhat similar reaction has been postulated for the determination of bilirubin with oxidized 2-hydrazinobenzothiazole.[56] The method depends on the oxidation of the hydrazine with ferric chloride to a diazonium salt which splits the bilirubin molecule and couples with the 2 dipyrrylmethene molecules

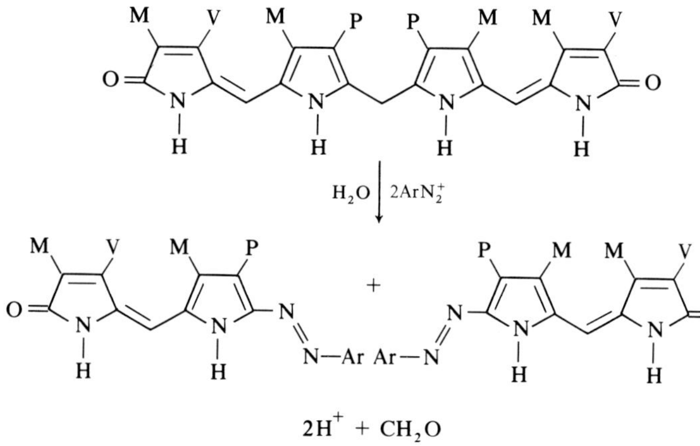

$2H^+ + CH_2O$

FIG. 12. Byproduct formation of formaldehyde in the determination of bilirubin with a diazonium salt.[55]

with the liberation of formaldehyde. The hydrazine does not react with the liberated formaldehyde since the reaction takes place in dimethylsulfoxide solution. The same type of reaction takes place with combined bilirubin.

In a somewhat similar fashion urinary urobilinogen, VII, can be determined with *p*-methoxybenzenediazonium tetrafluoroborate.[517]

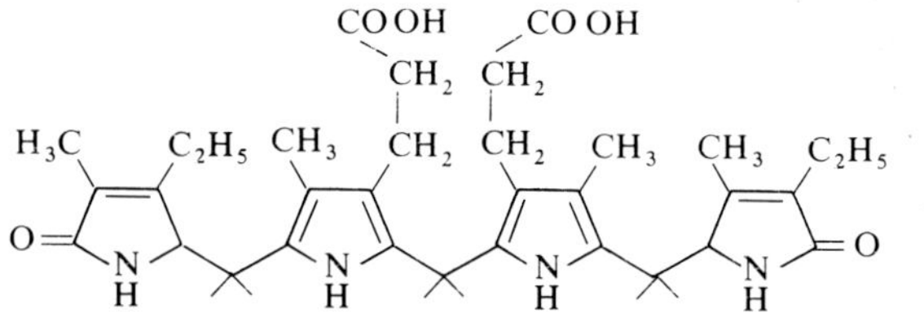

VII

Again, the byproduct is formaldehyde.

VIII. α-CHLOROMETHYL ETHERS

In these types of formaldehyde precursors the question that faces thinking people is whether any amount of these carcinogens in the human environment is dangerous and unallowable or whether many of these valuable chemicals (one thinks of DDT and malaria) should be available for its beneficial uses but kept below some environmental level. Industrial spokesmen have pointed out that if ethyleneimine, used to make an animal dewormer, were suddenly unavailable, the health and safety of $26 billion to $52 billion worth of beef cattle would be jeopardized.[582] And a ban on the chloromethyl ethers, used to make ion exchange resins, would aggravate the energy crisis according to these

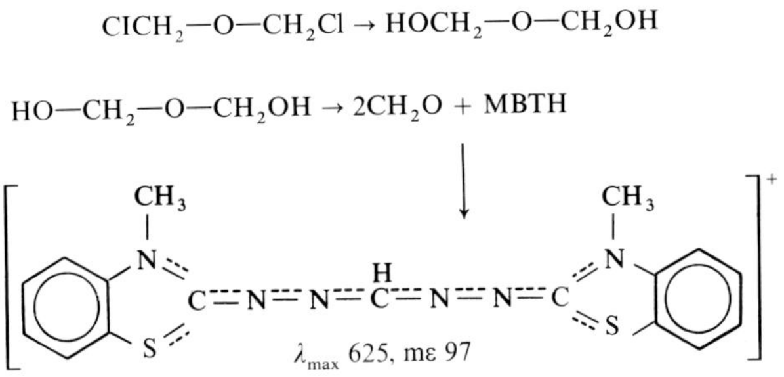

$$ClCH_2{-}O{-}CH_2Cl \rightarrow HOCH_2{-}O{-}CH_2OH$$

$$HO{-}CH_2{-}O{-}CH_2OH \rightarrow 2CH_2O + MBTH$$

λ_{max} 625, mε 97

FIG. 13. Reactions in the determination of bis-chloromethyl ether with MBTH.

spokesmen. Because of the ultrapure water produced by use of these resins, 10% less fuel is needed to produce electrical power now than in the 1950s. The problem here is whether the effects of various environmental carcinogens are additive and whether cofactors are present that could enhance the activity of the carcinogen considerably. In the case of bis-chloromethyl ether which appears to be a potent animal[583, 584] and human[585–587] lung carcinogen, a knowledgable human in continual contact with such a carcinogen would worry and/or break contact. Rats exposed to 0·1 ppm in air gave a high incidence of squamous carcinoma of the lung as well as nasal tumors extending into the brain.[584]

Bis-chloromethyl ether can be hydrolyzed to formaldehyde and determined with MBTH, Fig. 13, by the following procedure.

MBTH determination of bis-chloromethyl ether.[59] Following separation, extraction or distillation collect the ether in warm water. To 1 ml of this aqueous test solution add 1 ml of aqueous 0·8% MBTH·HCl followed by 2 ml of 0·5% aqueous ferric chloride. Read the absorbance at λ_{max} 625 nm at 14 to 40 min after the addition of the last solution. Bis-chloromethyl ether gives an mε of 97 at λ 625 nm while chloromethyl methyl ether gives an mε of 58·0. The α-chloroalkyl ethers also react.

Bis-chloromethyl ether can be synthesized in reactions between formaldehyde and hydrochloric acid (or their precursors) in aqueous solution[588–590] or in moist air.[590] Reactants used to prepare the ether in solution include HCl and aqueous formaldehyde; HCl and a sulfuric acid solution of paraformaldehyde; paraformaldehyde in concentrated HCl with chlorosulfonic acid; HCl and hexamethylenetetramine; and phosphorous trichloride, zinc chloride and paraformaldehyde. The ether is a byproduct in the zinc chloride—catalyzed reaction of phthalyl and acyl chlorides with paraformaldehyde.[588] Vapors above formaldehyde slurries of Friedel–Crafts chloride salts contain the ether.[590] These authors suggested that any acidic solution which contains chloride ion and formaldehyde should be considered a potential source of bis-chloromethyl ether.

Chloromethyl methyl ether has also been determined in industrial atmospheres.[541] Chromotropic acid in sulfuric acid was used to determine the derived formaldehyde.

A permeation tube containing bis-chloromethyl ether has been used as the gas standard in the atmospheric studies.[590]

These reactions involving the formation of the chloromethyl ethers have not been studied too thoroughly yet. Coexisting pollutants will be found which will catalyze and accelerate the formation of these types of carcinogens.

Since the chloromethyl ethers are formaldehyde precursors they can be analyzed or studied *in vivo* with the help of many of the reagents discussed in this and the formaldehyde sections.

Atmospheric bis-chloromethyl ether and other chloromethyl alkyl ethers would be determined as formaldehyde in the chromotropic acid procedure used in the analysis for atmospheric formaldehyde.

IX. LIGNIN PRODUCTS

The liberation of formaldehyde from spruce lignin and lignin model compounds on 4 hr refluxing with 0·2 M HCl in dioxane–water (9:1) indicates that the formaldehyde produced from lignin arises primarily from structural elements of the arylglycerol-β-aryl ether, phenylcoumaran, and 1,2-diaryl-1,3-propanediol types.[58] A proposed reaction route for the formation of homovanillin and formaldehyde from guaiacylglycerol-β-(2-methoxyphenyl)ether is shown in Fig. 14.[58] The formaldehyde was determined by the chromotropic acid procedure.

X. CELLULOSE AND DERIVATIVES

A. Carboxymethylcellulose

The determination of this compound in food products consisted of the following steps:[71] digestion of proteins in the sample with papain,

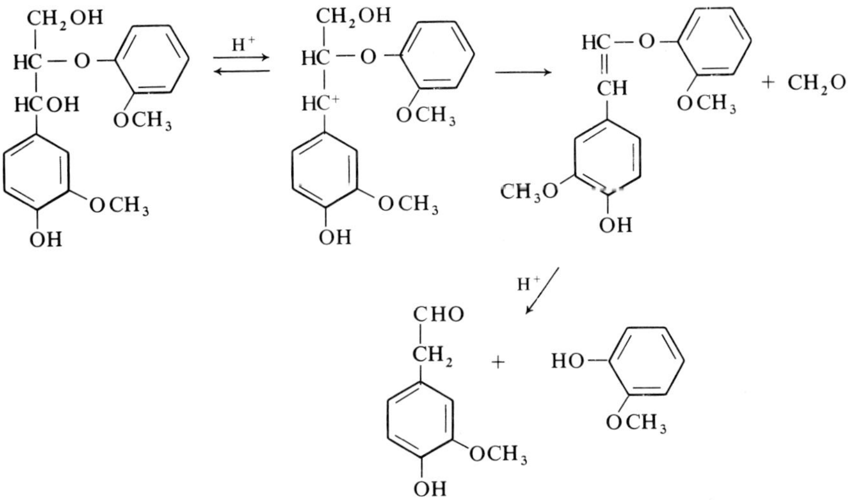

FIG. 14. Reaction route for the formation of homovanillin from guaiacylglycerol-β-(2-methoxyphenyl)-ether on acidolysis.[58]

precipitation of algin and pectic acid with calcium chloride, precipitation of sulfated hydrocolloids with cetylpyridinium chloride in the presence of 0·5 M NaCl, addition of Celite 535 or Hyflo Super-Cel, filtration, washing of residue with 0·01% cetylpyridinium chloride–0·01 M NaCl, dilution of combined filtrate and washings with water to give a solution 0·2 M in NaCl, precipitation of carboxymethylcellulose by cetylpyridinium chloride, filtration over a Celite 535 or Hyflo Super-Cel column, washing of residue with 0·5 M Na_2SO_4 and then with 0·01% cetylpyridinium chloride–0·01 M NaCl until the washings gave a negative reaction to phenol–H_2SO_4 reagent, washing of residue on column with hot 30% H_2SO_4, and determination of the hydrolyzed carboxymethylcellulose at 540 nm after reaction with naphthalene-2,7-diol–H_2SO_4 reagent.

With chromotropic acid as reagent in place of naphthalene-2,7-diol readings are made at 570 nm.[70] Down to 0·025% of carboxymethylcellulose can be determined with little interference from chocolate ingredients. Recoveries are greater than 80% in the presence of all food gums and 75% in the presence of starch.

B. Cellulose

To determine the aldehyde end-groups in cellulose, these groups are conveniently reduced to alditol end-groups by reduction with sodium borohydride.[74] After acid hydrolysis of the cellulose and removal of the mineral acid, ethanol is added and part of the D-glucose is removed by crystallization. The solution is chromatographed on a column (4·5 × 930 mm) of a cation exchange resin (Dowex 50W-X8, 14–17 μm) in the lithium form with elution by 85% (w/w) ethanol at 75°. The eluate is analyzed automatically in a two-channel analyzer. In one channel, the sugars are determined by the orcinol method,[591] and in the second channel, the alditols are determined by the periodate—formaldehyde method.[592]

Alditols are stable during acid hydrolysis while glucose gives rise to appreciable amounts of 5-hydroxymethyl-2-furaldehyde, levulinic acid, 1,6-anhydro derivatives and trace amounts of several monosaccharides. Under the applied working conditions, the artifacts do not interfere with the determination of the alditols.[74]

XI. CHLOROACETATES

Hydrolysis of these compounds in sulfuric acid forms glycolic acid which then can be made to disproportionate to formaldehyde, carbon monoxide and water. The glycolates will be considered more fully in a later section.

XII. 3-(4-CHLOROPHENYL)-1-METHYLUREA

A mixed function oxidase system which N-demethylates 3-phenyl-1-methyl ureas is found in the microsomal fraction of cotton leaf and requires molecular

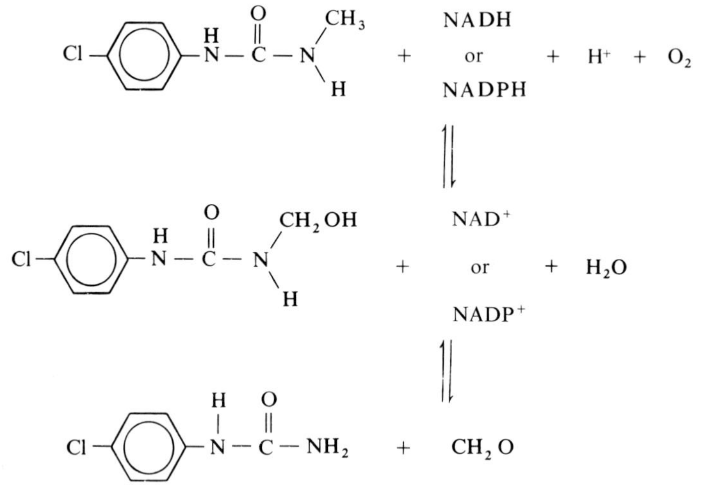

$$Cl-C_6H_4-NH-CO-NH-CH_3 + \text{NADH or NADPH} + H^+ + O_2$$

$$\rightleftharpoons$$

$$Cl-C_6H_4-NH-CO-NH-CH_2OH + \text{NAD}^+ \text{ or } \text{NADP}^+ + H_2O$$

$$\rightleftharpoons$$

$$Cl-C_6H_4-NH-CO-NH_2 + CH_2O$$

FIG. 15. Formaldehyde formation during the metabolic demethylation of N-(*p*-chlorophenyl)-N′-methylurea.[85]

oxygen and reduced pyridine nucleotides as cofactors.[85] The overall reaction for the enzymatic N-demethylation of 3-(4-chlorophenyl)-methylurea is shown in Fig. 15. The method could probably be used to determine 3-phenyl-1-methylureas.

XIII. CHOLESTEROL

This steroid could be determined with MBTH or some other reagent selective for formaldehyde. Utilizing a procedure similar to that of Sahagian and Levine,[593] cholesterol could be precipitated with digitonin. This latter reagent contains glucose and galactose. The digitonide is washed, oxidized with periodate, and the resultant formaldehyde determined with MBTH or some other appropriate reagent.

An alternative enzymatic method has been described for the determination of total cholesterol in serum through the determination of derived formaldehyde.[594] The cholesterol esters in the serum are hydrolyzed quantitatively into free cholesterol and fatty acids by cholesterol esterase (sterol-ester hydrolase, EC 3.1.1.13). The free cholesterol is oxidized by oxygen with the help of cholesterol oxidase (cholesterol: oxygen oxidoreductase, EC 1.1.3.6) into Δ^4-cholestenone. The hydrogen peroxide produced in the reaction oxidizes methanol to formaldehyde in the presence of catalase (hydrogen-peroxide: hydrogen-peroxide oxidoreductase, EC 1.11.1.6). The formaldehyde is then reacted with 2,4-pentanedione and ammonium ion to form 3,5-diacetyl-1,4-dihydrolutidine, which is determined colorimetrically at 405 nm. The accuracy and precision of the method are stated to be very good. The method correlates very well with an acetic anhydride–sulfuric acid method for cholesterol.[595] Beer's law is followed from 2·5 to 25·86 m mol cholesterol/l. serum. Serum proteins, bilirubin, creatinine, hemoglobin and drugs do not interfere.

XIV. 2-DEOXYRIBOSE

This sugar can be oxidized by periodic acid to formaldehyde and the latter determined with 2,4-pentanedione colorimetrically at λ_{max} 410 nm or fluorimetrically at F410/510 or with MBTH at λ_{max}623 nm. The 2 procedures which need to be further optimized are as follows.

2,4-*Pentanedione determination of* 2-*deoxyribose*. Let stand for 20 min at room temperature 1·4 ml of aqueous test solution and 0·2 ml of 0·5% H_5IO_6 in 1:280 H_2SO_4. Add 0·4 ml of 2% $NaAsO_2$ in 1:20 HCl followed by 2 ml of the reagent (15 g ammonium acetate, 0·3 ml of glacial acetic acid and 0·2 ml of 2,4-pentanedione diluted to 100 ml with water). Heat for 10 min at 60°. Cool and either read the absorbance at 410 nm or the fluorescence intensity at *F*410/510.

MBTH determination of 2-deoxyribose. To 0·5 ml of aqueous test solution add 0·25 ml of 0·025 M H_5IO_6 in 0·125 N H_2SO_4. Heat at 37° for 30 min. Add 0·2 ml of 2% $NaAsO_2$ in water followed by 2 ml of 0·4% aqueous MBTH•HCl. Heat at 100° for 3 min. Cool and add 10 ml of 0·2% aqueous ferric chloride. Read the absorbance at λ_{max} 623 nm. An mε of 48 is possible at this wavelength.

XV. 1,1-DICHLOROETHYLENE OR VINYLIDENE CHLORIDE

This compound has been shown to be mutagenic to *E. coli* K 12 strain[596] carcinogenic to rats,[597] and hepatotoxic to rodents.[598]

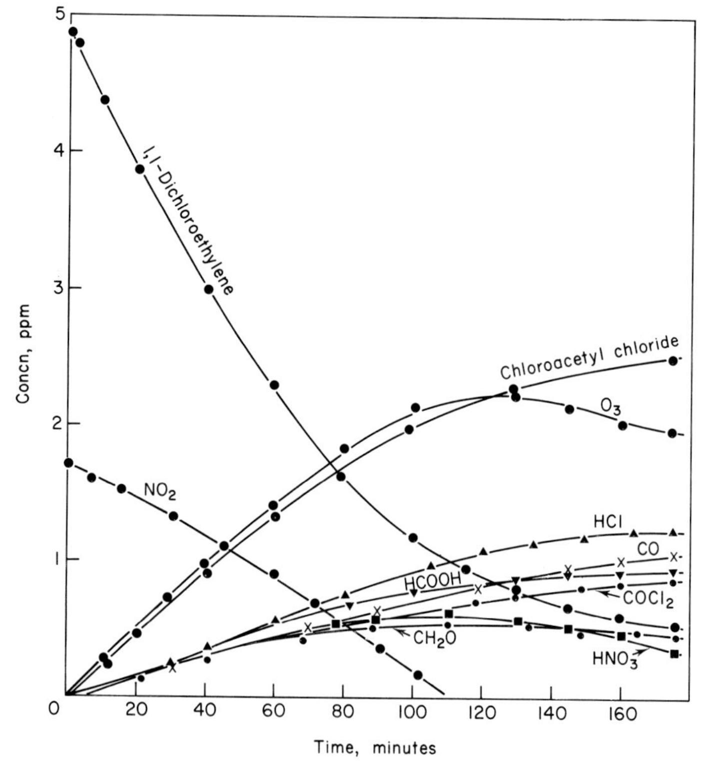

FIG. 16. Products and their variation with time from the irradiation with ultraviolet light of a mixture of 1,1-dichloroethylene and NO_2.[545]

Ultraviolet irradiation of this compound in the presence of nitrogen dioxide produces large amounts of chloroacetyl chloride and ozone, Fig. 16.[545] In addition to formaldehyde, hydrochloric acid, formic acid, phosgene, carbon monoxide and nitric acid are produced. On the other hand irradiation of 1,2-dichloroethylene has resulted in the production of hydrochloric acid, ozone, carbon monoxide, formic acid and nitric acid, but no formaldehyde.

XVI. DIMETHYLNITROSAMINE

A. Introduction

A large number of alkylating agents, many of which are mutagens and/or carcinogens, are known.[599–601] Qualitative tests have shown the presence of alkylating agents of unknown structure in various combustion products and in organic airborne particulates.[59,602] Since dialkylnitrosamines are carcinogenic alkylating agents, their presence in the human environment has to be determined. They may be present in cigarette smoke condensate[603,604] and have been found in wheat flour, smoked fish, and a product known as "liquid smoke".[605] Since they are formed by the reaction between a secondary amine, nitric oxide and nitrogen dioxide,[604] they could be formed in atmospheres or lungs containing these three components.

We will arbitrarily pick for discussion the simplest nitrosamine (which is also a well-known formaldehyde precursor), dimethylnitrosamine, as an example of the large class of nitrosamines many of which are carcinogenic and mutagenic and in some cases even teratogenic. We will discuss very briefly the concentrations of this carcinogen in the environment, its formation, its stability and its physiological activity. We will briefly discuss some aspects of the metabolism of this molecule but will consider this aspect of enzymatic activity in a later section on methylamines. Our main interest in this compound will be in terms of its ability to form formaldehyde.

Since the measurement of formaldehyde has proved to be vital to the study of (i) the metabolism of Me_2NNO and (ii) the relationship between dimethylnitrosamine N-demethylase activity and the toxic and genotoxic effects of Me_2NNO and since nitrosamine aldehyde formation (to be discussed in a subsequent volume) may play an important role in the genotoxic pathways of the nitrosamines, we will discuss the genotoxicity of Me_2NNO (and other nitrosamines) at some length.

B. Source and environmental concentrations

Let us consider 8 ways in which dimethylnitrosamine (and other nitrosamines) could arise in the environment or in living tissue. (i) Leakage from an industry into the ambient air. Examples of this type are found in the following studies, especially for some of the values obtained in Baltimore. In 1975 utilizing a cold trap for collection, a gas chromatograph for separation, and a chemiluminescence detector for the analytical finish, Fine *et al.*[606] reported the presence of dimethylnitrosamine in the urban atmospheres of Philadelphia, Baltimore, Md. and Belle, West Virginia. In this first study levels of 0 to 0·03 μg/m³ were found in Philadelphia, 0·01–0·9 μg/m³ in Baltimore, and 0·01–0·05 μg/m³ in Belle, West Virginia.

In a limited study lasting several weeks dozens of samples were taken at and near a plant (utilizing dimethylnitrosamine) by 2 different groups using 2 different sampling–analytical procedures—one group used the cold trap–GC–chemiluminescence detector procedure,[607] the other collected on Tenax GC and analyzed by GC–MS–COMP.[608] The plant process consisted of reacting dimethylamine with nitric oxide and nitrogen dioxide to form dimethylnitrosamine, and this latter compound reduced to the rocket fuel, 1,1-dimethylhydrazine. Values ranging from 0·4 to 13 μg $(CH_3)_2NNO$/m³ were found in the parking lot of this plant. Values ranging from 10 to 36 μg/m³ were obtained on the plant grounds.

(ii) Synthesis from the reactions of aliphatic secondary, tertiary and quaternary amines with NO_x ($x = 1$ or 2) in the air. Some of the values obtained in Belle, W. Virginia, Philadelphia, New York City and Baltimore could be due to this type of phenomenon. This reaction has been postulated for measurements made a few years ago in Germany, e.g. measurements of the air in plants (where dimethylamine is produced or processed and nitrogen oxides are present in relatively higher concentrations) showed the presence of the formaldehyde precursor, dimethylnitrosamine, in concentrations between 0·001 and 0·43 ppb.[609]

(iii) Industrial runoff of the dimethylnitrosamine in water effluents. On the grounds of a Baltimore plant utilizing dimethylnitrosamine, 196 ppb of dimethylnitrosamine were found in the soil and 6000 ppb in a rain puddle. The nitrosamine could have been washed out of the air into the soil. Dimethylnitrosamine is stable at room temperature for more than 14 days in aqueous solutions at neutral and alkaline pH in the absence of light.[610] It is slightly less stable at strongly acid pH at room temperature and is especially sensitive to ultraviolet light. It is reported to be stable for 3·5 months in lake water.[611] It disappears slowly from sewage but a minimum of 50% remains after 14 days. It appears to be stable for about 30 days in soil before it starts to disappear slowly.[611,612] However, once leached into puddles, rivers or lakes its persistence might be considerably shortened if the waters are naturally oligotrophic and are irradiated with sunlight.[613]

(iv) Synthesis from aliphatic secondary, tertiary and quaternary amines reacted with nitrite in water, sewage, soil or food.

Nitrosamines have been found in a large variety of foods, beverages, cigarette smoke, the aquatic environment, ambient atmospheres and in other miscellaneous sources. They are usually not detected in most foods. In the vast majority of foods in which nitrosamines have been found the amounts range up to 10 μg/kg, e.g. 1 μg of dimethylnitrosamine per kg of frankfurters,[614] 0 to 8 μg of dimethylnitrosamine per kg of bologna,[615] and 0 to 10 μg of dimethylnitrosamine per kg of cider distillate.[616] Occasionally fairly large amounts of nitrosamines have been found, e.g. 10 to 207 μg of nitrosopyrrolidine per kg of fried bacon (confirmed by mass spectrometry),[617] 15 to 75 μg of nitrosopyrrolidine per kg of fried bacon,[618] 2300 to 421 000 μg of dimethylnitrosamine per kg of winter herring,[619] and 40 to 1000 μg of dimethylnitrosamine per kg of salted white herring (some values confirmed by mass spectrometry).[620] The nitrosamines are usually formed in food through reaction of an appropriate amine and nitrous acid. The formation of dimethylnitrosamine in fish meal from fish preserved with nitrites has been shown to be proportional to the concentration of dimethylamine and to the square of the concentration of nitrite.[621] It has also been postulated that nitrosation of simple aliphatic tertiary amines also leads to the formation of dimethylnitrosamine which in this instance is proportional to the concentration of the amines and to the cube of the concentration of nitrites.[380, 622, 623] Thus, in both of these cases the concentration of the nitrite is the limiting factor.

The range in concentration of environmental nitrosamines has been reported.[624] Probably one of the most exceptionally high concentrations of a nitrosamine was that reported for N′-nitrosonornicotine in raw commercial U.S. tobacco.[625, 626] Values as high as 90 ppm were reported in chewing tobacco.

There is a possible microbial contribution to nitrosamine formation in sewage and soils.[627] Dimethylnitrosamine has been formed in soils treated with nitrogen compounds[628] and in samples of treated sewage and lake water containing methylamines.[629] Amines occur naturally in sewage and dimethylamine and diethylamine are found in fish products.[630, 631] Since these types of amines are in gaseous and liquid effluents from various types of industrial pollution sources, they would end up in streams, rivers and in the runoff water from sewage. Nitrates are widely distributed in nature, particularly in plants and some well waters; nitrites are readily formed from them, especially in the presence of bacteria containing nitrate reductase enzyme systems. Fairly large amounts of nitrate are found in runoff water from sewage, agricultural land and feedlots.

Fine and Rounbehler[632] found dimethylnitrosamine present in the ocean around Curtis Bay near Baltimore at the 0·1 to 0·3 μg/liter of water level. They believed that the total amount of dimethylnitrosamine present in the water to be too large to be derived from air leaks from the industry utilizing dimethylnitrosamine. Since the water effluent from an adjacent sewage treatment facility contained dimethylnitrosamine at the 3 μg/l level, they believed that there was a strong possibility that the source of the dimethylnitrosamine in the ocean could, in part, be the sewage treatment facility. Whether the nitrosamine is synthesized in the sewage, is an industrial effluent or is derived in some other fashion is still to be determined.

Fine and his group isolated the nitrosamines from water either by chloroform extraction or by adsorption onto carbon followed by extraction with chloroform and alcohol. The volatile nitrosamines were then analyzed by GC–chemiluminescence detection while the non-volatile nitrosamines were analyzed by high performance liquid chromatography–chemiluminescence detection.[632,633]

(v) Synthesis from reaction of aliphatic secondary, tertiary or quaternary amines with nitrate, nitrite or NO_x in living tissue (e.g. respiratory, digestive or urinary tracts). A large number of examples of this type of synthesis is available in the literature. For example, the formation of nitrosamines in the digestive tract of experimental animals has been widely demonstrated[634–636] and has led to the induction of tumors characteristic of nitrosamines.[637–639] The factors affecting the formation of nitrosamines have been discussed.[640] Because of the possibility that such a lethal synthesis can take place in living tissue nitrosamine artifact formation during environmental sampling is highly suggestive of possible physiological problems. However, in the preliminary sampling investigations of Fine *et al.*[607] and Pellizari *et al.*[608] artifact formation was attempted without success.

In attempts to explain the high frequency of cancer of the uterine cervix in southern Africans[641] pooled cervical and vaginal discharges were analyzed for nitrosamines.[642] Dimethylnitrosamine was identified. However, the importance of this report cannot be ascertained until it is confirmed and quantitated.

Many compounds can act as catalysts for nitrosamine formation. For example, readily oxidisable phenolic compounds can catalyze the formation of nitrosamines from secondary amines and nitrite at gastric pH.[643] Phenols postulated as playing this role include *o*- and *p*-dihydroxybenzene derivatives and *o*- and *p*- aminophenols. Positive cocarcinogenic results were obtained with 4-methylcatechol or chlorogenic acid, a major soluble constituent of coffee. The authors postulate that coffee and other foodstuffs containing readily oxidized phenolic materials may significantly increase human exposure to carcinogenic nitrosamines by catalyzing their formation in the digestive tract.[643] The ease of nitrosation of the amine can be influenced by the basicity of the amine,[636] pH,[621] substrate concentration,[621] the thiocyanate ion,[644] the chloride anion,[622,645,646] the acetate ion[646] and formaldehyde.[647]

Because of the importance of pH in the formation of nitrosamines it should be emphasized that large differences in pH can be found in the living system. For example, the difference between the H ion concentrations at the 2 outer cell boundaries of the gastric mucosa is approximately a millionfold.[648]

Drugs containing dimethylamino groups have been shown to react with nitrite to produce dimethylnitrosamine. Thus, anyone taking these drugs and ingesting large amounts of nitrate, nitrite or NO_x could be in a potentially high risk group. This synthesis could take place in the stomach where suitable acidic conditions and a significant intake of dietary nitrite are possible. Thus, in the reaction between aminopyrine and nitrite at pH 1–2 or pH 3–4 per cent yields of 31 and 71 of Me_2NNO were obtained.[649] Oxytetracycline reacts with nitrite at pH 3–4 to give a 63 % yield of Me_2NNO. Similar results have been reported for aminopyrine and oxytetracycline previously.[650]

The coadministration of aminopyrine and sodium nitrite by the oral route produced liver tumors in rats identical to those produced by dimethylnitrosamine.[651,652]

Data has been presented which indicates that thiocyanate increases both the rate of reaction and the final yield *in vitro* of dimethylnitrosamine from aminopyrine over the whole range of activity.[653] At pH 1·0 the yield is increased from 30 % to 50 % of theoretical and at pH 3·1 from 50 % to 80 % of theoretical. Because of the thiocyanate these high yields are obtained at very much lower concentrations of aminopyrine ($3{\cdot}3 \times 10^{-4}$ M) and nitrite ($3{\cdot}3 \times 10^{-4}$ M) than utilized by Lijinsky.[654]

(vi) Formation through combustion or oxidation of 1,1-dialkylhydrazines. Some evidence has been presented that dimethylnitrosamine might have been formed during the burning of 1,1-dimethylhydrazine as a rocket fuel for Surveyor III recovered in the Apollo 12 mission.[655] Another problem is whether dimethylnitrosamine could be formed from dimethylhydrazine in the environment or in living tissue.

(vii) Formation of a carcinogenic nitrosamine from the reaction of a noncarcinogenic nitrosamine with a secondary amine in the environment or *in vivo*. Transnitrosation has been shown to take place when noncarcinogenic N-nitrosodiphenylamine reacts with N-methylaniline to form carcinogenic N-methyl N-phenyl nitrosamine.[656]

(viii) Pollution of the environment with a secondary amine contaminated with a nitrosamine. Examination of a commercial dimethylamine has shown the presence of trace amounts of dimethylnitrosamine.[608]

C. Carcinogenic activity

The physiological, chemical and physical properties[657] and the analysis and formation[658] of dimethylnitrosamine have been reviewed. Approximately 100 nitrosamine compounds have been tested for carcinogenic activity; the majority are carcinogenic. Seventy nitrosamines have been tested in rats and guinea pigs.[659] The nitrosamines include some of the most potent carcinogens known. They are organ-specific in inducing selectively malignant tumors in bladder,[660] fore-stomach,[661, 662] kidney,[663, 664] liver,[664–668] lung,[664, 669] nasal sinus,[670] oesophagus[671] and ovary[670] in various animal species. Symmetrical dialkylnitrosamines induce predominantly liver cancer, whereas the assymetrical compounds produce carcinoma of the esophagus.[672]

After the inhalation of methyl n-butyl nitrosamine, test animals developed cancer of the esophagus after a total dose of 70–200 mg/kg body mass given over a period of 23 weeks.[673] This compound was carcinogenic even when inhaled at concentrations as low as 0·05 ppm.[672] Another experiment that has demonstrated the potency of this chemical as an air pollutant was shown in an animal room.[674] Mice were treated with methyl n-butyl nitrosamine and obtained tumors. The same tumors were found on untreated mice which were in an adjacent cage. The most likely possibility is that volatile methyl n-butyl nitrosamine was inhaled in very slight concentrations by the untreated mice.

Let us discuss dimethylnitrosamine a little more fully since it has been found in the ambient atmosphere, ocean, foods, cigarette smoke and a variety of other places. The compound is a formaldehyde precursor. In contrast to N-nitrosomethylurea and N-nitrosoethylurea,

$$R{-}\underset{\displaystyle NO}{\underset{|}{N}}{-}\overset{\displaystyle O}{\overset{\|}{C}}{-}NH_2$$

where R is CH_3 or C_2H_5

which are potent teratogens, mutagens and direct carcinogens to the rat, dimethylnitrosamine is not teratogenic to the rat and is only mutagenic and carcinogenic after metabolism to an active molecule.[675]

No animal species appears to be resistant to the nitrosamines. Thus, dimethylnitrosamine is a potent carcinogen in mouse, hamster, rat, guinea pig, rabbit and rainbow trout while diethylnitrosamine is carcinogenic in 12 animal species such as the mouse, hamster, rat, guinea pig, rabbit, dog, pig, rainbow trout, the aquarium fish, *Brachydanio rerio*, grass parakeet and the baboon.[676] Lung tumors in mice[677] and kidney tumors in rats[678] can result from a single injection of Me_2NNO, but liver tumors do not arise in adult intact mice or rats by a similar procedure.[679, 680]

The effect of small amounts of Me_2NNO on tumor induction has been studied. Among rats surviving a single intraperitoneal LD_{50} dose (6 mg) 20 % developed kidney tumors while much higher incidences were seen in rats on diets containing 200 ppm Me_2NNO for 1 week (21 mg) or 100 ppm for 4 weeks

(42 mg).[663] A 100 % incidence of kidney tumors was obtained in rats when they were given 1·5 mg of Me_2NNO daily for 6 days only.[681] Rats on a dietary level of 2 and 50 ppm Me_2NNO for 60 weeks got 1/26 and 8/74 liver tumors, respectively.[682] At dietary levels of 20 and 50 ppm liver tumors were observed in more than 66 % of the animals. A concentration of 5 ppm daily in the diet for one year can be regarded as carcinogenic and corresponds to a total intake of 54 mg per animal. A concentration of 50 ppm Me_2NNO in drinking water for one week is sufficient to induce tumors in the kidney and lungs of mice.[683] As little as 75 μg of Et_2NNO/kg body weight per day given to rats for more than 600 days gave 11 out of 20 rats benign or malignant tumors of the liver, esophagus or the nasal cavity.[684]

Let us very briefly consider some of the cocarcinogenic aspects in nitrosamine carcinogenesis. Forty-two and sixty hours after a non-lethal hepatotoxic dose of carbon tetrachloride a single dose of Me_2NNO (20 mg/kg body weight) resulted in an increase in kidney and liver tumors and a decrease in kidney tumors with a further increase in liver tumors, respectively.[685] Inhaled ferric oxide particles (geom. mean diam 0·11 μ with a percent standard deviation of the geom. mean diameter of 69) enhanced Et_2NNO tumorigenicity of the peripheral lung.[686] Following subcutaneous injection of Et_2NNO the cocarcinogenic effect of ferric oxide manifests itself only in the lower respiratory tract. Of importance in understanding some types of human carcinogenesis is this enhancing effect of ferric oxide. The number of animals with lung tumors was almost double and the number of lung tumors almost triple in Et_2NNO-treated hamsters exposed to ferric oxide dust compared with Et_2NNO-treated hamsters breathing filtered air. The authors explain the localized action by the accumulation and persistence of the inhaled particles in large quantities in the alveolar parenchyma. The possibility of other carcinogenic pathways beside the demethylation one is shown by the report that pretreatment of rats with 3-methylcholanthrene led to lung tumors after prolonged administration of Me_2NNO (which is not considered a lung carcinogen) even though there was no detectable dimethylnitrosamine demethylase activity as measured by formaldehyde.[687]

D. Formaldehyde formation and biological activity

1. *Introduction.* In this portion of the section we will discuss the relation between the demethylation of Me_2NNO (i.e. the formation of formaldehyde) and its physiological activity. Dimethylnitrosamine is a potent carcinogen,[663, 681, 682, 688, 689] mutagen[690] and hepatotoxin.[691] This nitrosamine methylates protein, RNA and DNA *in vitro*;[692, 693] this alkylation is presumed to play roles in the genotoxicity and acute toxicity of the compound. One of the genotoxic pathways appears to involve alkylation of DNA.

Investigation of the metabolism of Me_2NNO has led to the hypothesis that enzymatic demethylation by a hydroxylation mechanism yields formaldehyde and an unstable methyldiazonium salt which decomposes to the carbonium ion which then methylates electron-rich groups.[667] Mammalian metabolic activity seems to be necessary for mutagenic,[690, 694] carcinogenic[687, 695, 696] and hepatotoxic[691] activity of Me_2NNO. Frantz and Malling[695] have suggested that the effect of diet and other factors suggests that Me_2NNO carcinogenicity can also be influenced by altering mammalian metabolism.[697, 698]

2. *Methylation of DNA*. The relationship of the Me_2NNO N-demethylating enzyme activity (as measured by the formaldehyde formed from the N-demethylation of Me_2NNO) to the methylation of DNA was studied.[699] The following procedure was used in following the N- demethylation of Me_2NNO.

Determination of the N-*demethylation of* Me_2NNO *by liver microsomal preparations.*[699] At different intervals (0, 1, 3, 4·5, 6, 8, 10 and 17 weeks after the start of the experiment), the DMNA-N-demethylating activity of liver microsomes was assayed in a standard incubation system containing 3 mg microsomal protein, DMNA (3 mM), NADPH (2 mM), $MgCl_2$ (5 mM) and phosphate buffer (0·13 M, pH 6·9) in a final volume of 2 ml. The reaction mixture was incubated for 30 min. at 37° under an atmosphere of 95 % O_2 and 5 % CO_2. The reaction was stopped by the addition of 1·5 ml 2·5 % $ZnCl_2$ and 0·5 ml 0·5 N NaOH. The formaldehyde present in the supernatant was determined by the 2,4-pentanedione colorimetric procedure for formaldehyde.[556]

The pretreatment of mice with Me_2NNO led to a strong depression of the dimethylnitrosamine N-demethylating enzyme in the liver and to an increased half-life of unchanged Me_2NNO in the blood as compared with non-pretreated controls. In the course of 6 weeks of Me_2NNO ingestion, the N-demethylation of Me_2NNO by the isolated liver microsomes was progressively inhibited to reach some 20 % of the normal level at the end of this period.[669] Higher levels of unchanged Me_2NNO were observed in the blood of pretreated as compared with control animals. Engelse and Emmelot conclude that the liver N-demethylating enzyme activity governs (i) the rate of methylation of liver DNA, and (ii) the amount of systemic Me_2NNO available for metabolic demethylation and DNA methylation in lung and kidney. These changes are reversible and nearly normal values were obtained for the methylation of DNA and the Me_2NNO N-demethylating enzyme activity some weeks after the administration of unlabelled Me_2NNO had been stopped.

3. *Toxicity*. There appears to be a relationship between the metabolism of Me_2NNO as shown by the formation of formaldehyde and the hepatotoxicity of

the molecule. The hepatotoxic effects of carbon tetrachloride and dimethylnitrosamine can be affected by various factors.[700] A low protein diet gave some protection to rats from carbon tetrachloride[701] and Me_2NNO.[702] Drugs that decreased the activity of microsomal enzymes had a similar action[703] while those that increased the activity tended to increase the hepatotoxic effects.[701, 704, 705] Similarly, a single intragastric dose of carbon tetrachloride (0·01–2·5 ml/kg) protected male Sprague–Dawley rats against the toxic effects of Me_2NNO and reduced the levels of dimethylnitrosamine N-demethylase.[706] In addition, a dose of CCl_4 protected rats against a lethal dose

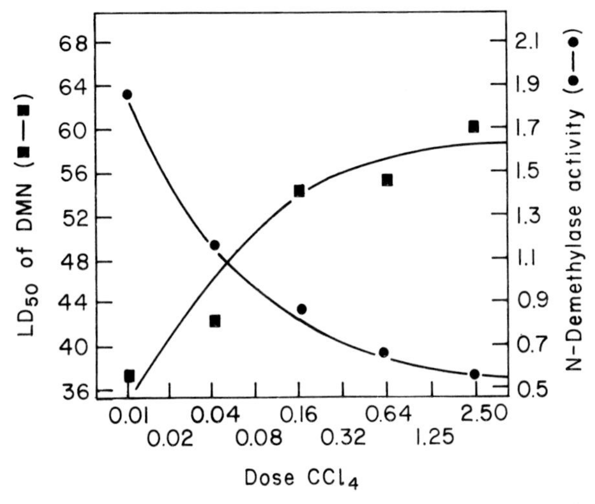

FIG. 17. Variation of LD_{50} of dimethylnitrosamine (mg/kg) and of microsomal dimethylnitrosamine demethylase with dose of CCl_4 (ml/kg) 12 hr after dosing with carbon tetrachloride. N-Demethylase activity is measured in nanomol formaldehyde per mg protein per 30 min.[706]

of Me_2NNO, the protection probably being due to a reduction in the activity of the N-oxidative demethylation of Me_2NNO.[707] The values of the LD_{50} of Me_2NNO and of the dimethylnitrosamine N-demethylase activity 12 hours after dosing with CCl_4 are plotted against the dose of CCl_4 in Fig. 17.[706] The effect of CCl_4 on the dimethylnitrosamine N-demethylase activity is shown in Fig. 18.[706] In a study involving mice a 70% reduction in liver dimethylnitrosamine N-demethylase activity was observed within 20 minutes of dosing with 0·6 ml CCl_4/kg. Dimethylnitrosamine N-demethylase activity was determined by measuring the amount of formaldehyde produced from Me_2NNO by a microsomal preparation from whole rat liver. In the rat study the increase in the LD_{50} of Me_2NNO and the reduction in demethylase levels were observed within 20 minutes of dosing with CCl_4. Maximum results were obtained in 12 hours, persisted until 48 hours, and began to recover between 48 and 60 hours. The LD_{50} of Me_2NNO and the level of dimethylnitrosamine N-demethylase returned to normal between 120 and 144 hours. Carbon tetrachloride can also depress the microsomal enzymes involved in the demethylation of aminopyrine for 4 days with recovery by 7 days[708—710] and can also prevent the development of the hepatocellular necrosis which

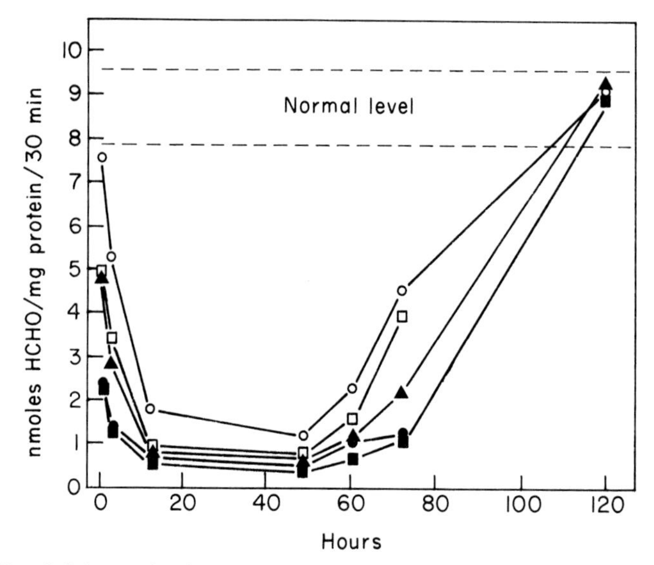

FIG. 18. Dimethylnitrosamine demethylase activity of a rat liver microsome preparation at times after a single dose of carbon tetrachloride. Doses of CCl_4 (ml/kg); ■—■ 2·5; ●—● 0·6; ▲—▲ 0·15; □—□ 0·04; ○—○ 0·01. Each point is the mean of 4 determinations.[706]

normally follows the administration of toxic doses of Me_2NNO.[706] These data support the suggestions that the hepatoxic effects of CCl_4 and Me_2NNO are derived from a metabolite formed in the enzymatic N-demethylation of this compound.[129, 701, 705, 711] In these cases the same derivatives that are involved in the mutagenic and carcinogenic pathways combine with other key components in cells to cause the cell damaging effects.

Consistent with the reports that the hepatotoxicity and carcinogenicity of Me_2NNO require prior enzymatic demethylation of the compound[712] are findings that the repression of hepatic Me_2NNO demethylation by 3-methylcholanthrene[687, 696, 713, 714] is paralleled by the decrease of toxicity[687] and inhibition of hepatocarcinogenicity[688] of Me_2NNO owing to methylcholanthrene administration and that the substantial lowering of Me_2NNO metabolism by aminoacetonitrile[713, 715—717] is paralleled by the inhibition of Me_2NNO toxicity[717, 718] and carcinogenesis[697, 717] by aminoacetonitrile. Argus *et al.*[712] have also pointed out that dibenamine [N-(2-chloroethyl)dibenzylamine] decreases Me_2NNO N-demethylation substantially[719] and also reduces the hepatotoxicity of Me_2NNO,[720] as well as the hepatocarcinogenicity of diethylnitrosamine.[721]

4. *Mutagenesis.* Let us discuss the relation between the demethylation of Me_2NNO (and the consequent formation of formaldehyde) and mutagenesis of the nitrosamine. It has been shown that the oxidative demethylation of Me_2NNO and mutagen formation by mouse liver microsomes *in vitro* follow similar kinetics.[722] Dimethylnitrosamine, like other nitrosamines, is not mutagenic by itself, but has to be converted by mammalian enzymes to highly mutagenic products, e.g. $CH_3N_2^+$, CH_3^+, $AcOCH_2{-}N(CH_3)NO$, and $OHC{-}N(CH_3)NO$ have been postulated as ultimate carcinogens. Evidence presently available seems to indicate that formaldehyde does not contribute to the mutagenic activity of the reaction products.[701] In attempts to correlate the activity of Me_2NNO N-demethylase with the formation of mutagenic metabolites from Me_2NNO, *Neurospora crassa* and *Salmonella typhimurium* have been used as indicators for mutagenic activity.[701] The ad-3 forward-mutation system in *Neurospora crassa* was used to detect specific locus mutations; mutants in this system can range from multi-locus deletions to leaky deletions. The mutation system is based on detection of forward-mutations in 2 adenine genes, *ad-3A* and *ad-3B*.[723] Mutants in the two genes block purine biosynthesis in two sequential steps, and lead to accumulation of purple pigments in the mycelium. The mutants are detected as purple colonies among white non-mutated colonies. The induction of mutations in *S. typhimurium* is detected as induction of histidine revertants of the histidine-requiring strain G46. This reverse mutation essentially involves reversion to histidine independence by a base-pair substitution.[724]

The activation of Me_2NNO is microsomal, is inhibited by SKF 525A[723] and requires oxygen, Mg^{+2} and NADPH. The activating enzyme is induced in mice by pretreatment with phenobarbital, 3-methylcholanthrene and butylated hydroxytoluene. The mutagenic activity of the reaction products is directly correlated with the metabolic formation of formaldehyde with and without induction by 3-methylcholanthrene and across strains of mice.[701] Dimethylnitrosamine and other nitrosamines require metabolic conversion to an active form to produce their cellular and biochemical effects,[725] as shown by the non-mutagenicity of Me_2NNO to *S. typhimurium* in the *in vitro* plate test and the mutagenicity to that microorganism in the host-mediated assay wherein the host mouse contains the necessary microsomal enzyme system for activation.[726] In marked contrast, nitrosoureas such as N-methyl-N-nitrosourea or the cancer chemotherapeutic agent 1,3-bis-(β-chloroethyl)-1-

nitrosourea do not require enzymatic activation[727] since they are readily hydrolyzed non-enzymatically to the alkylating agent.

Sodium nitrite has been shown to inhibit mouse liver microsomal aminopyrine demethylase (as shown by decreased formation of CH_2O) and aniline hydroxylase in a noncompetitive manner and also to reduce the acute toxicity of dimethylnitrosamine in mice.[728] Doses of sodium nitrite as low as 25 mg/kg lowered the frequency of observed mutants produced by Me_2NNO in *S. typhimurium* G46 in the host-mediated assay.[729] Sodium nitrite had no effect on the mutagenicity of 1,3-bis-(β-chloroethyl)-1-nitrosourea which was mutagenic to *S. typhimurium* both *in vitro* and in the host mediated assay. From this evidence the authors concluded that sodium nitrite interferes with the transformation of Me_2NNO to its active form.[729]

5. *Mutagenesis and carcinogenesis.* Let us consider the relation between demethylation of Me_2NNO (as shown by formaldehyde formation) and mutagenic and carcinogenic activity. All studies do not support a simple mechanism of metabolic demethylation of Me_2NNO to CH_2O and an alkylating agent which attacks some cellular products initiating a variety of toxic, mutagenic and carcinogenic pathways. It is most probable that other key metabolites are formed which help to initiate other toxic, mutagenic and carcinogenic pathways. Thus, a variety of studies in various strains of mice and rats treated with phenobarbital or 3-methylcholanthrene and Me_2NNO or Et_2NNO do not support a simple correlation between metabolic dealkylation and tumor incidence.[128, 688, 689, 730–732] However, the correlation of dimethylnitrosamine N-demethylase activity (as shown by CH_2O production) with toxic and genotoxic effects certainly needs much more study as shown by the available data. The consequent potentiality with increased knowledge will lead to a much better understanding of the mechanisms of toxicity and genotoxicity. For example, consider that rats fed a low-protein, fat-supplemented diet have a lower incidence of liver tumors from Me_2NNO than do rats on a normal diet.[698] It is suggested[695] that this decrease in liver tumors is probably due to a decrease in hepatic Me_2NNO activation, as a low-protein diet leads to a decreased hepatic cytochrome P-450 content[722] and, more specifically, to decreased Me_2NNO N-demethylase activity[733, 734] and liver nucleic acid methylation.[693] A low protein diet in mice similarly decreases the activation of Me_2NNO to a mutagen.[733, 735] Acetoaminonitrile treatment of rats decreases Me_2NNO-induced hepatic tumors[697] presumably by decreasing hepatic Me_2NNO metabolism.[715] Further evidence that the mutagenic pathway for Me_2NNO involves microsomal hydroxylation is the report that SKF 525-A (an inhibitor of microsomal hydroxylation reactions at a 1 mM concentration[736]) completely inhibits mutagenesis and bacterial killing. SKF 525-A is 2-diethylaminoethyl 2,2-diphenylvalerate HCl. Frantz and

Malling[695] have shown that with an increasing concentration of Me_2NNO (2–200 mM) there is an increasing number of revertants of the *S. typhimurium* G 46 indicator organisms. In this same experiment, a linear relation was obtained between the reaction product, formaldehyde, and the mutation frequency, Fig. 19.

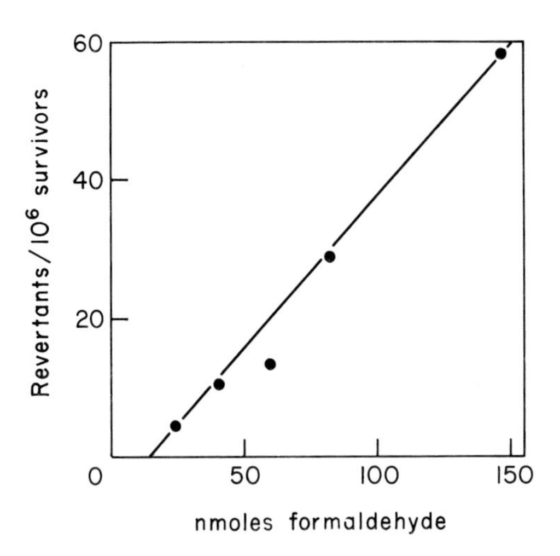

FIG. 19. Correlation between mutagenicity and demethylation (formaldehyde formation) of dimethylnitrosamine at varying concentrations of dimethylnitrosamine. In these studies 2 mg of calcium-precipitated microsomes from livers of 16-week-old male C3H mice, 8 mg of NADPH, and varying concentrations of DMN were incubated for 15 min. Survival of treated bacteria did not differ significantly from control.[695]

In the *Salmonella typhimurium* experiments the linear correlation between Me_2NNO N-demethylase activity and metabolism of Me_2NNO to a mutagen is consistent across inbred strains of mice tested with and without induction.[695] Phenobarbital or 3-methylcholanthrene given intraperitoneally increased liver microsomal Me_2NNO N-demethylase activity and Me_2NNO mutagenic activity in mice and rats. In rats acetoaminonitrile prolongs the half-life of Me_2NNO in blood,[715] decreases Me_2NNO methylation of nucleic acids in both liver and kidney,[715] decreases the *in vivo* mutagenicity of Me_2NNO,[735] and decreases the number of liver tumors from Me_2NNO with a possible increase of other tumors.[697] In line with a relationship between aldehyde plus activated metabolite formation and carcinogenicity of Me_2NNO is the report that Me_2NNO and Et_2NNO induce tumors in various organs of the rat in proportion to the ability of the organ to metabolize the carcinogens, as measured by $^{14}CO_2$ production upon incubation with [^{14}C]Me_2NNO or

[^{14}C]Et_2NNO.[737] Frantz and Malling[695] hypothesize that there is a strong relationship between the metabolism of Me_2NNO to a mutagen and its metabolism to a carcinogen and that the active products of Me_2NNO mutagenicity and carcinogenicity are probably identical. These same authors have also emphasized that the relationship has inconsistencies that will have to be resolved.

The importance of genetic factors involved in tumor etiology[738] in mammals has been demonstrated[739] through a correlation between the degrees of susceptibility to Me_2NNO-induced tumors of certain organs in mouse strains, Ha/ICR, BALB/cJ, RF/J, and C3H/HeJ, and the differential ability of these organs to activate Me_2NNO to mutagenic metabolites *in vitro* (utilizing *Salmonella typhimurium*, his$^-$, G-46). Preliminary *in vitro* assays with microsome fractions from stomach tissue, liver, lung, kidney, spleen and testes from random bred HA/ICR mice conducted with Me_2NNO showed that except for stomach tissue and the testes and probably the spleen, the other 3 organs were capable of metabolizing Me_2NNO to a mutagenic intermediate. The extremely high Me_2NNO activation exhibited by the kidney microsomes of this strain was not observed in any of the other three inbred lines; the ICR mice, unlike BALB/c and RF mice, develop kidney tumors following exposure to Me_2NNO.[740–743] Similarly, C_3H/He mice show high levels of Me_2NNO activation[744] and are susceptible to Me_2NNO induced kidney tumors.[741] Weekes and Brusick[744] conclude that tissues or organs which are susceptible to tumor induction such as the liver, lung and kidney activate Me_2NNO to its mutagenic form whereas tissues and organs which are not susceptible to the carcinogenic effect of Me_2NNO such as the testes or spleen are not effective in transforming the compound. These data provide additional support for proposals linking mutagenicity, carcinogenicity and Me_2NNO metabolism.

E. Human health effects

The question that arises is what relevance do all these dimethylnitrosamine and nitrosamine studies have to man? From all the animal studies it appears evident that the nitrosamines could be genotoxic to man. In fact, Preussmann, a prominent research worker in this field, has stated, "I do not hesitate to say, therefore, that it is practically certain that N-nitroso compounds are also carcinogenic in man."[676] The American Conference of Governmental Industrial Hygienists have specified a "zero" threshold limit in air for nitrosamines when defining an acceptable working environment.[745] ACGIH standards further prohibit exposure by any route in view of the high presumed carcinogenic potential for man. A comparative *in vitro* study has shown that Me_2NNO is metabolized in human liver slices at about the same rate as in rat liver slices.[746] In contrast to observations in rat and guinea pig[747, 748] it has been found that in early human intrauterine development microsomal fractions of the whole embryo as well as of embryonal liver show activity of dimethylnitrosamine (and aminopyrine) N-demethylases (as shown by formaldehyde formation).[557] The availability of this enzyme system in the human embryo indicates that prenatal carcinogenesis by Me_2NNO is already possible in the early prenatal phases in man. As a result of an industrial accident similar toxic symptoms were found to arise in the human liver as with experimental studies on animals. Exposure to Me_2NNO in industry gives rise to acute necrosis and, after 2 years, to cirrhosis in man.[749,750] Centrilobular necrosis is similarly produced in most animal species treated with high doses of nitrosamines in acute toxicity experiments.[676] Some of the circumstantial evidence for the carcinogenicity of nitrosamines to humans has been summarized.[640] Lacking the epidemiological evidence obtained from contact of humans for 20 to 40 years with nitrosamines, the circumstantial evidence is speculative. There is one report of the death from cancer of a chemist very active in the syntheses of nitrosamines.[751] With the increasing xenobiotic chemical burden in the environment, the genotoxic danger from nitrosamines may become all too real in the future.

It has been estimated that the marginal effect dose of Et_2NNO in the rat is about 0·5–0·75 mg/kg in the diet.[752] With the generally accepted international practice in establishing permissible doses, a safety factor of 100 is allowed when extrapolating animal data to man.[676] Thus a level of 5–10 μg/kg is suggested as the "permissible" dose for low-molecular-weight nitrosamines. Establishing a permissible dose in ambient air is a more difficult problem. Thus, in a polluted air sample where the atmospheric concentration of Me_2NNO is about 1 $\mu g/m^3$ a person breathing this air could take in about 7 mg per year. At somethat higher concentrations methylbutylnitrosamine produced carcinoma of the esophagus in rats when inhaled at a concentration of 0·05 ppm.[672] With a safety factor of 100 the permissible dose of a nitrosamine would be calculated to be about 0·5 ppb. However, because of the potency of the nitrosamines their atmospheric level should be preferably kept at as low a level as possible with no acknowledged permissible level.

F. Enzymatic reactions and formaldehyde formation

We will discuss a few enzymatic reactions involving Me_2NNO, but the majority of these reactions will be discussed in the methylamines section.

Treatment of rats with [^{14}C]dimethylnitrosamine followed by hydrolysis of liver DNA with perchloric acid and analysis of the free bases by column chromatography on Dowex-50 showed that the thymine fraction was labeled.[131] The radioactive component is not 3-methylthymine and evidence suggests that it is not a product formed by methylation of the nucleic acid. The radioactive compound is probably formed from the labeled formaldehyde arising from dimethylnitrosamine, but the radioactivity is not in thymine, deoxyribose or in any amino acid tested. It probably represents an uncharged metabolite of formaldehyde which becomes bound more extensively to DNA than to RNA.

Dimethylnitrosamine could be analyzed by reduction to dimethylhydrazine, followed by oxidation, tautomerism, hydrolysis to formaldehyde and then determination. The procedure used for 1,1-dimethylhydrazine could be used here (see hydrazine section).

In the presence of microsome and cell sap preparations dimethylnitrosamine forms formaldehyde.[128, 129]

Oxidative N-dealkylation of dimethylnitrosamine *in vitro* with rat liver microsomal preparations was shown to form formaldehyde.[753]

In early human intrauterine development microsomal fractions of the whole embryo as well as of embryonal liver show N-demethylase activity.[557] This activity consists of hydroxylation of the molecule at the α-C followed by splitting off of the corresponding aldehyde. Thus, in the incubation of dimethyl or diethylnitrosamine with the liver homogenates of adult rats formaldehyde or acetaldehyde is detected; with embryonic tissue before the twenty-first day after impregnation the corresponding aldehyde is not formed.[747] In line with this is the report that when dialkylnitrosamines are given to pregnant rats on the fifteenth day of pregnancy cancers are not produced in the offspring.[754] This result shows the N-demethylase enzyme system necessary for the activation of the dialkylnitrosamines and the formation of aldehydes is not present in the fetus at that time. In addition, the preactivated intermediates formed in the maternal organism are much too unstable and short-lived to reach the fetus. However, enzymatic-derived aldehyde formation and a strong carcinogenic reaction to the dialkylnitrosamines becomes evident in the rats after birth.

With human embryonic tissue and dimethylnitrosamine as the substrate N-demethylase activity had tripled 12 weeks after impregnation.[557] Five and twelve weeks after impregnation 10·3 and 28·2 μM of formaldehyde were obtained for each gram of tissue in 30 min. The authors emphasize that since the presence of such an enzymatic system is necessary for the activation of nitrosamines, it seems possible that prenatal carcinogenesis by these substances is possible already in man's early prenatal phases.

Dimethylnitrosamine was incubated with the extracted microsome fraction by standard procedure.[755] The formaldehyde formed during enzymatic oxidation was collected and stabilized with semicarbazide, reacted with 2,4-pentanedione and measured at λ 415 nm.

G. Other nitrosamine reactions

Although we have only discussed one metabolic pathway (i.e. the one involving the formation of formaldehyde from Me_2NNO)[756] there are other possible pathways[757] some of which could be important in genotoxicity. One of the simpler situations is shown for Me_2NNO (Fig. 20) where one does not have the complications of β-hydroxylation, ω-hydroxylation, ring opening (as in the heterocyclic nitrosamines) and methylation with non-methylated nitrosamines. In the pathways shown in Fig. 20 the acyloxy compound, the nitrosamine aldehyde, the methyldiazonium cation could all be potential ultimate carcinogens.

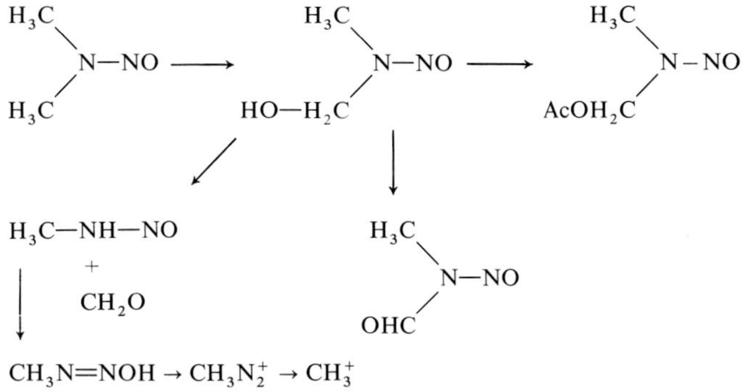

$(H_3C)_2N{-}NO \longrightarrow (H_3C)(HO{-}H_2C)N{-}NO \longrightarrow (H_3C)(AcOH_2C)N{-}NO$

$(H_3C)(HO{-}H_2C)N{-}NO \longrightarrow H_3C{-}NH{-}NO + CH_2O$

$(H_3C)(HO{-}H_2C)N{-}NO \longrightarrow (H_3C)(OHC)N{-}NO$

$H_3C{-}NH{-}NO \longrightarrow CH_3N{=}NOH \rightarrow CH_3N_2^+ \rightarrow CH_3^+$

FIG. 20. A few of the postulated pathways in the metabolism of dimethylnitrosamine to hypothesized ulticarcinogens.

Nitrosamines decompose in ultraviolet light to aldehydes and other derivatives. The reaction has been shown to take place in solution,[758] essentially $(RCH_2)_2NNO \rightarrow RCHO$. In a preliminary study of alternative pathways and in an attempt to account for some puzzling metabolic observations Olah *et al.*[757] have reported 3 distinct modes of protolytic dialkylnitrosamine fragmentation in magic acid (equimolar amounts of $HSO_3F:SbF_5$) solution. Dimethylnitrosamine was cleaved to the protonated Schiff base of formaldehyde and methylamine. The other modes involved the formation of the *tert*-butyl cation or denitrosation to the dialkylammonium ion.

XVII. EPOXIDES

The properties and uses of ethylene oxide and other epoxides have been described.[759, 760] Ethylene and propylene oxides are used extensively for the sterilization of a large number of materials. Persistence of these epoxides can be due to absorbance into plastics and other materials, chemical combination with water to form the corresponding glycols, combination with HCl to form the corresponding chlorohydrins, combination with compounds containing active hydrogen, etc. For example, residue ethylene oxide in plastic tubes sterilized with ethylene oxide takes 4 or 5 months of storage at room temperature to dissipate.[761]

Epoxides containing the grouping $RCH\overset{O}{\diagup\diagdown}CH_2$ can be formaldehyde precursors either through enzymatic means or through hydrolysis-oxidative methodology in laboratory containers, e.g..

$$R{-}\underset{\diagdown O \diagup}{CH{-}CH_2} \longrightarrow \underset{OH}{RCH}{-}\underset{OH}{CH_2} \xrightarrow{(O)} RCHO + CH_2O$$

Atmospheric ethylene oxide and propylene oxide can be determined following collection, hydrolysis, oxidation with periodate and determination of the formaldehyde with chromotropic acid.[549]

Epoxides are probably widely prevalent through the human environment. Meager investigations of the atmosphere indicate the presence of ethylene oxide, propylene oxide, styrene oxide and 1-octene oxide.

In addition, a very large number of precursors of epoxides are also in the environment and especially in the atmosphere; these are the olefins which are present in the air in huge amounts. Olefins are emitted into the air in huge amounts from stationary and mobile sources of pollution. They are believed to undergo autoxidation and photochemical oxidation in air to epoxides, hydroperoxides, peroxides and other oxygenated aliphatics.[762] In addition, metabolic oxidation can take place. Thus, one of the major metabolic pathways (for compounds containing an unsaturated C=C bond) in mammalian liver is now believed to be oxidation to the corresponding epoxides by microsomal epoxidase followed by conversion of the products to glycols by hepatic microsomal epoxide hydrolase or to α-hydroxyalkyl-S-glutathione conjugates by soluble epoxide-S-glutathione transferase.[534, 763] Like 1-hexadecene many of the larger olefins could be converted to epoxides in the environment and in living tissue. It has been shown that 1-hexadecene in the presence of an epoxide hydrolase inhibitor forms carcinogenic 1,2-epoxyhexadecane[536] on incubation with rabbit liver microsomes and appropriate cofactors.[534] Thus, the presence in the environment of the appropriate 1-alkene carcinogenic epoxide precursors and epoxide hydrolase inhibitors could be dangerous to the individuals in contact with such a mixture.

Many of the epoxides are mutagenic. For example, the following carcinogenic epoxides have been shown to be mutagenic to *Escherichia coli*, glycidaldehyde, 1,2,3,4-diepoxybutane, 1,2,4,5-diepoxypentane, 1,2,5,6-diepoxyhexane, 1,2,7,8,-diepoxyoctane and vinylcyclohexene-3 diepoxide.[764] The following epoxides have been shown to be mutagenic to *Neurospora*—ethylene oxide, glycidol, epichlorohydrin and propylene oxide[765] and di-(2,3-epoxypropyl)ether.[766] 1,2,3,4-Diepoxybutane has been shown to be mutagenic to *Drosophila*[767] and to *Salmonella typhimurium*[768] and to cause chromosomal aberrations in mammalian cells.[769] The mutagenicity of many of the epoxides and analogous compounds have been discussed.[770]

Since formaldehyde can be formed during the metabolism of 1-alkene oxides and some nitrosamines, its possible reactions in living tissue are of some importance. Formaldehyde is stated to be a mutagen to *Drosophila*.[771, 772] The mutagenic effects of formaldehyde are enhanced in the presence of hydrogen peroxide, ultraviolet light or catalase inhibitors.[773, 774] The genetic effects of formaldehyde are postulated as arising from an intermediate oxidation product of formaldehyde, probably a free radical peroxide.

A small number of epoxides have been tested for carcinogenicity.[536, 775–782] The following epoxides have been shown to be carcinogenic to mouse skin—glycidaldehyde, 1,2-epoxyhexadecane, styrene oxide, dl-1,2,3,4-diepoxybutane, meso-1,2,3,4-diepoxybutane, 1,2,4,5-diepoxypentane, 1,2,5,6-diepoxyhexane, 1,2,6,7-diepoxyheptane, 1,2,7,8-diepoxyoctane, 1-ethyleneoxy-3,4-epoxycyclohexane, and 3,4-epoxy-6-methyl-cyclohexylmethyl-3,4-epoxy-6-methylcyclohexanecarboxylate.[776] Glycidaldehyde, dl-1,2,3,4-diepoxybutane and 1,2,5,6-diepoxyheptane have also been shown to be carcinogenic to mice and rats following subcutaneous injection. Propylene oxide has been shown to bind covalently with DNA.[783]

Many carcinogens with unsaturated groupings, e.g. the polynuclear aromatic hydrocarbons, safrole, isosafrole, pyrrolizine alkaloids, aflatoxins, can be metabolized to epoxides which could be ultimate carcinogens in living tissues. One of the ways to detoxify these possibly dangerous metabolites is through hydrolysis by epoxide hydrolase to the much less toxic glycol. Some of the 1-alkene oxides have been shown to inhibit this detoxification. Thus, styrene oxide

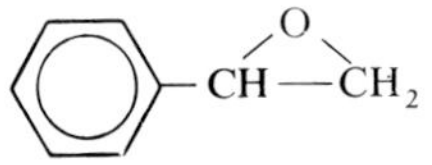

O
CH —CH$_2$

stabilizes benzo[a]pyrene 4,5-oxide,[784] which could be a potential ultimate carcinogen. Many alkene oxides have an inhibitory effect on epoxide hydrolase activity and are also poor substrates for epoxide hydrolase. A most potent inhibitor of this type is 1,2-epoxydecane.[534] These inhibitors can stabilize a carcinogenic epoxide in living tissue so that this alkylating agent will have a longer lifetime in which to reach the target area where the carcinogenic pathway could be initiated.

Several examples of such analytical procedures which have been used to determine ethylene oxide and could be used for other 1-alkene oxides are presented. Atmospheric ethylene oxide is collected in 6 ml of 40% aqueous sulfuric acid, hydrolyzed to ethylene glycol, oxidized with periodate to formaldehyde, and this aldehyde determined with chromotropic acid.[150] The equation is essentially given in Fig. 9, except that phenylhydrazine is used to determine the formaldehyde.

Ethylene oxide can be adsorbed onto plastic and so can be a problem. The following procedure has been used to determine ethylene oxide in plastics. Standards were prepared by dissolving a measured volume of the gas in water. Ethanediol cannot be an interference since it is retained in the flask. 2-Chloroethanol cannot be determined by this method.

Determination of ethylene oxide in plastics.[785] Draw air through 100 ml of an aqueous solution containing 1·7 g of hydroxylammonium chloride and 3·3 ml of triethanolamine to remove aldehydes, then through a capillary tube in the neck of a 3-necked 1-liter flask, thence through a reflux condenser and into two bubblers with Vigreaux points, each containing 50 ml water at 0°, and finally through a check bubbler containing water. Boil the contents of the flask in a stream of air (5 to 6 liters per hour), then add the sample or aqueous standard through the third neck of the flask and continue distillation for 45–60 min. Hydrolyze the contents of the bubblers separately and the reagent blank with 1 ml of 0·5 N H_2SO_4 and maintain the solution at 100° for 1 hr. Cool and neutralize with 1 ml of 0·5 N NaOH. To this solution add 2 ml of 0·1 M $NaIO_4$, shake intermittently for 15 min, then add 2 ml of 5·5% aqueous Na_2SO_3 solution and dilute to 100 ml. To 3 ml of each of these solutions, cooled in ice, add dropwise 5 ml of reagent solution (dissolve 100 mg of sodium chromotropate in 2 ml water plus 50 ml concentrated H_2SO_4). Heat the mixture for 10 min at 100°, cool, dilute to 10 ml and measure the absorbance at 570 nm. Recoveries of 96–99% are obtained.

XVIII. ETHANOLAMINE AND ETHYLENIMINE

Ethanolamine or its derivatives can be absorbed from the air into a dilute acid solution, oxidized by periodic acid to formaldehyde and then determined with chromotropic acid.[143] Sensitivity is about 0·04 μg of ethanolamine. Formaldehyde and its precursors interfere. Amines do not interfere. (See also 1,2-Aminoalkanol Section.)

Atmospheric ethylenimine can be analyzed by hydrolysis to ethanolamine, oxidation of the latter with periodic acid and determination of the formaldehyde with chromotropic acid.[165] The limit of determination is 40 ng of ethylenimine in the test sample.

To obtain a 55% dissociation equilibrium of this compound at 20·5°, 3 days at pH 10, 1 day at pH 11·2 or 7 hours at pH 12·2 are necessary.

XIX. ETHYLENE

This compound is the dominant olefin in auto exhaust and is phytotoxic. The production of formaldehyde from photo-oxidized ethylene has been studied in the smog chamber.[149] This compound is discussed more fully in the 1-alkene section.

A colorimetric method is available for the determination of ethylene through its derived formaldehyde.[148] Formaldehyde and 1-alkenes could interfere. Essentially the ethylene is collected on active carbon, the carbon treated with anhydrous acetic acid, a solution of a periodate in 5% H_2SO_4, and 2% potassium permanganate, and stirred for 10 min. The mixture is decolorized with concentrated sodium sulfite solution, two drops of this solution are added in excess, the mixture is centrifuged, and the formaldehyde determined in the supernatant solution with chromotropic acid at 574 nm.

XX. ETHYLENEDIAMINES, N-PHENOTHIAZINYL

Ethylenediamine (see Section VB), like ethylene glycol, is readily oxidized by periodate to CH_2O and then determined. However, some more complicated derivatives of this type can be oxidized to formaldehyde, e.g. VIII, where R = H or CH_3.[151]

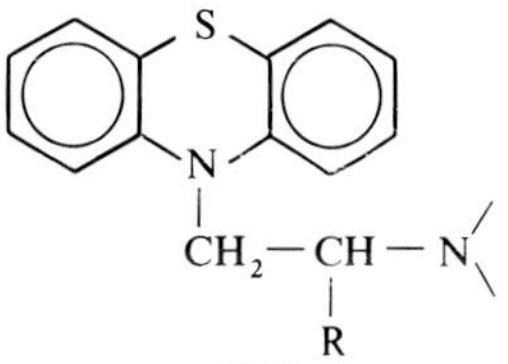

S

N

$CH_2—CH—N<$

R

VIII

Oxidation with persulfate of compounds containing the $>N—CH_2—CH—N<$ chain could give 2 equivalents of formaldehyde while the $>N—CH_2—CH(CH_3)—N<$ chain would give 1 equivalent of formaldehyde and 1 equivalent of acetaldehyde. Antihistamines, such as promethazin would give positive results while tranquilizers such as promazine would give negative results.

XXI. ETHYLENE GLYCOL

This topic is covered more fully in the 1,2-glycols part of this section. Ethylene glycol and epichlorohydrin have been determined in industrial atmospheres through oxidation to formaldehyde[786] and colorimetric determination with chromotropic acid.[137] Interference of formaldehyde and its precursors has to be overcome. The method is sensitive to 1 μg of ethylene glycol or 2 micrograms of epichlorohydrin in a 1 ml sample.

Ethylene glycol can also be determined with an Auto Analyzer.[158] Essentially the ethylene glycol is oxidized with periodate to give 2 molar equivalents of formaldehyde which latter compound is then determined fluorimetrically with 2,4-pentanedione. The ratio of formaldehyde formed to the formaldehyde of the standard for this oxidation is consistent with previous work in the literature.[787–789] In these oxidations as the pH was decreased the reaction rate increased.[158, 778]

Surface waters can also be analyzed for ethylene glycol with the periodate method.[157] However, surface waters without ethylene glycol show fairly high blank values indicating the presence of aldehyde precursors (periodate) whereas permanganate oxidation indicates little or no aldehyde precursors. The authors emphasize the inherent weakness of the periodate method and indicate that monitoring a surface water would require sample blank values before contamination but they emphasize the extreme sensitivity of the periodate method. An mε of 105 was obtained at the wavelength maximum. The following procedure was used.

MBTH determination of ethylene glycol in surface waters.[157] *Reagents.* 2 N H_2SO_4. 0·04 M *Potassium periodate solution*—dissolve 0·92 g of potassium periodate in 100 ml of 2 N H_2SO_4. 1 M *Sodium arsenite solution*—dissolve 13 g of reagent in 100 ml water. MBTH.HCl, 2%. *Ferric chloride–sulfamic acid solution*—2% $FeCl_3 . 6H_2O$ and 3% sulfamic acid in aqueous solution. *Standard ethylene glycol solution* (10% w/v)—immediately prior to use dilute 1:20 with water and then dilute 2 ml of this solution to 500 ml to give a working solution of 2 mg l^{-1}.

Procedure. To 10 ml of the clear test solution add 1 ml of the periodate reagent. Mix and heat at 100° for 2 min. Add 1 ml of the sodium arsenite solution followed by 1 ml of the MBTH reagent solution. Immerse the flasks at 100° for a further 6 min. Cool to room temperature, add 1 ml of the ferric chloride solution and dilute to 25 ml with water. After 20 min read the absorbance at 630 nm.

Methods of analysis are given for ethylene glycol and many of its derivatives in industrial atmospheres.[549] The methods involve periodate oxidation followed by chromotropic acid analysis. Some of the compounds thus analyzed include *p*-dioxane, 2,2′-dimethyldioxane, 2-methoxyethanol, a diglycid ester of ethylene glycol, epichlorohydrin and ethylene oxide. Obviously, the methods lack specificity and can only be useful as screening tests or in analyzing for large amounts of one of the compounds in the presence of much lower amounts of other formaldehyde precursors.

Since ethylene glycol is a toxic substance which has been responsible for several deaths due to its accidental consumption,[790–792] methods for its estimation in biological material have been developed.[154, 156, 793] The method using Schiff's reagent[154] is stated to be unreliable because of the reagent, the non-removal of interfering substances and the lack of obedience of Beer's law.[793] In the 2,4-pentanedione method[156] a high blank was obtained. A chromotropic acid method is recommended for the estimation of ethylene glycol in any biological material.[793] Recovery of known amounts of ethylene glycol added to blood and liver homogenate is 98–100%. Tungstate is unsuitable for protein precipitation since tungstate filtrates give a deviation in colour from those of standards which is not the case with trichloracetic acid filtrates. Beer's law is followed from 1·0–10 µg of ethylene glycol. The following procedure is used.

Estimation of ethylene glycol in biological material.[793] *Reagents. Trichloroacetic acid*—10% in water. 20% *Copper sulphate*—dissolve 20 g $CuSO_4.5H_2O$ in 100 ml water. 25% calcium hydroxide suspension in water. 0·025 M *sodium metaperiodate*—dissolve 0·503 g in 100 ml water. Stable at room temperature. 0·5 M *sodium arsenite*—dissolve 2·25 g NaOH and 5 g As_2O_3 in 100 ml water. Store at room temperature. *Chromotropic acid*—dissolve as much of the 2 g of the acid in 200 ml of water as possible. Add cold H_2SO_4–ice cold water (add 600 ml to 300 ml). Filter through glass wool and store in a brown bottle. *Ethylene glycol standard solutions*—stock = 1%; stock standard, 1 ml = 100 µg; working standard, 1 ml = 20 µg.

Procedure. Prepare fresh 10% tissue homogenate in water. Collect blood in a heparinized container. Add 2 ml of the 10% homogenate to 8 ml of TCA (1 ml of blood to 9 ml TCA). Mix well and centrifuge for 5 min. Transfer 5 ml of the supernatant to a glass-stoppered centrifuge tube. Add 2 ml of 20% $CuSO_4$ followed by 2 ml of 25% calcium hydroxide suspension and 1 ml water to make up a total volume of 10 ml. Stopper the tubes and mix vigorously, continuously for the first 5 min, and then intermittently for 10 min. This will remove the interfering substances such as hexoses, pentoses, glucosamines, ascorbic acid, glycolic acid, and glyceric acids.[794] Centrifuge the contents for 5 min and transfer 1 ml of the aliquot to another tube. Add 1 ml periodate, mix and allow to stand for 10 min at room temp. Reduce the excess periodate by the addition of 1 ml arsenite. After 10 min, add 1 ml of water, mix and centrifuge. Pipette a 1-ml aliquot into a 20 × 5 cm Pyrex tube (lesser volume of the aliquot is taken if the EG concentration is high and the volume is made up to 1 ml with water) and add 5 ml of chromotropic acid. Mix the contents and keep in a boiling water bath for 30 min protected from excessive light (a copper water bath is preferred). After cooling to room temperature measure the absorbance at 550 nm in a spectrophotometer.

Blank. Substitute 2 ml water for the homogenate and carry through the procedure.

Standards. Both direct standards and standards carried through the procedure give the same absorbance since ethylene glycol is not affected by the $CuSO_4$–$Ca(OH)_2$ treatment.

Direct standardization. Treat standards (1 ml = 20 μg) in the range of 1–10 μg with 0·25 ml periodate. After 10 min, add 0·25 ml arsenite. After standing for 10 min, dilute to 1 ml. Add 5 ml chromotropic acid and proceed similar to test.

XXII. EUGENOL

In the determination of atmospheric ozone use has been made of the oxidation of eugenol by ozone to formaldehyde which is then determined by reaction with sodium dichlorosulfitomercurate and acidified pararosaniline to give a chromogen absorbing at 560 nm.[165] Increased speed and precision can be achieved by modifying the method for use in a parallel analysis technique of the centrifugal photometric analyzer.[168] Formaldehyde, 1 to 10 μg/ml, can be determined with an average relative standard deviation of 1·63% and ozone can be determined with a difference of 2% from the neutral potassium iodide method. The authors claim that interference from nitrogen dioxide can be removed with sulfamic acid while a possible interference from formaldehyde could be subtracted by determining formaldehyde separately.

XXIII. FORMALS

Some evidence has been presented for the presence of formaldehyde precursors in auto exhaust.[795, 796] These are probably formal-type molecules. These types of structures will be discussed in the methylene section.

XXIV. FORMATES

Formate is present in air[797–799] and in irradiated auto exhaust.[800] The chemistry of formic acid and its simple derivatives has been discussed.[801]

Formic acid is usually detected or determined by reduction to formaldehyde with magnesium in acid solution followed by reaction with chromotropic acid in sulfuric acid.[171–173, 176] The precision of the method appears to be within 5%, although the yield of formaldehyde is only about 30%.[173] This means that a molar absorptivity of only about 4700 is obtained. The main interferences in this procedure are formaldehyde and compounds yielding formaldehyde on treatment with sulfuric acid and heat. The identification limit for formic acid in the spot test is 1·4 μg.[176]

Other indirect tests which can be used in the detection of formic acid are the HBT, MBTH, and J-acid tests,[177] which are all more sensitive than the chromotropic acid spot test. Positive results are shown in the J-acid test by a yellow fluorescent color and in the HBT and MBTH tests by a blue color. The J-acid test is the most sensitive of this group; an identification limit of 0·05 μg is obtained. Formaldehyde is the main interference in these tests.

Formic acid can be considered an aldehyde in that it contains the CHO group. Consequently analytical methods making use of this aldehyde group were believed to be feasible and so were developed.[177]

A comparison of these spot tests is shown in Table 7.

Table 7. A comparison of spot tests for detection of formic acid[177]

Reagent	Color[a]	Identification limit (μg)
1-Methylquinaldinium methosulfate	B	0·08
1-Methylquinaldinium toluene-*p*-sulfonate	G	0·04
1-Ethylquinaldinium iodide	G	0·5
1-Ethylquinaldinium toluene-*p*-sulfonate	B	0·08
2-Hydrazinobenzothiazole	B	0·7
3-Methyl-2-benzothiazolinone hydrazone	B	0·7
J-acid	Y (fluor)	0·05

[a] B, blue; G, green; Y, yellow.

Procedures for the detection of formate as given in this section could be applied to the determination of formic acid if a more efficient and soluble reducing agent were used.

J-acid spot test.[177] To a 10- by 75-mm test tube add ca. 5 mg of powdered magnesium, one drop (0·02 ml) of aqueous test solution, and one drop of concentrated hydrochloric acid. Immediately cover the mouth of the tube with glass fiber filter paper (No. 934-AH, H. Reeve Angel and Co., Clifton, N.J.) treated with a drop of 0·2% J-acid in sulfuric acid. Heat the tube gently. A positive test is shown by the brilliant yellow fluorescence of the spot under ultraviolet radiation. With a drop of water the spot turns blue.

The various spot tests are compared in Table 1.

HBT spot test.

To 5 mg of powdered magnesium on a white porcelain spot plate add a drop of aqueous test solution and a drop of 1% 2-hydrazinobenzothiazole in concentrated hydrochloric acid. After 30 sec, add a drop of aqueous 1% potassium ferricyanide and a drop of aqueous 40% sodium hydroxide solution. The blank is yellow. A positive test is shown by a blue color.

MBTH spot test

To a few mg of powdered magnesium on white porcelain spot plate add one drop of aqueous test solution, one drop of 1:10 hydrochloric acid, and one drop of aqueous 2% 3-methyl-2-benzothiazolinone hydrazone hydrochloride. After 3 min, add one drop of aqueous 3% ferric chloride solution. The blank is pale yellow-green. A positive test is shown by a blue color.

The reaction with J-acid takes place in the following fashion, Fig. 21.

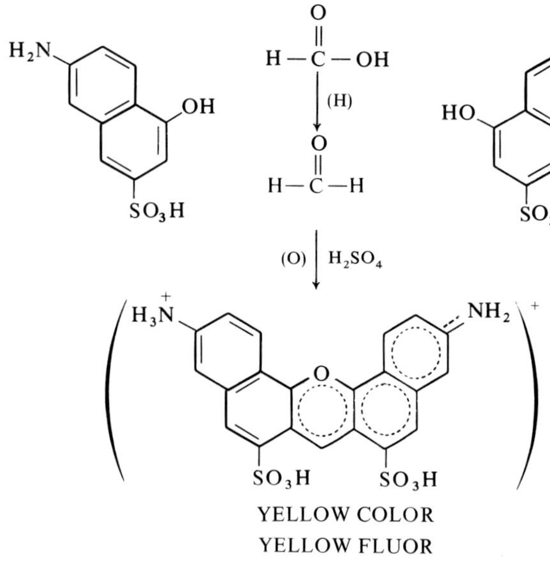

H_2N OH SO_3H H—C(=O)—OH (H) H—C(=O)—H HO NH_2 SO_3H

(O) H_2SO_4

$H_3\overset{+}{N}$ O NH_2 + SO_3H SO_3H

YELLOW COLOR

YELLOW FLUOR

FIG. 21. Reactions in the determination of formic acid with J-acid.

XXV. FORMYL AND ISONITRILE GROUPS

The ultramicromethod for the determination of the formyl and isonitrile groups is based on the basic hydrolysis (1 N NaOH) of the sample.[179] The sample is present in either 1 N aqueous or ethanolic NaOH. The hydrolyzate is acidified, and the quantitatively formed formic acid is distilled in the closed system and then reduced to formaldehyde using magnesium turnings and hydrochloric acid.[67, 802] The amount of formaldehyde is then measured with chromotropic acid.[173, 179] The test sample should contain less than 20 μg of potential formic acid. Beer's law is followed from 1 to 20 μg of formic acid/5 ml. The isonitrile group can also be determined in this fashion, e.g.

$$RNC + H_2O \xrightarrow{OH^-} RNH{-}CHO \xrightarrow{H_2O} RNH_2 + HCOOH$$

$$HCOOH \xrightarrow[H^+]{Mg} CH_2O$$

Some of the compounds undergoing this reaction are shown in Table 8.

Table 8. Chromotropic acid determination of formyl and isonitrile derivatives[179]

Substance	MW	Formyl group (%) Theory	Found
Formylhydrazine (practical grade)	60·06	48·32	47·11
N-Formylaniline (formanilide)	121·13	23·96	23·76
N-Formylindoline	147·17	19·72	19·63
N-Formyldiphenylamine	197·23	14·72	14·59
N-Formyldiethylaminomalonate	203·19	14·28	14·18
Folinic acid	473·44	6·13	6·04
Vincristine sulfate	923·02	3·14	2·76
		Isonitrile group	
$C_{16}H_{24}O_8(NC)_2$	396·39	13·13	12·47

XXVI. GLUCARIC ACID

This acid can be separated from mammalian urines using ion-exchange column chromatography, oxidized with periodic acid to glyoxylic acid, and reacted with phenylhydrazine to give the intensely colored 1,5-diphenylformazan which is then determined colorimetrically.[180, 181] See the Glyoxylic Acid Precursor Section for further methodology.

XXVII. GLUCOSE AND DERIVATIVES

A. Glucose

Periodate oxidation forms formaldehyde which can then be determined, as discussed in the 1,2-glycols part of this section. Treatment with hot sulfuric acid and pyrolytic decomposition will also form formaldehyde. The formation of formaldehyde during the thermal decomposition of D-glucose labelled with ^{14}C at various positions has been studied.[184] (This study is a continuation of the investigations of the pyrolitic reactions of carbohydrates).[803] The results show that 65% of the formaldehyde originates from the C-6 of the glucose skeleton. Elimination of the C-6 atom as formaldehyde can occur either directly from the hydroxymethyl group of the D-glucose unit, or more probably from intermediates leading to the formation of 5-hydroxymethylfurfural, or from 5-hydroxymethylfurfural itself. Carbon atoms C-1 and C-2 contribute about 15% and 5%, respectively to the formation of formaldehyde.

Glucose can be oxidized with periodate to give one molar equivalent of formaldehyde which can then be assayed automatically and fluorimetrically with 2,4-pentanedione.[158]

B. Glucose phenylosazone

Glucose reacts with phenylhydrazine to form the 1,2-phenylosazone. The substituted portions of the phenylosazones of the reducing sugars are resistant to periodate oxidation under mild conditions[189] while the unsubstituted portions are oxidized, as shown for glucose phenylosazone, Fig. 22. Thus, besides formic acid and formaldehyde, mesoxalaldehyde bisphenylhydrazone is obtained.

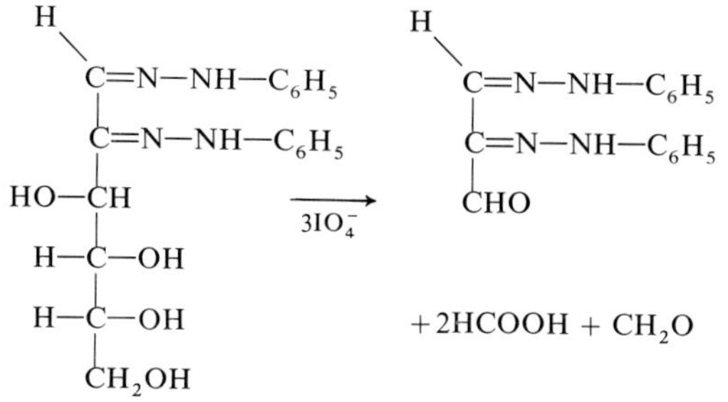

$$
\begin{array}{l}
\mathrm{H} \\
\quad \backslash \\
\quad\ \mathrm{C{=}N{-}NH{-}C_6H_5} \\
\quad\ \ | \\
\quad\ \mathrm{C{=}N{-}NH{-}C_6H_5} \\
\quad\ \ | \\
\mathrm{HO{-}CH} \\
\quad\ \ | \\
\mathrm{H{-}C{-}OH} \\
\quad\ \ | \\
\mathrm{H{-}C{-}OH} \\
\quad\ \ | \\
\quad\ \mathrm{CH_2OH}
\end{array}
\xrightarrow[3IO_4^-]{}
\begin{array}{l}
\mathrm{H} \\
\quad \backslash \\
\quad\ \mathrm{C{=}N{-}NH{-}C_6H_5} \\
\quad\ \ | \\
\quad\ \mathrm{C{=}N{-}NH{-}C_6H_5} \\
\quad\ \ | \\
\quad\ \mathrm{CHO} \\
\\
\mathrm{+2HCOOH + CH_2O}
\end{array}
$$

FIG. 22. The oxidation of glucose phenylosazone.

XXVIII. GLYCEROL DERIVATIVES

A. Glycerides

This topic is covered more thoroughly in the triglycerides part of this section. Hydrolysis of a glyceride forms glycerol which can be oxidized with periodate to formaldehyde, which after condensation with 2,4-pentanedione and ammonia is determined colorimetrically or fluorimetrically as 3,5-diacetyl-1,4-dihydrolutidine.[219–221, 485, 556, 804] The procedure is made possible through the destruction of the excess interfering periodate which can be reacted with potassium iodide to form iodine which is then destroyed with thiosulfate.[485] In the fluorimetric method excess thiosulfate can cause some interference. When rhamnose is used to destroy the excess periodate, the fluorogen is stable for at least 2 hr, Fig. 23.[192] The following procedure is used.

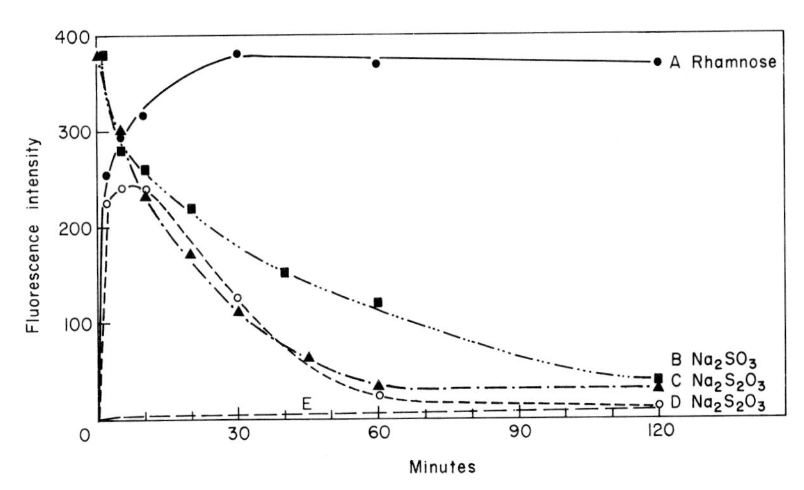

FIG. 23. The influence of rhamnose (A), sodium sulphite (B) and sodium thiosulphate (C) upon the fluorescence of 3,5-diacetyl-1,4-dihydrolutidine at pH 5 and 60°. Curve ε = upper limit of blank values.[192]

Determination of glyceride glycerine.[192] Heat the test solution containing 3×10^{-10} to 3×10^{-8} moles of a glyceride (e.g. a phospholipid-free serum extract or a mono-, di- or triglyceride) with 100 µl of isopropanol and 20 µl of 0·005 N KOH an hour at 60°. Cool and add 100 µl of 0·5 N HCl followed by 20 µl of 0·005 M $NaIO_4$ and oxidize at room temperature for 10 min. Destroy the excess oxidant with 50 µl of 0·1 M rhamnose. After 5 min add 200 µl of reagent (0·1 ml of freshly distilled 2,4-pentanedione in 100 ml of 2 M ammonium acetate solution and bring to pH 5·6). After 30 min at 60° dilute with 500 µl of isopropanol and read the fluorescence intensity at *F*405/485.

Glycerol and glycerides are one class of 1,2-glycols which have been analyzed extensively as formaldehyde precursors. Some of the procedures which have been used are the following.

Chromotropic acid determination of glycerol. To 0·4 ml of aqueous test solution containing 0·05 to 0·3 μmole of glycerol add 0·1 ml of 2·5 N sulfuric acid and 0·2 ml of 0·1 M periodic acid. Allow the solution to stand 5 min. Add 0·2 ml of 1 M sodium arsenite and then allow to stand an additional 15 min. Dilute to 2 ml with 1·1 ml of water. Add 0·5 ml of the aliquot to 5 ml of CA reagent (Add 80 ml of 24 N H_2SO_4 to a clear solution of 200 mg of disodium 4,5-dihydroxy-2,7-naphthalenedisulfonate in 20 ml of water. Stable 2 weeks). Heat on the water bath for 30 min. Cool to room temperature and read at 570 nm.

Glycerol can be separated from other polyols by column chromatography through water-washed Celite using ethyl acetate followed by benzene–n-butanol (1:3) as eluting solvents.

MBTH determination of glycerides.[227] Evaporate 5 ml of aqueous test solution; add 1 ml ethanol and 0·1 ml 2·5% aqueous potassium hydroxide to the residue. Reflux at 60° for 30 min. Cool and add 0·2 ml 6% acetic acid. Evaporate the ethanol on a water bath and add 10 ml of methylene chloride. Reflux again for 30 min. Cool and add water to give an aqueous phase of 5 ml.

To 0·02 to 0·16 ml of this aqueous solution containing 0·92 to 4·6 μg of glycerol add water (so that total volume is > 0·1 ml) and 0·1 ml 0·002 M periodic acid. After 10 min add 0·1 ml of 2% aqueous potassium bicarbonate and 0·03 ml of 0·01 M of sodium arsenite. After 10 min add 0·05 ml of 2 N hydrochloric acid, 0·1 ml of saturated aqueous MBTH·HCl and 0·1 ml 5% ferric chloride. After 10 min add water to 2 ml and after a further 30 min read at 630 nm.

Phenylhydrazine determination of serum glycerides.[223] Place half a gram of silicic acid in a glass-stoppered tube and wet with 0·5 ml isopropyl ether. Add 40 μl serum on the top of the silicic acid followed by an additional 3·5 ml isopropyl ether and then centrifuge 5 min at 2000 rpm. Transfer a 0·5 ml aliquot of the supernatant (equivalent to 5 μl serum) to a clean glass-stoppered tube and evaporate to dryness at 70° under a nitrogen stream.

Treat a blank and a tripalmitin standard similarly. To each tube add 0·25 ml alkali (Dilute 1 ml of aqueous 5 M KOH to 100 ml with 95% ethanol). Stopper tubes firmly and saponify contents at 60° for 20 min. Cool and add 0·05 ml of 4 N phosphoric acid and 0·1 ml of 0·345% (0·15 M) aqueous potassium metaperiodate. After 15 min add 0·2 ml of 1% aqueous phenylhydrazine hydrochloride, and after a 10-min interval 0·2 ml 1% aqueous potassium ferricyanide. Immediately place all tubes into an ice bath for exactly 5 min. Removing the tubes one at a time, add 1 ml of 60% (v/v) sulfuric acid–glacial acetic acid (1:1, v/v) and 1 ml of isopropanol to each tube, mixing after each addition. After 20 min read at 520 nm.

In the oxidation of glycerol 2 molecules of periodate are consumed and 2 molecules of formaldehyde are formed. In the case of glycerol, the oxidation proceeds as follows:[158]

$$\begin{matrix} CH_2OH \\ | \\ CHOH \\ | \\ CH_2OH \end{matrix} \xrightarrow{IO_4^-} \begin{matrix} CH_2O + IO_3^- \\ + \\ CHO \\ | \\ CH_2OH \end{matrix} \xrightarrow{IO_4^-} \begin{matrix} HCOOH \\ + \\ CH_2O + IO_3^- \end{matrix}$$

The results indicate that the initial reaction of periodate with glycerol is rate-limiting instead of the secondary reaction with glycolaldehyde being that way; these 2 reactions occur concomitantly. Formaldehyde is formed at least twice as fast with glycolaldehyde as with glycerol. On an equimolar basis, twice as much formaldehyde is obtained from glycerol as from glycolaldehyde. The rate at which formaldehyde is formed indicates that glycolaldehyde and glycerol are completely oxidized in about 3 and 11 min, respectively.

Using a fluorimetric 2,4-pentanedione automated system of analysis[489] glycerol, glucose, glycerophosphate, glycolaldehyde and ethylene glycol gave formaldehyde equivalents of 1·97, 1·01, 1·06, 1·01 and 1·94, respectively.[158] Of these compounds and formaldehyde only glycolaldehyde was degraded by alcoholic potassium hydroxide treatment to non-reactive products in the test. In the periodate oxidation of ethylene glycol, glycerol, and glycolaldehyde, the reaction rate increased as the pH was decreased.[158, 598]

B. α-Glycerophosphates

These compounds can be determined in the presence of the β-glycerophosphates. The α-derivatives, having adjacent hydroxyl groups, are readily oxidized to formaldehyde and a glycolaldehyde phosphate.[239] The formaldehyde can then be determined by one of the many available reagents.

$$HOCH_2{-}CHOH{-}CH_2OPO(OH)_2 \xrightarrow{IO_4^-} CH_2O + OHC{-}CH_2OPO(OH)_2$$

C. Glycerophosphatides

In the following procedure recoveries of lecithin added to serum ranged from 95 to 105%. The coefficient of variation for a serum containing 132 mg of triglycerides and 141 mg of glycerophosphatides per 100 ml were 0·83% and 3·75%, respectively, and for a serum containing 82·8 mg and 88·6 mg per 100 ml were 2·61% and 1·62%, respectively. An Auto Analyzer can be used for the oxidation and color-development stages.

Determination of glycerophosphatides through glycerol.[240] (a) For total glycerol.—To prepare a lipid extract, shake 1 ml of blood serum with 8 ml of methanol, mix with 16 ml of $CHCl_3$ and shake the mixture with 15 ml of 0·05% H_2SO_4; separate and discard the aqueous phase. Mix 2 ml of the extract with 0·1 ml of phenyl ether and evaporate the $CHCl_3$ on a water bath. Add 2 drops of H_2O, then stopper the tube and heat it to 255° to 260°; maintain this temperature for 10 min. Cool, dissolve the residue in 1 ml of isopropyl alcohol and warm for 20 min at 65° with 1 ml of 5% KOH soln. in 75% isopropyl alcohol. Cool, and add 0·5 ml of 0·535% $NaIO_4$ solution in 12% acetic acid and 4 ml of color reagent (15 ml of acetylacetone and 50 ml of isopropyl alcohol are added to 1 liter of 30·82% aqueous ammonium acetate and the solution is diluted to 2 liters). Heat for 15 min at 65°, set aside for 30 to 60 min at 0°, centrifuge to separate the ether, and measure the extinction at 412 nm against a reagent blank. (b) For triglycerides.—Stir 3 to 4 ml of the lipid extract from (a) with silica gel (0·1 g per ml), centrifuge and evaporate 2 ml of the supernatant solution to dryness. Dissolve the residue in 1 to 2·5 ml of isopropyl alcohol and proceed as for (a) but omitting the thermal hydrolysis stage. The glycerophosphatide content is determined by calculating the difference in the results obtained from (a) and (b).

XXIX. GLYCIDYL STERATE

Because of the possible carcinogenicity of epoxides, their possible presence in the environment and their assay could be of some importance. One possible pathway to their analysis is through their cleavage and oxidation to aldehydes, as has been done with the use of periodate for ethylene oxide,[805] for detection of epoxides,[806] and in a direct cleavage method to demonstrate the location of the oxirane group.[242] In the latter procedure 2 ml of aqueous 62·5% paraperiodic acid was stirred with 10 ml of a dioxane solution of the epoxide for 15 min, poured into 100 ml of water, extracted with petroleum ether, evaporated and assayed.

Glycidyl stearate cannot be assayed in this fashion but can be oxidized in good yield to formaldehyde and formylmethylstearate on treatment with a large excess of periodic acid dissolved in acetic acid and chloroform.[241]

XXX. GLYCINE

This compound has also been discussed in Section VD.

Since hyperglycinemia is a common feature of a number of inborn errors of metabolism (glycinosis, ketotic and nonketotic hyperglycinemia, methylmalonic aciduria) leading some times to serious or fatal illness early in life or even to mental retardation, there have been many attempts to develop a rapid screening method for glycine. The most generally used colorimetric method for the determination of glycine in mixtures containing other amino acids consists of the oxidation of glycine by ninhydrin to formaldehyde which is then measured with chromotropic acid.[243, 247] Essentially, formaldehyde, carbon dioxide, ammonia and hydrindantin are formed on oxidation. In recent work the formaldehyde has been condensed with ammonia and 2,4-pentanedione to form the highly fluorescent 3,5-diacetyl-1,4-dihydrolutidine.

Table 9. Relative fluorescence of amino acids and sugars analyzed by the fluorimetric glycine method[246]

Compound	Rel. fluor. intensity	Compound	Rel. fluor. intensity
Histidine	2·9	Leucine	3·2
Norleucine	0·7	Serine	3·1
Threonine	1·9	Valine	2·1
Taurine	2·2	Lysine	3·3
Alanine	1·5	Hydroxyproline	2·3
Aspartic acid	1·6	Arginine	3·6
Tryptophan	4·7	Tyrosine	4·9
Proline	0·6	Phenylalanine	1·1
Methionine	3·7	Homocystine	2·2
Glutamine	3·8	Cystine	1·8
Glutamic acid	0·7	Cysteine	4·2
Asparagine	3·4	Galactose	0·05
Glycine	100·0	Glucose	0·02

The ninhydrin reaction with glycine to form formaldehyde has 2 advantages: (a) there is no need to remove excess oxidizing agent and (b) ninhydrin complexes ammonia and other free amino acids. Quantities of glycine as low as 3 nmoles/ml can be detected by the automated fluorimetric procedure.[246] No serious interferences were found in a study of the automated procedure carried out in the presence and absence of glycine with each of the 23 common protein amino acids in 30-fold excess of normal plasma levels and the common blood sugars in concentrations as high as 1000 mg %, Table 9. Glycine added to whole blood was recovered to the extent of 96 ± 5 %. A linear relationship was found between the fluorescence intensity and concentration for 0·2 to 4 mg % of glycine. The glycine levels of urines obtained from 32 subjects selected at random gave a mean value of 61·8 ± 38·3 μg/mg creatinine (±I.S.D.). The method is applicable to biological fluids and can be adapted to the analysis of glycine in blood spotted on filter paper.

An alternative manual method for the determination of glycine in biological fluids through the determination of the derived formaldehyde differs in that the oxidizing agent is chloramine-T.[251] A fluorimetric procedure is used for concentrations of glycine from 0·1 to 3 μg while colorimetry is used for the higher ranges (5 to 50 μg). Beer's law is obeyed in both cases. Glycine added to blood or urine was recovered in 95 to 102 % or 94 to 98 % amounts, respectively. Other amino acids added to the extent of 2 μg (fluorimetry) or 20 μg (colorimetry) did not interfere in the determination of glycine.

The method described is based on the decarboxylation and deamination of α-amino acids treated with an alkali hypohalogenite such as chloramine T to form an aldehyde with one carbon less than the original amino acid.[807]

Determination of glycine in biological fluids.[251] *Reagents.* 2,4-*pentanedione* (pH 6)—dilute 150 g of ammonium acetate, 3 ml of glacial acetic acid and 2 ml of 2,4-pentanedione to 1 liter with water. Stable for 2 weeks in the refrigerator. *Glycine stock standard*—1 % aqueous glycine. Stable 1 month in refrigerator. *Glycine working standards*—0·1 to 3 μg/ml for fluorimetric determination and 5 to 50 μg/ml water for colorimetric determination.

Preparation of calibration curve. To 0·5 ml of the working standard add 0·1 ml of 1 % chloramine-T and 0·5 ml of 0·2 N H_2SO_4. Mix and place in oil bath (135 to 140°) for exactly 1 min. At end of incubation add 0·9 ml water and 2 ml of 2,4-pentanedione reagent, mix and incubate tubes in water bath at 58° for 10 min. Then cool to room temperature and read the fluorescence intensity at *F*405/510 or the absorbance at 412 nm, dependent on the concentration of glycine.

Determination of glycine in blood. Mix 0·5 ml of heparinized blood with 4 ml of 0·08 N H_2SO_4 and 0·5 ml of 10 % sodium tungstate in a centrifuge tube and centrifuge. To 0·5 ml of the supernatant add 0·1 ml of 1 % chloramine-T and 0·5 ml of 0·2 N H_2SO_4 and carry through procedure as described above.

Determination of glycine in urine. Dilute 1 ml of urine to 50 ml with water. To 0·5 ml of this diluted sample add 0·1 ml of 1 % chloramine T and 0·5 ml of 0·2 N H_2SO_4 and carry through as above.

Some difficulty has been experienced with the application of this fluorimetric procedure to some urine samples, so that the procedure has been modified.[252] To cancel the inhibitory effect of some physiological constituents in serum or urine the test samples are purified by passage through a column of ion-exchange resins before analysis. Results for urinary glycine obtained by a single-column amino acid analyzer system compare very favorably with those obtained by the proposed chromotropic acid and 2,4-pentanedione colorimetric methods.

Interferences in the colorimetric method were investigated. Without chromatographic purification urea in concentrations above 200 μg per sample aliquot interferes with the recovery of glycine; with chromatographic purification 20 000 μg of urea did not interfere. Ammonia (ammonium sulfate) interferes with the estimation of glycine in concentrations above 60–100 μg per sample aliquot. Resin treatment eliminates the ammonia interference. Bilirubin in aqueous solutions interferes at concentrations greater than 16 mg/dl. In the presence of albumin no interference was encountered in bilirubin concentrations below 32·5 mg/dl.

Glycine is the only amino acid that on deamination and decarboxylation is converted to formaldehyde. Ethanolamine and sarcosine gave positive results in the 2,4-pentanedione procedure. In equal concentrations ethanolamine and sarcosine produced 32 % and 84 % of the color produced by glycine. However, on column treatment all of the sarcosine was recovered while only 31 % of the ethanolamine was obtained. Recovery of glycine added to urine ranged from 94 to 102 % for the 2,4-pentanedione procedure. Glycine was determined in a single urine specimen (9·4 mg glycine/dl) during the course of 2 weeks with a variation of ±4·5 %.

The measurement of glycine in both urine and blood is invaluable in distinguishing hyperglycinemia in a number of syndromes.[808]

In the following procedures the formaldehyde is formed by the following reactions:

$$p\text{-}CH_3{-}C_6H_4SO_2NClNa + H_2O \xrightarrow{H^+} CH_3{-}C_6H_4SO_2NH_2 + NaOCl$$

$$H_2NCH_2COOH + NaOCl \xrightarrow{H^+} CH_2O + NH_3 + CO_2 + NaCl$$

Determination of glycine in serum and urine.[252] *Reagents.* Chloramine-T, 1 g/dl and 2 g/dl; prepare daily. Sodium tungstate, 10%. 2,4- *Pentanedione reagent* (pH 6·0)—dissolve 150 g ammonium acetate in about 600 ml of water, add 3 ml glacial acetic acid, 2 ml of 2,4- pentanedione, and dilute to a liter with water; keep refrigerated; stable for 1 month. *Chromotropic acid reagent*—add 300 ml of concentrated H_2SO_4 to 150 ml of ice-cold water, cool, and add a filtered solution of 1% aqueous chromotropic acid (4,5-dihydroxy-naphthalene–2,7-disulfonic acid); refrigerate in amber-colored bottle; stable for 2 months. *Lithium hydroxide saturated solution*—add 12 g of $LiOH.2H_2O$ to 100 ml of water with heat, cool, and use after overnight standing. Phenolphthalein indicator, 0·1% in 95% ethanol. Dowex 50W-X2 cation exchange resin, 50 to 100 mesh, H^+ form. Glycine stock standard, 0·1% in 0·1 M HCl. *Glycine working standards*—dilute stock tenfold with water and from this solution make 0·002%, 0·004% and 0·008% working solutions.

Chromatographic purification (if necessary) through removal of formaldehyde precursors. Suspend 2 g of Dowex 50W-X2 resin in water and pour into a 10 × 100 mm coarse glass-sintered chromatographic column fitted with a stopcock, wash with water, with 10 ml of 10 M NaOH, with water until eluate is neutral, with 15 ml of 8 M HCl and then with water until eluate is neutral.

Isolation of glycine in serum and urine. Introduce either 2 ml of urine, 4-ml of protein-free supernatant fluid, 4 ml each of glycine working standards or 4 ml water sample (for blank) into the column and allow to pass through. Wash the resin with 15 ml of distilled water and discard. Elute the column with 15 ml of 4 M NH_4OH. Evaporate the eluate to dryness in an evaporating dish. Reconstitute the sample up to a volume of 4 ml with 0·1 M HCl.

Free glycine in urine. Dilute the chromatographically purified glycine standards and urine eluates fivefold. To separate test tubes add 1 ml of the treated urine and standard samples followed by 0·2 ml of the chloramine-T solution, and 1 ml of 0·2 M H_2SO_4. Heat the tubes at 60° for 10 min. Cool and proceed with color development.

Total glycine in urine. Add 2 ml of concentrated HCl to 2 ml of chromatographically purified urine in a polyethylene-lined screw-capped tube. Aerate with nitrogen to remove oxygen. Screw cap on tightly. Heat at 125° for 4 hr. Neutralize with LiOH with phenolphthalein as indicator. Dilute about twentyfold and proceed with color development.

Free glycine in plasma or urine. Add 0·5 ml of serum or plasma to 4 ml of 0·1 M H_2SO_4 followed by 0·5 ml of the sodium tungstate reagent. The glycine working standards suitably diluted with water are taken through the same process. Mix, centrifuge, and decant the supernatant fluid. Treat 4 ml of the protein-free filtrate as described under "isolation of glycine". To 2 ml of the chromatographically purified protein-free supernate add 0·2 ml of chloramine-T, followed by 0·2 ml of 1 M H_2SO_4. Heat for 10 min at 60 or 100°, cool, and develop color.

2,4-*Pentanedione color development.* Add 4 ml of the dione reagent to the formaldehyde-containing tubes. Heat for 10 min at 60°. Cool and read the absorbance at 415 nm. Calculate the concentration of glycine in the samples from the Beer's law curve.

Chromotropic acid color development. To 0·5 ml of the formaldehyde-containing solution add 4·5 ml of the chromotropic acid reagent. Mix, and heat at 100° for 30 min. Cool, and read the absorbance at 570 nm. Calculate the concentration of glycine in the samples from the Beer's law curve.

Other methods of oxidizing glycine and N-alkylated glycines to formaldehyde have been described. Periodiate oxidation of glycine and N-substituted glycine proceeds by an initial electrophilic attack of oxidant on N to give formaldehyde (where R = H)[809]

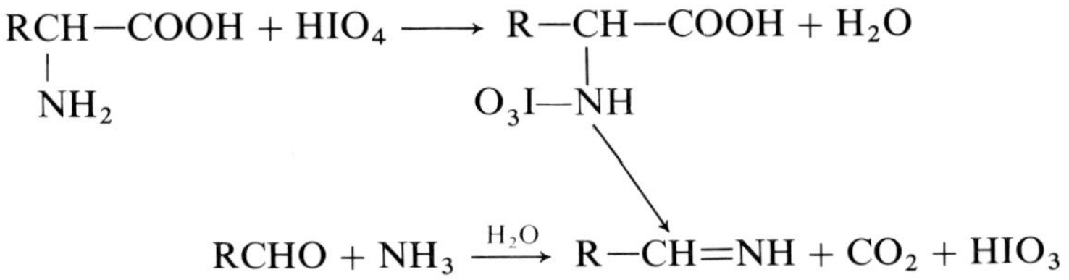

$$\underset{\displaystyle NH_2}{RCH}-COOH + HIO_4 \longrightarrow R-\underset{\displaystyle O_3I-NH}{CH}-COOH + H_2O$$

$$RCHO + NH_3 \xrightarrow{H_2O} R-CH=NH + CO_2 + HIO_3$$

A method of analysis for the antibiotic, bacitracin,[810] a compound containing aminoacarboxylic acid groups, could probably also be used to determine glycine. The postulated mechanism is shown in Fig. 24. Phloroglucinol was the reagent used in analysis. The chromogen is probably a dibenzoxanthylium cation.

$$\underset{\displaystyle NH_2}{RCH}-COOH + NaBrO \rightarrow RCHO + NH_3 + CO_2 + NaBr$$

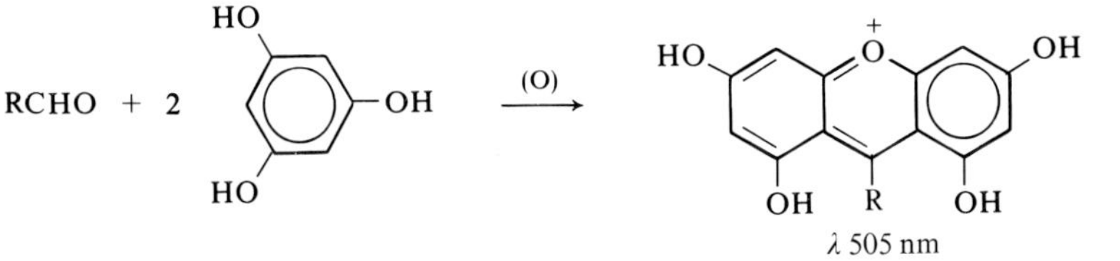

RCHO + 2 (phloroglucinol: HO, HO, OH) $\xrightarrow{(O)}$ (dibenzoxanthylium cation: HO, $\overset{+}{O}$, OH, OH, R, OH)

λ 505 nm

FIG. 24. Oxidative deamination of bacitracin and phloroglucinol determination of resultant aldehyde.

XXXI. GLYCOLIC ACIDS

A. Introduction

These compounds and their precursors, such as the aryloxyacetic acids and tartronic acid, disproportionate to formaldehyde, carbon monoxide and water, as shown in Fig. 25.

A variety of reagents dissolved in concentrated sulphuric acid have been used for the determination of glycolic acid and its precursors. Some of the common formaldehydogenic methods available in the literature are described below.

B. Glycolic acid

1. 2,7-*Dihydroxynaphthalene. Determination of glycolic acid.*[265] To 2 ml of aqueous test solution add 20 ml of 0·01 % 2,7-dihydroxynaphthalene in 95 % sulfuric acid. Mix and heat at 100°C for 30 min. Allow to cool to room temperature and then dilute to 50 ml with distilled water. Read at wavelength maximum.

Alternative methods with this reagent have been described.[255, 256] It has been used to determine glycolic acid in irradiated aqueous acetic acid.[255] Since formaldehyde also reacts, a correction was made for this aldehyde.

In the following method the final color is stable for several hours. Beer's law is followed from about 2 to 100 nanomoles of glycolic acid. The reagent has to be prepared every 2 weeks. The presence of more than 5 % by volume of water in the reaction mixture significantly interferes with the color development.

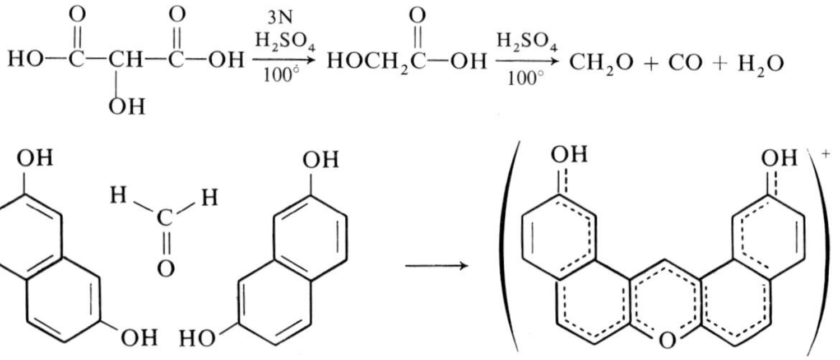

$$HO-C(=O)-CH(OH)-C(=O)-OH \xrightarrow[100^\circ]{3N\ H_2SO_4} HOCH_2C(=O)-OH \xrightarrow[100^\circ]{H_2SO_4} CH_2O + CO + H_2O$$

FIG. 25. Determination of tartronic acid with 2,7-naphthalenediol.

Formaldehyde, acetaldehyde, anisaldehyde and salicylaldehyde also react with the reagent and so are interferences. The method has been utilized for the determination of glycolic acid liberated from the γ-carboxymethyl ester of L-glutamic acid[810] and α-N-glycolyl-D-arginine.[811]

2,7-*Dihydroxynaphthalene determination of glycolic acid.*[256] To dried test substance (or less than 50 μl of solution) containing about 0·1 to 10 μg of sodium glycolate add 1 ml of 0·01 % 2,7-dihydroxynaphthalene in concentrated H_2SO_4. Cover tube with a polyethylene or glass stopper and heat at 100° for 20 min. Centrifuge down any condensed water on sides of tube. Read absorbance of cooled solution at 540 nm.

2. *Chromotropic acid. Chromotropic acid determination of glycolic acid.*[265] To 2 ml of aqueous test solution add 1 ml of 5 % aqueous sodium chromotropate followed by about 37 ml of concentrated sulfuric acid. Mix and heat at 100°C for exactly 30 min. Allow to cool to room temperature and then dilute to 50 ml with concentrated sulfuric acid. Read at 580 nm.

A method has also been devised to determine glycolic acid in mixtures obtained by irradiating methanol–water–carbon dioxide.[812] The irradiated mixture contained principally ethylene glycol and formaldehyde. In the following procedure the volatile components are completely expelled during the evaporation. Ethylene glycol and formic, tartaric and oxalic acids do not interfere in concentrations <10 mM.

Microdetermination of glycolic acid in irradiated solvents.[812] To the test solution add 0·1 ml of 0·1 M sodium acetate, evaporate almost to dryness at 100°, dissolve the residue in 1 ml of 0·5% disodium chromotropate, add 8 ml of concentrated H_2SO_4, heat for 30 min on the water bath, cool, and measure the absorbance at 610 nm. Beer's law is followed from 0·56 to 1·3 mM.

3. β-*Naphthol* can also be used for the determination of glycolic acid. Maximum absorbance was obtained after 10 min with only a slight change in absorbance after 20 hr. The β-naphthol solution is stable for 48 hr if refrigerated and protected from light. Under test conditions, dextrose, lactic and tartaric acid equal in weight to glycolic acid interfere with the assay to the extent of 9, 7·5, and 2·5% errors, respectively. Formaldehyde and acetaldehyde do not interfere since both compounds are removed during the evaporation step. Beer's law is followed from 0·2 to 15 μg of glycolic acid.

Heating a sulfuric acid solution of glycolic acid gives formaldehyde, carbon monoxide and water. Equimolar amounts of glycolic acid and formaldehyde reacted with β-naphthol give the same spectra and identical absorbances.

β-*Naphthol determination of glycolic acid.*[72] Make the test solution (containing less than 15 μg of glycolic acid) alkaline with 0·01 N NaOH. Evaporate to dryness in an oven at 125 to 130°. Add 1 ml of 0·1% β-naphthol in 92·5% H_2SO_4 and mix with a vortex mixer. Cover test tubes with glass marbles and heat at 100° for 20 min. Cool to room temperature and read absorbance at λ_{max} 480 nm.

C. Phenoxyacetic acids

J-Acid determination of phenoxyacetic acids.[84] To a residue in beaker or test tube add 3 ml 0·3% J-acid in concentrated sulfuric acid. Heat at 165° ± 2°C for exactly 5 min. Cool to room temperature. Dilute to 25 ml with 20% acetic acid. Read at 580 nm within 3 hr.

Phenyl J-acid determination of phenoxyacetic acids.[84] As in the J-acid procedure, but 0·3% phenyl J-acid in sulfuric acid is used. Dilute to 25 ml with water. Read at 635 nm within 6 hr.

Methoxone or [(4-chloro-*o*-tolyl)oxy]acetic acid is discussed in the Methoxone section.

D. Alkoxyacetic acids

Because of the concern about the problem of accelerating eutrophication due to the input of man's wastes (e.g. phosphates) into bodies of natural waters, possible biodegradable substitutes for phosphates are being investigated. A promising substitute is trisodium carboxymethyloxysuccinate (CMOS)

$$\begin{array}{l} CH_2{-}COONa \\ | \\ CH{-}COONa \\ | \\ O{-}CH_2{-}COONa \end{array}$$

The compound can be readily hydrolyzed to glycolic acid which can be oxidized to formaldehyde, the latter being then determined with β-naphthol.[72]

The color obtained in the procedure was stable for several hours. Beer's law was followed from 1 to 30 μg of CMOS. Recoveries of CMOS from primary sewage effluent, river water, pretreated sewage effluent, and pretreated river water were 97%, 96·4%, 94·5% and 93·7%.

Investigation of the precision showed that the average slope of the Beer's law curve (calculated by least squares) was 0·0321 with a standard deviation of 0·0003. The standard deviation of a single measurement was 0·0129. The distilled water blank gave an absorbance value usually less than 0·04.

Interferences were also investigated. Without the evaporative step glycolic acid, formaldehyde, acetaldehyde, and CMOS in 0·000968 M amounts in the test solution gave values equivalent to 20·5, 20·5, 2·7, and 20 μg of CMOS, respectively.

With the evaporative step formaldehyde and acetaldehyde were removed and thus showed no interference in the procedure. With this step 20 μg of dextrose, glycolic acid, lactic acid, and tartaric acid were equivalent to 0·3, 0·24, 0·30 and 0·94 μg of CMOS, respectively.

CMOS can also be determined by the fluorimetric β-naphthol method used for malic acid[813] or the chromotropic acid method for formaldehyde.[70] About 98 % formaldehyde was recovered from CMOS with the chromotropic acid method.

Without the evaporative step of the following procedure CMOS, formaldehyde and glycolic acid give mε values of 34, 35 and 35, respectively. With the evaporative step CMOS, formaldehyde and glycolic acid give mε values of 34, 0 and 0·41, respectively.

Colorimetric assay for carboxymethyloxysuccinate (CMOS).[72] *Pretreatment procedure.* Briefly centrifuge the test samples from sewage effluents or river water, Millipore-filter through a 0·22 μ membrane and acidify with concentrated HCl to approximately 1 N HCl. After 30 min centrifuge again, if necessary. Heat 1 ml of each supernatant solution in a 12 ml heavy-duty conical centrifuge tube at 100° for 30 min. Cool and add 0·3 ml 0·5 % phenylhydrazine or 2,4-dinitrophenylhydrazine in 2 N HCl to each sample. Heat again at 100° for 30 min. Add 0·5 ml of 10 % $BaCl_2$ solution. After 30 min mix in 0·15 ml concentrated NH_4OH immediately after delivery and add 6 ml of absolute ethanol to each sample. Store at 4° for about 20 hr to insure complete precipitation. Centrifuge for 15 min at 7000 g and remove and discard the clear supernatant fluid. Bake precipitates in an oven at 130° for 30 min to remove all traces of moisture. Add 1 ml of β-naphthol reagent (0·05 % in 92·5 % H_2SO_4, stable for 72 hr if refrigerated and protected from light) and then treat according to prescribed method.

Assay procedure. Acidify test solution containing not more than 30 μg CMOS with 0·1 ml of concentrated HCl per ml. Evaporate to dryness in an oven maintained at 125–130°. Add 1 ml of colorless β-naphthol reagent, stir vigorously on a vortex mixer, cap test tubes with glass marbles and heat at 100° for 60 min. Cool, add 3 ml of 80 % H_2SO_4, and mix on a vortex mixer until the solution is homogeneous. Let stand 20 min and then read the absorbance at 480 nm.

XXXII. 1,2-GLYCOLS

A. Introduction

These types of compounds are widely prevalent in nature. Since they are readily oxidized to formaldehyde, almost any of the reagents used in the analysis of formaldehyde can be used in their analysis. Periodate oxidation of these types of molecules has been discussed.[276] An example of the determination of

$$HOCH_2-CH_2OH+IO_4^- \longrightarrow 2CH_2O$$

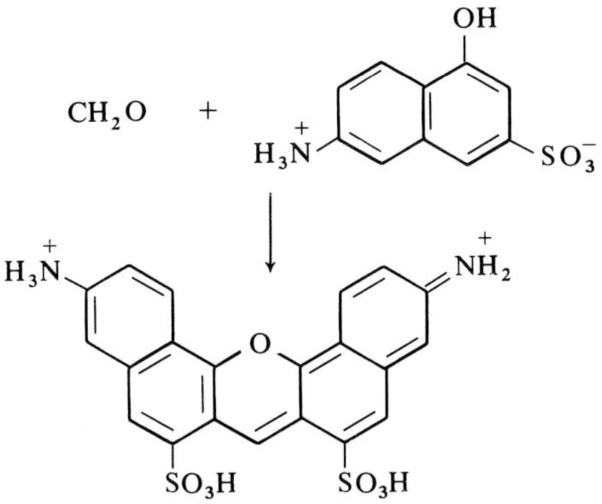

FIG. 26. J-Acid determination of ethylene glycol.

ethylene glycol with J-acid is given in Fig. 26. The fluorogen formed in this reaction can be assayed fluorimetrically.

Sugars and other polyols can be analyzed similarly. An example is shown with MBTH as the reagent, Fig. 27. Gluconic acids can be oxidized and analyzed in a similar manner. However, in their case glyoxylic acid is also produced on oxidation.

$$HOCH_2 - (CHOH)_n - CH_2OH + 2IO_4^- \dashrightarrow 2CH_2O + 2IO_3^- + nHCOOH + H_2O$$

$$CH_2O + MBTH$$

$$\downarrow \; 1.\ \Delta \quad 2.\ (O)$$

$$\left(\text{C}_6\text{H}_4\text{-N(CH}_3\text{)-S-C} = N - N = \overset{H}{C} = N - N = \text{C-N(CH}_3\text{)-S-C}_6\text{H}_4 \right)^+$$

FIG. 27. MBTH determination of sugar alcohols.

In the case of the more complicated sugar derivatives a larger variety of products are formed on oxidation as shown in Fig. 28 for the disaccharide, palatinose and Fig. 29 for lactulose.[324]

Formaldehyde can also be formed in the oxidation of other complicated molecules containing the α-glycol group, e.g. the pseudouridine phosphates, as shown in Fig. 30.[444]

Corticosteroids containing the 1,2-glycolic group (see Table 3) also can be oxidized to formaldehyde and characterized or determined with one of the appropriate reagents.

A large number of oxidative procedures are available for the determination of the 1,2-glycols of various types, e.g. ethylene glycol, glycerol and glycerides, mannitol and other sugar alcohols, amino sugars, sialic acids, sugar acids, etc. Some of the more pertinent ones are presented.

A very large number of inhalant allergens have been studied over the years.[814–816] A group of allergens isolated from ragweed pollen have been purified and appear to be proteins,[817–819] but a much larger number (which have been investigated and have not, as yet, been highly purified) appear to be glycoproteins.[814–816] A common structure has been postulated for the glycoprotein allergens.[820, 821] Essentially their characteristic feature is postulated to be a bond involving the ε-amino group of lysine linked to carbon atom 1 of a 1-deoxy-2-ketose sugar in the 1,2-enol form.

Monosaccharides of plant and animal origin are present in "purified" house dust allergen.[423] They are highly branched, suffering little attack during

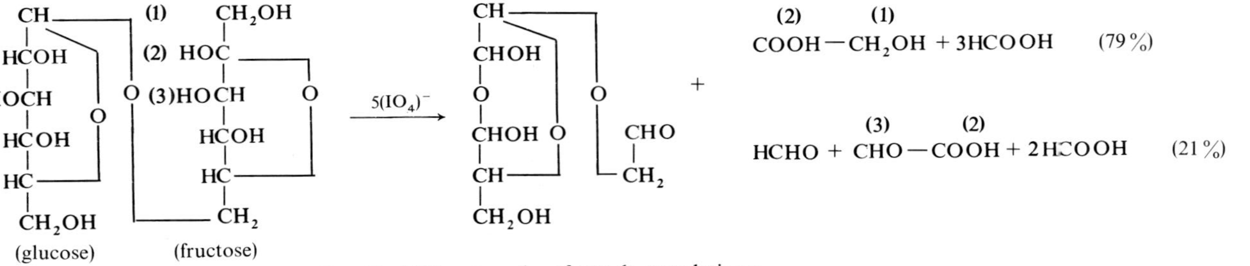

FIG. 28. Different modes of attack on palatinose.

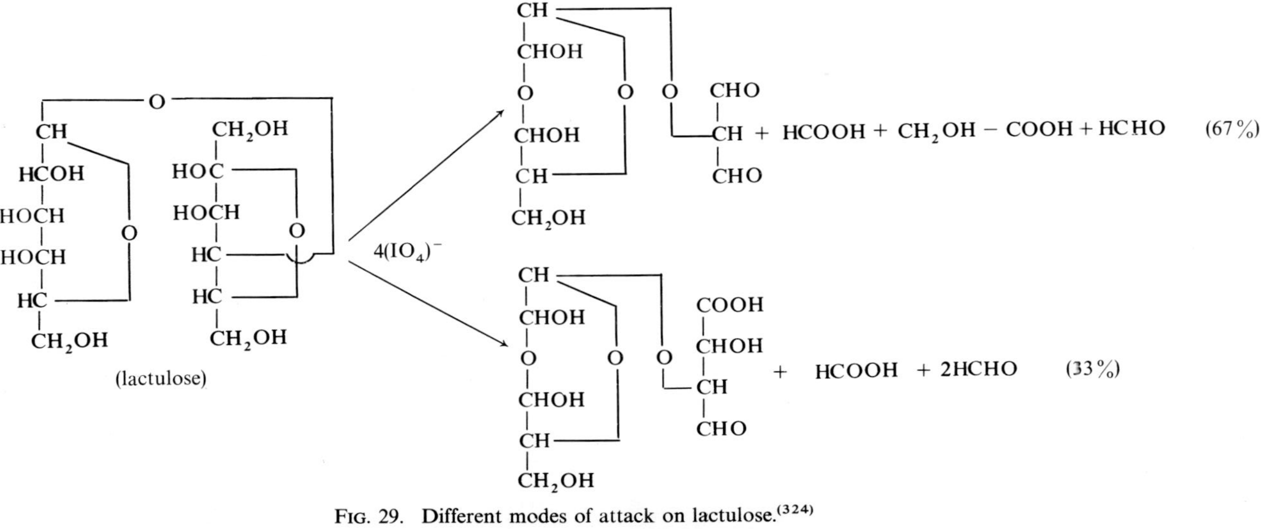

FIG. 29. Different modes of attack on lactulose.[324]

periodate oxidation. Furanose configuration for a proportion of the sugars (pentoses) is indicated because of liberation of pentose by very mild hydrolysis and release of formaldehyde following periodate treatment. About 25% of the sugars are oxidized. The formaldehyde is determined by the chromotropic acid method.[3]

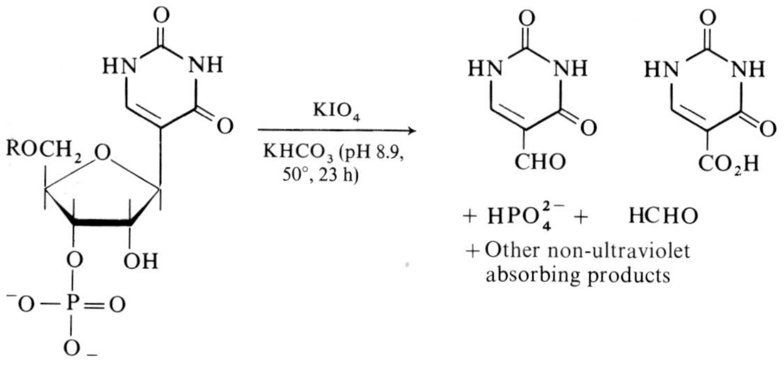

KIO_4, $KHCO_3$ (pH 8.9, 50°, 23 h)

CHO CO_2H

+ HPO_4^{2-} + HCHO
+ Other non-ultraviolet absorbing products

A, R = H
B, R = $-PO_4^{2-}$

A. Pseudouridine 3′-phosphate
B. Pseudouridine-3′,5′-diphosphate

FIG. 30. The oxidation of the pseudouridine phosphates.[444]

An alternative method of oxidation of 1,2-glycols to formaldehyde utilizes a cobaltous salt in the presence of oxygen,[277, 278] e.g.

$$CH_3(CH_2)_7CHOH{\cdot}CH_2OH \xrightarrow[\text{Co (II) laurate}]{O_2} CH_3(CH_2)_7COOH + CH_2O$$

Since the reaction takes place at 100° over several hours, the method needs further study to optimize conditions before it can become practical.

Another method of oxidation uses hydrogen peroxide plus a ferric salt.[822] When diethyleneglycol and propylene glycol are oxidized and coupled with resorcinol only the reaction product of the former produces a greenish yellow color on basification. The diethyleneglycol is possibly oxidized to formaldehyde which couples with resorcinol in strongly acid solution to form a dibenzoxanthenyl anion on alkalinization. If so, this anion is fluorescent and could be determined fluorimetrically. Beer's law is followed up to 12 ppm of diethyleneglycol. The accuracy is stated to be about ±2%. The following procedure was used.

Determination of diethylene glycol in propylene glycol.[822] To 5 ml of 80% H_2O_2 solution add 1 ml of the test solution followed by 0·1 ml of ferric ammonium sulphate solution (1 mg/ml) and 2 ml of 36 N H_2SO_4. Heat for 20 min at 170°, add 1·5 ml of resorcinol solution and heat at 170° for 30 min.

Dilute to 100 ml with distilled water. Pipet a 2 ml aliquot of the solution, adjust pH to 10·5 and dilute to 50 ml, keeping pH at 10·5. Measure absorbance at 490 nm.

The method needs to be made more sensitive and optimized.

B. Atmospheric α-glycols

Methods have also been developed for the determination of atmospheric α-glycols.[2, 8] In this assay airborne particulates are collected from polluted air onto a filter with the help of a high volume sampler. The particulate is extracted with water and then an aliquot of the extract is analyzed. Airborne pollen, fungi, house dusts, grain dusts and various air pollution source effluents can be analyzed by one of these procedures. The oxidative MBTH procedure is one of the most convenient ones to use here.

The details of the procedures are given in the following pages.

Oxidative MBTH procedure for acetols and 1,2-*glycols.*[2] Two-tenths ml of 0·5 % periodic acid in water–sulfuric acid (280:1) was added to 1 ml of an aqueous test solution. After 15 min 0·4 ml of 2 % sodium arsenite in water-concentrated hydrochloric acid (20:1) was added; after 2 min 1 ml of 0·8 % aqueous MBTH salt was added. The mixture was then heated in a boiling water-bath for 3 min and then cooled to room temperature under the tap. Two ml of 0·5 % aqueous ferric chloride was added and the absorbance was read in 15–25 minutes against the blank at 654 nm.

C. Miscellaneous reagents

2,4-*Pentanedione, dimedone and J-acid procedures for* 1,2-*glycols.*[8] To 1 ml of aqueous test solution add 0·2 ml 0·5 % periodic acid in concentrated sulfuric acid–water (1:280, v/v). After 15 min add 0·4 ml 2 % sodium arsenite in concentrated hydrochloric acid–water (1:20, v/v). Wait 2 min and add the reagent solution and proceed according to the reagent procedure.

In the procedure using 2,4-pentanedione the reagent contains 15 g ammonium acetate, 0·2 ml 2,4-pentanedione and 0·3 ml of distilled acetic acid diluted to 100 ml with water. After the 2 min wait, add 2 ml of the pentanedione reagent and heat the mixture at 57° for 10 min. Cool and read the fluorescence intensity at *F*405/510.

In the procedure using dimedone the reagent contains 25 g ammonium acetate, 0·3 g dimedone and 0·4 ml distilled acetic acid diluted to 100 ml with water. After the 2 min wait add 2 ml dimedone reagent and heat the mixture for 10 min at 100° with the help of a cold finger. Cool under the tap and read at *F*392/462.

In the procedure using J-acid the reagent contains 0·01% J-acid in concentrated sulfuric acid. After the 2 min wait add 5 ml of the J-acid solution. Fifteen minutes later dilute to 10 ml with concentrated sulfuric acid. Cool under the tap and read the fluorescence intensity at *F*462/520.

Oxidative J-acid procedure for acetols and 1,2-*glycols.*[2] To 1 ml of aqueous test solution was added 0·2 ml of 0·5% periodic acid (H_5IO_6) in aqueous sulfuric acid (1 ml H_2SO_4 diluted to 280 ml with water). After 15 min 0·4 ml of 2% sodium arsenite in aqueous hydrochloric acid (1 ml concentrated hydrochloric acid diluted to 20 ml with water) was added, followed in 2 min by the addition of 5 ml of 0·01% J-acid in concentrated sulfuric acid. The mixture was then heated for 5 min in a boiling water-bath and then cooled in ice water. The fluorescence was read 10–25 minutes later at *F*462/520.

Chromotropic acid determination of 1,2-*glycols.*[139] Mix 0·2 ml of aqueous test solution with 0·2 ml of 0·3 M periodic acid and 0·2 ml of 1 M sodium bicarbonate. Allow to stand 1 hr at 25°. Add 1·5 ml of 0·5 M sulfuric acid and 0·5 ml of 1 M sodium arsenite (13% $NaAsO_2$). Gently agitate till formed iodine disappears (within 5 min). To a 1 ml aliquot (less than 0·62 μmoles of formaldehyde) of the mixture add 10 ml of 1% sulfuric acid solution of chromotropic acid. Heat 30 min at 100°C, stoppering when tubes have become hot. After cooling, read absorbance at 570 nm.

The periodic acid–Schiff stain is used extensively in the histochemical studies of polysaccharides, mucopolysaccharides, mucoproteins, and glycoproteins. The analogous Feulgen stain is also used extensively in the cytophotometric analysis of DNA in individual cells. Both of these reactions will be discussed in the aliphatic aldehydes precursor section.

The oxidative cleavage of 1,2-diols with cobalt acetate is carried out in an aprotic, polar solvent with aldehyde yields of 60 to 80%.

D. Chromatographic localization

Sensitive location agents are available for the detection of the various types of 1,2-glycols, e.g. sugars, amino sugars, sialic acids, sugar acids, sugar alcohols, etc. An acetylacetone (or 2,4-pentanedione) procedure has been described for the location and detection of these types of molecules on chromatograms.[33]

Acetylacetone detection of 1,2-*glycols on chromatograms.* For paper chromatography, the dry chromatogram is dipped through a reagent containing 2 ml 40% periodic acid, 2 ml pyridine and 100 ml acetone. When the paper is seen to be dry, it is further dipped through a solution containing 15 g ammonium acetate, 0·3 ml glacial acetic acid and 1 ml acetylacetone in 100 ml methanol. A visible yellow colour appears in 30–60 min, but yellowish green fluorescent spots can be detected within 10–15 min. For thin-layer chromatography, the method is essentially the same as that described for paper, but the reagents are used as spray rather than dip. The plates are left, after application of the first reagent, until no smell of acetone can be detected. Both the described reagents should be freshly prepared. It is also advisable to use redistilled acetylacetone.

Some of the compounds detected by this procedure are shown in Table 10.

A J-acid procedure can also be used to detect these compounds on chromatograms. A lavender color with a detection limit of about 3 μg/cm^2 is obtained for these types of molecules with the exception of an orange color with a detection limit of about 2 μg for 2-deoxy sugars.

Table 10. Location and detection of sugars, amino sugars and sialic acids on chromatograms

	Detection Limits[b] μg/cm^2		Fluor. Color[c] Detection Limits μg/cm^2	
Compound[a]	paper	si gel	paper	si gel
N-Acetyl-D-glucosamine	(5)	(5)	(10)	(3)
Arabitol	(3)	(3)	(1)	(3)
2-Deoxy-D-galactose	(5)	(5)	(5)	(5)
2-Deoxy-D-glucose	(5)	(5)	(5)	(10)
2-Deoxy-D-ribose	(3)	(3)	(5)	(3)
Dulcitol	(5)	(3)	(3)	(3)
1-Erythritol	(3)	(3)	(1)	(3)
D-Galactonic acid	(10)	(5)	(5)	(5)
Galactosamine	(5)	(3)	(5)	(3)
Glucose	(5)	(3)	(5)	(3)
Glucosamine HCl	(5)	(3)	(10)	(3)
Mannitol	(5)	(3)	(3)	(3)
Mannonic acid	(5)	(5)	(5)	(5)
Perseitol	(5)	(3)	(5)	(3)
Ribitol	(5)	(3)	(1)	(3)
D-Ribose	(5)	(10)	(10)	(20)
Sedoheptulose, anhydride	—	—	(10)	(5)
Sorbitol	(5)	(3)	(1)	(3)
D-Xylitol	(3)	(3)	(1)	(3)

[a] The following compounds were negative at 20 μg. (1) Cellobiose (2) Deoxycytidine (3) Deoxyguanosine 1·5 H_2O (4) Deoxyinosine (5) Dextrose pentaacetate (6) Fucose (7) Glucuronic acid (8) Glucuronic acid lactone (9) Inositol (10) Maltose (11) Raffinose (12) Rhamnose (13) Trehalose.

[b] All compounds gave a yellow color.

[c] All compounds gave a green fluorescence except sedoheptulose anhydride which gave a purple fluorescence.

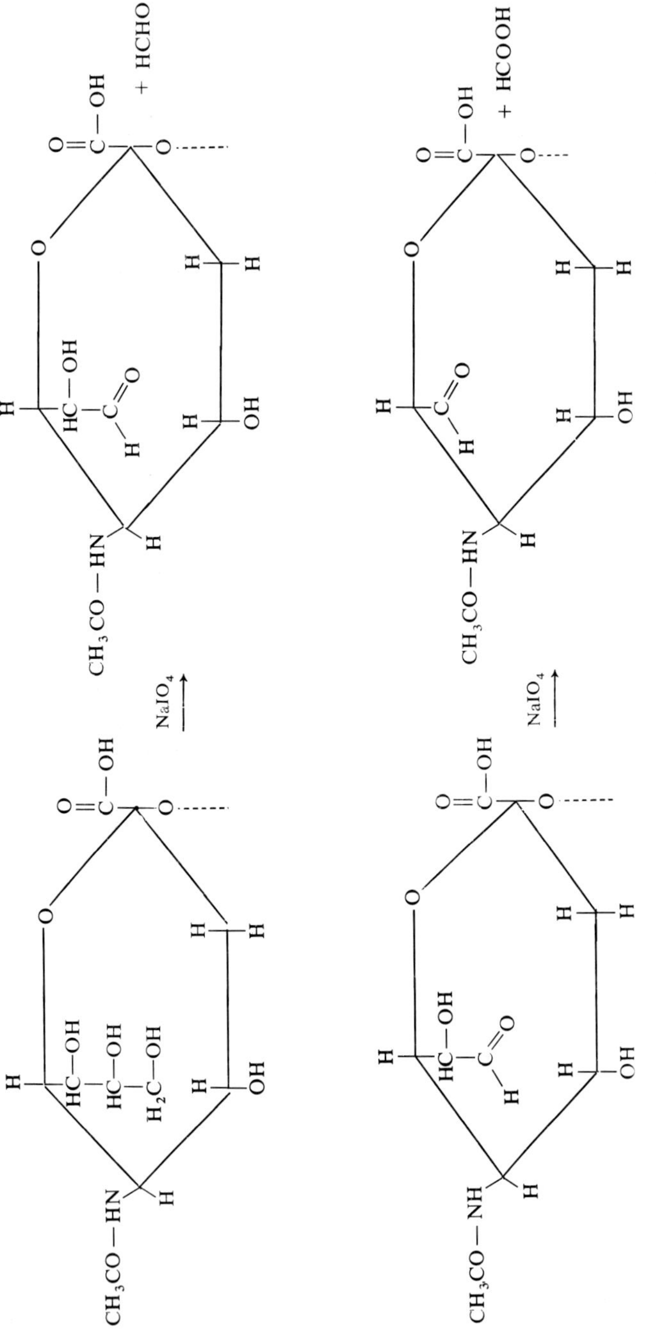

FIG. 31. Periodate oxidation of N-acetylneuraminic acid in a nonreducing terminal position of a glycoprotein.[458]

J-Acid detection of 1,2-glycols on chromatograms. Dip or spray in sequence with reagent 1 (4 ml of 40% periodic acid and 4 ml of pyridine in 100 ml of acetone) and then reagent 2 (5% of the pure sodium salt of J-acid in concentrated sulfuric acid brought up to 100 ml with ethanol, allowed to stand for 30 min and filtered). Dry the chromatogram partially with an air dryer. The colors develop at room temperature.

E. Sialoglycoproteins

N-Acetylneuraminic acid (NANA)-containing glycoproteins are widely distributed in blood group specific substances, serum glycoproteins, hormones and cell membrane glycoproteins. Under very mild conditions,[823] formaldehyde is only liberated from the cleavage between C9 and C8 of NANA residues and a 8-carbon aldehyde analog of NANA is formed,[824] Fig. 31. Some of the biological properties of the glycoproteins often involve the 3 carbon polyhydroxy side chain of NANA, so that the measurement of formaldehyde liberated during periodate oxidation of a sialoglycoprotein is postulated as a good way of approaching the establishment of a correlation between structure and biological activity. On further oxidation a 7-carbon aldehyde analog of NANA is formed.[458]

The formaldehyde formed in the oxidation of NANA is determined with MBTH. This reagent reacts with formaldehyde and glyoxalic acid in the same way, e.g. mε = 65,[825] but the latter compound does not interfere since it is not formed during periodate oxidation of a sialoglycoprotein.

The following procedure was used in the determination.

MBTH determination of a sialoglycoprotein.[458] *Reagents.* 0·01 M periodate solution, prepare daily. 2% Sodium bicarbonate solution. 0·1 N sodium arsenite solution. *Formaldehyde solution* (10^{-4})—prepare daily by decomposition of paraformaldehyde, the solution being analyzed by the dimedone gravimetric method. 2 N HCl. MBTH, 0·154 M in 0·1 N HCl.

Procedure. To 1 ml of aqueous test solution add 0·1 ml of 2 N HCl, 0·1 ml of MBTH reagent and 0·1 ml of 5% $FeCl_3 \cdot 6H_2O$. Let stand 0·5 hr, add 2 ml of methylene chloride, shake tubes vigorously and centrifuge. Separate the methylene chloride layer and add a few milligrams of anhydrous sodium sulfate to it. Read the absorbance of the organic layer at 630 nm. Colors are stable for 17 hr.

XXXIII. GLYCOPROTEINS AND POLYSACCHARIDES

These compounds can be detected directly after electrophoresis; periodate oxidation forms formaldehyde which is reacted with phenylhydrazine to form the hydrazone, and this latter derivative is coupled with diazotized *o*-anisidine in the presence of cupric acetate to form a formazan, IX[281]

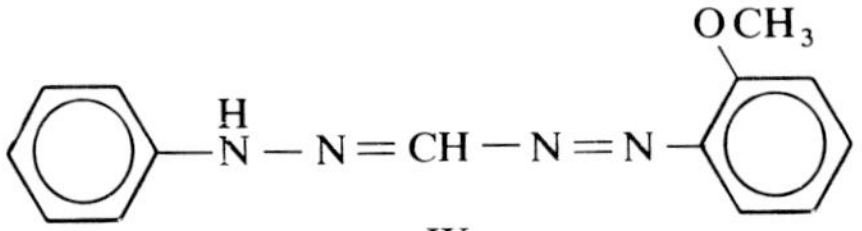

OCH$_3$

H
N—N=CH—N=N

IX

Polysaccharides are, in most cases, potential formaldehyde precursors. An example of such a precursor is guaran, the galactomannan from guar seeds. The high degree of substitution in guaran shows a regular periodicity in that every second mannose residue in the main chain carries a galactose residue as a side-chain. Hydrolysis of guaran gives D-galactose and D-mannose in the ratio of 36:64. When the mannose residue (in guaran) carrying a galactose residue at

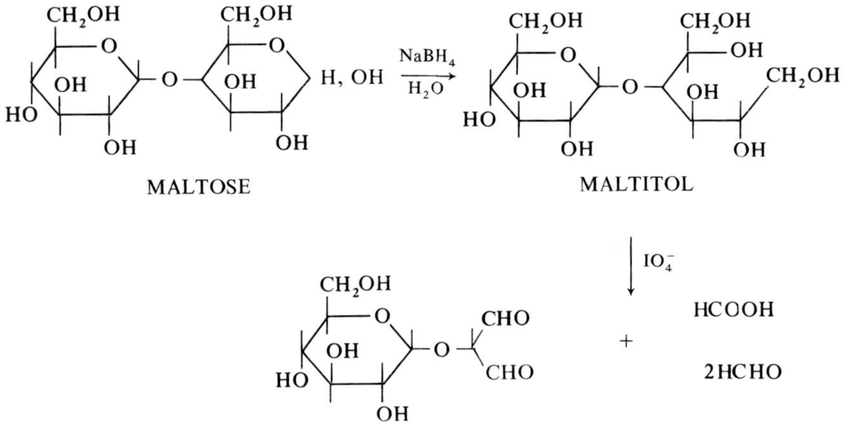

CH$_2$OH CH$_2$OH
O O
OH O OH H, OH
HO
OH OH
MALTOSE

NaBH$_4$
H$_2$O

CH$_2$OH CH$_2$OH
O OH
OH O OH CH$_2$OH
HO
OH OH
MALTITOL

IO$_4^-$

CH$_2$OH
O CHO
OH O
HO CHO
OH

\+

HCOOH

2HCHO

FIG. 32. The reduction → oxidation of maltose to a dialdehyde and formaldehyde.[897]

position 6 is oxidized, the liberated aldehyde groups immediately form stable hemiacetals with the closest hydroxyl groups on the 2 adjacent mannose residues, thus preventing their subsequent oxidation.[283] To obtain a direct measure of the fraction of singly oxidized galactose residues present at any time the oxidized samples were reduced with borohydride and then oxidized with periodate with measurement of the consumption of additional periodate and of the liberated formaldehyde. The aldehyde was determined by a chromotropic acid method.[324]

Maltose can be reduced to maltitol which then can be oxidized to a dialdehyde, formic acid and formaldehyde, Fig. 32.

XXXIV. GLYOXYLIC ACID

Precursors of this acid will be discussed in more detail in the glyoxylic acid section in another volume. However, we would like to emphasize in this section that in most cases, the same chromogen is obtained with formaldehyde. Phenylhydrazine is the reagent used most often in the analysis of this acid and its precursors. Thus, allantoin, allantoic acid, ureidoglycolate can be analyzed by the procedure, Fig. 33.

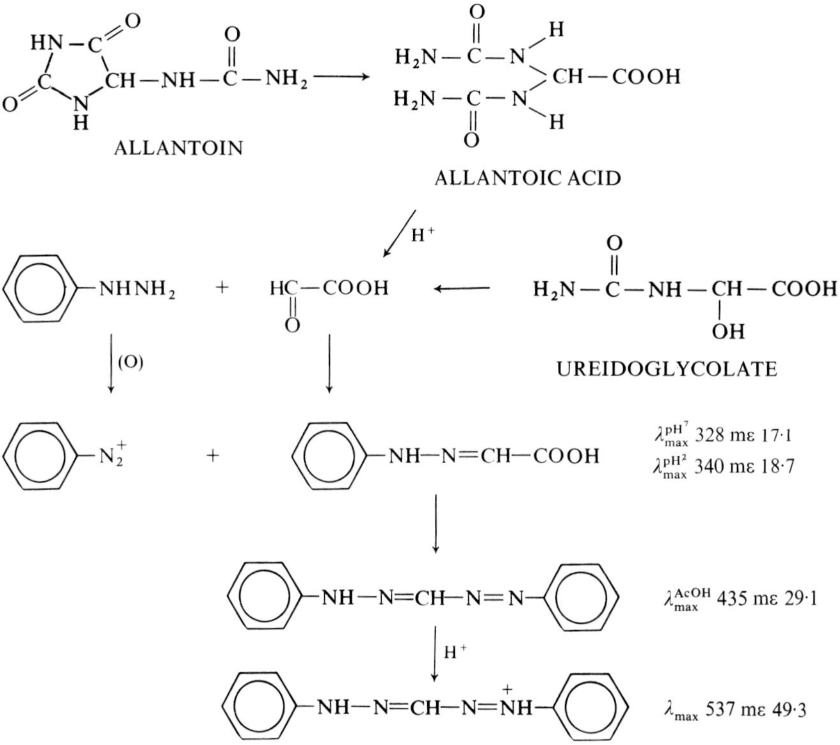

ALLANTOIN

ALLANTOIC ACID

UREIDOGLYCOLATE

$\lambda_{max}^{pH^7}$ 328 mε 17·1

$\lambda_{max}^{pH^2}$ 340 mε 18·7

λ_{max}^{AcOH} 435 mε 29·1

λ_{max} 537 mε 49·3

FIG. 33. The determination of allantoin with phenylhydrazine.

One reagent that has been used in the analysis of glyoxylic acid is pyrogallol. The following procedure has been described.

Pyrogallol determination of glycolaldehyde or glyoxylic acid.[254] To 0·5 ml of aqueous test solution (containing 0·05 to 0·5 μmole of test substance) add 7 ml of 0·043 % pyrogallol in concentrated sulfuric acid. Heat for 30 min at 100°C. Cool and measure at 600 nm.

Compounds forming formaldehyde in hot sulfuric acid will probably also react.

A pyruvic oxidase in pig heart muscle is stated to cleave glyoxylate to formaldehyde; this enzyme is stimulated by CoA·SH and NAD^+.[826] On the other hand, lactic dehydrogenase from *Aspergillus niger*, rabbit muscle and pig heart is known to catalyze a dismutation of glyoxylate to glycolate and oxalate.[827–829] Furthermore, a NADH-dependent reduction of glyoxylate to glycolate is known to occur in *L. plantarum*.[830, 831] Glycolate can be determined by conversion to formaldehyde in H_2SO_4 and then reaction with 2,7-naphthalenediol while oxalate can be reduced with the help of zinc dust and acid to glycolic acid which can then be determined colorimetrically at λ_{max} 530 nm after reaction with naphthalene 2,7-diol. These analyses have been utilized to determine glycolate and oxalate in urine after chromatography on a column of Dowex AG2-X8 resin (Cl^- form).[832]

XXXV. HEPTITOLS, HEXITOLS, TETRITOLS AND OTHER SUGAR ALCOHOLS

Sugar alcohols are widely distributed among the different plant families. The separation and determination of threitol and erythritol are important in structural studies on plant and bacterial polysaccharides using periodate oxidation, reduction and acid hydrolysis; the production of threitol and erythritol characterizes a 1,4-linked galacto and 1,4-linked glyco-pyranoside unit respectively. Erythritol and threitol can be present in mixtures obtained from the partial oxidation of hexitols or oxidation of polysaccharides and be automatically determined following chromatography on Dowex 1 (molybdate).[295] Allitol and ribotol can be separated from their isomers while propane-1,2-diol, glycolaldehyde, glycerol, erythritol and xylitol are separated by chromatography on Dowex 1 (sulfate), Fig. 34. Detection and assay is by periodate oxidation to formaldehyde which is then reacted with ammonia and 2,4-pentanedione to form the yellow chromogen, λ_{max} 420 nm. Beer's law is followed from about 30 to 160 μg of D-threitol or erythritol.

Partition chromatography on anion- and cation-exchange resins in aqueous ethanol can also be used for the separation of aldoheptitols and various oligomeric sugar alcohols.[833] A similar method has been used for the separation of oligomeric sugars.[834] A straight line relationship was shown to exist between the logarithm of the distribution coefficient and the degree of polymerization of oligomers with the same mode of glycosidic bonds. The distribution coefficients listed in Table 11 show the good separations obtained with the two investigated resins. At high ethanol concentration the solutes are eluted in order of increasing size while at low ethanol concentration the order is

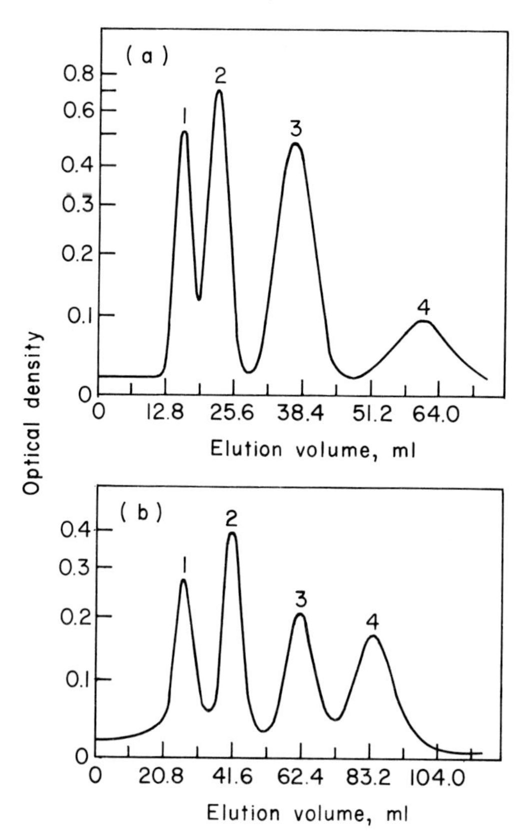

FIG. 34. Separations of sugar alcohols: (a) (1) = propane-1,2-diol, (2) = ethylene glycol, (3) = glycerol, (4) = D-threitol; (b) (1) = glycolaldehyde, (2) = glycerol, (3) = erythritol, (4) = xylitol. Sugar alcohols determined by formaldehyde assay (absorbance measured at 420 nm).[295]

reversed. Two resins were used,[833] one cation-exchanger in the lithium form (Aminex A-6, 15–19 μ) and one anion-exchanger in the sulfate form (Technicon T5C, 10–17 μ). Eluates were analyzed automatically in a two-channel Technicon Autoanalyzer by the orcinol[835] and periodate–2,4-pentanedione[592] methods. The volume distribution coefficients (D_v) were calculated from the peak elution volumes,[836] where:

$$D_v = \frac{\text{amount of solute per cm}^3 \text{ of exchanger bed}}{\text{amount of solute per cm}^3 \text{ of solution}}$$

The type of separation of a small group of sugars and sugar alcohols on a sulfate resin is shown in Fig. 35, as is the usefulness of the 2-channel analyzer for precursors of the furfurals (orcinol test) and formaldehyde (periodate–2,4-

pentanedione test).[833] A chromatogram illustrating the separation of a mixture of 7 sugar alcohols from glucose on a lithium resin, Fig. 36, is ideal for the analysis of complex mixtures of sugars and polyhydric alcohols. At pH 2 the sugar alcohols give a strong response in the formaldehyde channel whereas aldoses give a very weak response. Like ketoses[837] the glycosides of sugar alcohols are precursors of the furfurals and formaldehyde and so give a response

Table 11. Distribution coefficients of some oligomeric sugar alcohols separated by partition chromatography on ion-exchangers[833]

Sugar alcohols	SO_4^{2-} resin, Technicon T5C (10–17 μ) 70% EtOH	85% EtOH	Li^+ resin, Aminex A–6 (15–19 μ) 85% EtOH
Xylitol	1·41	5·31	4·38
Glucitol	2·15	10·5	6·50
Mannitol	2·54	14·1	5·83
D-Glycero-D-galacto-heptitol	4·18	31·6	11·3
D-Glycero-D-gulo-heptitol	3·25	21·6	9·47
D-Glycero-D-manno-heptitol	3·60	25·4	8·06
4-0-β-D-Xyl*p*-D-xylitol	1·81	13·4	4·70
3-0-β-Gal*p*-D-arabinitol	2·81	26·6	9·23
4-0-β-D-Gal*p*-D-glucitol	3·78	46·1	18·3
4-0-β-D-Gal*p*-D-mannitol	3·80	46·8	13·4
6-0-α-D-Gal*p*-D-glucitol	3·99	49·1	20·9
3-0-α-D-Glc*p*-D-glucitol	4·80	64·9	12·1
3-0-α-D-Glc*p*-D-mannitol	4·75	63·5	12·7
4-0-α-D-Glc*p*-D-glucitol	3·97	51·5	15·0
4-0-β-D-Glc*p*-D-glucitol	4·05	52·0	12·0
6-0-α-D-Glc*p*-D-glucitol	4·26	57·0	15·6
6-0-β-D-Glc*p*-D-glucitol	4·78	68·0	11·8
6-0-α-D-Glc*p*-D-mannitol	4·94	73·9	14·6

in both the orcinol and periodate channels. The periodate → formaldehyde response is higher for 4-0-glycosides of hexitols (such as 4-0-β-D-glucopyranosyl-D-glucitol, 4-0-α-D-glucopyranosyl-D-glucitol, and 4-0-β-D-galactopyranosyl-D-glucitol) than for those with the glycosidic bond at C-6 (such as 6-0-α-D-galactopyranosyl-D-glucitol).[833] Two moles of formaldehyde are formed from each mole of 3-0- and 4-0-glycosides of hexitols and 3-0-glycosides of pentitols (such as 3-0-β-D-galactopyranosyl-D-arabinitol), whereas only one mole is formed when another site is blocked by a glycosidic bond.

An excellent separation can also be obtained of the β-(1 → 4)-linked-D-xylopyranosides of D-xylitol on an anion-exchanger in sulfate form in 75%

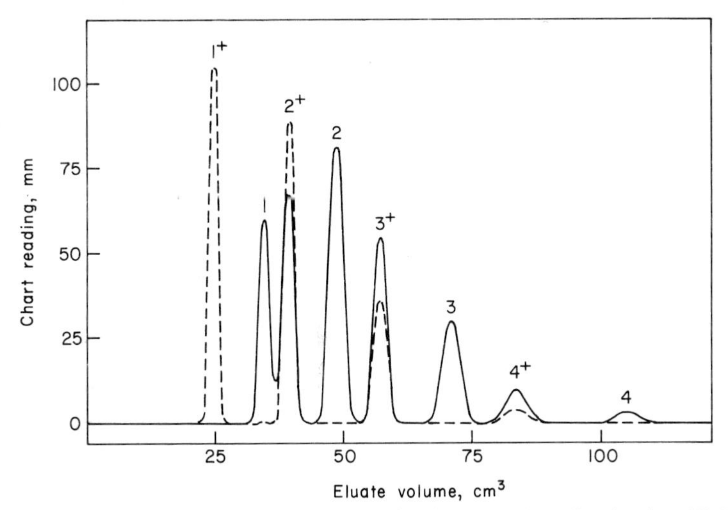

FIG. 35. Separation of 50 μg of glucose (1), 100 μg of maltose (2), 50 μg of maltotriose (3), 50 μg of glucitol (1^+), 100 μg of maltitol (2^+), 100 μg of maltotriitol (3^+) and traces of maltotetraose (4) and maltotetraitol (4^+) in 70% aqueous ethanol (w/w) at 75°C on anion-exchanger in sulphate form. Resin bed: 4·4 × 720 mm. Nominal linear flow: 3·2 cm min^{-1} (Calc. for unpacked column). Orcinol method, full line; periodate–formaldehyde method, broken line.[833]

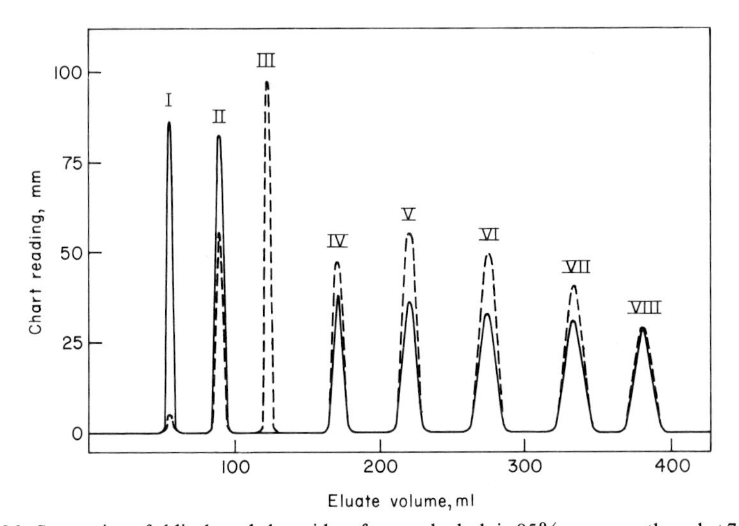

FIG. 36. Separation of alditols and glycosides of sugar alcohols in 85% aqueous ethanol at 75°C on cation-exchanger in its Li^+-form. Resin bed: 4·4 × 1170 mm. Nominal linear flow: 3·9 cm min^{-1}. Orcinol method, full line; periodate–formaldehyde method, broken line.
I, glucose (100 μg); II, 4-0-β-D-Xyl*p*-D-xylitol (100 μg); III, glucitol (70 μg); IV, 3-0-β-D-Gal*p*-D-arabinitol (200 μg); V, 4-0-β-D-Glc*p*-D-glucitol (200 μg); VI, 4-0-α-D-Glc*p*-D-glucitol (200 μg); VII, 4-0-β-D-Gal*p*-D-glucitol (200 μg); VIII, 6-0-α-D-Gal*p*-D-glucitol (200 μg).[833]

aqueous ethanol at 75°.[833] Xylitol and 6 members of the series 0-β-D-xylopyranosyl-(1,4)-[0-β-D-xylopyranosyl-(1,4)]$_{DP-2}$-D-xylitol, where the DP values of 1 to 7 refer to the degree of polymerization include the xylitol moiety. The relative responses in the orcinol and periodate channels can be useful in characterization of the oligomeric sugar alcohols and for the estimation of the number of monomeric units (DP) of the lower oligomers.

Table 12. Recommended ethanol concentrations at 75°C for various oligomeric series[833]

Oligomeric series	Resin form	Recommended conc. of ethanol % (w/w)
β-(1 → 4)-linked	Li^+	80–85
D-xylose-D-xylitol	SO_4^{2-}	72–77
β-(1 → 3)-linked	Li^+	82–87
D-glucose-D-glucitol	SO_4^{2-}	70–75
β-(1 → 4)-linked	Li^+	78–83
D-glucose-D-glucitol	SO_4^{2-}	68–73
α-(1 → 6)-linked	Li^+	70–75
D-glucose-D-glucitol	SO_4^{2-}	60–65

Havlicek and Samuelson[833] have emphasized that with complex mixtures containing members of different oligomeric series separations can easily be carried out at an eluent composition corresponding to the critical concentration of one series of oligomers present in the sample. If for instance glucopyranosides of glucitol are chromatographed at 75° C on a sulphate resin in 59% ethanol the β-(1,4)-linked oligomers appear in one peak at $D_v = 1{\cdot}28$, the β-(1,3)-linked oligomers appear earlier on the chromatogram and the members of the α-(1,6)-linked series appear afterwards and in the order of increasing molecular weight. The fractions containing overlapping compounds are then re-chromatographed at higher ethanol concentration. The ethanol concentrations recommended for separations of oligomers belonging to only one series are shown in Table 12.

XXXVI. HEXAHYDRO-1,3,5-TRINITRO-S-TRIAZINE

Treatment of this compound with concentrated H_2SO_4 forms formaldehyde[287] which can then be determined by one of the available reagents.

XXXVII. HEXAMETHYLENETETRAMINE (or hexamine or methenamine)

This food preservative can be determined in food samples by steam-distilling the sample in the presence of citric acid and magnesium sulfate.(293) The formaldehyde in the distillate is then determined.

This compound and its mandelate can be determined in tablets.(289) A semiautomated chromotropic acid method has been collaboratively studied.

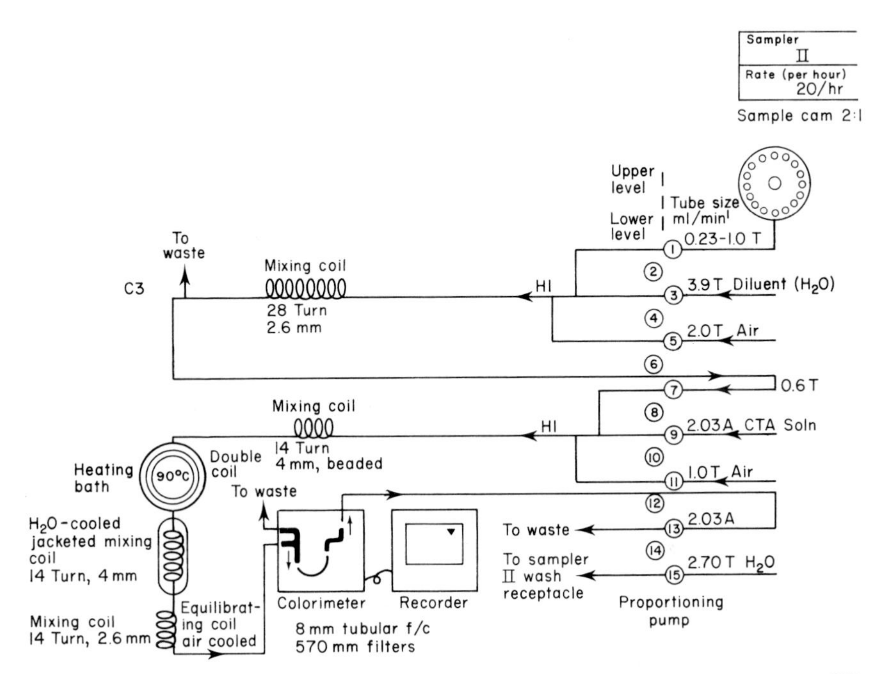

FIG. 37. Flow diagram for automated analysis of methenamine and methenamine mandelate.(289)

The coefficients of variation were found to range from 0·54 to 2·48%. Pure hexamethylenetetramine was used as the standard material. The reaction does not go to completion in this procedure; the percent hydrolysis ranged around 47·3 to 47·6%, but the method was considered satisfactory and reproducible. The flow diagram is shown in Fig. 37. Recoveries ranged from 98 to 101%.

Hexamine in vulcanizates could also be determined with an appropriate formaldehyde reagent by modification of the following detection procedure.

Detection of hexamine in vulcanizates.(291) Extract the finely divided vulcanizate (5 g) with chloroform in a Soxhlet apparatus. (Ultrasonic or Polytron extraction would be faster and cleaner). Concentrate the extract to

25 ml, and re-extract the hexamine into 0·5 ml water. Apply 5 μl of this solution to an activated Silica gel G plate and develop to a distance of 15 cm with methanol–concentrated aqueous ammonia (9:1). After drying the plate in air, detect the hexamine by spraying with 60% H_2SO_4–10% chromotropic acid (5:1) giving a blue-violet spot at R_f 0·45 against a pale blue background.

The various methylene compounds which can be readily hydrolyzed to formaldehyde are discussed much more fully in the Methylene Section.

XXXVIII. HEXAMETHYLMELAMINE

This compound, X, has minimal activity against animal tumors but

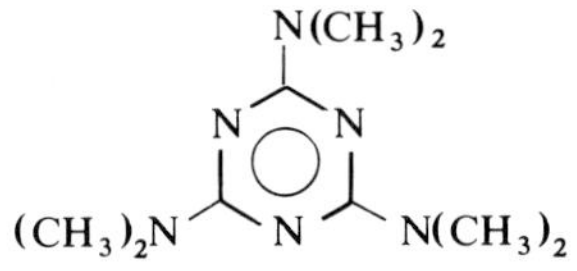

$N(CH_3)_2$

$(CH_3)_2N$ — triazine ring — $N(CH_3)_2$

X

has significant activity in man, particularly in the treatment of lung and ovarian cancer.(838) It is also highly active against human tumors, (such as those of the lung,(839) kidney, ovaries and colon) transplanted into immunologically-deprived mice. Many of the anticarcinogens used in the treatment of human cancer are carcinogenic bifunctional alkylating agents. The requirement for at least two alkylating entities in the molecule suggests that their cytotoxicity is due to a cross-linking reaction. However, because the alkylating agents are so reactive they react with all types of entities in normal living tissue before they ever interact with the tumor cell and its DNA. Thus, the anticarcinogenic activity of such a drug will depend on the balance between the damage caused to the cancer cell and the damage to the normal cell. There is an apparent correlation between demethylation and anticancer activity(838) since the hexamethyl-, N,N,N′,N″-tetramethyl-, N,N-dimethyl-, N-methyl-, N,N′,N″-trimethyl- and the hexaethyl-melamines have antitumor activities of 10·3, 1·8, 0, 0, 0, and 0, respectively.(839) Some of these compounds could be metabolized to bis-methylol derivatives which could be bifunctional alkylating agents.

Hexamethylmelamine (or Hemel or 2,4,6-trisdimethylamino-s-triazine) and trimethylolmelamine, XI, are mutagenic to *Musca domestica*(840) and Drosophila,(841) respectively.

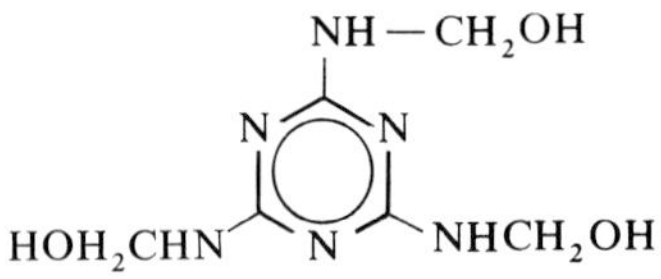

$NH-CH_2OH$

HOH_2CHN — triazine ring — $NHCH_2OH$

XI

Trimethylolmelamine is also useful in human tumour therapy.[842] This compound could also be readily hydrolyzed to formaldehyde and thus analyzed. It has been suggested that the intermediate methylol could be the active metabolite. N-Methylols have been postulated as cross-linking agents responsible for the mutagenicity of formaldehyde.[843] Hexamethylmelamine is metabolized to N-methylols enzymatically and these could be hydrolyzed to formaldehyde. Formaldehyde and its precursors could be analyzed by many of the methods described in this section and in the formaldehyde section in Volume 1. The role of formaldehyde and its precursors in the anticarcinogenic activity of hexamethylmelamine needs more thorough study.

XXXIX. HUMAN FOLLICLE-STIMULATING HORMONE (HFSH) AND HUMAN CHORIONIC GONADOTROPHIN (HCG)

Studies of the carbohydrate moieties of high-activity HFSH and HCG have been reported.[11,844,845] These compounds are hydrolyzed, chromatographed on Dowex resin AG-1 (× 8, sulphate form, 200–400 mesh) at 50° with 86% ethanol.[295] The column eluate is then continuously monitored automatically for formaldehyde after oxidation with periodate.[11]

XL. HYDRAZINES, TRIAZENES AND TETRAZENES

A. Introduction

The potential risk of contact with these direct and indirect alkylating agents has been emphasized.[672] The carcinogenicity of the hydrazines and triazenes has been discussed.[50,672,846,847] Druckrey[50] has discussed the organic biochemistry and carcinogenicity of 3 groups of indirect carcinogens, e.g. dialkylnitrosamines; 1,2-dialkylhydrazines, azo- and azoxy-alkanes; and 1-aryl-3,3-dialkyltriazenes. Many of these compounds have been found to be organotropic and transplacental carcinogens and teratogens in BD- rats. It was demonstrated that enzymatic dealkylation (probably initiated by α-C-hydroxylation) and subsequent degradation of the resulting monoalkyl compounds, eventually yielding the respective alkyldiazonium salt as the proximate alkylating carcinogen or teratogen, is a common feature of these three groups. Aldehydes are byproducts in all these reactions, formaldehyde being formed from the methyl derivatives.

B. Methylhydrazines

1,2-Dimethylhydrazine and cycasin can induce colon cancer in rats.[848] Cycasin is active *per os* only after hydrolysis by a bacterial β-glucosidase in the lower intestinal tract while the dimethylhydrazine requires metabolic activation with formaldehyde as one of the products.[563] 1,2-Dimethylhydrazine is carcinogenic to the colon and rectum of rat.[849, 850] Hydrazine, 1,1-dimethylhydrazine, and 1,2-dimethylhydrazine are mutagenic.[851] 1,1-Dimethylhydrazine is carcinogenic primarily in mice and hamsters.[852–854] This compound forms formaldehyde on metabolism with a microsomal fraction of liver from various species.[854] The rate of formaldehyde formation is increased in the presence of enzyme inducers.[855, 856]

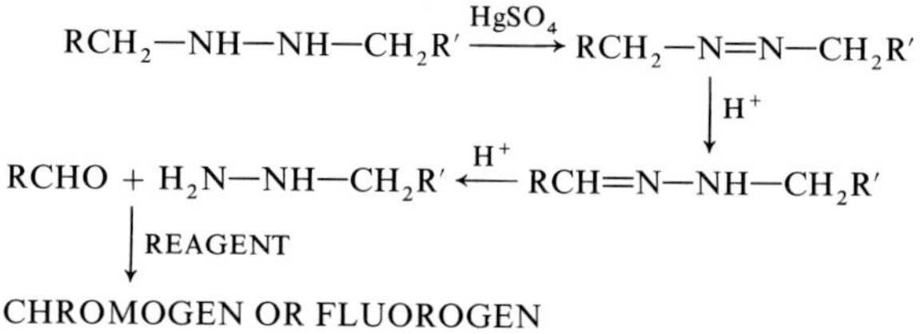

$$RCH_2{-}NH{-}NH{-}CH_2R' \xrightarrow{HgSO_4} RCH_2{-}N{=}N{-}CH_2R'$$

$$\downarrow H^+$$

$$RCHO + H_2N{-}NH{-}CH_2R' \xleftarrow{H^+} RCH{=}N{-}NH{-}CH_2R'$$

$$\downarrow \text{REAGENT}$$

CHROMOGEN OR FLUOROGEN

FIG. 38. Determination of 1,2-dialkylhydrazine through the derived aldehyde.

Polymethylated derivatives of these functional groups can form formaldehyde as shown in Table 1. The symmetrical dimethylhydrazine through oxidation followed by tautomerism and hydrolysis forms an equivalent amount of formaldehyde which can then be determined with an appropriate reagent, Fig. 38. Asymmetrical dimethylhydrazine can be analyzed as shown in Fig. 39.

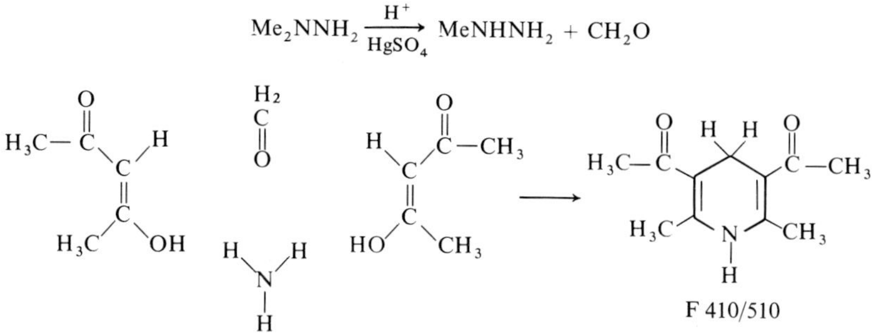

$$Me_2NNH_2 \xrightarrow[HgSO_4]{H^+} MeNHNH_2 + CH_2O$$

$$H_3C{-}C({=}O){-}CH{=}C(OH){-}CH_3 + H_2C{=}O + NH_3 + HO{-}C(CH_3){=}CH{-}C({=}O){-}CH_3 \longrightarrow$$ 3,5-diacetyl-2,6-dimethyl-1,4-dihydropyridine (ring with $H_3C{-}C(=O)$, H, H, $C(=O){-}CH_3$, H_3C, N–H, CH_3)

F 410/510

FIG. 39. Determination of 1,1-dimethylhydrazine with 2,4-pentanedione.[127]

A chromotropic acid procedure for the determination of methylalkylhydrazines, polymethylhydrazines and tetramethyltetrazene is of value because of the physiological and carcinogenic activity of some of the molecules of this type.

CA determination of methylalkylhydrazines.[49, 127] To 0·1 to 2 μmoles of the hydrazine in a 50 ml distilling flask add 4 ml of 0·1% mercuric sulfate in 10% aqueous sulfuric acid. The distillation and the rest of the analysis was as in the azomethane procedure.

1,2-Dimethylhydrazine is oxidized easily to azomethane and then to azoxymethane. A side reaction is the tautomerization of azomethane to formaldehyde methylhydrazone which readily forms formaldehyde and toxic but noncarcinogenic methylhydrazine. However, under appropriate conditions methylhydrazine could be oxidized to the proximate electrophilic

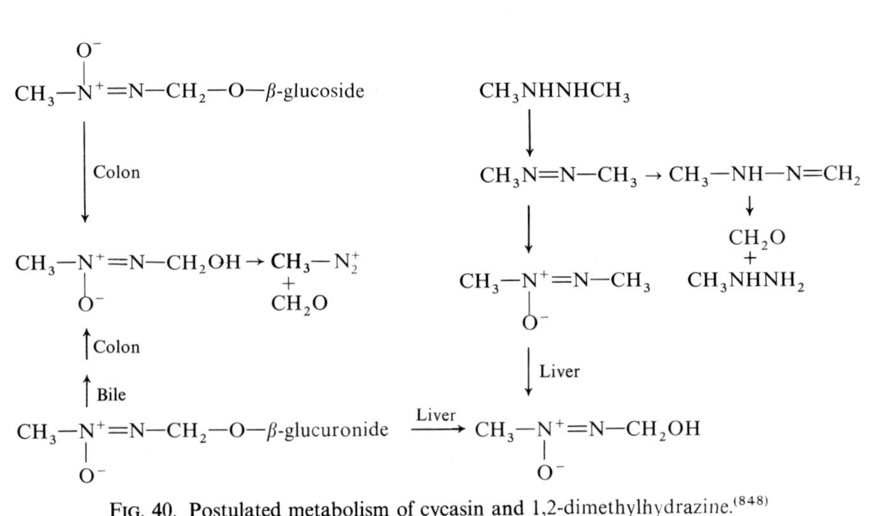

FIG. 40. Postulated metabolism of cycasin and 1,2-dimethylhydrazine.[848]

carcinogen, methyldiazonium salt. Azoxymethane is hydroxylated by enzymes in liver endoplasmic reticulum, and further conjugated there to a glucuronide which is secreted in bile and hydrolyzed by β- glucuronidase in colon. These various reactions are depicted in Fig. 40.

On the basis of these reagents, Weisburger[848] postulates that colon cancer formation in man may involve: 1. ingestion of carcinogens with diet, with liberation of the active agent by bacterial enzymes; 2. ingestion of procarcinogens, absorption, metabolism by liver, secretion in bile, and freeing of active agent in gut; 3. production of active agent by bacterial flora in gut by conversion of select food precursors; and 4. production of a mycotoxin by an intestinal microorganism.

The role and analytical usefulness of "byproduct" formaldehyde in these biological reactions has not as yet been satisfactorily evaluated.

C. Alar

Alar, the pesticide, is an asymmetrical dimethylhydrazide which can be hydrolyzed to dimethylhydrazine and then oxidized to formaldehyde with the aldehyde determined with 2-hydrazinobenzothiazole, Fig. 41[10] or some other appropriate reagent. Its analysis has been described.

$$HO-\overset{\displaystyle O}{\overset{\|}{C}}-CH_2-CH_2-\overset{\displaystyle O}{\overset{\|}{C}}-NH-N\begin{matrix}CH_3\\CH_3\end{matrix}$$

$$\Big\downarrow \; NaOH, \Delta$$

$$HO-\overset{\displaystyle O}{\overset{\|}{C}}-CH_2-CH_2-\overset{\displaystyle O}{\overset{\|}{C}}-OH + H_2N-N\begin{matrix}CH_3\\CH_3\end{matrix}$$

$$\Big\downarrow \; \text{1. } SeO_2 \;\; \text{2. } H_2SO_4 \;\; \text{3. DISTILL}$$

$$H_2N-NHCH_3 + CH_2O$$

FIG. 41. Hydrolysis of Alar to 1,1-dimethylhydrazine which is then oxidized to formaldehyde.[10]

Determination of Alar (N-dimethylaminosuccinamic acid) residues in apples.[10] Mix the shredded sample (50 g) with water (50 ml) and distill and discard 30 ml of liquid. Add 250 ml of 50% sodium hydroxide and 40 g of solid sodium hydroxide to the sample and collect 30 ml of distillate. Acidify with 3 to 4 drops of 10 N sulfuric acid and concentrate to 4 ml. After the addition of 1 ml of 5% aqueous selenium dioxide heat for 30 min at 50°, add 10 ml of 7·5 N sulfuric acid and collect 6 ml of distillate in 1·25 ml of 0·5% 2-hydrazinobenzothiazole in concentrated hydrochloric acid–water (1:9, by volume). Allow solution to stand for 18 min with intermittent shaking. Add 1·25 ml of 1% aqueous potassium ferricyanide and allow solution to stand for 25 min with intermittent shaking. Add 2·5 ml of dimethylformamide followed by 2·5 ml of 10% aqueous potassium hydroxide. Transfer contents of tubes to 25 ml volumetric flasks and dilute to volume with water. Read absorbance at 582 nm. The recovery ranged from 83 to 106% for 0·5 to 20 ppm of Alar added to apples.

D. Triazenes

The potent carcinogens, 1-phenyl-3,3-dimethyltriazene and 1-(3-pyridyl)-3,3-dimethyltriazene and several other triazenes undergo oxidative enzymatic dealkylation to form formaldehyde and arylmonomethyltriazenes, the latter being probable proximate carcinogens.[427] Many triazenes are carcinogenic even after application of a single dose. In rats phenyldimethyltriazene is a powerful teratogen.[857, 858] Some of the aryldialkyltriazenes have been shown to be strong mutagens in several test systems.[859–863] Examination of 13 1-aryl-3,3-dialkyltriazenes in BD-rats demonstrated a preference for tumor formation in the brain, nervous system and kidney.[847]

It has been reported that only phenyltriazenes which can be converted enzymatically to 1-phenyl-3-methyltriazene have antitumour properties against the mouse TLX5 transplanted lymphoma.[864]

$$C_6H_5-N=N-N(R)(R')$$

R	R′	Anti-tumour activity[864]
CH_3	CH_3	+
CH_3	C_2H_5	+
CH_3	$CH(CH_3)_2$	+
CH_3	$n\text{-}C_4H_9$	+
CH_3	$C(CH_3)_3$	−
C_2H_5	C_2H_5	−
$n\text{-}C_3H_7$	$n\text{-}C_3H_7$	−

The monomethyl derivative $C_6H_5{-}N{=}N{-}NHCH_3$, is a powerful carcinogen in the rat, causing carcinoma of the oesophagus and forestomach on oral administration and neurogenic malignant tumours and local carcinoma at the site of subcutaneous injection.[563] It is a powerful alkylating agent forming the methyl cation, nitrogen and aniline. The monomethyl derivative methylates guanosine and nucleic acids to form 7-methylguanine.

4(5)-3,3-Dimethyl-1-triazenoimidazole-5(4)-carboximide (DIC) causes complete regression of animal tumours which are resistant to alkylating agents and yet one of the metabolic pathways through which it acts is probably through the formation of an alkylating agent.[865] Although it is inactive, it is metabolized by liver microsomes to the monomethyl derivative (the other methyl group coming off as formaldehyde) which can react directly with nucleophiles or which can break down to the methyl carbonium ion. The remainder of the molecule, namely 4(5)-aminoimidazole-5(4)-carboxamide is found in the urine of rodents and man.[865, 866] DIC is also a powerful carcinogen. Thus, in one of the genotoxic pathways DIC would form the highly reactive monomethyl derivative and formaldehyde. Thus, the metabolism of these active triazenes could be followed through the determination of the derived formaldehyde. In a similar fashion the action of triazine N-demethylase inducers and inhibitors could be followed.

An example of such an analysis is the following. Formaldehyde obtained from these triazenes after enzymatic dealkylation by rat liver microsomal fraction *in vitro* is distilled from the mixture containing the incubated material and 10% aqueous sulfuric acid, and then determined in the distillate by the chromotropic acid procedure.[427]

The formaldehyde is believed to be formed by the following mechanism.

$$C_6H_5N{=}N{-}N(CH_3)_2 \longrightarrow C_6H_5{-}N{=}N{-}N(CH_3)CH_2OH$$

$$\downarrow$$

$$C_6H_5NH_2 + ({}^+N{=}N{-}CH_3) \xleftarrow{H^+} C_6H_5{-}N{=}N{-}NHCH_3 + CH_2O$$

$$\downarrow H^+$$

$$C_6H_5N_2{}^+ + H_2NCH_3$$

Aniline and formaldehyde have been identified as products of the reaction. The possibility has been discussed that triazenes react *in vivo* according to their chemical reactivity by two different mechanisms, either as diazotizing and/or alkylating agents or as alkylating agents after enzymatic activation.[847]

E. Hydrazones of formaldehyde

These compounds can be determined with MBTH, Fig. 42. The spectra of the chromogen and the blank is shown in Fig. 43. Since an alcoholic solvent is used

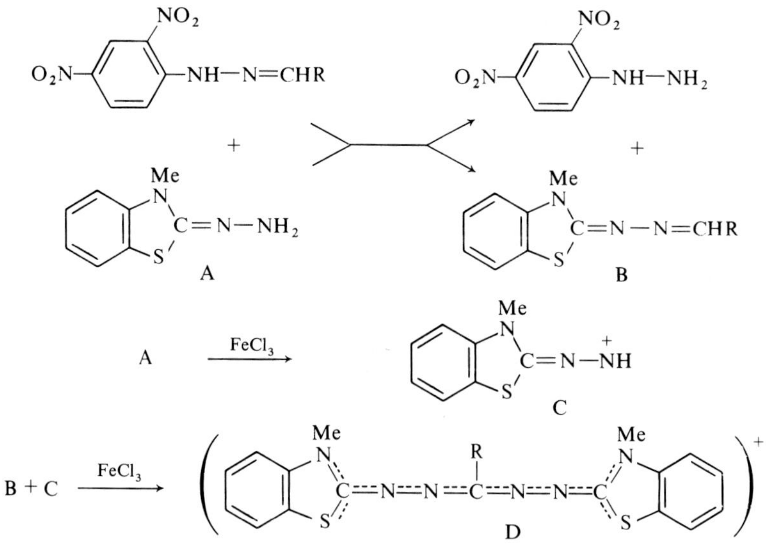

FIG. 42. Determination of aldehyde 2,4-dinitrophenylhydrazones with MBTH.

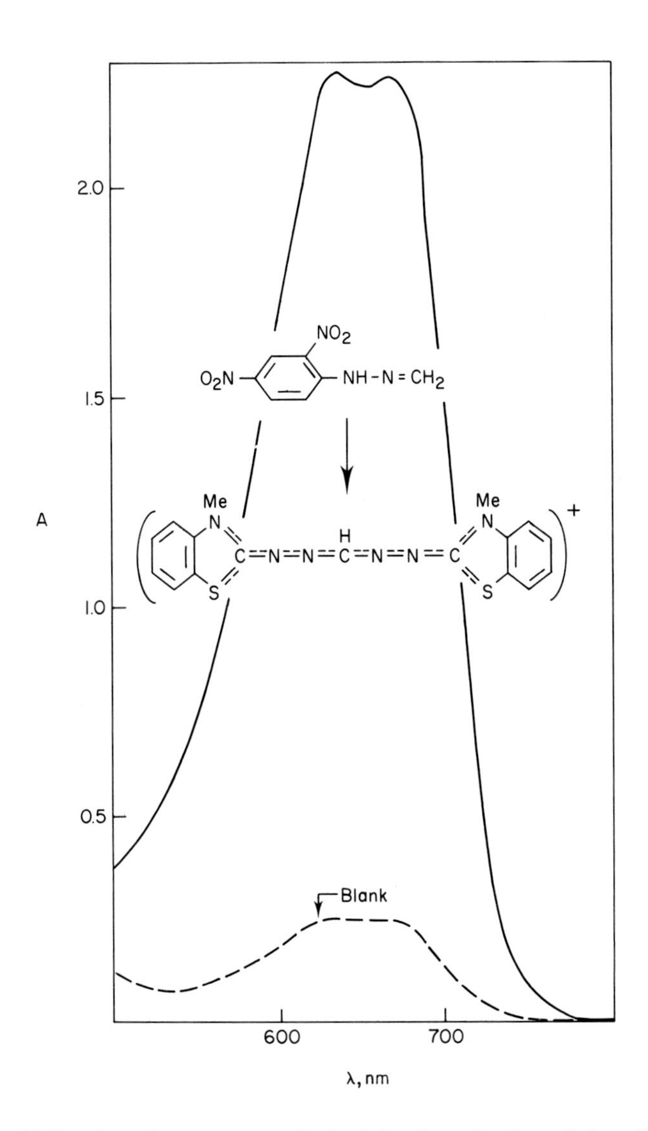

FIG. 43. Visible absorption spectrum obtained in determination of formaldehyde 2,4-dinitrophenylhydrazone with MBTH.

as the test solution, aldehydes in the solvent can give a high blank, unless the solvent is freshly purified.

XLI. HYDROCARBONS, UNSATURATED

(See $R_2C{=}CH_2$ and Ethylene Sections also)

Photochemical smog appears to be a recent phenomenon. Its characteristic symptoms are reduced visibility, a brownish colored atmosphere, plant damage, eye irritation, and respiratory distress. It first became prominent

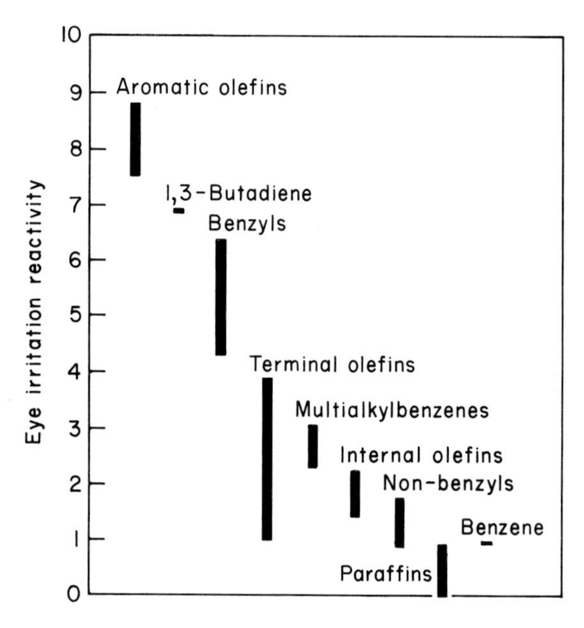

FIG. 44. Effect of hydrocarbon structure on eye irritation.[867]

during the middle 1940s in Los Angeles. The prime ingredients in smog formations appear to be hydrocarbons and nitrogen oxides derived from automobiles and some stationary sources. Some of the most potent precursors of eye irritation are shown in Fig. 44.[867] Lachrymatory effect was found to be most closely correlated with hydrocarbon structure. All the unsaturated hydrocarbons produced formaldehyde to a varying degree and some also produced peroxyacetyl nitrate. There are probably quite a few other pollutants (and factors) which contribute to eye irritation beside acrolein, formaldehyde and PAN.

The effects of gasoline aromatic and lead content on exhaust hydrocarbon reactivity has been investigated in a large irradiation chamber.[296] Formaldehyde was determined by the chromotropic acid method with an automated analyzer. Low aromatic gasolines produced the most formaldehyde, 0·81 ppm. Adding monocyclic arenes to the gasoline decreased the yield by as much as 20% for the benzene and toluene blends. This has been borne out in pure hydrocarbon studies which have shown that methylated benzenes produce

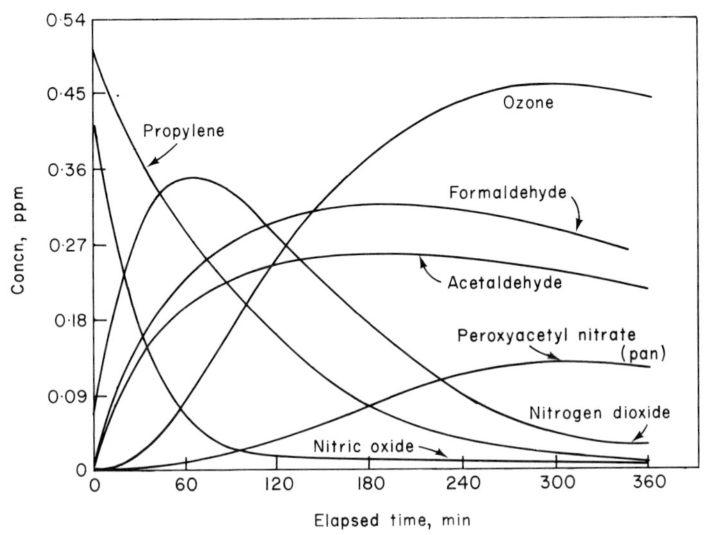

FIG. 45. Results of a typical smog chamber experiment. Products and their variation with time for the irradiation with ultraviolet light with a mixture of 0·5 ppm propylene, 0·45 ppm NO, and 0·05 ppm NO_2 in 760 torr of highly purified air.

less formaldehyde than do paraffins or olefins. The formaldehyde yields from the commercial gasolines ranged from 0·55 to 0·72 ppm and generally decreased with the aromatic content of the gasoline. The lead content (0 or 2·3 g Pb per gallon) had no effect on the emission of formaldehyde.

Classic examples of smog chamber oxidations of olefins are shown in Figs. 2 and 4. As can be seen from the data depicted in Fig. 45 an irradiated mixture of propene, nitric oxide and air forms the expected acetaldehyde and formaldehyde in equivalent amounts.[868]

XLII. p-HYDROXYBENZYL ALCOHOLS

Treatment of these types of molecules with aqueous sodium metaperiodate give formaldehyde and the corresponding benzoquinone,[869] as shown in the postulated oxidation of p-hydroxybenzyl alcohol, Fig. 46.[870] The cyclic glycol periodic ester decomposes in the normal way[871] to give formaldehyde and *p*-benzoquinone.

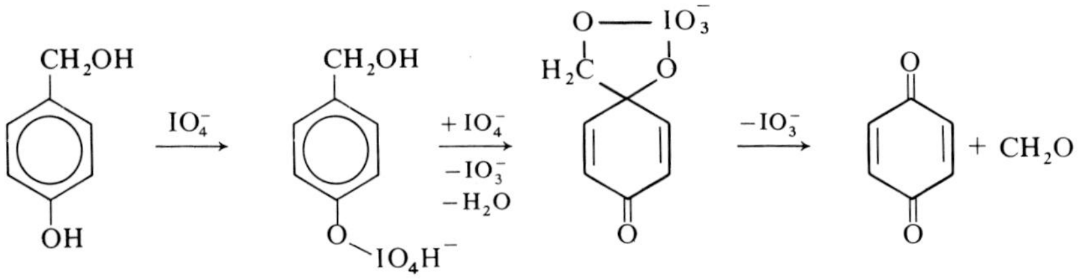

FIG. 46. Periodate oxidation of *p*-hydroxybenzyl alcohol to *p*-benzoquinone and formaldehyde.[870]

Although periodate and bismuthate are generally considered to be equivalent as glycol-cleaving agents,[872] they differ in their oxidation of 4-hydroxybenzyl alcohol. With bismuthate an intermediate oxirane is formed which then hydrolyzes to hydroquinone and formaldehyde. The postulated reaction is shown in Fig. 47.[870] The pure intermediate oxirane, 1-oxaspiro[2.5]octa-4,7-dien-6-one, is rapidly converted to formaldehyde and *p*-benzoquinone in acid solution.

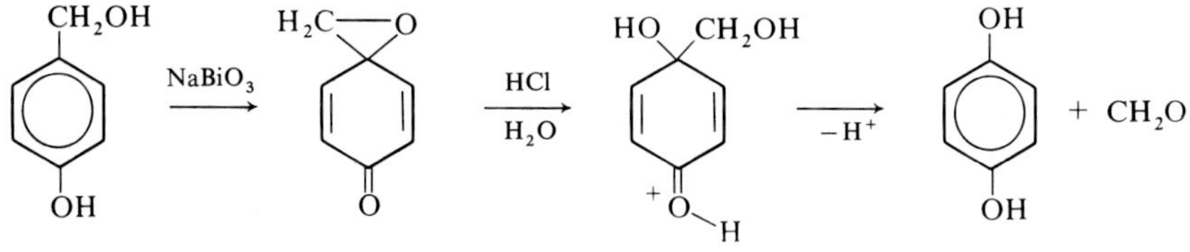

FIG. 47. Bismuthate oxidation of *p*-hydroxybenzyl alcohol to hydroquinone and formaldehyde.[870]

Similar losses of a 4-carbinol substituent has been shown to take place in some diazotization reactions of phenols, in 2,5-cyclohexadienone intermediates arising in certain modes of radical coupling involved in the biosynthesis of lignin,[873–876] in the oxidative degradation of *p*-hydroxybenzyl alcohols with dipotassium nitrosodisulphonate,[877] in the formation of oligomers of the polyphenylene oxide type from 3,5-disubstituted 4-hydroxybenzyl alcohols,[878, 879] and in the color reaction of 4-hydroxybenzyl alcohols with N-chloroquinone imide.[880, 881] Many of these reactions might be capable of modification and application analytically to the determination of the various precursors through formaldehyde.

XLIII. HYDROXYLYSINE

This amino acid has been discussed more thoroughly in the Ethanolamine Section; it is readily oxidized to equimolar amounts of α-aminoglutaric γ-semialdehyde, formaldehyde and ammonia, Fig. 48.[882] The semialdehyde (discussed in Volume 3) is a crucial intermediate in the *in vivo* interconversions between proline, glutamic acid, arginine and ornithine[883–888] and cyclizes intramolecularly to Δ′-pyrroline 5-carboxylic acid, but can exist as the free aldehyde at low pH values.[889]

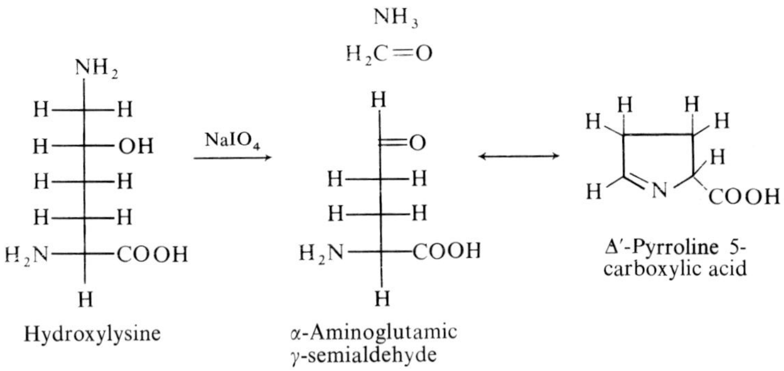

FIG. 48. Conversion of hydroxylysine to Δ′-pyrroline-5-carboxylic acid and formaldehyde.[882]

XLIV. HYDROXYMETHYLKETONES OR ACETOLS

Corticosteroids (Table 3), hydroxy- and dihydroxy-acetone can be analyzed in terms of formaldehyde. Formaldehyde can be obtained through oxidation by bismuthate or periodate,[97] Fig. 49. Androstane derivatives such as allotetrahydrocortisol, XII, corticoids, such as 5α-dihydrocortisone, XIII, and pregnane derivatives, such as 3β,11β,21-trihydroxy-5α-pregnane-20-one, XIV, would be expected to react to form formaldehyde following oxidation.

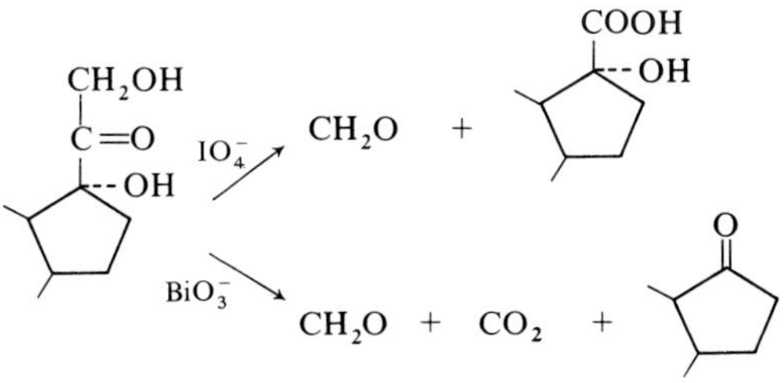

FIG. 49. Oxidation of 17,21-dihydroxy-20-ketosteroids.

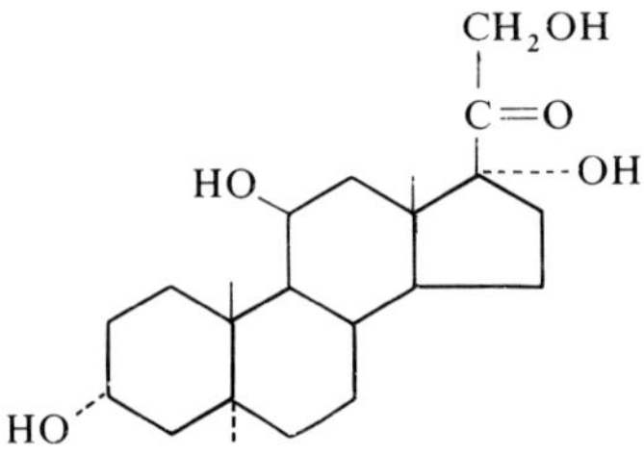

XII 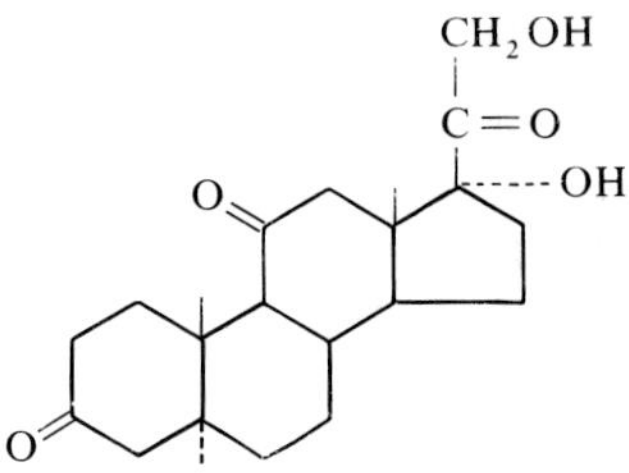 XIII

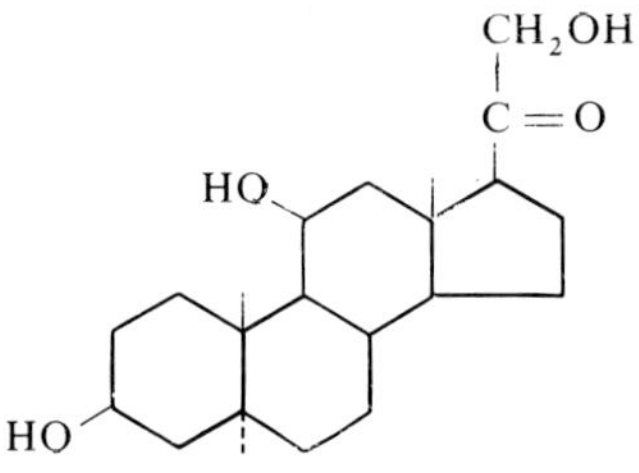

XIV

These types of molecules could be readily analyzed through their derived formaldehyde by many of the methods described in these volumes.

XLV. 4-HYDROXYSTYRENES

The mechanism of reaction between phenols with a *p*-vinyl grouping and oxidized MBTH is unknown.[890] However, a chromogen is formed which has the same absorption spectrum as obtained for the chromogen obtained in the determination of formaldehyde with MBTH. Thus, this type of phenol may be a formaldehyde precursor. Isolation and purification of the dye should prove whether this is so. The types of compounds which can be analyzed are shown in Table 13. Two types of procedures are available, one for water-soluble compounds, as described for chlorogenic acid and one for insoluble compounds as described for the 4-hydroxystyrenes in the Table and in the procedure.

MBTH determination of chlorogenic acid. To 1 ml of 0·4% aqueous 3-methyl-2-benzothiazolinone hydrazone hydrochloride add 1-ml of aqueous test solution. Add 5 ml of 0·2% aqueous ferric chloride. After 5 min. dilute to 10·0 ml with acetone. Read at 630 nm. A millimolar absorptivity of 43 is obtained at λ_{max}630. The relative standard deviation is $\pm$2·84%.

Table 13. Determination of 4-hydroxystyrene derivatives

Compound	λ_{max}, nm	mε
4-Methoxystyrene	667	62·0
	634	61·0
4-Hydroxystilbene	667	55·0
	624s	48·0
4-Methoxystilbene	667	52·0
	624s	45·6
iso-Safrole	667	65·0
	624s	60·0
iso-Eugenol	662	64·0
	624s	58·0
3-Methoxy-4-hydroxypropenylbenzene	662	62·0
	624s	57·0
4-Propenylveratrole	667	70·0
	624s	63·0
4-Hydroxybenzylideneacetophenone	667	23·2
	630	22·6

MBTH determination of 4-hydroxystyrenes.[890] To 1 ml of the methanolic test solution add 1 ml of 0·2% aqueous MBTH • HCl followed by 2 ml of 1·3% aqueous ferric chloride. Let stand 15 min and dilute to 10 ml with methanol. Read at the wavelength maximum, Table 13.

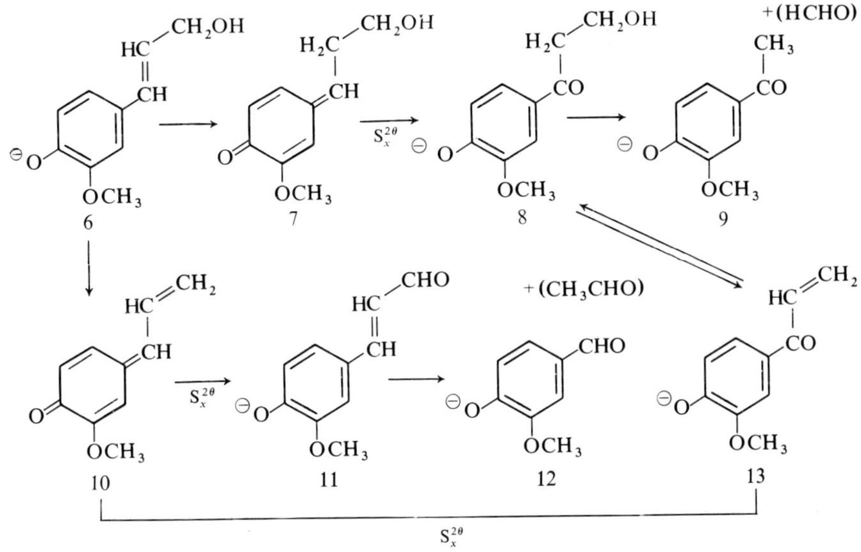

FIG. 50. Postulated oxidation of *p*-hydroxystyrenes to aldehydes.[891]

Color is stable 1 hr. Beer's law is obeyed from 0·39 ($A = 0{\cdot}1$) to 7 μg per ml of final solution. Negative results were obtained with phenol, polynuclear phenols, aromatic aldehydes, stilbene and *p*-anisoin.

p-Hydroxyarylglycerol-β-aryl ether substructures are the most important site of lignin degradation in kraft pulping (heating of wood chromogen aqueous sodium sulphide–sodium hydroxide to about 170°).[891] These structures fragment to *p*-hydroxystyrene derivatives, (e.g. *p*-hydroxycinnamyl alcohol) and phenols with formation of $S°$ in the form of polysulphide.[892–894] Brunow and Miksche[891] investigated the reaction of model *p*-hydroxystyrenes, such as *p*-hydroxycinnamyl alcohol, with polysulphide liquor at high temperatures. The postulated mechanism is shown in Fig. 50. The formation of formaldehyde, acetaldehyde and vanillin fits in with results obtained in other investigations.[895, 896]

The *p*-hydroxystyrenes derived from lignin would be expected to react in the MBTH procedure to give the aliphatic aldehyde-derived blue chromogen.

XLVI. LACTOSE AND MALTOSE

These carbohydrates can be analyzed by the usual periodate oxidation to formaldehyde followed by analysis of formaldehyde. However, it has been shown that the expected 2 moles of formaldehyde are not obtained[3] because of the recombination of formaldehyde with active hydrogen containing structures.[323] This effect can be overcome by carrying out periodate oxidations in the presence of freshly distilled benzaldehyde or *p*-hydroxybenzaldehyde.

Structural studies in carbohydrate chemistry have benefited through the use of sodium borohydride reduction followed by periodate oxidation. A convenient way of investigating the structure of a polysaccharide is to degrade it and identify the di-, tri- and other oligosaccharides produced. Thus, in the case

Table 14. Expected results of initial and rapid periodate oxidation of borohydride-reduced oligosaccharides containing reducing aldohexopyranoside end-groups[897]

Internal glycosidic linkage	Moles IO_4^- consumed	Moles HCHO liberated	Moles HCOOH liberated
(1 → 2)–	3	1	2
(1 → 3)–	3	2	1
(1 → 4)–	3	2	2
(1 → 5)–	3	1	2
(1 → 6)–	4	1	3

of maltose periodate oxidation would not give a clearcut answer while reduction to maltitol followed by oxidation would give 1 molar equivalent of formic acid and 2 molar equivalents of formaldehyde, Fig. 32.[897] With a knowledge of the periodate consumed and the amount of formic acid and formaldehyde liberated a correct assignment of the position of glycosidic linkages in oligosaccharides of the reducing type can be obtained. In this respect Dryhurst[897] has shown that in oligosaccharides containing a terminal reducing aldohexopyranoside group the expected initial rapid results of oxidation would be as shown in Table 14.

XLVII. MANNITOL

Serum mannitol can be determined in the following fashion.

Chromotropic acid determination of serum mannitol.[330] Transfer 50 μl of serum from venous or capillary blood to a small disposable tube containing 1 ml of 0·087 M $ZnSO_4$. Add 1 ml of 0·083 M $Ba(OH)_2$ and centrifuge. The pH of supernatant must not exceed 7·0 for complete deproteinization. Transfer 0·3 ml of supernatant to a glass stoppered test tube and then add 0·1 ml of 0·02 M potassium periodate in 1 N H_2SO_4. Ten minutes later add 0·1 ml of 0·2 M sodium arsenite. After another 5 min add 3 ml of CA reagent (add 100 ml of 1% of the disodium salt of CA to a mixture of 300 ml concentrated H_2SO_4 and 150 ml of distilled water. Prepare 2 days before use. Store at 4°C; stable for at least 2 months). Heat at 100°C for 30 min. After 1 min of boiling place glass stoppers loosely on the tubes. Cool to room temperature and read at 570 nm.

MBTH can also be used for the determination of mannitol or sorbitol. The following modification has been reported.

MBTH determination of mannitol and sorbitol.[331] To 1 ml of test solution containing 10^{-8} to 10^{-7} mole of mannitol or sorbitol add 0·1 ml of 0·025 M HIO_4 and after 10 min 0·1 ml 0·25 M $NaAsO_2$, 0·2 ml 2 N HCl, 0·2 ml of 2·76% MBTH in 0·1 N HCl, and 0·2 ml of 5% aqueous $FeCl_3 \cdot 6\,H_2O$. After 30 min read the absorbance at 620 nm.

Since pure D-mannitol can be readily prepared, it can serve the function of providing, on periodate oxidation, a standard source of formaldehyde which can be compared with the formaldehyde produced by similar oxidation of other formaldehyde precursors. Pure D-mannitol was prepared by the following procedure.[898] Studies on the proof of purity were also reported.

Preparation of D-mannitol.[898] To a cooled solution of recrystallized D-mannose (30 g) in water (90 ml) add dropwise, with magnetic stirring, a solution of sodium borohydride (10 g) in water (150 ml) during 0·5 hr. Stir the mixture for a further 0·75 hr; a negative test for reducing sugar is then obtained with boiling

Fehling solution. Add washed, regenerated, Amberlite IR-120 (H^+) ion exchange resin gradually and, when the effervescence has ceased, add ~200 ml of this resin, followed by Norit. Stir the suspension for 1 hr at room temperature, and then filter through Celite that has been thoroughly prewashed with water. Concentrate the filtrate to a syrupy suspension by evaporating under diminished pressure with three 400-ml portions of methanol, and then twice at atmospheric pressure with 500 -ml of methanol, to remove boric acid as methyl borate; a negative flame-test for methyl borate is then obtained. Dissolve the preparation, evaporate to dryness, in hot water (50 ml), add hot ethanol (500 ml), and allow the product to crystallize at room temperature and then in a refrigerator. Filter off the crystals of D-mannitol, wash twice with ethanol, and dry under vacuum at 60°C; yield 26·5 g (87%), mp 166·9–167·9°C. Three recrystallizations from 5:1 (v/v) ethanol–water (600 ml) give D-mannitol as globular clusters of colorless needles, wt 19·2 g, mp 167·1–168·0°C.

Prepare D-glucitol in a similar way by reduction of SRM D-glucose. After two recrystallizations from ethanol, and drying under vacuum at 55°C, the D-glucitol premelts at 87–89°C, and melts at 95–96°C.

XLVIII. METHANOL AND METHYL ESTERS

A. Methanol-chemical oxidation

Methanol and derived methanol can be determined by chemical oxidation to formaldehyde followed by chromogen formation as shown in Tables 1 and 5. An example of a method utilized for methanol is the following.

Determination of methanol with chromotropic acid.[899] To the test solution containing 1 to 15 μg of methanol add 0·2 ml of 0·3% (v/v) propionaldehyde solution and 0·4 ml of permanganate solution (dissolve 1·5 g of $KMnO_4$ in 30 ml of water and 7·5 ml of 85% H_3PO_4 and dilute to 50 ml). After 5 min add 0·2 ml of 20% Na_2SO_3 solution. Then add 0·3 ml of 2% aqueous chromotropic acid and 4 ml of 75% (v/v) H_2SO_4. Heat the mixture at 80–85° for 10 min, cool to room temperature, and read the absorbance at 575 nm.

Methanol in hair spray[364] and in alcoholic beverages[365] has been determined. In addition, the methyl ester content of pectin can be assayed.[366] Methanol in serum and cerebrospinal fluid can be oxidized by permanganate and H_2SO_4 to formaldehyde and then determined with chromotropic acid at 570 nm.[355] Interference by ethanol[265] and other hair spray constituents in the colorimetric determination of methanol in hair spray is cancelled by distillation of the spray, extraction with light petroleum of the sodium chloride saturated distillate, and the use of a standard graph.[364] Although the presence of ethanol reduces the intensity of the color, it does not affect the rectilinearity of the relationship between absorbance and methanol concentration. The interference of ethanol is eliminated by a procedure of standard additions of methanol. Chromotropic acid is used to determine the formaldehyde.

One method for the determination of methanol has also been applied to the measurement of pectin ester content, pectin methyl esterase activity, and methyl esters.[366] The following method for the determination of methanol obeys Beer's law from 2 to 40 μg. Large amounts of 5-hydroxymethylfurfural or EDTA are serious interferences. Traces of unreduced permanganate interfere considerably. A 14% increase in absorbance is obtained when 10 μg of methanol is assayed in the presence of 250 μg ethanol. High concentrations of ethanol inhibit chromophore formation.

2,4-Pentanedione procedure for determination of methanol.[366] *Reagents.* 0·02 M 2,4-pentanedione (freshly distilled) in a solution containing 2 M ammonium acetate and 0·05 M acetic acid. Premixed reagent. Used sometimes in blank determinations and prepared by mixing 9 parts of 2 N H_2SO_4 with 10 parts of filtered 2% aqueous potassium permanganate, 10 parts of 0·5 M sodium arsenite in 0·12 N H_2SO_4 and 21 parts of water. Probably stable indefinitely.

Procedure. To 1 ml of aqueous test solution cooled in ice-water bath add 0·2 ml of 2% aqueous potassium permanganate, taking care sides of tubes are not splashed. Mix gently, cool in ice-water for 15 min. Add 0·2 ml 0·5 M sodium arsenite in 0·12 N H_2SO_4 followed by 0·6 ml water, mix and let stand for 1 hr at room temperature. Add 2 ml of the pentanedione reagent, shake, cover tubes with marbles, heat at 58–60° for 15 min, and cool to room temperature. Read absorbance at 412 nm, using water as blank.

Procedure for pectin ester content. Add 0·25 ml of 1·5 N NaOH to 0·5 ml of aqueous pectin solution (50 to 200 μg) and swirl tubes. After 0·5 hr at room temperature acidify with 0·25 ml of 5·5 N H_2SO_4, and cool in ice-water bath for permanganate oxidation, and assay as already described. To provide control solutions, add 1 ml of premixed reagent to saponified, acidified pectin samples followed by pentanedione reagent as usual. Procedure for other methyl esters. Saponify, acidify and assay. Procedure for pectin esterase. Add 0·2 or 2 ml of the pectin esterase solution (100 μg/ml) to 2 ml of a 0·1% aqueous pectin solution and dilute digest to 4 ml with a 0·1 M sodium acetate buffer, pH 7·5. Remove 0·5 ml aliquots at interval and add 0·5 ml 2·0 N H_2SO_4. Cool in ice-water bath and oxidize with 0·2 ml of 2% permanganate and continue assay as described.

Methanol in industrial atmospheres can be determined through oxidation by permanganate to formaldehyde followed by analysis of the aldehyde with chromotropic acid.[549]

The use of methanol as a gasoline extender has been critically reviewed.[900] Vehicular emissions of aldehydes are not being currently regulated, although they do make a partial contribution to the hydrocarbon readings. However, aldehydes are as photochemically reactive as olefins. Formaldehyde can be produced from methanol by partial oxidation. Tests on a methanol blend showed 30% and 50% higher aldehyde emissions for the 1973 and 1967 cars, respectively. Aldehyde emissions on a catalyst-equipped car with or without the methanol were very much lower than was obtained with cars without a catalyst. However, the danger of obtaining copious amounts of formaldehyde from the use of methanol as a fuel needs careful appraisal.

B. Methanol—enzymatic oxidation

The microsomal NADPH-dependent ethanol oxidation system consisting of NADPH–cytochrome *c* reductase and catalase[901] can help oxidize methanol.[902] Essentially NADPH-dependent methanol oxidation to formaldehyde was observed in a system consisting of catalase (hydrogen-peroxide: hydrogen-peroxide oxidoreductase, EC 1.11.1.6), NADPH–cytochrome *c* reductase (NADPH: ferricytochrome oxidoreductase, EC 1.6.2.4), and a low molecular weight factor from trypsin digested hepatic microsomes. The formaldehyde was determined with 2,4-pentanedione.[556] The factor inhibited NADPH-dependent lipid peroxidation in microsomes (as measured by the TBA determination of malonaldehyde), but did not affect aniline hydroxylation and aminopyrine demethylation (as measured by the determination of formaldehyde). Menadione at concentrations as low as 1 μM could be substituted for the factor in the oxidation of methanol. FAD (flavin-adenine dinucleotide) or FMN (flavin mononucleotide) activated the oxidation of methanol to formaldehyde in the reconstituted system but they could not replace the activator.[902]

Several bacteria,[903, 904] yeast,[903, 904] fungi[905] and an actinomycete[906] are known to grow in methanol (or certain C_1 compounds) as sole source of carbon and energy. In this case the methanol is oxidised to formaldehyde before proceeding further on the metabolic pathway.

C. Methyl esters

Methyl esters can be hydrolyzed to methanol, the latter oxidized by permanganate to formaldehyde, which is then determined by the chromotropic acid method. Examples are the analysis of methyl acrylate or dimethyl terephthalate present in industrial atmospheres[549] and the pectin esters discussed previously.

Another example of a methyl ester is methyl methanesulphonate, $CH_3O_3SCH_3$. Just like the formaldehyde precursor, methylazoxymethanol ($CH_3{-}NO{=}N{-}CH_2OH$),[574] methyl methanesulfonate is mutagenic,[724,

907, 908) carcinogenic[909–911] and teratogenic.[912] Since this compound is very reactive toward nucleophilic centers, it shows a complex metabolic pattern.[913] In some ways it shows a similar metabolic pathway to that of methyl iodide. *In vivo* retention of the alkyl moiety suggests widespread alkylation of cellular material. The 5·1 % of the expired ^{14}C-carbon dioxide originates mainly by hydrolysis to methanol with subsequent oxidation to formaldehyde and entry into the one-carbon pool. Methyl methanesulphonate is of interest as a formaldehyde precursor in terms of analytical and metabolic investigations.

D. Indirect reactions

Methanol can also be used as a reagent or a substrate, to determine various compounds in biological fluids and tissues. Thus, with methanol as a reactant uric acid can be determined in biological fluids through the derived formaldehyde.[367] This reaction will be discussed in the Uric Acid Section.

In the assay of methanol peroxidase hydrogen peroxide was formed by addition of glucose and glucose oxidase, and the resulting formaldehyde[914] was measured by chromotropic acid.[915]

A simple selective method has been introduced for the enzymatic determinations of total and free cholesterol.[594] The method involves the oxidation of methanol to formaldehyde and is described in the Cholesterol Section. This manual method utilizing 2,4-pentanedione has been modified and automated.[916] To increase sensitivity the method has been adapted to fluorimetry since the final chromogen, 3,5-diacetyl-1,4-dihydrolutidine is also fluorescent. The method checks well over a wide range in concentration with a reference manual colorimetric method.[917]

Enzymatic oxidation of methanol can also be used to determine glucose indirectly as shown in Fig. 51. Chromotropic acid and 3-methyl-2-benzothiazolinone hydrazone (MBTH) have been used as chromogen-forming reagents. The procedure using the latter reaction is described.

Many methoxy compounds can be hydrolyzed, oxidized to formaldehyde (Table 1) and determined with an appropriate reagent.

MBTH enzymatic determination of glucose.[188] To 1 ml of aqueous test solution containing 1 to 20 μg of glucose, add 1 ml of enzyme reagent (0·4 g of glucose oxidase (GOD-III, Boeringer, Mannheim) dissolved in 100 ml of 20% methanol) and 2 ml of phosphate buffer (pH 5·6). Incubate mixture at 37–38° for 90 min. Then add 1 ml of 0·5% aqueous MBTH • HCl and allow the mixture to stand at room temperature for 60 min. Finally add 1 ml of oxidizing reagent (0·83 g $FeCl_3 \cdot 6\,H_2O$ in 10% aqueous potassium bisulfate to 100 ml volume). After 60 min read absorbance at 620 nm.

$$\text{GLUCOSE} + O_2 + H_2O \xrightarrow[\text{OXIDASE}]{\text{GLUCOSE}} \text{GLUCONIC ACID} + H_2O_2$$

$$CH_3OH + H_2O_2 \overset{\text{CATALASE}}{\rightleftarrows} CH_2O + 2H_2O$$

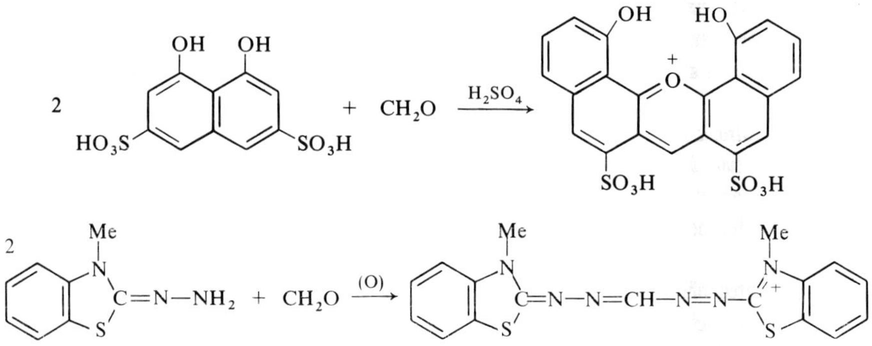

FIG. 51. Enzymatic determination of glucose with either chromotropic acid or MBTH.

Beer's law obeyed from 1–20 μg per ml of sample solution. Added glucose recovered in 100% yield. Somogyi precipitation method and 20 μl of serum were necessary for the glucose estimation.

XLIX. METHIONINE SULFOXIDE

It is difficult to determine methionine or its sulfoxide in a protein since hydrolysis of the protein results in changes in these 2 amino acids. Two methods are discussed. The first involves hydrolysis but in the second the methionine sulfoxide content of intact proteins is determined. In both cases the specificity of the reaction is based on the reaction between sulfoxides and acetic anhydride to give a formaldehyde precursor.

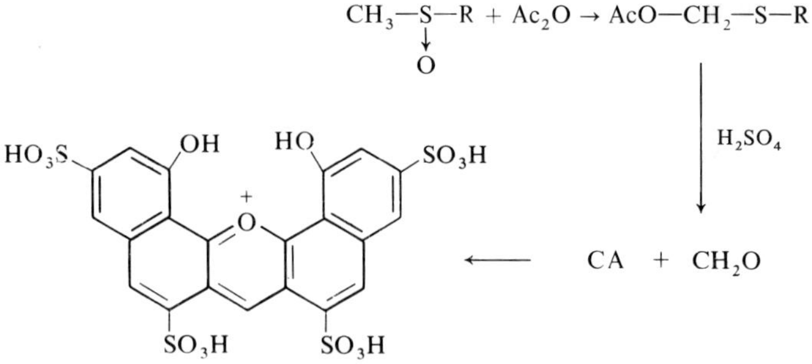

$$CH_3\text{—}\underset{\downarrow \atop O}{S}\text{—}R + Ac_2O \rightarrow AcO\text{—}CH_2\text{—}S\text{—}R \xrightarrow{H_2SO_4} CA + CH_2O \rightarrow \text{dye}$$

FIG. 52. Determination of methyl sulfoxides with chromotropic acid.[130]

Some compounds can undergo a synthetic reaction to form a hydrolyzable methylene which can then be hydrolyzed to formaldehyde and assayed. An example is the methyl sulfoxides which can be assayed in the following manner, Fig. 52. Methionine sulfoxide or methionine hydantoin sulfoxide can be assayed in a somewhat similar fashion, but with chromotropic acid as the reagent.

Chromotropic acid determination of methionine sulfoxide or methionine hydantoin sulfoxide.[130] Dilute an aqueous solution of the test substance or the residue to about 10 to 60 μmol per ml with purified acetic acid (treatment with phenylhydrazine and acetic anhydride, refluxing and distillation). To a 0·5 ml aliquot add 7·5 ml of purified acetic anhydride (treatment with a few drops of concentrated sulfuric acid and distillation) and reflux the mixture for 30 min on an oil bath. Cool with water and dilute to 100 ml with distilled water. To a 1 ml aliquot add 10 ml chromotropic acid reagent (1 g of the disodium salt of CA dissolved in 100 ml of water followed by the addition of 400 ml of 24 N H_2SO_4). Heat at 100° for 30 min. Allow to cool to room temperature and read at 580 nm.

In the following method protein is analyzed directly for methionine sulfoxide. The acetic anhydride in the reaction appears to only react with the methionine sulfoxide since other protein residues remain unchanged. Beer's law is followed.

Determination of methionine sulfoxide content of intact protein.[370] Suspend a carefully weighed sample of 1 to 3 g of protein in 5 to 10 ml acetic acid. Add 10 to 20 ml acetic anhydride. Reflux 30 min at 140° in an oil bath. Cool 5 min; stop the reaction by addition of 50 ml distilled water introduced through the condenser. This step is very important and must be carefully followed in order to avoid further reaction to give homocysteine. Add 200 ml 4 N H_2SO_4 and distill 200 ml into a volumetric flask in 30 min. Add a 1 ml aliquot of the distillate to 10 ml of reagent (Dissolve 1 g chromotropic acid in 100 ml water, filter, and dilute to 500 ml with 12·5 M H_2SO_4. Keep in dark). Heat at 100° for 30 min. After 30 min standing, cool to room temperature. Read absorbance at 570 nm using 3 cm cuvets.

L. METHOXONE OR (4-CHLORO-*o*-TOLYL)OXYACETIC ACID

$$Cl-C_6H_3(CH_3)-O-CH_2COOH$$

This phenoxyacetic acid has been determined in industrial atmospheric samples.[371] The compound was hydrolyzed and oxidized with sulphuric acid and determined with chromotropic acid. The method was sensitive to 10 μg.

Atmospheric formaldehyde could be an interference. However, if the collected sample were evaporated to dryness, the residue dissolved in a little water and the solution treated with a solution of chromotropic acid in sulphuric acid, a fairly selective and sensitive method for methoxone would be available after appropriate optimization of the procedure.

LI. METHOXYFLURANE

The metabolism of the anesthetic, methoxyflurane has been of some interest. Its metabolic degradation has been studied in man.[918] It has been reported that the enzymatic metabolism of methoxyflurane *in vitro* yields a compound which, when treated with acid, produces inorganic fluoride and formaldehyde,[919] e.g.

$$H_3COCF_2CHCl_2 \longrightarrow \text{metabolite} \xrightarrow{H^+} F^- + CH_2O + ?$$

Further study is needed of this metabolism and the analytical usefulness of the derived formaldehyde needs to be explored.

LII. METHOXYMETHYL METHANESULFONATE, $CH_3SO_2OCH_2OCH_3$

This compound is a powerful alkylating agent[920] whose physiological effects need extensive study. As a formaldehyde precursor it could be determined by chromotropic acid, MBTH, 2,4-pentanedione, etc. The metabolism and genotoxic properties of this formaldehyde precursor need thorough investigation.

LIII. N-METHYL COMPOUNDS

A. Introduction

A large number of these types of methyl derivatives have been investigated in terms of their metabolism and their utilization as substrate molecules in studying demethylating enzymes. The induction and suppression of these enzymes by xenobiotics plays important roles in (i) the physiological well being and survival of all organisms and (ii) the increase of, and the magnification of, acute toxic and/or genotoxic effects. A list of many of the N-methyl compounds which have been utilized in a wide variety of metabolic studies is given in Tables 15 and 16. Most, if not all, biochemical oxygenations of N-methylated carcinogens, pesticides and other xenobiotics are carried out by low-specificity, mixed-function oxidases often localized in microsomes or similar structures.

Oxidative dealkylation at hetero atoms can lead to the production of formaldehyde. The N-methylated compounds listed in Tables 15 and 16 are demethylated to formaldehyde with the help of an N-demethylase. Essentially these secondary and tertiary amines are oxidatively dealkylated to formaldehyde and the dealkylated amine catalyzed by hepatic cytochrome P-450.[1039,1040] Of the more than 900 pesticides supposedly used in the USA,[1041] many are aldehyde precursors. Their fate in the environment and their metabolic transformation have been discussed.[1042]

Table 15. Effect of various detergents on formaldehyde formation from ethylmorphine through N-demethylation by rat liver microsomes[a(939)]

Detergent	Relative amount of CH_2O
None	100
Triton B-1956	97
Cholate	92
Triton QS-15	90
Triton X-67	89
Brij-35	76
Triton CF-32	58
Cetyl trimethylammonium bromide	38
Triton DF-12	32
Triton N-101	25
Triton X-100	22
Triton N-57	22
Triton CF-10	20

[a] The final 5-ml volume of incubation mixture contained 1·8 mg of microsomal protein from phenobarbital-treated rats, 0·2 mg of the detergent, 2 mM ethylmorphine, an NADPH-generating system and phosphate buffer, pH 7·4. The mixture was incubated at 37° for 10 min. The formaldehyde formed during the reaction was determined by the 2,4-pentanedione method.[556]

In studies of the various types of carcinogenesis, and especially hepatocarcinogenesis, very little attention has been paid to the metabolic and genotoxic roles of the formed aldehydes. This applies also to the metabolic and genotoxic roles of the formation, interconversions, and utilization of formaldehyde and the other labile 1-carbon compounds in the synthesis of the biologically active macromolecules[1043–1045] and in the regulation of cellular differentiation.[1046,1047] Poirier *et al.*[1048] have pointed out that a series of interrelations appear to exist between carcinogens and 1-carbon compounds. Thus choline,[1049–1052] methionine,[967] ethanolamine,[1053] folic acid, and vitamin B_{12},[967,1051,1054] or their derivatives alter chemical carcinogenesis.

Table 16a. Some N-methylated formaldehyde precursors utilized in metabolic studies

<table>
<tr><th rowspan="2">Compound</th><th colspan="2">Effect on N-demethylation</th><th rowspan="2">Ref.</th></tr>
<tr><th>Stimulator</th><th>Inhibitor</th></tr>
<tr><td>Acetylmethadol</td><td></td><td></td><td>921, 922</td></tr>
<tr><td>4-Aminoantipyrine</td><td></td><td></td><td>923</td></tr>
<tr><td>Aminopyrine</td><td></td><td></td><td>736, 924–927</td></tr>
<tr><td></td><td></td><td>Adrenalectomy + castration[a]</td><td>928</td></tr>
<tr><td></td><td>Age in days[b]</td><td></td><td>929, 930</td></tr>
<tr><td></td><td>Alcohols[c]-n-propanol</td><td></td><td>931</td></tr>
<tr><td></td><td></td><td>Methanol + ethanol</td><td>931</td></tr>
<tr><td></td><td colspan="2">BaP[d]</td><td>932, 933</td></tr>
<tr><td></td><td></td><td>CS_2</td><td>561</td></tr>
<tr><td></td><td></td><td>CCl_4</td><td>708, 710, 934</td></tr>
<tr><td></td><td colspan="2">Castration[d]</td><td>935</td></tr>
<tr><td></td><td></td><td>*p*-Chloromercuribenzoate</td><td>936</td></tr>
<tr><td></td><td colspan="2">Cold exposure[e]</td><td>937</td></tr>
<tr><td></td><td>DDT</td><td></td><td>938</td></tr>
<tr><td></td><td></td><td>Desoxycholate</td><td>936</td></tr>
<tr><td></td><td></td><td>Detergents</td><td>939, 940</td></tr>
<tr><td></td><td></td><td>Ethionine</td><td>941</td></tr>
<tr><td></td><td></td><td>Menthol</td><td>942</td></tr>
<tr><td></td><td colspan="2">3-Methylcholanthrene[d]</td><td>932, 933</td></tr>
<tr><td></td><td></td><td>MtT[f]</td><td>943</td></tr>
<tr><td></td><td></td><td>Nicotinamide</td><td>944</td></tr>
<tr><td></td><td>Phenobarbital</td><td></td><td>928, 932, 933</td></tr>
<tr><td></td><td>Phenylbutazone[g]</td><td></td><td>932, 945</td></tr>
<tr><td></td><td>Prednisolone[h]</td><td></td><td>928</td></tr>
<tr><td></td><td></td><td>Pregnancy</td><td>946</td></tr>
<tr><td></td><td></td><td>Propylene glycol</td><td>947</td></tr>
</table>

		Selenium deficiency	948
		TCDD[i]	949, 950
		Terpin hydrate	942
	Testosterone		935
	Testosterone propionate		928
		Vit. E deficiency	948, 954, 955
DL-Amphetamine		pH and partition[j]	956
Azo dyes—			
N,N-Dimethyl-4-phenyl-	DBacA, DBahA[k]		957, 958[l]
azoaniline	Phenobarbital[m]	Ethionine	959, 960
	3-Methylcholanthrene[n]		961
N,3-Dimethyl-4-phenyl-			736, 962
azoaniline	3-Methylcholanthrene		963–965
	BaP		965
	β-Naphthoflavone		963
	PAH		932, 958
	Phenobarbital		932, 965, 966
	TCDD[o]		963
N-Methyl-4-phenyl-			967, 968
azoaniline			
Benzphetamine			969, 971[p]
		BaP	972
	Griseofulvin[q]		973
	3-Methylcholanthrene		977, 978
	Phenobarbital		977, 978
	Pregnenolone-16		977, 978
	α-carbonitrile	TCDD[r]	948, 949
Bidrin			979
Carbaryl			980
Chlorcyclizine			977
p-Chloro-N-methylaniline		SKF 525-A[s]	736
		Ethanol[d]	981

Table 16a—*continued*

Compound	Effect on N-demethylation		Ref.
	Stimulator	Inhibitor	
Chloropromazine[t]			982
Codeine			923
	Phenobarbital		928
Cotinine			989
Desimipramine			990
6-Dimethylaminopurine	Phenobarbital[u]		991, 992
N,N-Dimethylaniline		pH and partition[j]	956
		$HgCl_2$[v]	377
	Mg^{+2} and Hg^{+2}		995[w]
		Triton X-100	996[x]
N,N-Dimethylaniline N-oxide		CO, SKF 525-A	123, 996
N,N-Dimethylbenzylamine		pH and partition[j]	956

[a] In rats.

[b] Changes in aminopyrine N-demethylase during the neonatal period in rat liver microsomes. Days after birth 1% aminopyrine demethylation: 1/17, 4/25, 8/36.

[c] Methanol and ethanol inhibit 51 and 61%, respectively while 2-propanol stimulates 83%.

[d] No effect.

[e] Rats and mice at 5° to 7° for a week had no effect on aminopyrine demethylase.

[f] Effect of a pituitary mammotropic tumor on hepatic demethylation in the rat.

[g] Stimulates liver microsomal metabolism in rats(932) and accelerates metabolism in man.(945)

[h] Stimulates aminopyrine N-demethylation.

[i] 2,3,7,8-Tetrachlorodibenzo[p]dioxin is a teratogen in the rat.(951–953)

[j] Rate of demethylation decreased with decreasing pH and with decreasing partition between the microsomes and the aqueous incubation media.

[k] Among more than 60 polynuclear aromatic hydrocarbons studied with N-demethylation of an aminoazo dye as a test system, the optimal size for enzyme induction ranged from 75 to 150 A°, with coplanar hydrocarbons more potent than noncoplanar.

[l] Both dibenz[a,c]anthracene and dibenz[a, h]anthracene induce aminoazo dye N-demethylase but only the latter has carcinogenicity.

[m] Barbital has no promoting effect on metabolism and excretion of N,N-dimethyl 4-phenylazoaniline.[961]

[n] Benzo[a]pyrene found to be generally less potent than 3-methylcholanthrene to induce azo dye N-demethylase.[957]

[o] Unlike PAH (e.g. 3-methylcholanthrene and naphthacene) and β-naphthoflavone, TCDD (2,3,7,8-tetrachlorodibenzo-*p*-dioxin) administration to so-called genetically "non-responsive" mice stimulates the hydroxylase, O-deethylase, O-demethylase, and N-demethylase activities and the new formation of cytochrome P_1-450 to the same extent as that found in genetically "responsive" mice.

[p] Cytochrome P-450 stimulated N-demethylation of benzphetamine while cytochrome P-448 had much less effect.[971]

[q] Large amount causes enlarged liver and multiple hepatomas in mice but not rats'[974,975] Formaldehyde formed in demethylation measured by a modified 2,4-pentanedione method.[556,976]

[r] The suppression of benzphetamine N-demethylase in male rat liver microsomes was age related: the suppression was seen only in adult animals while in the very young (10 days old) the enzyme was actually induced by TCDD.[948]

[s] 2-Diethylaminoethyl 2,2-diphenyl valerate.

[t] The phenothiazines are a major group of neuroleptic agents used in the treatment of mental illness, of which chlorpromazine is probably the most widely used in the treatment of schizophrenia. There are about 168 possible metabolites of chlorpromazine.[983] Many metabolites have been identified in plasma[984] and urine[985,986] samples of patients receiving the drug. Some of the metabolites which have been identified in incubation studies with liver microsomes from rabbit, rat[987] and human[988] are mono-N-demethylchlorpromazine, di-N-demethylchlorpromazine, the corresponding hydroxylated compounds and sulphoxides and formaldehyde.

[u] Pretreatment of rats with phenobarbital stimulates the liver microsomal demethylation of 6-dimethylaminopurine, 6-methylaminopurine, puromycin and puromycin aminonucleoside, also. 6-Dimethylaminopurine and 6-methylaminopurine are constituents of liver RNA.[993,994]

[v] Other inhibitors of microsomal N-oxidation of, as well as formaldehyde formation from, dimethylaniline are hexobarbital, desmethylimipramine, metapyrone, 2,2′-bipyridyl, SKF-525A, piperonyl butoxide, ethylmaleimide, mercaptoethanol, and CO/O_2 (98%/2%).

[w] Lung N-oxidase enzyme activity is 3 times higher than liver N-oxidase at the optimum pH (8·9) while the activities are more nearly the same at physiological ranges of pH. Changes in dimethylaniline metabolism with age in rabbits from 4 days old to adult show a steady increase in lung demethylase activity (as shown by the increase in formaldehyde formation). Liver demethylase had a sharp increase in activity between 2 weeks and one month. Activities of N-demethylase in liver and lung of newborn rabbits were 10 to 20% of adult levels.

[x] Authors report two modes of formaldehyde formation from dimethylaniline, the first through oxidation of the C-atom of the methyl group, without participation of dimethylaniline N-oxide, and the second from dimethylaniline-N-oxide by means of transfer of oxygen from the N- to the C-atom.

Table 16b. Some N-methylated formaldehyde precursors utilized in metabolic studies

Compound	Effect on N-demethylation		Ref.
	Stimulator	Inhibitor	
Dimethylnitrosamine			557, 997[a]
		Aminoacetonitrile	697, 713, 715–717, 998
	Amino acids		712, 734
		Aroclor 1254[b]	713
	BaP(999)[c]	BaP	713
		Carbohydrates	712, 734
		CCl_4	706, 1000
		Dibenamine	719
	Fasting	Low protein diet	999
		3-Methylcholanthrene[b]	687, 713, 714, 1001
		β-Naphthoflavone	713
		PAH	712
	Phenobarbital(999)[d]	Phenobarbital[c]	713
		Pregnenolone-16α-carbonitrile	713
	Starvation[e]	Starvation[e]	712
N,N-Dimethylphenethylamine		pH and partition[f]	956
N,N-Dimethylphentermine		pH and partition	956
N,N-Dimethylphenylpropylamine		pH and partition	956
Diphenamide			1003
Ephedrine			990
Ethylmorphine			719, 736, 1004–1006
		Antipyrine	1007
		Aroclor 1254[g]	1008
		Cobalt (II)	1014

	Corticosterone[h]		1015, 1016
		Cortisone acetate[c]	1017
	Detergents[i]		939
		Dibenamine	719
	α,α′-Dipyridyl		1014
	EDTA(1014)	EDTA[j]	1018
		Ethanol[c]	981
		Fe (II)	1014
		3-Methylcholanthrene[c,k]	1008, 1019
	o-Phenanthroline		1014
	Phenobarbital		964, 965, 975, 978, 1007
		Phenylbutazone	1020
	Pregnenolone-16a-carbonitrile		975, 978, 1017
		SKF-525A[1]	736
	Spironolactone[m]		1017
		Storage at −20°	1023
	Stress		1016
		TCDD[n]	948, 949
Hexobarbital			1024
	Phenobarbital		928
		Benzo[a]pyrene	1025
Meperidine[o]		Benzo[a]pyrene	1026
Methadone			1027 1028
dl-Methamphetamine			990
N-Methylaniline			377, 378, 943, 990
N-Methylbenzylamine			990
N-Methyl-m-chlorobenzylamine			990
N-Methyloctylamine			990
Monuron			1029
Natulan[p]			1030

F

Table 16b—*continued*

Compound	Effect on N-demethylation		Ref.
	Stimulator	Inhibitor	
Nicotine			989
Nortriptyline			990
Phenylephrine			990
Propylhexedrine			990
Pseudoephedrine			990
Zectran			1038

[a] The relative activities of various rat tissues in demethylating dimethylnitrosamine

Organ	Demethylating activity (%)	Relative demethylating activity per organ (%)
Liver	100	100
Kidney	79	26
Lung	15	4
Brain	13	5
Spleen	12	1
Muscle	10	

[b] Represses Me_2NNO in rat but no effect in mice.[712]

[c] No effect.

[d] The mutagenic activity of the reaction products from dimethylnitrosamine is directly correlated with the metabolic formation of formaldehyde with and without induction by 3-methylcholanthrene and in various strains of mice. Formaldehyde is a weak mutagen in several organisms,[772, 1002] but it could be much more potent when formed at an appropriate tissue site.

[e] Starvation has a substantial inducing effect on Me_2NNO N-demethylase in rats, and results in a moderate induction of the demethylase in mice.

[f] Rate of demethylation and formaldehyde formation decreases with decreasing pH and with decreasing partition between the microsomes and the aqueous incubation media.

[g] Contamination of human adipose tissue and of bovine and human milk with PCB has been reported.[1009–1012] Recent studies indicate that ingestion of PCB by rhesus monkeys results in hyperplastic, dysplastic and invasive changes suggestive of eventual neoplastic transformation of gastric mucosa.[1013]

[h] Single intravenous injection of corticosterone increased formaldehyde formation from ethylmorphine.
[i] See Table 15.
[j] At $>1000\ \mu M$ EDTA.
[k] 3-Methylcholanthrene reported as stimulating ethylmorphine N-demethylation.[975,978]
[l] β-Diethylaminoethyl-diphenylpropyl acetate.
[m] Other so-called "catatoxic" steroids were also shown to enhance microsomal N-demethylation and aliphatic hydroxylation in female rat liver.[1021,1022]
[n] 2,3,7,8-Tetrachlorodibenzo-*p*-dioxin.
[o] Also decrease of N-demethylation of Benadryl following BaP pretreatment.
[p] Natulan, p-$(CH_3)_2CH{-}NH{-}CO{-}C_6H_4{-}CH_2NHNHCH_3$, is rapidly metabolized in man, dog and rat to the azo derivative and finally to N-isopropylterephthalamic acid in blood. Alternatively, methylhydrazine derivatives may be degraded to azomethine, N-hydroxymethyl compounds and formaldehyde.[1031–1032] Natulan is active against chronic leukemia, lymphosarcoma and Hodgkin's disease in man[1033,1034] and is carcinogenic[1035,1036] and teratogenic[1037] in rats.

Alternatively, chronic administration of a hepatocarcinogenic formaldehyde precursor appears to alter 1-carbon metabolism.[1048, 1055–1058] Some of the enzymes involved in the metabolism of formaldehyde and other 1-carbon compounds and their precursors include dihydrofolate reductase (H_2–folate reductase), formylase, formiminoglutamic acid, histidase, methylene–H_4–folate dehydrogenase, serine hydroxymethylase and urocanase.[1048]

In the microsomal metabolic studies involving formaldehyde formation it is usually necessary to determine the drugs as well as formaldehyde and other metabolites, many of which are labile to air oxidation. To prevent decomposition of these materials during initial isolation and analysis sodium bisulphite has been used as an antioxidant. The influence of bisulfite on

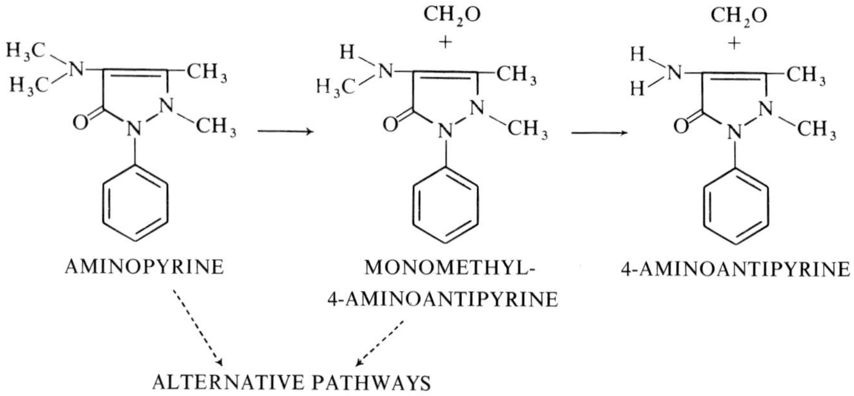

CH_2O + ; CH_2O +

AMINOPYRINE → MONOMETHYL-4-AMINOANTIPYRINE → 4-AMINOANTIPYRINE

ALTERNATIVE PATHWAYS

FIG. 53. Enzymatic N-demethylation of aminopyrine.

microsomal activity has been determined[924] through a study of the N-demethylation of aminopyrine to formaldehyde, 4-methylaminoantipyrine, and 4-aminoantipyrine,[1059] Fig. 53.

The above reaction indicates that demethylation activity may be determined by measuring the formation of either formaldehyde or 4-aminoantipyrine. However, the disappearance of aminopyrine or N-methyl-4-aminoantipyrine cannot be used as a measure of dealkylation since appreciable amounts of these substrates disappear by alternative pathways. Aminopyrine is just one of the many model substrates that can be used to study N-dealkylation of drugs, acute toxicants and the genotoxicants and the method described can be applied to the many substrates which will be discussed in this section.

Sulphite has been reported as inhibiting[556, 1060] the 2,4-pentanedione determination of formaldehyde.[556] This method has been used extensively for the determination of formaldehyde formed during the demethylation of N-methyl substrates with the help of an N-demethylase. The postulated mechanism for the bisulphite inhibition is the following

$$CH_2O + NaHSO_3 \rightleftharpoons HOCH_2SO_3Na$$

$$HOCH_2SO_3Na + NH_3 \longrightarrow H_2NCH_2SO_3Na + H_2O$$

Such a so-called sulphomethylation reaction has been reported for these reactants.[1061] Because of this interference the 2,4-pentanedione procedure has been modified for use in the presence of bisulphite. The interference of sulphite can be cancelled through the oxidation of sulphite to sulphate by iodine. The standard curve obtained in this manner is identical to that derived in the absence of sulphite and iodine. Beer's law is followed from about 8 to 90 μM of formaldehyde. Smith and Erhardt[924] suggest the use of sodium bisulphite as an effective alternative to semicarbazide as a formaldehyde-trapping agent in biological work since bisulphite appears to be compatible with the microsomal metabolic function. The difficulty with the use of semicarbazide as a formaldehyde trapping agent is that excess reaction times are needed for the 2,4-pentanedione determination.[166] When bisulphite is used under typical drug-microsomal incubation conditions, formaldehyde losses associated with volatility and/or further metabolic oxidation to formate and carbon dioxide[1062, 1063] are completely prevented.[924] The following mechanism has been postulated for the determination of formaldehyde with 2,4-pentanedione in the presence of bisulphite, Fig. 54.[924]

1) $CH_2O + H_2O \rightleftharpoons HOCH_2OH^b$

2) $HOCH_2OH + NaHSO_3 \rightleftharpoons HOCH_2SO_3Na + H_2O$

3) $NaHSO_3 + H_2O + I_2 \rightarrow NaHSO_4 + 2HI$

4) $CH_2O + NH_3 + 2CH_2(COCH_3)_2 \rightarrow$ 3,5-diacetyl-1,4-dihydrolutidine (ring structure with $CH_3C{=}O$, $CCH_3{=}O$, H_3C, N–H, CH_3)

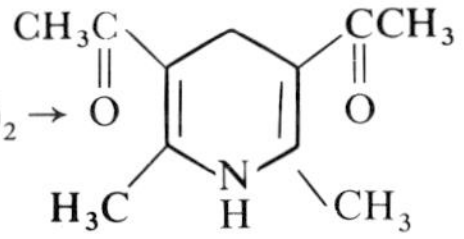

FIG. 54. Chemical events in the Nash determination of formaldehyde in the presence of bisulfite.[a]

[a] Altered chemical equilibria peculiar to the protein precipitation–acidification step have been excluded for diagrammatic simplicity.

[b] Dilute aqueous CH_2O solutions largely contain methylene glycol in equilibrium with small amounts of unhydrated CH_2O.[1064]

Determination of formaldehyde in the presence of bisulphite.[924] *Reagents.* Standard solutions—dilute formaldehyde solution to provide 246, 123, 62, 31, and 15 μM standard solutions. 2,4-Pentanedione (acetylacetone) solution—prepare fresh daily a 100 ml aqueous solution containing 30 g ammonium acetate and 0·4 ml 2,4-pentanedione. Iodine solution—freshly prepare 28 mM I_2 solution in acetone.

Procedure. Add 2 ml of pentanedione solution to 5 ml of test solution and incubate at 60° for 30 min. Cool to room temperature and read absorbance at 415 nm. Run determinations in triplicate.

Procedure for bisulphite treated test solutions. Treat 5 ml of a formaldehyde test solution containing 0·05 % sodium bisulphite with 1 ml of the iodine solution and then react as above.

Microsomal studies. Decapitate, exsanguinate, and hepatectomize male Sprague–Dawley rats weighing about 250–300 g. Blot livers dry and homogenize (glass tube, Teflon pestle) at 0° in 2 volumes of 0·25 M sucrose. Centrifuge the homogenate at 1000 g (av.) at 0° for 15 min and retain the supernatant fraction. Model incubations contained 1 ml 0·2 M Tris buffer (pH 7·4, 37°), 1 ml of 10 000 g supernatant (≡0·5 g. liver), 1 ml Tris buffered co-factor solution containing 2 μmol NADP, 10 μmol glucose 6-phosphate, 25 μmol $MgCl_2$, 1 ml standard formaldehyde solution, and 1 ml aqueous $NaHSO_3$ (0·0025 g). Prepare blanks containing water in place of formaldehyde and/or water in lieu of $NaHSO_3$ similarly. Perform incubations in the presence of air and proceed for 0, 5, 15, and 30 min at 37° in a Dubnoff metabolic shaker. Effect protein precipitation by adding 0·5 ml of 3 N CCl_3COOH, mix for 2 min, and centrifuge (2000 g, 10 min). Neutralize 3-ml portions of supernatants with 0·2 ml of 3·6 N KOH and react with 2-ml aliquots of the 2,4-pentanedione solution. For microsomal incubations containing 0·05% sodium bisulphite, treat supernatant aliquots with 0·6 ml of freshly prepared 28 mM I_2–acetone solution prior to formaldehyde determination. Perform neutralization prior to this step so as to eliminate the possibility of acid catalyzed aldol condensation between formaldehyde and acetone,[1064] and to promote the reverse reaction of bisulphite addition product to free formaldehyde and bisulphite. Run all determinations in triplicate and carry appropriate blanks through the entire microsomal study and analysis. Formaldehyde recoveries were determined from average values obtained in these experiments compared to appropriately diluted standard formaldehyde solutions which were analyzed without prior incubation.

In this section we will discuss in a concise fashion a large number of $N-CH_3$ compounds from the viewpoint of formaldehyde formation, analysis of the precursors and the involved N-demethylases as well as some of the physiological byproducts of this metabolism. We have tried to discuss a representative group of these phenomena but feel that many of the important aspects of this important problem either could not be covered or were overlooked. For example, consider hexamethylphosphoramide or hempa, $[(CH_3)_2N]_3PO$. Inhalation of 400 ppb of this compound by rats in about 8 months results in squamous cell carcinoma in the nasal cavity, these tumours originating from the epithelial lining of the nasal turbinate bones.[1065] It is also mutagenic to fruit flies. This compound and its metabolites are probably formaldehyde precursors and possibly even precursors of formylphosphoramides, for in the rat and the mouse hempa is metabolized to 3-demethylated products, viz., pentamethyl, tetramethyl and sym. trimethylphosphoramide.[1066] Degradation of the phosphoramides by rat liver slices *in vitro* indicates that their metabolism occurs by a process of oxidative dealkylation. Thiohempa would be similarly metabolized since it forms hempa in the rat and the mouse. The role of aldehyde formation in the physiological activity of these and other N-methyl compounds needs a much more thorough examination.

B. Aliphatic amines

A purified microsomal mixed function oxidase isolated from pork liver has been shown to catalyze the NADPH- and oxygen-dependent N-oxidation of a variety of secondary and tertiary amines, including tranquilizers, antihistamines, narcotics, tropine alkaloids, ephedrine and ephedrine-like compounds.[990] The tertiary amines are oxidized only to the amine oxide. Secondary amines, which do not form a stable N-oxide, yield the primary amine and an aldehyde, and probably an hydroxylamine. Tertiary amines are oxidized at a faster rate than secondary amines, and N-methyl alkylamines at a faster rate than N-ethyl alkylamines. Monoamine oxidase activity is completely absent. N-Alkyl amides, carbamates and heterocyclic nitrogen compounds containing a polar group α or β to the amino group are not substrates for the enzyme. Some of the N-methyl compounds[990] which give formaldehyde under these conditions are listed in Tables 16a and 16b.

Another formaldehyde precursor, acetylmethadol, is more effective than methadone in the treatment of opiate dependence.[1067–1071] From studies of laboratory animals it would seem that some of the metabolites of acetylmethadol are responsible for the time-action characteristics of some of the pharmacological effects of acetylmethadol.[1072–1074] Some of the metabolites have been identified in the urine of acetylmethadol maintenance subjects,[922,1075,1076] and in the biofluids of rats given acetylmethadol.[1077] Preliminary studies indicate that biotransformation is a prerequisite for the elimination of acetylmethadol and that N-demethylation is quantitatively more important than deacetylation for acetylmethadol in the human.[921] The metabolism of this compound is shown in Fig. 55.

$$C_6H_5-CH_2-\underset{\displaystyle CH_3}{\underset{|}{CH}}-\overset{\displaystyle CH_3}{\overset{|}{N}}-CH_2-C_6H_5$$

XV

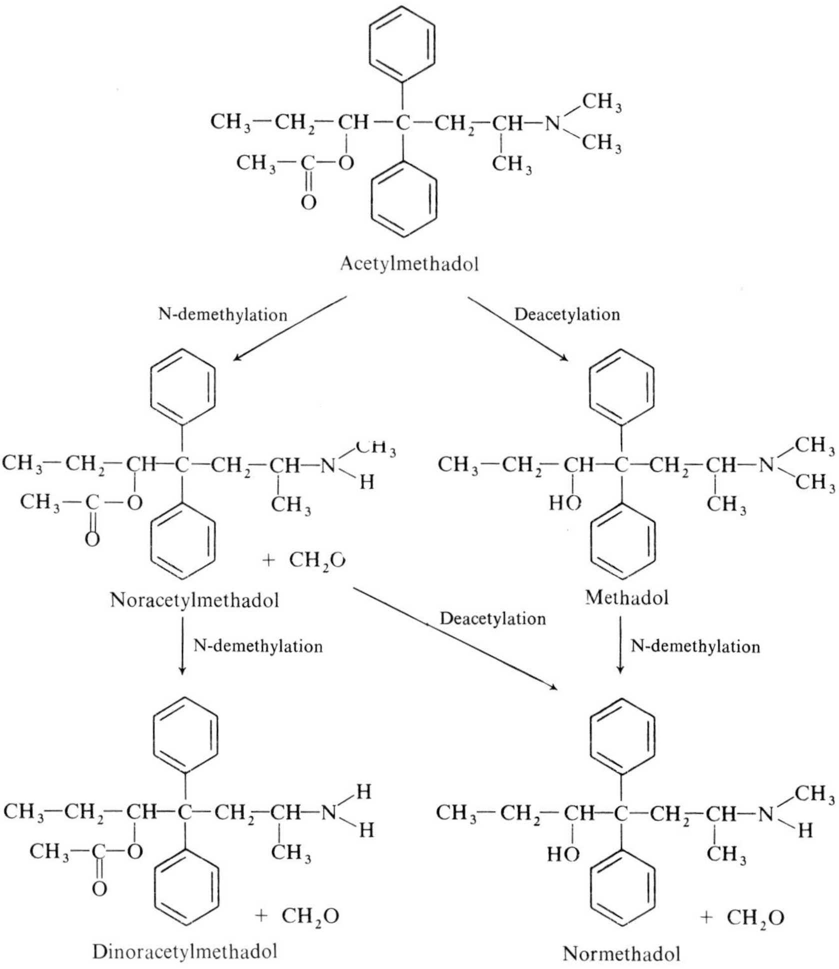

Acetylmethadol

N-demethylation

Deacetylation

Noracetylmethadol

$+ CH_2O$

Methadol

Deacetylation

N-demethylation

N-demethylation

Dinoracetylmethadol

$+ CH_2O$

Normethadol

$+ CH_2O$

FIG. 55. Enzymatic N-demethylation of acetylmethadol and some of its metabolic products.[921]

Benzphetamine, XV, is one of the substrates utilized in investigating N- demethylase activity.

This activity is assayed by following the release of formaldehyde.[556, 1078] With the help of this analytical step it was found that after treatment with 2,3,7,8-tetrachlorodibenzo-*p*-dioxin, XVI, the N-demethylation of benzphetamine, aminopyrine and ethylmorphine was suppressed in hepatic microsomes from male but not from female rats, with the suppression of benzphetamine N-demethylase in male rat liver being age related.[948]

XVI

Semicarbazide can be used to trap the benzphetamine-derived formaldehyde[970] for the determination.[969, 976] Cytochrome P-450 is active in benzphetamine demethylation but is inactive in supporting benzo[a]pyrene hydroxylation even at high concentrations; with cytochrome P-448 the opposite results are found, Fig. 56.[971] Of interest is the report that human liver microsomes are lower in cytochrome P-450 content than are microsomes from commonly used animals.[968]

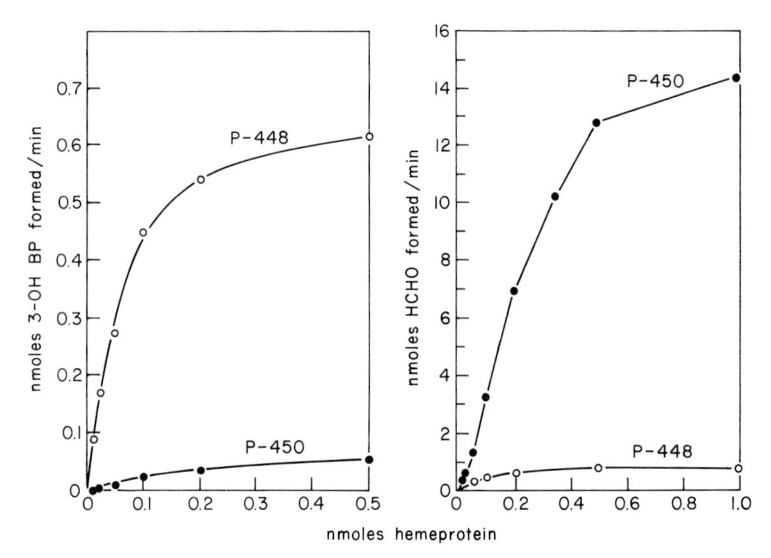

FIG. 56. Hydroxylation of benzo[a]pyrene (left) and N-demethylation of benzphetamine (right) by highly purified cytochrome P-448 and P-450. The final BaP concentration was 95 μM, and the final benzphetamine concentration was 1·0 mM.[971]

Bidrin or 3-hydroxy-N,N-dimethyl-cis-crotonamide dimethyl phosphate is an aliphatic amide and a formaldehyde precursor. It is an insecticide that exhibits systemic action, wide spectrum insecticidal activity and moderate persistence; it is fairly toxic for mammals and is teratogenic in hen eggs, the cis-crotonamide isomer being the active teratogen component. In rats Bidrin is N- demethylated, O-demethylated, and the amide and alkyl phosphate groups are hydrolyzed, Fig. 57.[977] Bidrin, its metabolism and the N-demethylase systems concerned in the metabolism could be studied through the determination of formaldehyde.

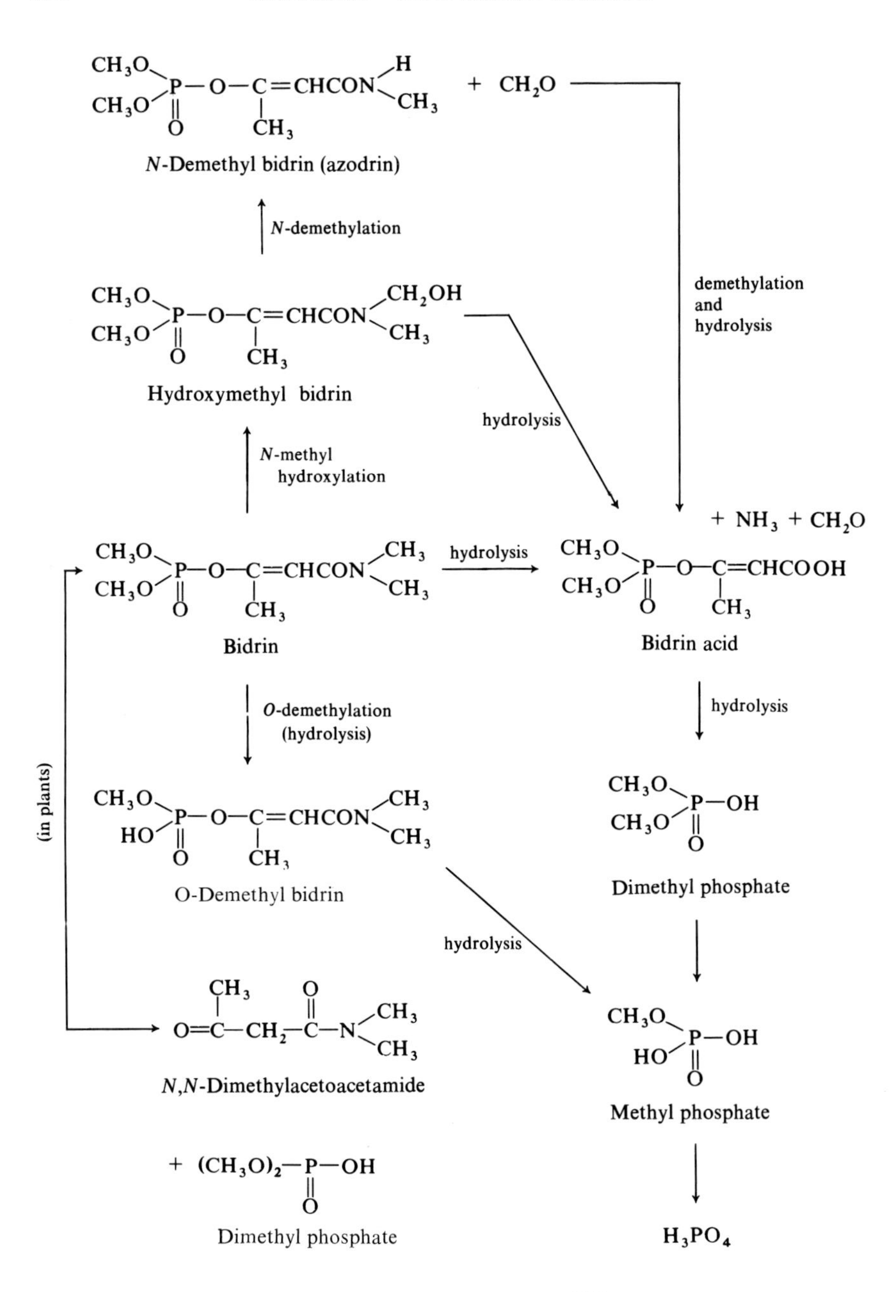

FIG. 57. Metabolism of Bidrin.[977]

Another formaldehyde precursor of importance which can be considered a dimethylalkylamine is chloropromazine. The N-demethylated metabolites of chloropromazine have the ability to prevent the formation of psychotogenic substances by inhibiting the activity of indolethylamine N-methyltransferase; this is postulated as a mechanism by which chloropromazine exerts its antischizophrenic action.[982] Some of the metabolites of chloropromazine are shown in Fig. 58.

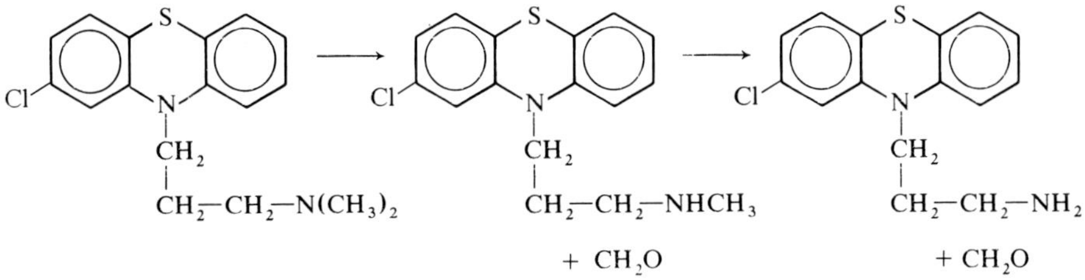

FIG. 58. Enzymatic N-demethylation of chloropromazine.

Another aliphatic amide whose mammalian metabolism has been studied is N,N-dimethylurea.[1079]

Methadone has been used in the treatment of opiate dependence. Thin-layer chromatographic methods have been described for the characterization of methadone metabolites.[1027, 1028] One possible series of metabolic reactions are shown in Fig. 59. The N-methyl and N-demethyl heterocyclic compounds

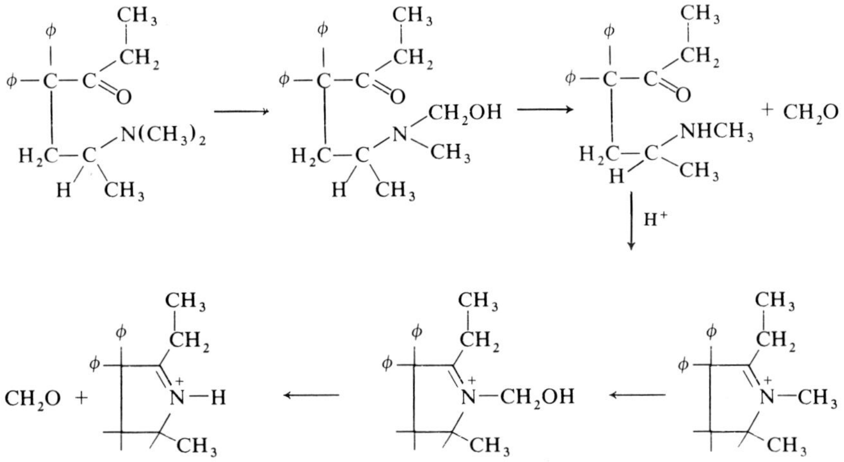

FIG. 59. Enzymatic N-demethylation of methadone.

have been isolated. The byproduct formation of formaldehyde could be utilized in various studies of the compound and its effects.

Methylamines can also be oxidized to formaldehyde in plants with the help of an amine oxidase.[1080] This oxidation is the first step in methylamine utilization with formaldehyde being incorporated into metabolites on the C_1 pathways.

Monoamine oxidase (monoamine: oxygen oxidoreductase (deaminating), EC 1.4.3.4) can act on a primary amine to form an aldehyde, e.g.

$$CH_3NH_2 + H_2O + O_2{=}CH_2O + NH_3 + H_2O_2$$

This reaction can be utilized to determine methylamine (or ethyl- or propylamine),[1081,1082] e.g.

$$RCH_2NH_3Cl + H_2O + PMS \longrightarrow RCHO + NH_4Cl + PMSH_2$$

$$PMSH_2 + DCPIP \rightleftharpoons PMS + DCPIPH_2$$

where PMS is phenazine methosulphate, and DCPIP is 2,6-dichlorophenol-indophenol.

Nitrosamines and aldehydes can be formed from the reaction of trialkylamines with tetranitromethane,[1083,1084] e.g.

$$R_2NCH_2R' + C(NO_2)_4 \longrightarrow R_2NNO + R'CHO + CH(NO_2)_3$$

when $R' = H$, formaldehyde is formed.

Since many alkaloids, other natural products, and numerous drugs are tertiary amines and many tertiary amines have been shown to react smoothly with nitrite ion at pH 3 to 6 to produce carcinogenic nitrosamines,[380,622,645,1085] the presence of these precarcinogens (tertiary amines and nitrite) in the human environment has made necessary the investigation of this possibly lethal synthesis[658] and the analysis of these precursors in environmental samples.

Let us emphasize the importance of the nitrosamines as carcinogens, even though they have not been proven unequivocally to be carcinogenic to humans.[1086,1087]

The mechanisms by which nitrite and trisubstituted amines could give rise to secondary nitrosamines have been divided into 3 categories,[380] all of which could involve aldehyde as byproduct. For example consider the postulated mechanism[1085] in Fig. 60.[380] When $R' = H$ in the tertiary amine formaldehyde is the byproduct. Of course, other aldehydes could also be formed here. In the second type cleavages of the N—R bond are those involving both oxidative and non-oxidative steps, e.g. the N-methylmorpholine reaction with nitrite under mildly acidic and essentially neutral conditions results in a smooth production of N-nitrosomorpholine.[380] Formaldehyde would be a byproduct

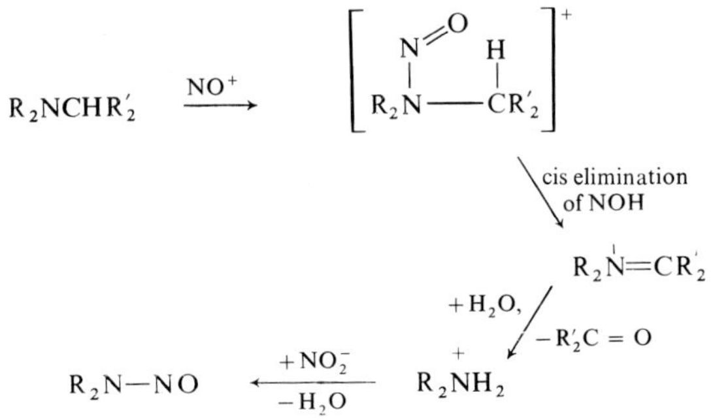

$$R_2NCHR'_2 \xrightarrow{NO^+} \left[\begin{array}{cc} N{=}O & H \\ | & | \\ R_2N & \!\!-\!\!-\; CR'_2 \end{array}\right]^+ \xrightarrow[\text{of NOH}]{\text{cis elimination}} R_2\overset{+}{N}{=}CR'_2 \xrightarrow[-R'_2C=O]{+H_2O,} R_2\overset{+}{N}H_2 \xrightarrow[-H_2O]{+NO_2^-} R_2N{-}NO$$

FIG. 60. Proposed mechanism for the nitrosative dealkylation of tertiary amines.[380, 1085]

here. A third type of cleavage involves nonoxidative cleavage of one N—R bond, where the R group is in an oxidation state appropriate to its removal by hydrolysis. Compounds like alkoxymethylene amines would be expected to hydrolyze to formaldehyde and to nitrosate smoothly under mildly acidic conditions. An example of this type is S-piperidinomethylthiobenzoate where one N-substituent is in the oxidation state of formaldehyde; this compound would be easily hydrolyzed even at pH 4–5 to formaldehyde and in the presence of nitrite is converted to nitrosopiperidine in 67% yield after 22 hr of reaction at 37°.[380] The following reaction takes place, Fig. 61.

When members of the methylamine series—CH_3NH_2, $(CH_3)_2NH$, $(CH_3)_3N$, and $(CH_3)_3NO$—are compared in terms of reaction with nitrite, they all give the same ultimate carcinogen, $CH_3N_2{}^+$ or $CH_3{}^+$, and except for CH_3NH_2 they

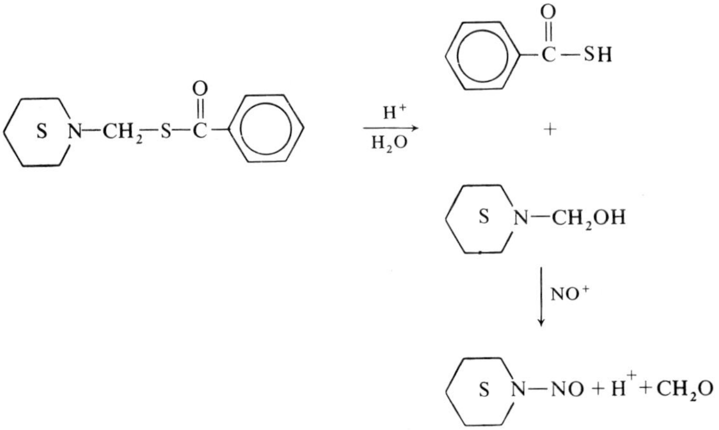

$$\text{(piperidino)}N{-}CH_2{-}S{-}\overset{O}{\overset{\|}{C}}{-}C_6H_5 \xrightarrow[H_2O]{H^+} C_6H_5{-}\overset{O}{\overset{\|}{C}}{-}SH + \text{(piperidino)}N{-}CH_2OH \xrightarrow{NO^+} \text{(piperidino)}N{-}NO + H^+ + CH_2O$$

FIG. 61. Hydrolysis and nitrosation of S-piperidinomethylthiobenzoate.[380]

give formaldehyde as a byproduct and their reaction with nitrite is catalyzed by formaldehyde. And yet the mechanism of reaction is different for each one.

While tertiary amines are optimally reactive toward nitrous acid in the pH range 3 to 6 with the reaction suppressed almost completely under more acidic conditions, secondary amines are nitrosated at rates that decrease rapidly as the pH is increased above the relatively acid optimum, e.g. 3·4 for dimethylamine.[621] Consequently, trimethylamine converts more readily to dimethylnitrosamine at pH 6·4 than does dimethylamine.[1086] On this basis it has been suggested that tertiary amines are more important than secondary amines as environmental precursors of nitrosamines in the limit of mildly acidic conditions.[380]

Thus, on this basis many dimethylamines, $(CH_3)_2NR$, present in air, water, natural products and in human tissue could be considered to be possible precursors of dimethylnitrosamine and eventually of formaldehyde. Examples of these precursors are the analgesic drug, aminopyrine; betaine, a natural precursor of trimethylamine; the antihistamines, chlortrimeton and benadryl; the antibiotics, tetracycline, erythromycin and puromycin; captamine, the cutaneous depigmenter; aminopentamide, a smooth muscle antispasmodic; cyclopentolate, the cyclopegic and mydriatic; tetracaine, the local anesthetic; propoxyphene, a narcotic analgesic; clophedianol, an antitussive; chlorpromazine, a tranquilizer; amitriptyline, a stimulant; and bufotenin and psilocybin, psychotomimetic agents.

Trimethylamine is a decomposition product of betaine and other natural products and occurs widely in sea weed[1088, 1089] and is present at high levels in some lichens.[1090] Trimethylamine would react with nitrous acid to give dimethylnitrosamine and formaldehyde. Thus, this reaction is a potentially lethal synthesis which could take place in the environment or in living tissue.

A preparation of the trimethylamine dehydrogenase from *H. vulgare* has been applied to the enzymatic estimation of trimethylamine.[1091] The primary electron acceptor for the enzyme is phenazine methosulphate (PMS) the reoxidation of which is nonenzymatically coupled to the reduction of 2,6-dichlorophenolindophenol (DCPIP), e.g.

$$(CH_3)_3NH^+ + PMS + H_2O \longrightarrow (CH_3)_2NH_2{}^+ + PMSH_2 + CH_2O$$

$$PMSH_2 + DCPIP \rightleftharpoons PMS + DCPIPH_2$$

The trimethylamine could also be estimated through the derived formaldehyde.

It has been shown that strong acid treatment liberates formaldehyde from trimethylamine oxide, a compound which is a normal constituent of most sea fish.[514] A somewhat similar reaction is the formation of formaldehyde in the muscle of gadoids when the fillets are frozen and stored at temperatures above

$-25°$.[511,512] This formaldehyde has its origin in the trimethylamine oxide of the muscle[510,513,1092] which decomposes through enzyme action to form equimolar amounts of dimethylamine and formaldehyde.

Compounds such as trimethylamine oxide, tropine N-oxide, N-methylmorpholine N-oxide, and various N,N-dimethylalkylamine oxides form adducts with sulfur dioxide which are hydrolyzed with hot water to formaldehyde, Table 1. Any of the formaldehyde reagents can be used in the analysis.

In the reaction of trimethylamine-N-oxides with nitrite under weakly acidic conditions nitrosodimethylamine is obtained with formaldehyde a most likely byproduct.[380]

It would appear obvious that the huge assortment of methylamines present in the environment and in living tissue and used as drugs could be determined and studied through the derived formaldehyde.

C. Aromatic amines

Methylanilines, *p*-substituted methylanilines and some heterocyclic N-methylated compounds can be oxidized to formaldehyde and determined. In the determination of the *p*-substituted methylanilines with MBTH a variety of chromogens is obtained of which the formaldehyde chromogen absorbs at longest wavelengths, Fig. 62.

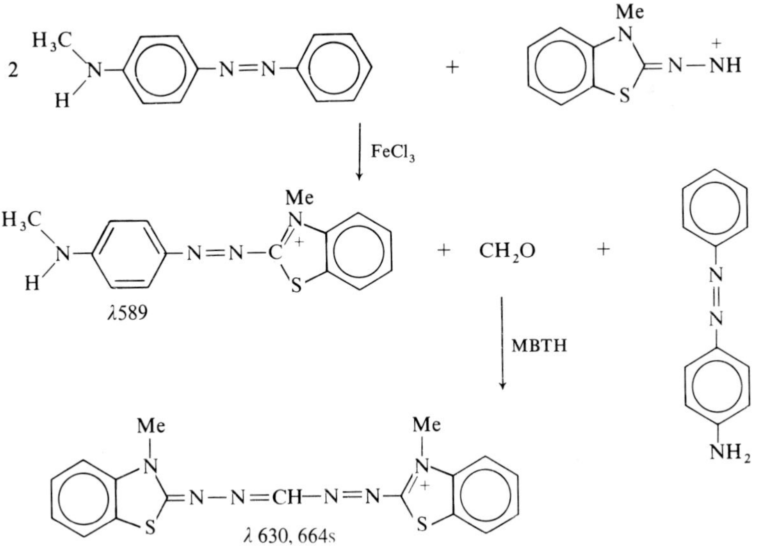

FIG. 62. MBTH determination of N-methyl 4-phenylazoaniline.

Two types of procedures are available, one for the simple *p*-substituted dimethylanilines and one for the N-methyl 4-phenylazoanilines discussed in Section E. These procedures could be utilized for the determination of para-substituted N-methyl aromatic amines. Some of the results are shown in Table 17.[120] The 2 bands near 630 and 670 nm are probably derived from formaldehyde, the band near 600 nm from a chromogen formed by attack of MBTH at the position para to the $N(CH)_3$ group to form an azo dye.

Table 17. Determination of *p*-Substituted N,N-Dimethylaniline Derivatives[120]

$$X-C_6H_4-N(CH_3)_2$$

X	λ_{max}, nm		$\varepsilon \times 10^{-3}$	
CH_3	635	665	19	19
Br	599	665	52	25
SCN	605	668	48	37
NO	630	670	72	72
NO_2	629	668	13	13
CHO	635	670	55	55
CH=CH—CHO	623	666	29	27
CO·CO·ϕ	635	670	13	13
CO·ϕ·$N(CH_3)_2$	630	666	46	43
N=N—ϕ[a]	630[b]	664s	34	32
N=N—ϕ	604	672	70	56
$CH=C(CN)_2$	637	667	56	56
CH=N—P[c]	637s[d]	677	90	102
CH=CH—ϕ	632	667	116	119

[a] $NHCH_3$ instead of $N(CH_3)_2$.
[b] Also a band at 589 with mε 36.
[c] P = 1-pyrenyl.
[d] Also a shoulder at 755, mε 49.

MBTH determination of p*-substituted N,N-dimethylanilines.*[120] Add 0·5 ml of 0·35% aqueous MBTH·HCl to 0·5 ml of the methanolic test solution. After 10 min add 1 ml of 0·6% aqueous ferric chloride solution. After 5 min dilute to 10 ml with methanol. Read at long wavelength maximum.

Various types of oxidizing agents can be used for the oxidation of N-methylanilines to formaldehyde. Thus, for example lead tetraacetate can be used to oxidize N-methyl aromatic amines to formaldehyde. The procedure has been used to prepare aliphatic aldehydes in 50 to 90% yields,[376] e.g.

$$ArN(R')CH_2R + Pb(OAc)_4 \xrightarrow{Ac_2O} ArN(R')Ac + RCHO$$

Dimethylaniline can be oxidized to formaldehyde in 25% yield with chromium trioxide.[124] The general equation for the formation of aliphatic aldehydes is postulated to be

$$3(RCH_2)_2\overset{+}{N}H{-}C_6H_5 + 32H^+ + 4\,Cr_2O_7^{-2} \longrightarrow$$

$$8\,Cr^{+3} + 16\,H_2O + 3\,O{=}C_6H_4{=}O + 3\,NH_4^+ + 6\,RCHO$$

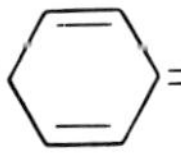

Aliphatic amines could be degraded to aliphatic aldehydes by the following procedure with the help of chromate,[119] Fig. 63.

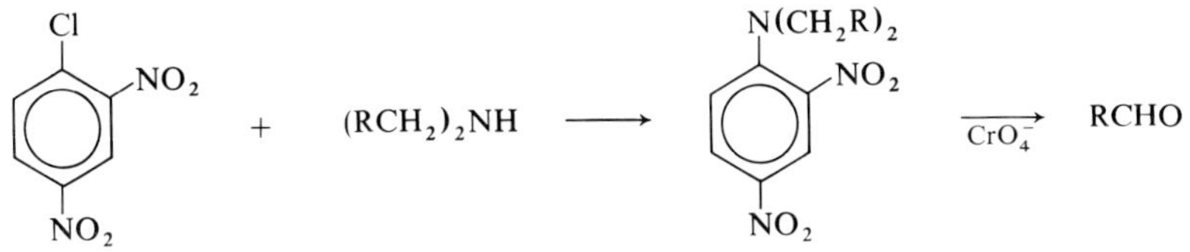

$$\text{1-Cl-2,4-}(NO_2)_2C_6H_3 + (RCH_2)_2NH \longrightarrow \text{1-}N(CH_2R)_2\text{-2,4-}(NO_2)_2C_6H_3 \xrightarrow{CrO_4^-} RCHO$$

FIG. 63. Conversion of dialkylamines to aldehydes.

Enzymatic reactions have been utilized in the formation of formaldehyde from N,N-dimethylamines. N-Demethylation is one of the reactions catalyzed by the mixed-function oxidases and since many of the substrates are bases, their ionization characteristics are important in the reaction. Cho and Mirva[956] examined the effect of pH on the N-demethylation of a series of arylalkylamines by liver microsomes. It was found that the rate of reaction decreased with decreasing pH for those amines that were ionized in the pH range studies. The rate is believed to be dependent on the ionization of the bases and its partition between the microsomes and the aqueous incubation media.

N,N-Dimethylaniline and probably other N,N-dimethyl aromatic amines could be determined by the following series of reactions with the help of dimethylaniline monooxygenase (N-oxide forming) (EC 1.14.13.8, N,N-dimethylaniline, NADPH: oxygen oxidoreductase (N-oxide-forming)) (122)

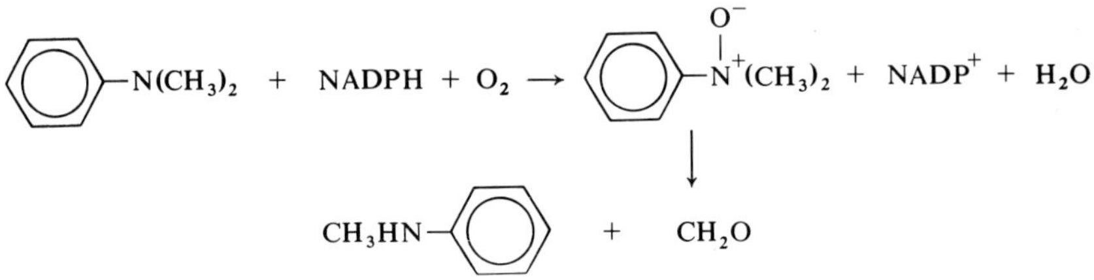

$$C_6H_5{-}N(CH_3)_2 + NADPH + O_2 \rightarrow C_6H_5{-}\overset{+}{N}(O^-)(CH_3)_2 + NADP^+ + H_2O$$

$$\downarrow$$

$$CH_3HN{-}C_6H_5 + CH_2O$$

FIG. 64. Enzymatic N-demethylation of N,N-dimethylaniline.

and dimethylaniline-N-oxide aldolase (EC 4.1.2.24, N,N-dimethylaniline-N-oxide formaldehyde-lyase)[123] Fig. 64.

It has been shown that N-methylaniline which can be formed from N,N-dimethylaniline through enzymatic demethylation can be further demethylated,[377,378] e.g.

$$2\,C_6H_5NHCH_3 \xrightarrow[O_2]{NADPH} C_6H_5NHOH + C_6H_5NH_2 + 2\,CH_2O$$

D. Heterocyclic amines

One heterocyclic dimethylamine, amidopyrine, has been chosen for brief discussion. We have discussed it before; in this section we will discuss it in terms of the effects of various substances in inhibiting or stimulating its metabolism to formaldehyde.

The oxidative demethylation of amidopyrine (given to healthy adult males exposed for 6 hr to graded CS_2 concentrations between 10 and 80 ppm) was inhibited.[561]

Immediately after administration of a small protective dose of CCl_4 to rats, activity of liver microsomal aminopyrine demethylase begins to fall. By 24 hr, activity of this enzyme system is one-fourth of control levels and rats become completely resistant to an LD_{95} of CCl_4.[710] The aminopyrine demethylase level as shown by the formaldehyde production remains depressed for 4 days, after which it gradually recovers. After sedimentation of the precipitated protein by centrifugation, the formaldehyde content was determined by the 2,4-pentanedione procedure.[556] On the other hand in a test period of 10 days the inhalation of carbon tetrachloride, methylene chloride, trichloroethylene and benzene in concentrations between 450 and 500 ppm produced an increase in liver contents of cytochrome P_{450} and in the activity of aminopyrine demethylase.[1093] The *in vitro* N-demethylation of aminopyrine was calculated on formaldehyde formed within 5 min.[1094] On the basis of this work it is believed that with prolonged exposure even small solvent concentrations in the respiratory air may lead to undesirable alterations in biotransformations of foreign substances and endogenous compounds in the liver.[1093] High and fairly high concentrations of solvent vapors can be present in industrial atmospheres and in ambient atmospheres adjacent to some industrial areas, respectively. Unfortunately, not much data is available on the organic vapor concentration of industrial and other highly polluted areas.

After oral treatment with 100 ppm DDT daily for 84 days the aminopyrine demethylase activity has increased 2·3 times and does not come back to normal till about 78 days after ending DDT treatment.[937] However, a single injection of phenobarbital causes an increase of about 50% in the amount of cytochrome

P-450, but doubles the rate of hexobarbital oxidation and the rate of formation of formaldehyde from amino pyrine.

Aminopyrine has an inhibitory effect on the rate of acetaldehyde formation from ethanol but has no inhibitory effect on the oxidation of 50 mM methanol in microsomes to formaldehyde.(1095) The authors suggest that these and other data provide support for the existence of a microsomal ethanol oxidizing

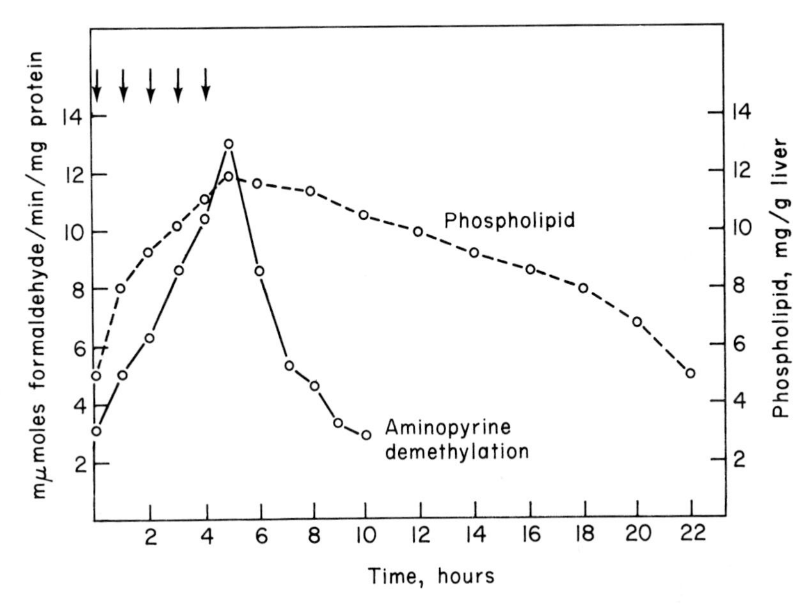

FIG. 65. Effect of phenobarbital administration to rats *in vivo* on the oxidative demethylation activity and phospholipid content of the liver microsomes. The arrows indicate the phenobarbital injections.(928, 1101)

enzyme system,(1096, 1097) in addition to the azide sensitive and H_2O_2 dependent pathway for ethanol oxidation.(901, 1098) Formaldehyde production was assayed by the 2,4-pentanedione method. Acetaldehyde was separated from reaction mixtures containing formaldehyde and acetaldehyde by diffusion into the side-arm of a Warburg vessel containing 0·5 ml of a 0·4% solution of MBTH • HCl. The aldehyde was then determined by the MBTH procedure.(538) Formaldehyde produced from aminopyrine or N-methylaniline can be trapped with semicarbazide(943, 1099, 1100) and determined with 2,4-pentanedione.(556, 976)

Phenobarbitol administered to rats results in a drastic increase in formaldehyde formation from aminopyrine, Fig. 65.(928, 1101) Adrenalectomized and castrated rats show a drastic decrease in testosterone hydroxylation and aminopyrine demethylation activities.[928] Treatment of these rats with nothing, prednisolone or testosterone propionate results in aminopyrine demethylation as shown by the values of mμ moles formaldehyde/min/mg protein of 0·91, 3·80 or 3·69, respectively.

Pregnancy in rats results in a decrease in the N-demethylation of aminopyrine as measured by the formation of formaldehyde with the lowest value at the 15th to 20th day after conception and an increase to normal non-pregnancy values 3 days after delivery.[946]

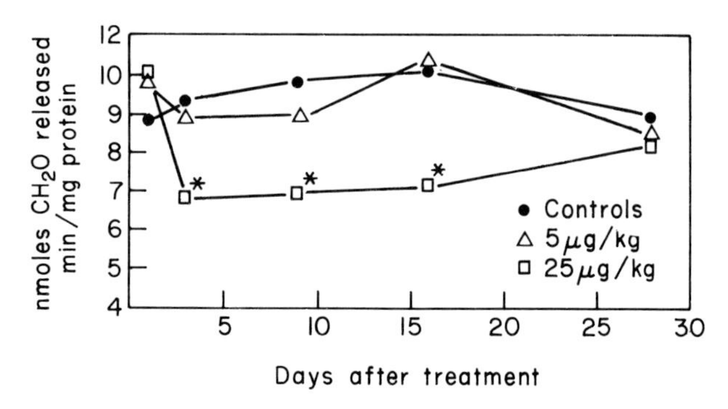

FIG. 66. Time–course effects of a single oral dose of TCDD on liver microsomal aminopyrine demethylation. An asterisk indicates that values are significantly different from controls at $P < 0{\cdot}05$. $N = 3$ male rats.[949]

The effect of a single dose of 2,3,7,8-tetrachlorodibenzo[p]dioxin on the demethylation of aminopyrine is shown in Fig. 66.[949]

Aminopyrine demethylation is impaired in liver microsomal fractions of rats deprived of vitamin E.[954, 955, 1102] When both vitamin E and selenium are withdrawn from rats, the initial rate and extent of aminopyrine demethylation is depressed to an extremely low value.[948]

E. Azo dye amines

MBTH can be used to determine azo dyes containing a para-substituted methylamino group. Analogs of the azo dyes containing —CH=CH— or —N=CH— groups instead of the —N=N— could also be determined by a somewhat similar procedure. It is possible that a large variety of aromatic $—N(CH_3)_2$ and $NRCH_3$ compounds could be determined through the derived formaldehyde. The procedure for determining the azo dyes and their analogs is as follows.

MBTH determination of N-methyl or N,N-dimethyl 4-phenylazoanilines and analogs.[121] Add 1 ml of 0·2% aqueous MBTH•HCl to 1 ml methanolic test solution followed by 2 ml of 1% aqueous ferric chloride. After 30 min dilute to 10 ml with methanol and read at long wavelength maximum.

However, MBTH can also attack the para position of aromatic amines forming brightly colored azo dyes.[120,121] In some of the compounds determined in Table 17 the para substituent is replaced to give an azo dye chromogen (as shown by the band near 600 nm for the bromo, thiocyano and phenylazo derivatives in addition to the formaldehyde-derived formazan cation.

Thus, this reaction can be used for the analysis of carcinogens such as N-methyl 4-phenylazoaniline,[967] N,N-dimethyl 4-phenylazoaniline,[967] and N,N-dimethyl 4-aminostilbene.[1103] Many of these compounds could probably also be determined enzymatically through N-demethylation to formaldehyde which can then be determined by one of the many methods described in these volumes. Thus, for example para-chloro N-methylaniline can be demethylated to formaldehyde by rat kidney subcellular fractions[1104] which can then be determined.

It has been emphasized that in the 4-phenylazoanilines at least one N-methyl group is a necessary but not sufficient condition for carcinogenesis.[967] These authors have studied the N-demethylation of these azo dyes. A dye which is much more readily demethylated than are the tertiary aminoazo dyes is more readily studied. Such a compound is 4-phenylazo N,2-dimethylaniline (or 3-methyl-4-methylaminoazobenzene).[962] Since the 3-methyl group almost completely hinders the reductive cleavage of the azo group, demethylation and formaldehyde formation can be studied independently of other reactions. For optimum demethylation activity NAD^+, $NADP^+$, oxygen, ATP and magnesium ions are required. It is felt that the reaction proceeds through a monomethylol derivative. Attempts to determine whether demethylation proceeds by an N-oxide intermediate have been inconclusive.[968] When the demethylation was carried out in the presence of semicarbazide, formaldehyde and 4-phenylazo-2-methylaniline were recovered in amounts accounting stoichiometrically for the 4-phenylazo-N,2-dimethylaniline metabolized.[962] Mouse liver homogenates contain two to three times as much of the enzyme system responsible for the derived formaldehyde as rat liver homogenates. The activity of mouse liver and rat liver homogenates can be increased about 250% and 50%, respectively, by feeding the animals diets containing certain meat products or organic peroxides for at least one week before they are killed.[1105,1106]

The formaldehyde resulting from the N-demethylation of 4-phenylazo-N,2-dimethylaniline (or 3-CH_3—MAB) has been measured by the modified chromotropic acid method.[915,1107]

The studies of Mannering *et al.*[964] have compared the N-demethylation of ethylmorphine and 3-CH_3—MAB. Two-substrate kinetic studies employing ethylmorphine and 3-CH_3—MAB supported the view that the microsomal system responsible for 3-CH_3MAB N-demethylation in rats treated with 3-methylcholanthrene is not the same as that found in untreated rats or rats treated with phenobarbital.[964] These studies also suggested that a single enzyme system is responsible for the N-demethylations of both ethylmorphine and 3-CH_3—MAB when microsomes from untreated and phenobarbital-treated rats were used, but that when microsomes from 3-methylcholanthrene-treated rats were employed, two N-demethylating systems are involved, one of

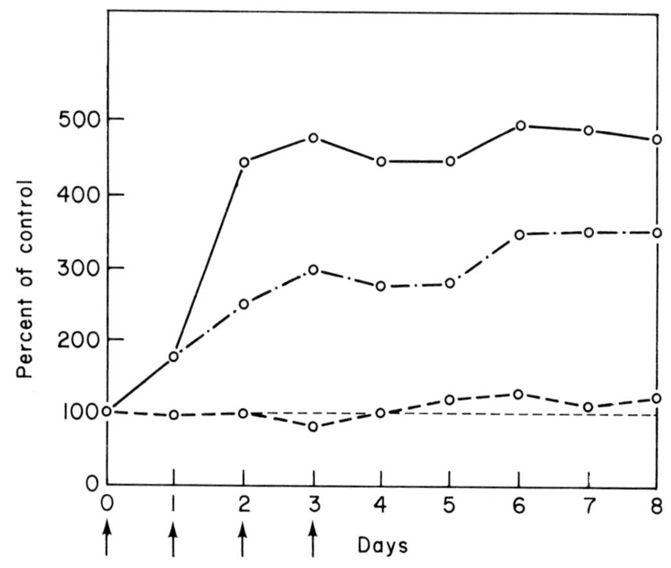

FIG. 67. Effect of concurrent 3-methylcholanthrene and thioacetamide administration on microsomal N-demethylation and cytochrome P-450 levels. Rats (male, 90–120 g) were injected with 3-methylcholanthrene, 20 mg/kg, plus thioacetamide, 50 mg/kg, daily for 4 days (arrows). Cytochrome P-450 levels (○ — · — ○); 3-CH_3-MAB N-demethylation (○——○); ethylmorphine N-demethylation (○------○). 9000 *g* supernatant microsomal fraction was used for the enzyme studies; washed microsomes were used for the cytochrome P-450 determinations. Each point represents the mean of 3 animals.[964]

which is incapable of reacting with ethylmorphine. Thioacetamide was used to provide evidence that the mechanisms by which phenobarbital and 3-methylcholanthrene stimulate drug metabolism may be different. Thioacetamide blocked phenobarbital induction, but had little or no effect on 3-methylcholanthrene induction. Some of the results are shown in Fig. 67.

N,N-Dimethyl 4-phenylazoaniline (or 4-dimethylaminoazobenzene or DAB) carcinogenesis in rats is delayed by the simultaneous administration of phenobarbital, but accelerated by DL-ethionine.[959] The authors believe that the inhibition of DAB carcinogenesis by phenobarbital administration may be related to its action in promoting N-demethylation and hydroxylation of the compound. In contrast, DAB plus DL-ethionine treatment markedly depressed the hepatic activity to N-demethylate DAB. Some of the possibilities for this effect of DL-ethionine on carcinogenesis are (i) an increase in proximate carcinogenic metabolites through depression of the N-demethylating activity, (ii) damage of the N-demethylating system in liver microsomes, and (iii) the action of DL-ethionine as a carcinogenic agent. The method of Matsumoto and Terayama[1108] was used in the measurement of formaldehyde. An example of a method of analysis is the following.

Determination of N-demethylation of amino azo dyes.[968] Prepare microsomes in 0·2 M potassium phosphate buffer (pH 7·5 at 20°C) and adjust to 2·6 mg/ml. Measure demethylation in a total volume of 2·0 ml containing 3·9 mg microsomal protein, 0·1 μmoles $NADP^+$, 120 μmoles of nicotinamide, 1·0 μmoles of ADP and NADPH regenerating system utilizing glucose-6-phosphate or isocitrate (vide supra). Incubate at 37°C aerobically, stopping the reaction at 0 and 20 min by the addition of 1/2 volume of 20% TCA. After standing at room temperature for 15 min centrifuge at 1000 × g for 15 min. Mix 2 ml of the supernatant fluid with 0·6 ml of the Nash Reagent,[556] incubate for 30 min at 37°C, and measure the absorbance at 415 nm against a tube run without $NADP^+$ and formaldehyde standards prepared in the same manner as the samples. Express the results in mμ moles of formaldehyde released/mg protein/20 min.

A separate series of experiments were undertaken in which the same system was used to measure potential demethylation of N-methyl and N,N-dimethyl *p*-phenylenediamine. This analytical procedure was used in a study of the N-demethylation of 4-phenylazoaniline and its N-methyl and N,N-dimethyl derivatives. It was found that the rate of formation of formaldehyde from the mono- and di-substituted dyes was identical. When propylene glycol was used as a solvent for these dyes it was effectively metabolized by the microsomal preparation to formaldehyde, and the reaction was NADPH dependent.[968] The dyes in the presence of propylene glycol were metabolized more rapidly in terms of released formaldehyde or increased demethylated products. Similarly the presence of 4-phenylazoaniline augmented formaldehyde release with both ethanol and propylene glycol.

A fairly large group of aromatic azo compounds are carcinogenic.[1109] Of these, quite a few contain $N{-}CH_3$ groups. These compounds and their carcinogenicity in living creatures could be studied through the derived formaldehyde. Similarly, the N-demethylase systems could be studied through the derived formaldehyde.

Frightening results needing thorough examination have been reported by Golub *et al.*[1110] Not only do some individuals pay in such an appalling way for their long contact with some carcinogenic mixture but their offspring may also have to pay for this unfortunate "transgression" of a parent. The oncogenic action of carcinogenic azo dyes, such as *o*-aminoazotoluene and N,N-dimethyl 4-phenylazoaniline, and aromatic amines, such as *o*-tolidine and 3,3′-dichlorobenzidine, was demonstrated in the progeny of experimental BALB/c mice. Golub *et al.*[1110] have emphasized that the results of such experiments make it imperative to take into consideration the possible carcinogenicity of azo dyes and aromatic amines both on persons in direct contact with them and on their progeny through placental transmission and/or postnatal transmission through the milk of lactating mothers. Transplacental carcinogenesis has been accomplished with quite a few carcinogens.[1111–1114] There are even examples where initial chemical carcinogenesis can be passed on through several generations.[564, 1115] In humans the only known example of transplacental carcinogenesis occurred following the treatment of pregnant women with diethylstilbestrol.[1116, 1117] Some of the female offspring eventually developed an adenocarcinoma of the vagina. Diethylstilbestrol was once given as an example of a chemical "safely" used in women for over 10 to 12 years and causing cancer in animals only when given in fairly large doses.

The significance of the report that some aromatic amines and azo dyes are transplacental carcinogens is that any individual in contact with these carcinogens, some of which cause human cancer, can get cancer and/or pass cancer on to his offspring. This could apply to any individual in contact with such dyes, aromatic amines and any mixtures, such as "coal tar pitch volatiles" which contain these carcinogens. Thus, the sins of the parents or even the grandparents can be visited on the children.

F. Heterocyclic

The compounds we are concerned with in this section are the $N-CH_3$ derivatives, such as heroin, XVII, dextromethorphan, XVIII, morphine, XIX, dihydromorphinone, XX, codeine, XXI, lincomycin, XXII, caffeine, XXIII, methysergide, XXIV, and demerol, XXV.

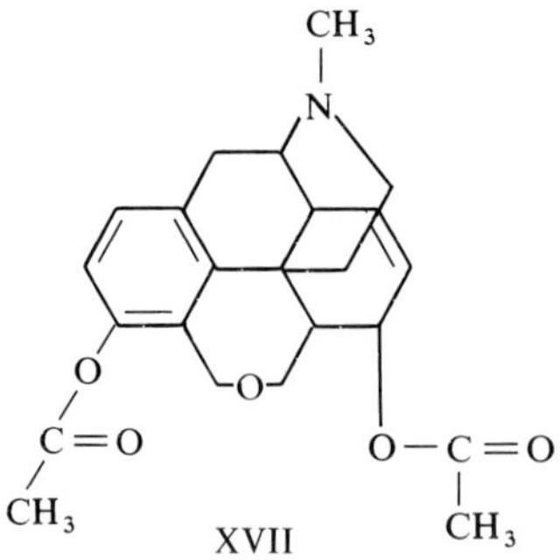

CH_3 N O O $C=O$ CH_3 $O-C=O$ CH_3

XVII

CH_3 N O CH_3

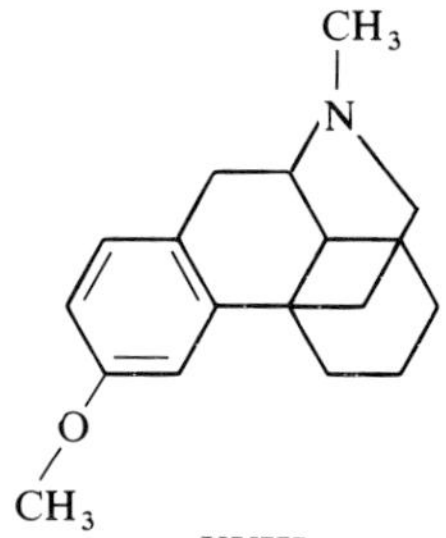

XVIII

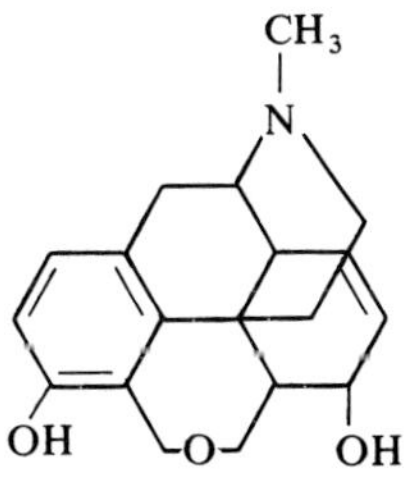

CH_3 N OH O OH

XIX

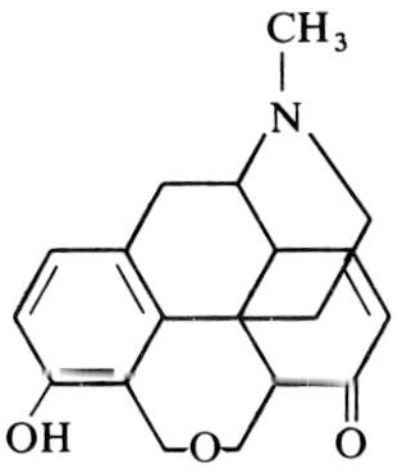

CH_3 N OH O O

XX

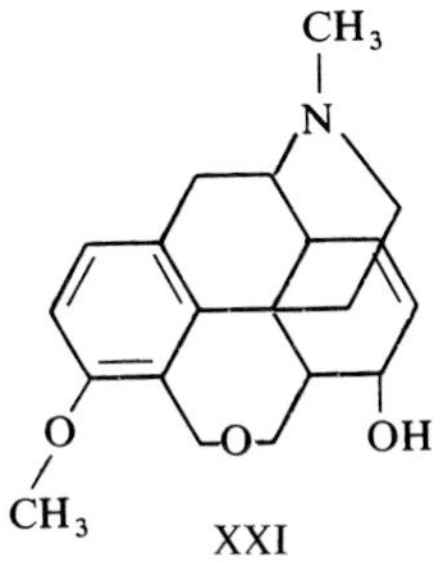

CH_3 N O CH_3 O OH

XXI

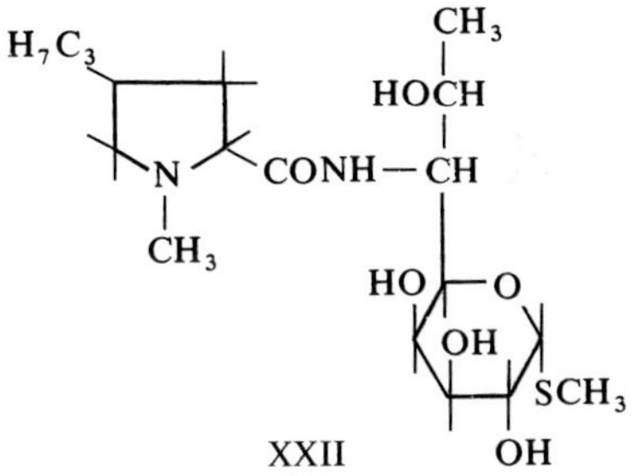

H_7C_3 N CH_3 CONH — CH CH_3 HOCH HO O OH SCH_3 OH

XXII

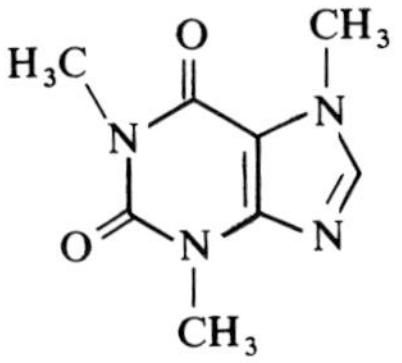

H_3C O CH_3 N N O N N CH_3

XXIII

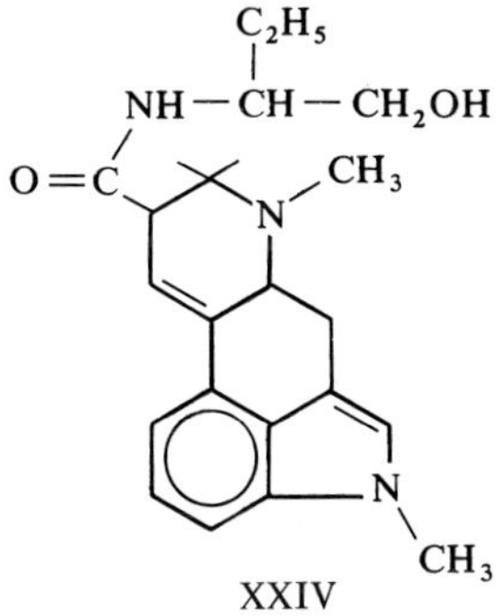

C_2H_5 NH — CH — CH_2OH O=C N CH_3 N CH_3

XXIV

$COOC_2H_5$ N CH_3

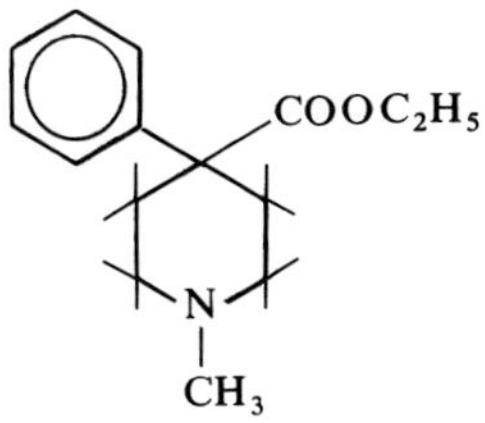

XXV

Table 18. General properties of the liver microsomal hydroxylation system[939]

General reaction catalyzed	$RH + NADPH + H^+ + O_2 \rightarrow ROH + NADP^+ + H_2O$
Specific reactions catalyzed	aromatic hydroxylation, aliphatic hydroxylation, N-dealkylation, O-dealkylation, S-dealkylation, sulfoxidation, deamination, epoxidation, desulfuration, dehalogenation
Endogeneous substrates	steroids, fatty acids, bile acids, and heme
Exogenous substrates	drugs, carcinogens, insecticides and herbicides, and other foreign compounds
Factors affecting activity	species, strain, sex, age, hormonal and nutritional conditions, stress, inducers and other environmental conditions
Inducers which stimulate activity	phenobarbital, 3-methylcholanthrene, benzo[a]pyrene, DDT, chlordane, anabolic steroids and hundreds of other compounds.
Nomenclature	
General	mono-oxygenase, hydroxylase, mixed-function oxidase
According to type of substrate	drug-metabolizing enzyme, steroid hydroxylase, fatty acid hydroxylase, aryl hydrocarbon hydroxylase, etc.
According to specific substrate	aniline hydroxylase, 3,4-benzpyrene hydroxylase, testosterone 6β-, 7α- and 16α-hydroxylases, benzphetamine N-demethylase, ethylmorphine N-demethylase, *p*-nitroanisole O-demethylase, etc.

The continual introduction of large numbers of xenobiotics into the human environment and the increase in concentration of a large variety of environmental pollutants has resulted in a tremendously increased interest in the metabolism of these compounds in hepatic tissue and in extrahepatic tissues such as lung, skin, and intestine.[1118–1121] The study of the detoxication–toxication mechanism in extrahepatic tissues needs much more thorough study since some of these tissues can be the portal of entry for environmental pollutants and knowledge about the fate of environmental contaminants or drugs at the first lines of defense in the body is necessary for an understanding of the mechanisms of the genotoxic pathways in living tissue.

Table 19. Ethylmorphine N-demethylase activity in liver and intestine of rodents[a][(1005)]

Species	Liver
Guinea pig	38 ± 3, 6[b, c]
Rabbit	60 ± 23, 3[d]
Mouse	101 ± 10, 4[e]
Hamster	111 ± 15, 4[e]
Rat	189 ± 9, 4[e]

[a] Activity is expressed as nanomoles of formaldehyde formed per min per mg of microsomal protein ±SE.
[b] Number of animals or separate tissue pools after comma.
[c] In intestine 8·8 ± 0·3, 4.
[d] In intestine 11·2 ± 1·6, 3.
[e] Activity not detected in intestine.

Knowledge about the microsomal mixed function oxidase system is desirable for this understanding. Lu and Levin[939] have summarized the properties of the liver microsomal hydroxylation system, Table 18.

Table 19 compares the ethylmorphine N-demethylase activity in hepatic and intestinal microsomes from various species. Environmental contaminants come in contact with the small intestine either through ingestion or inhalation (by swallowing) so that intestinal microsomes could play an important role in detoxication–toxication, as indicated in the rat with benzo[a]pyrene.[1122–1125]

The demethylation of ethylmorphine is a typical reaction undergone by heterocyclic N—CH_3 compounds, Fig. 68. Ethylmorphine is an ideal substrate for kinetic studies since it is believed to be metabolized *in vitro* only by dealkylation.[1126] The small amount of acetaldehyde formed by deethylation

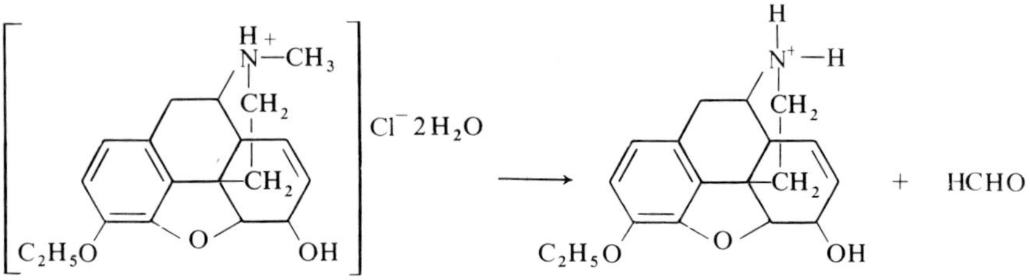

ETHYLMORPHINE HYDROCHLORIDE

$$[\text{ethylmorphine-}NH^+CH_3]\,Cl^-\,2H_2O \longrightarrow \text{norethylmorphine-}NH_2^+ + HCHO$$

FIG. 68. N-Demethylation of ethylmorphine.

does not interfere with the formaldehyde determination by 2,4-pentanedione or chromotropic acid.

SKF-525A or β-diethylaminoethyl-diphenylpropylacetate, XXVI, has been used extensively in drug metabolism studies.

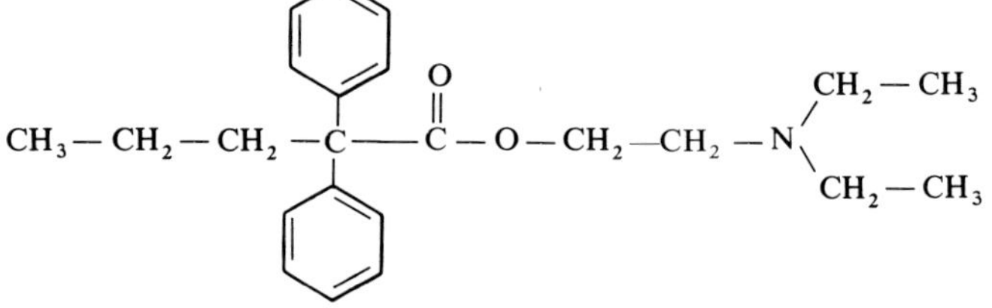

$$CH_3-CH_2-CH_2-C(C_6H_5)_2-\overset{O}{\overset{\|}{C}}-O-CH_2-CH_2-N(CH_2-CH_3)_2$$

XXVI

It inhibits the metabolism of many substrates *in vivo* and *in vitro*, e.g. N- and O-demethylation, side chain oxidation of barbiturates, hydrolysis of procaine and the formation of morphine glucuronide. Some of the reactions which are not inhibited include the N-demethylation of N-methylaniline, O-dealkylation of phenacetin, sulphoxidation of chloropromazine and the reduction of nitroarenes and azo dyes. The kinetics of the inhibition of the N-demethylation of ethylmorphine by SKF-552A and related compounds has been investigated.[736]

There is a close inverse relationship between lipid peroxidation (as shown by malonaldehyde formation) and activities of drug-metabolizing enzymes (as shown by formaldehyde formation) in liver microsomes of rats.[558] The effects of lipid peroxidation on the activity of ethylmorphine N-demethylase have been studied in various animal species.[1014] The activity of ethylmorphine N-demethylase was significantly increased by addition of EDTA and decreased by ferrous ion, with an inverse relation to changes in lipid peroxidation in rat and mouse liver microsomes. No significant effects of EDTA and ferrous ion on activity of ethylmorphine N-demethylase could be found in guinea pig and rabbit liver microsomes. The values of the activity of ethylmorphine N-demethylase per content of cytochrome P-450 was highest in rats, followed by mice, guinea pigs and rabbits, especially when ethylmorphine N-demethylase was assayed in the presence of EDTA.[1014]

We have described a large number of formaldehyde-forming substrates which have been and are being used to study the process of N-demethylation in living tissue. In most cases 2,4-pentanedione has been used for the determination of the derived formaldehyde. Many other methods for the determination of formaldehyde are described in this section and in Volume 1 of the aldehyde series of books.

A recently described tryptophan method shows promise for the determination of formaldehyde and its precursors. It has been applied to the estimation of formaldehyde produced from ethylmorphine metabolism by rat liver.[1006] Essentially, formaldehyde is determined by reaction with tryptophan in the presence of sulphuric acid and iron, nickel, or cobalt. A violet chromogen is obtained with absorption maximum at 575 nm. The mechanism of this type of reaction has been discussed in previous volumes concerned with the determination of aldehydes by indole-type molecules, and especially by 3-substituted indoles.

Tryptophan determination of formaldehyde and its precursors (capable of hydrolysis by acid or oxidation by ferric chloride to formaldehyde).[1006] Mix 1 ml of 0·1% tryptophan in 50% ethanol with 1 ml of the test solution containing 0·4 to 12 μg or 13·3 to 400 nanomoles of formaldehyde. Then carefully add 1 ml of 90% sulfuric acid (sp. gr. 1·81) and mix thoroughly. Add 0·2 ml of 1% $FeCl_3$, mix thoroughly, and incubate in a water bath for 90 min at 70°. After cooling, read the violet color at 575 nm against the blank.

Estimations of formaldehyde in biological mixtures (tissue homogenates) is possible following precipitation of protein with a twofold volume of 1% HCl in absolute ethanol. Following centrifugation of the precipitated protein, 1 ml of the supernatant can be analyzed for formaldehyde by the above procedure.

Sugars, glyoxal, amino acids, acetaldehyde, pyruvic acid, hydroxypyruvic acid, oxalacetic acid, acetic acid and formic acid give negative results in the procedure. The method is ten times as sensitive as the 2,4-pentanedione procedure; at λ_{max} 575 nm it gives an mε of 30. The final colored solution can be stored for several days at room temperature with no decrease in color.

Trichloroacetic or perchloric acid utilized for precipitation of protein slow the color development, shift the wavelength maximum of the chromogen, and in some cases produce cloudy supernatant solutions.

Since the method does use sulfuric acid and ferric chloride, there is the possibility of formaldehyde formation from easily hydrolyzable or oxidizable precursors during the analytical procedure. Thus, the absorbance of blanks of tissue homogenates from liver incubated under O_2 increases with incubation time[1006] because of this phenomenon. Similar difficulties with blank values have been noted for the Hantzsh reaction. However, 10^{-3} M sodium azide obviates these difficulties in the blanks of the tryptophan method and has no effect on demethylation of ethylmorphine.[1006]

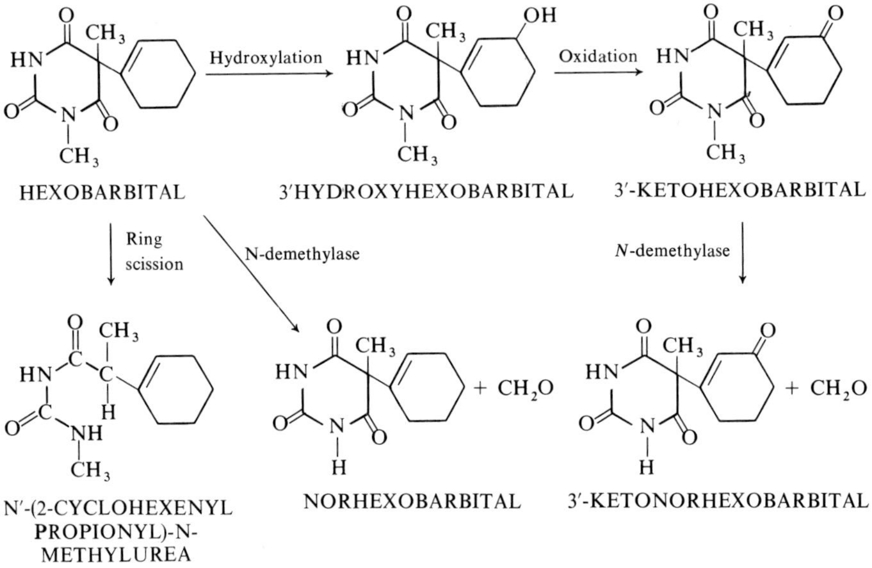

FIG. 69. Metabolism of hexobarbital.[1024]

Tryptophan determination of formaldehyde formed during metabolic studies.[1006] Incubation mixture for drug metabolism contained 21 μmoles of TPN, 90 μmoles of glucose-6-phosphate, 240 μmoles of $MgSO_4$, 250 μmoles of semicarbazide, 100 μmoles of ethylmorphine as the substrate, and an amount of 9000 g liver supernatant equivalent to 1·7 g of liver, all in a total volume of 25 ml q.s. with 0·1 M phosphate buffer (pH 7·35). Incubate the reaction mixture at 37°C under O_2 and take 2 ml samples for analysis at various periods by the tryptophan method for CH_2O.

Other incubation mixtures contained sodium azide, potassium cyanide, or sodium fluoride in a final concentration of 10^{-3} M. Sodium cyanide inhibited the formation of formaldehyde, sodium azide had no effect, while sodium fluoride enhanced the formation of formaldehyde.

The enzymatic metabolism of hexobarbital has been studied.[1024] Some of the metabolic reactions are shown in Fig. 69. The decrease in sleeping time response caused by hexobarbital can be used as an indicator of increased hepatic microsomal metabolism activity. Hexobarbital could be determined through the derived formaldehyde.

In the pig nicotine can be metabolized by enzymatic oxidation or demethylation.[989] Formaldehyde would be a byproduct in these reactions, Fig. 70. The rat can also metabolize the nicotine metabolite, cotinine, to demethylcotinine.[1127]

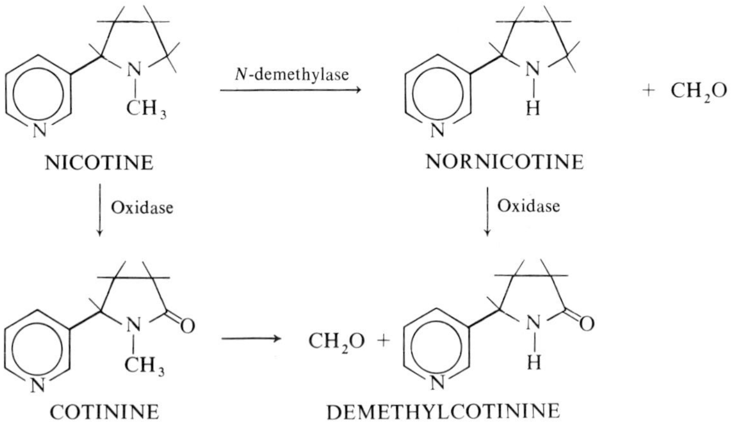

FIG. 70. Metabolism of nicotine.

G. Nitrosamines

Dimethylnitrosamine interacts with cytochrome P-450 as shown by the difference spectra, and the stimulation of lipid peroxidation probably depends upon metabolism of dimethylnitrosamine at this site.[1128] With the stimulation of lipid peroxidation, carcinogenic malonaldehyde precursors and malonaldehyde could be formed from the unsaturated lipids. Dimethylnitrosamine needs metabolic activation for its genotoxic action. Less than 3% of the injected dimethylnitrosamine is found in urine or expired as such.[691] A second possible genotoxic pathway would be through a methyl acyloxynitrosamine, a third through methylformylnitrosamine. A fourth possible genotoxic pathway would lead to oxidative demethylation to an unstable monomethylnitrosamine, formaldehyde[667,753] and carbon dioxide.[737,1129] The unstable monomethylnitrosamine has been postulated as producing[1130] carcinogenic diazomethane,[659] hydrazine, hydroxylamine,[1130] methylamine and mutagenic[1131] nitrous acid (in the presence of water). It would appear evident that a thorough knowledge of all the compounds derived from dimethylnitrosamine metabolism is a necessary step in clarifying the mechanism of action of this genotoxicant.[(1132)] Utilizing an *in vitro* system containing rat liver microsomes and pH 5-cellular soluble enzymes, formaldehyde, formic acid, methylamine, N-methylhydrazine, N-methylhydroxylamine and N,N-dimethylhydrazine are produced.[(1132)]

The metabolites formed at pH 7·0, both in the presence and absence of cofactors, are listed in Table 20;[(1132)] their amount is 3-fold higher than pH 7·4 metabolites. The postulated reactions are shown in Fig. 71. This modified figure taken mainly from Grilli and Prodi[(1132)] is an attempt at depicting the possible genotoxic pathways for dimethylnitrosamine. The role of formaldehyde in some of these pathways is unclear, although formaldehyde could be part of a genotoxic pathway.

pH 5-enzymes are capable by themselves of forming N,N-dimethylhydrazine through reduction of dimethylnitrosamine.[(1132)] The carcinogen, N,N-dimethylhydrazine, can be N-demethylated enzymatically by microsomal enzymes, which are slightly inducible by treatment with 3-methylcholanthrene,

Table 20. Metabolism of ^{14}C-dimethylnitrosamine by subcellular fractions at pH 7·0 *in vitro*[(1132)]

Metabolites	Peak fraction	Normal microsomes and pH 5-enzymes, complete medium	
HCOOH	5	1·6[b]	0·12[c]
CH_2O	8	30·4[b]	2·58[c]
$(CH_3)_2NNO$[a]	17	926·5[b]	992·9[c]
$CH_3NHOH{\cdot}HCl$	23	8·3[b]	1·2[c]
$CH_3NH_2{\cdot}HCl$	27	10·9[b]	0·70[c]
$(CH_3)_2NNH_2 \cdot 2HCl$	31	10·1[b]	1·5[c]
$CH_3NHNH_2 \cdot 2HCl$	35	12·2[b]	1·0[c]

^{14}C-DMNA (2 μCi) was incubated at 37° in air for 80 min in the dark at pH 7·4 or pH 7·0, in a total volume of 2·5 ml; both purified and non-purified DMNA were used. After incubation and removal of enzymic fractions, 1·8-ml samples were dissolved in 1 N HCl and chromatographed immediately on Dowex AG50-WX8 column (52 × 1 cm), developed with 1 ~ 4 N HCl gradient in the dark. The radioactivity detected is given as the % of the DMNA incubated; mean of two experiments. All the radioactivity was always recovered.

[a] Stable under the following conditions: 1 N HCl at 20° in the dark for at least 96 hr ± Tris ± NADPH ± GSH.

[b] Values obtained in the presence of cofactors.

[c] Values obtained in the absence of cofactors.

the reaction rate being 3-fold higher than that reported for oxidative demethylation of dimethylnitrosamine.[1133] Of the isolated metabolites the most important appear to be N-methylhydroxylamine, which can react with cytosine, breaking the pyrimidine ring and removing the base from the background[1134] and is mutagenic in transforming DNA of *B. subtillis*[1135] and *Neurospora*,[1136] and N-methylhydrazine which causes (i) inactivation of transforming DNA and (ii) chromosome breaks.[1134] The methyl, ethyl and dimethyl hydrazines are powerful carcinogens for liver and lung.[563, 852, 1137–1140]

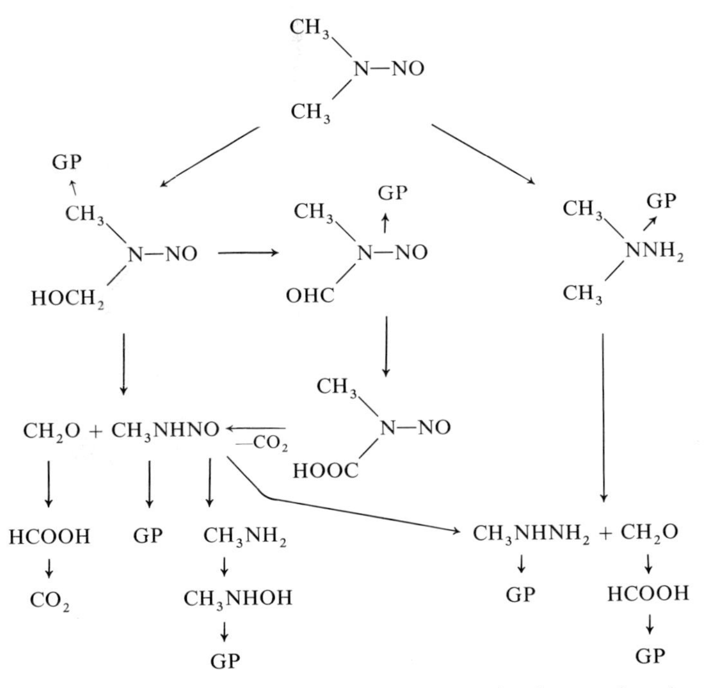

FIG. 71. Metabolism of dimethylnitrosamine. GP = postulated genotoxic pathway.

The degradation compounds obtained from dimethylnitrosamine by physicochemical agents coincide with the metabolic derivatives of this compound.[1141] Both photoirradiation with sunlight or ultraviolet ray and reduction reactions under acid conditions (similar to what would occur in the stomach) lead to the formation of formaldehyde, formic acid, N-methylhydrazine, methylamine, dimethylamine and N-methylhydroxylamine, Table 21.[1141] The metabolites, N-methylhydroxylamine and N-methylhydrazine are genotoxic[1134, 1139] formaldehyde precursors.

Paik and Kim[1142] postulate that methylations by dimethylnitrosamine of various side chains of nuclear proteins[1143] (and probably DNA also) occur by two pathways. One involves the formation of formaldehyde by dealkylation of dimethylnitrosamine and the subsequent incorporation of the formaldehyde into S-adenosyl-L-methionine through tetrahydrofolate, vitamin B_{12}, and homocysteine. The other is the direct methylation by the postulated alkylation intermediate arising from the demethylation of the parent compound.

Table 21. Photolysis of DMNA by sunlight and UV irradiation under neutral conditions[1141]

Compound detected[a]	Eluted fraction (ml)	Sunlight[b]	UV ray[c] (344 nm)
HCOOH	5	4·0	1·6
HCHO	8	4·9	10·5
$(CH_3)_2NNO$	17	969·3	393·9
CH_3NHOH	23	16·0	275·0
CH_3NH_2	28	4·8	294·0
$(CH_3)_2NH$	30	1·0	25·0

^{14}C-DMNA (2 μCi) was irradiated in 3 ml of 0·33 M Tris-HCl buffer (pH 7·0), both for sunlight at 20° for 19 days and for a nearly monochromatic UV ray of 334 nm at 10° ± 0·2° for 14 hr, 1·5-ml samples were chromatographed over a column, and radioactivity was counted. The results are expressed as % of the incubated nitroso compound. All the radioactivity was recovered from the column.

[a] Also identified by R_f values on paper and thin-layer chromatograms.

[b] DMNA decomposition, $1{\cdot}28 \times 10^{-9}$ mol.

[c] DMNA decomposition, $2{\cdot}35 \times 10^{-8}$ mol; quantum yield, <0·01.

Preliminary contact with a subcarcinogenic dose of one carcinogen (such as N-2-fluorenylacetamide) results in an increased susceptibility to cancer on contact with a subcarcinogenic dose of a carcinogen (such as dimethylnitrosamine).[1144] The possibility of increased demethylation (and formaldehyde formation) under such conditions needs investigation. Similarly, a single subcarcinogenic dose of dimethylnitrosamine, which induced a 100% yield of hepatocellular carcinomas in partially hepatectomized rats, induced none in either quiescent or regenerating normal livers.[1144] This result is fairly similar to that reported for 3′-methyl-4-dimethylaminoazobenzene when it is administered some time after an initial subcarcinogenic exposure.[1145] Similar results have also been reported with a 2-carcinogen system consisting of radiation and carbon tetrachloride.[1146]

Study of amino acid induction and carbohydrate repression of dimethylnitrosamine N-demethylase in rat liver has shown that the dietary control of the demethylase is similar to the dietary regulation of a number of other enzymes. [734] Polycyclic hydrocarbons, phenobarbital and the polychlorobiphenyl mixture, Aroclor 1254, are potent inducers of liver tissue proliferation and of azo dye N-demethylase and potent repressors of Me_2NNO demethylase in rats.[712] However, in mice there is also an inducing effect on liver tissue proliferation and a maintenance of azo dye N-demethylase activity but no effect of Aroclor administration on Me_2NNO N-demethylase activity.

Many attempts have been made to correlate the metabolism of a carcinogen with its activity. Thus, the metabolism of dimethylnitrosamine to mutagenic intermediates by kidney microsomal enzymes has been correlated with reported host susceptibility to kidney tumors.[739]

To understand this relationship the enzymatic activities of a wide variety of tissues need to be studied. In this respect the dimethylnitrosamine N-demethylating activities of some tissues have been found to decrease in the order—hamster bronchus, hamster trachea, hamster lung, GRS/A mouse lung, C3Hf/A mouse lung, human lung, Sprague–Dawley rat lung.[1147] The extent of enzymatic activation or binding of activated intermediates to cellular macromolecules does not always correlate with the degree of susceptibility to these carcinogens.[677, 730] From this evidence it can be hypothesized[1147] that cancer is a multi-stage process in which a number of successive factors/events (e.g. metabolic (in)activation, interaction with DNA, repair processes, hormonal environment, antigenic structure of tumour cells, etc.) determine whether a tumour will ultimately develop or not. However, there are many other factors involved in cancer, such as the environmental chemical background of the exposed individual and the accumulated and inherited genetic memories acquired from past contacts with genotoxicants by the individual and his recent ancestors.

Since *in vitro* metabolism of dimethylnitrosamine by hamster lung slices is relatively high compared to mouse lung slices and hamster lung DNA is methylated to an appreciable extent by dimethylnitrosamine,[1147] the resistance of the hamster to dimethylnitrosamine[1148–1152] is remarkable. It is all the more remarkable in that tumours in the hamster respiratory system were observed not only after administration of diethylnitrosamine,[1153] di-n-propylnitrosamine,[1154] dibutylnitrosamine,[1155] and a number of cyclic nitrosamines[1156, 1157] but also after administration of methylnitrosourea[1158] which theoretically can form the same ulticarcinogen as dimethylnitrosamine, but much more readily.

These types of anomalies can be seen in the mutagenesis tests with *Salmonella typhimurium.* In one report it was found that irrespective of the animal susceptibility to 2-acetylaminofluorene carcinogenesis, mutation frequency was always increased in the *Salmonella typhimurium* TA 1538 tester strain.[1159] In fact, the greater response was found in the presence of liver from cotton rats, a species which is resistant to 2-acetylaminofluorene-induced carcinogenesis.[1160] Although mutation plate tests are very important and useful, they are performed in highly artificial conditions which do not take into account any species differences in cell membrane permeability, transport mechanisms, important differences in excretion rates, the mammalian cell's ability to repair damage, or the fidelity inherent in the process.[1159] In addition, it has been shown that dimethylnitrosamine activated by rat or mouse liver enzymes is non-mutagenic in the *E. coli* plate test, even though this compound is highly carcinogenic to rats and mice.[1161]

Contact with subcarcinogenic doses of one carcinogen (e.g. 2-acetylaminofluorene) can lead to increased susceptibility to contact with subcarcinogenic doses of other carcinogens (e.g. the formaldehyde precursor, dimethylnitrosamine).[1162]

Utilizing the 2,4-pentanedione procedure for formaldehyde it is possible to measure dimethylnitrosamine metabolism and to observe the effects of diet and inducers on formaldehyde formation from dimethylnitrosamine in microsomes and post-mitochondrial supernatant fractions.[999]

Estimation of the metabolism of dimethylnitrosamine by measuring the rate of formaldehyde production.[999] Add a 5 ml portion of 10% post mitochondrial fraction of liver homogenate, or an equivalent quantity of microsomal suspension to an incubation mixture to give a final volume of 10 ml containing the following: $MgCl_2$, 75 μmoles; Na isocitrate, 30 μmoles; KH_2PO_4 buffer (adjusted to pH 7·2 with KOH), 500 μmoles; semicarbizide (neutralized to pH 7·2 with NaOH), 40 μmoles; $NADP^+$ 4·5 μmoles. Add isocitrate dehydrogenase (0·1 unit Sigma type 4) when microsomal functions were assayed.

Incubate the mixture at 37° in air in a stoppered 25 ml conical flask, and start the reaction by addition of dimethylnitrosamine, to give a concentration of 1·4 μmoles/ml. Take out 2·5 ml samples of the incubation mixture at zero time, and after 10 and 20 min incubation. Precipitate protein by addition of 1 ml saturated barium hydroxide, and neutralize the sample with 1 ml of a 20% w/v zinc sulphate solution.

Measure formaldehyde in the supernatant by the 2,4-pentanedione method with the following modification. Incubate 2·5 ml of sample with 1 ml of double strength Nash reagent at 60° for 30 min, cool the sample and extract the color into 1·5 ml amyl alcohol, and measure the absorbance at 421 nm. The amyl alcohol extraction removed the high and variable blank values that were due to turbidity from glycogen and other interfering substances.

The liver is the major site of metabolism of drugs, pollutants and other xenobiotics,[1162a, 1163] but other extrahepatic tissues can also biotransform some compounds. For example, the demethylation of *p*-chloro-N-methylaniline by rabbit lung microsomal preparations is 1·5 fold greater than that observed for liver [1164] while a similar demethylation by rat liver is 2·5 fold that of rat kidney.[1165]

Precursors of dimethylnitrosamine and some of the other methylnitrosamines, such as aliphatic amines and NO_x, could be analyzed through the derived formaldehyde. Thus, reactions forming dimethylnitrosamine could be of some value from the analytical viewpoint. Varying concentrations of dimethylamine (DMA) and trimethylamine (TMA) were reacted with sodium nitrite at pH 6·4, 100°C, for 2·5 hr.[1166] In the presence of equimolar concentrations of amines and nitrite, more dimethylnitrosamine (DMNA) was produced from DMA rather than TMA. When the molar ratio of amine to nitrite was increased, the amounts of DMNA from the 2 amines became nearly equal, and, at very high amine to nitrite ratios, more DMNA was formed from TMA than from an equimolar amount of DMA. The optimum pH for the conversion of TMA to DMNA at 100°C was 3·2–3·3. These results indicate that, as the TMA to nitrite ratio increased, a different reaction mechanism became operative. The tertiary amine undergoes nitrosative dealkylation to the corresponding secondary amine which reacts with nitrite to form the N-nitrosamine. It is suggested that the catalytic effect of formaldehyde produced by nitrosative dealkylation of TMA explains the increased yields of DMNA at the higher TMA levels used in this study. Because relatively large amounts of TMA can form in refrigerated seafood, the dietary intake of TMA may have to be reevaluated if *in vivo* formation of DMNA from TMA does occur.

The weak carcinogen, N-nitrososarcosine, is decarboxylated at room temperatures in solutions of pH > 9 (and probably even in the mildly alkaline intestinal region) to dimethylnitrosamine.[610, 1167–1170] The induction of lung adenomas in mice treated with N-methylaniline and nitrite is due to *in vivo* nitrosation (presumably in the stomach) with the formation of carcinogenic N-nitroso N-methylaniline[637] which in the body is probably metabolized to the ulticarcinogen, benzenediazonium cation, and formaldehyde. In the study of oxidative demethylation of dimethylnitrosamine with both reconstituted and unresolved liver microsomal cytochrome P-450 enzyme systems from rats and hamsters it was found that both cytochrome P-450 and NADPH–cytochrome c reductase fractions were required for the optimum formation of formaldehyde.[1171]

Dimethylnitrosamine can be formed from the reaction of the analgesic amidopyrine and nitrite extracted from foods.[1172] Treatment of laboratory animals with amidopyrine concurrently with nitrite results in liver damage and in tumours of the liver and lung typical for dimethylnitrosamine.[651, 1173]

Other compounds which react with nitrite at pH 5·6 to form dimethylnitrosamine include neurine, carnitine, betaine, choline, acetylcholine, 2-dimethylaminoethyl acetate and dimethylglycine.[1174]

Appropriate secondary and tertiary amines could be determined through their metabolically-derived formaldehyde. Alternatively the nitrosamine could be determined by gas chromatography.[1175]

H. Hydroxylamines

Methylhydroxylamines can undergo a series of reactions wherein formaldehyde is one of the products.[1176] Formaldehyde is produced at a rate 0·25 times the rate of NADPH oxidation. Formaldehyde in the reaction medium was measured by the 2,4-pentanedione method after complete extraction of methylbenzylhydroxylamine with ether, since N-methylhydroxylamines interfere with formaldehyde estimations. The postulated sequence of reactions is shown in Fig. 72. The use of secondary amine N-demethylation to measure cytochrome P-450-dependent activity may be misleading since part of the formaldehyde could be formed by the reactions

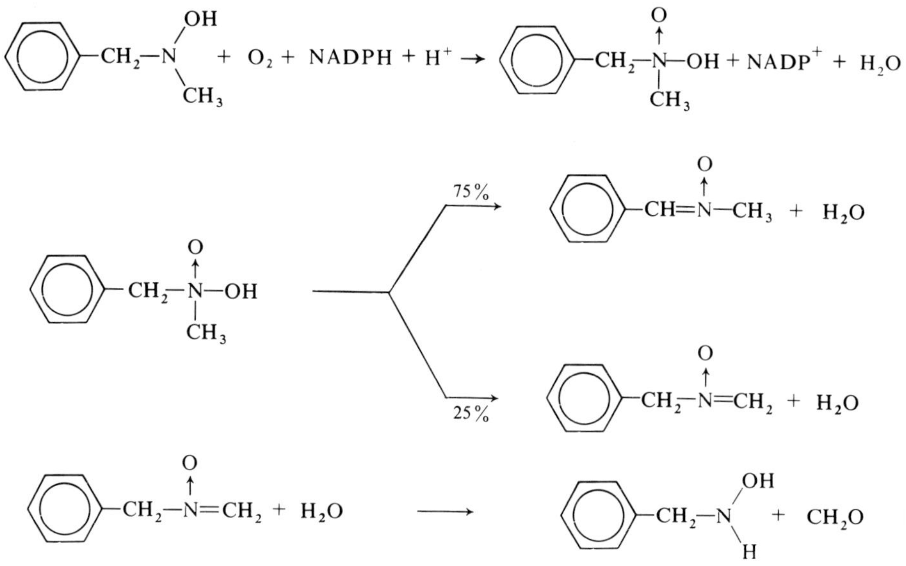

$$C_6H_5\text{—}CH_2\text{—}N(OH)\text{—}CH_3 + O_2 + NADPH + H^+ \rightarrow C_6H_5\text{—}CH_2\text{—}N(\rightarrow O)(OH)\text{—}CH_3 + NADP^+ + H_2O$$

$$C_6H_5\text{—}CH_2\text{—}N(\rightarrow O)(OH)\text{—}CH_3 \xrightarrow{75\%} C_6H_5\text{—}CH{=}N(\rightarrow O)\text{—}CH_3 + H_2O$$

$$C_6H_5\text{—}CH_2\text{—}N(\rightarrow O)(OH)\text{—}CH_3 \xrightarrow{25\%} C_6H_5\text{—}CH_2\text{—}N(\rightarrow O){=}CH_2 + H_2O$$

$$C_6H_5\text{—}CH_2\text{—}N(\rightarrow O){=}CH_2 + H_2O \longrightarrow C_6H_5\text{—}CH_2\text{—}N(OH)\text{—}H + CH_2O$$

FIG. 72. Proposed sequence of reactions taking place during the oxidation of N,N-disubstituted hydroxylamines catalyzed by an amine oxidase. Only reaction 1 is enzymic. The ratio of α-phenyl-N-methylnitrone to N-benzylnitrone formation (reactions 2 and 3) is based on the rates of α-phenyl-N-methylnitrone and formaldehyde formation relative to NADPH oxidized and O_2 reduced.[1176]

shown in Fig. 72.[1176] This could be true for N-methylarylamines which upon di-N-oxidation can dehydrate only to the arylnitrone which quantitatively yields formaldehyde upon hydrolysis.

LIV. O-METHYL COMPOUNDS

Methoxy compounds can be oxidatively cleaved by powerful oxidizing agents or through enzymatic methods. An example of the first type is the oxidative splitting of compounds containing OCH_3 or NCH_3 groups by benzoyl peroxide,[1177] e.g.

$$ArOCH_3 + (C_6H_5CO)_2O_2 \longrightarrow ArOH + (C_6H_5CO)_2O + CH_2O$$

A large number of enzymatic methods are known.[1178] Essentially the alkoxy compound is cleaved during metabolism to an alcohol or a phenol and an aldehyde, probably by initial hydroxylation at the α-carbon to give an unstable hemiacetal which can then undergo nonenzymatic cleavage. The initially formed alcohol or phenol and aldehyde are directly excreted or more commonly, undergo further oxidative, reductive and conjugative transformations before eventual elimination (usually). Microsomal O-demethylation, N-demethylation, and aromatic hydroxylation have been studied in the presence of bisulphite and dithiothreitol.[1179]

Another group of formaldehyde precursors are the aflatoxins. A review on these compounds is available.[1180] The carcinogenesis of these compounds has also been reviewed.[1181] The aflatoxin problem has broad geographical dimensions. It has been identified in peanut meals and can be produced by toxin-producing fungal strains whenever conditions permit their growth on practically any natural substrate. Aflatoxins have been shown to have carcinogenic activity in the rat, mouse, rainbow trout and duck.[1181] In the underdeveloped countries there is an extensive contamination of market foods and foodstuffs by small amounts of the aflatoxins. Results of a Thailand study suggest an association between aflatoxin consumption and the incidence of primary liver cancer.[1182] Dietary aflatoxin levels of 0·1 ppm and higher induce liver carcinoma in rats at an incidence greater than 50%, when feeding is continued up to 80 weeks.[1181] Rats injected with 2 μg of aflatoxins B_1 and G_1 for 44 weeks had a local tumor incidence of 5 out of 6.[1183] The carcinogenic activity of the aflatoxins appear to be in the order $B_1 > G_1 \gg B_2 > G_2$.[1181] The primary target organ for aflatoxin B_1 in most rat strains is the liver, in which it induces hepatocellular carcinomas, cholangiosarcomas and other lesions.[1184, 1185]

The metabolism of the aflatoxins can lead to various aldehydes. Two of the pathways in the metabolism of aflatoxin B_1 will be briefly discussed. This compound can be oxidized to the 2,3-oxide, XXVII; evidence for the formation of this compound in rat liver *in vivo* and by human liver microsomes *in vitro* has been presented.[1186] Being an acetal this compound could probably readily form the aldehyde by hydrolysis. Another compound which could be formed in this pathway or through the acid-catalyzed addition of water to B_1 is aflatoxin B_{2a}, the hemiacetal of B_1, XXVIII.

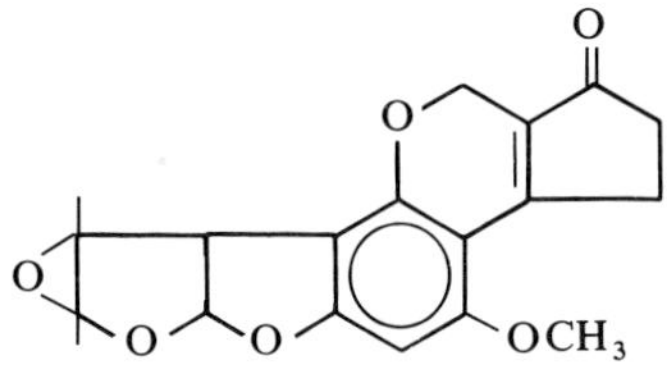

XXVII

This compound has been isolated from liquid cultures of aflatoxin-producing fungi.[1187, 1188] Its formation from B_1 has been applied as a confirmatory test for aflatoxin B_1.[1189] Since this hemiacetal and XXVII could be hydrolyzed to aldehydes, aflatoxin B_1, XXVII and XXVIII could be determined with MBTH.

These various products could also be demethylated enzymatically to give the phenol and formaldehyde.

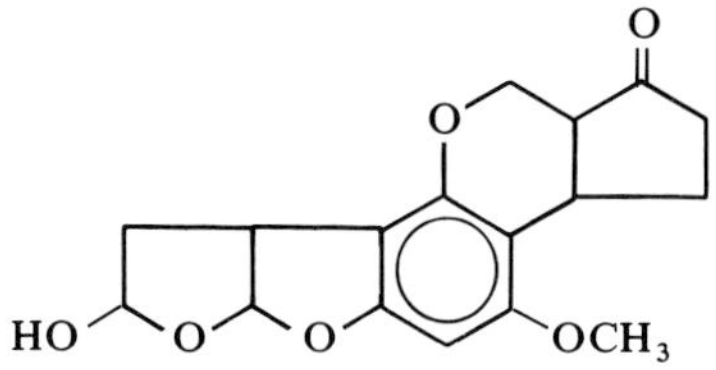

XXVIII

Another possible metabolic pathway involves demethylation of aflatoxin B_1, Fig. 73. Aflatoxin P_1, the product of O-demethylation of B_1, was recently isolated and identified as a major metabolic product from the urine of rhesus monkeys dosed with B_1.[1190] The oxidative metabolism of B_1 by mammalian liver slices and microsomes has been studied.[1191] The derived formaldehyde was estimated by the method of Cochin and Axelrod[976] as modified by Stitzel *et al.*[166] using double strength Nash reagent.[556]

The palmotoxins are of unknown structure but are believed to be related to the aflatoxins.[1192] Like the aflatoxins, they induce liver lesions.[1193, 1194] They are metabolized *in vitro* like the aflatoxins, undergoing demethylation and a possible hydroxylation.[1192] Incubation of Palmotoxins B_0 and G_0 with male rat liver microsomal fractions gave optimum amounts of formaldehyde at a pH between 7·5 and 7·6. Phenobarbitone was found to induce the palmotoxins B_0 and G_0 as well as the aflatoxin B_1-metabolizing enzymes. Carbon monoxide aeration inhibited the activity of these enzymes.

The microsomal dealkylation of anisoles has also been studied. The rate of demethylation for a series of ring-substituted anisoles is greater for para than for meta substitution.[1195] The rate of dealkylation of ArOR decreases as the size of the alkyl group increases through the series ethyl > n-propyl > isopropyl > n-butyl > n-hexyl.[1040] Compounds with chains larger than propyl are often metabolized by side-chain oxidation, rather than dealkylation.[1196]

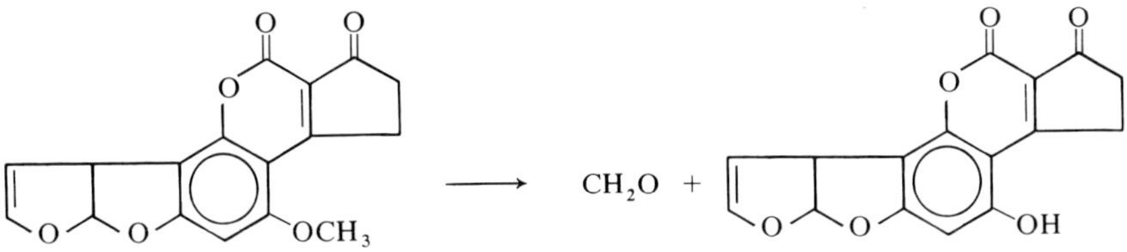

CH_2O +

FIG. 73. O-Demethylation of aflatoxin B_1.

Of the various anisoles *p*-nitroanisole has been one of the most extensively studied.[1195, 1197] In living tissue it is metabolized to *p*-nitrophenol and formaldehyde, Fig. 74. The products primarily excreted in the urine are the corresponding glucuronide and sulphate.[1198] The dealkylation (formaldehyde formation) of *p*-nitroanisole was enhanced after treatment with polynuclear aromatic hydrocarbons or phenobarbital,[1199] and was reduced in the Wistar rat during pregnancy.[946]

When genetically "responsive" inbred mouse strains are exposed to aromatic hydrocarbons, such as 3-methylcholanthrene, or to β-naphthoflavone, induction of *p*-nitroanisole O-demethylase, aryl hydrocarbon hydroxylase, 7-ethoxycoumarin O-deethylase, and 3-methyl-4-methylaminoazobenzene N-demethylase activities is associated with the same genetic region,[1200] designated the Ah locus[1201, 1202] for aromatic hydrocarbon "responsiveness".[963] These chemicals fail to induce these changes in genetically "nonresponsive" inbred strains even when administered chronically at high doses. However, 2,3,7,8-tetrachlorodibenzo-*p*-dioxin administration to so-called genetically "nonresponsive" mice stimulates the O-demethylase, N-demethyl-

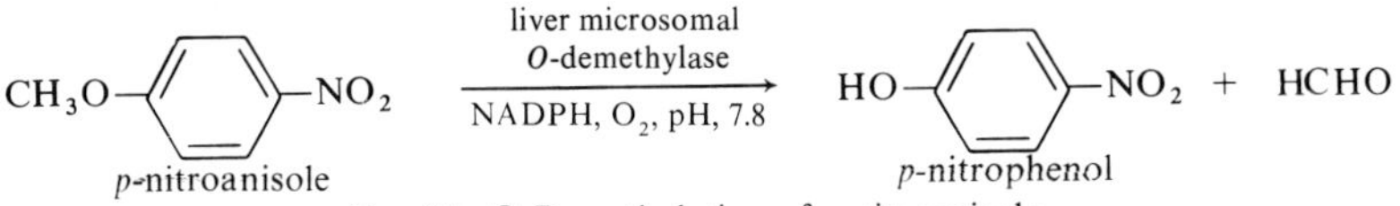

CH_3O–C₆H₄–NO_2 (*p*-nitroanisole) → [liver microsomal *O*-demethylase; NADPH, O_2, pH, 7.8] HO–C₆H₄–NO_2 (*p*-nitrophenol) + HCHO

FIG. 74. O-Demethylation of *p*-nitroanisole.

ase, O-deethylase and the hydroxylase activities and the new formation of cytochrome P_1-450 to the same extent as that found in genetically "responsive" mice.[963]

O-Dealkylation has been studied for compounds as dissimilar as Dicamba[1203] and griseofulvin.[971] Griseofulvin is an orally effective antifungal agent and is used widely in the treatment of fungal infections.[1204] It is effective against species of Microsporon, Epidermophyton and Trichophyton. Griseofulvin is also a carcinogen. Ingestion of very large amounts of griseofulvin can result in enlarged liver with the ultimate formation of multiple hepatoma in mice but not in rats.[974, 975] The metabolism of griseofulvin has been studied in animals and man *in vivo*,[1205, 1206] and in rat tissue slices *in vitro*.[1207, 1208] In liver microsomes of rats and mice griseofulvin is converted enzymatically to 4- and 6-desmethylgriseofulvin.[971] The O-demethylation of ^{14}C-griseofulvin by liver microsomes has been studied in rats and mice treated with phenobarbital, 3-methylcholanthrene, and griseofulvin. Phenobarbital treatment increased 4- and 6-demethylation of ^{14}C-griseofulvin in rats and mice. Treatment with 3-methylcholanthrene increased both 4- and 6-demethylation in mice but only 6-demethylation in rats. Prolonged intake of griseofulvin increased 6-demethylation but not 4-demethylation in rats and mice. Based on this differential stimulation of 4- and 6-demethylation by phenobarbital, 3-methylcholanthrene, and griseofulvin in rats and mice, it is suggested that 4-demethylation and 6-demethylation of griseofulvin in liver microsomes of rats and mice are probably catalyzed by two different enzymes.[971]

Methoxy compounds can also be oxidized to formaldehyde as shown for 4-methoxybenzoate.[374] The enzyme catalyzing this reaction is 4-methoxybenzoate monooxygenase (O-demethylating) [EC 1.14.99.15, 4-methoxybenzoate, hydrogen donor: oxygen oxidoreductase (O-demethylating)]

$$\text{4-Methoxybenzoate} + AH_2 + O_2 = \text{4-Hydroxybenzoate} + CH_2O + A + H_2O$$

This enzyme system also acts on 4-ethoxybenzoate (acetaldehyde formation) and on N-methyl-4-amino-benzoate and -toluate.

Methoxychlor can also be dealkylated to formaldehyde,[1209] as can 8-methoxykynurenic acid,[1210] Fig. 75.

An example of another enzymatic method is the one described for 8-methoxykynurenic acid which can be used to analyze for this compound and for the demethylating enzyme system. In the catabolism of tryptophan the indole ring is opened up by tryptophan 2,3-dioxygenase (L-tryptophan: oxygen 2,3-oxidoreductase (decyclizing), EC 1.13.11.11) to form L-formylkynurenine. Further reactions form metabolites, such as kynurenine, kynurenic acid, xanthurenic acid, 3-hydroxykynurenine, 3-hydroxyanthranilic acid, quinolinic acid, nicotinic acid, 2,6-dioxo-3-carboxyhexanoic acid, and 8-methoxykynurenic acid. The latter compound is a urinary metabolite of tryptophan in humans,[1211] swine[1212] and monkeys[1213] and is carcinogenic to the mouse lymphoreticular system[1214] and to the mouse bladder when implanted in that organ in cholesterol pellets.[1215, 1216]

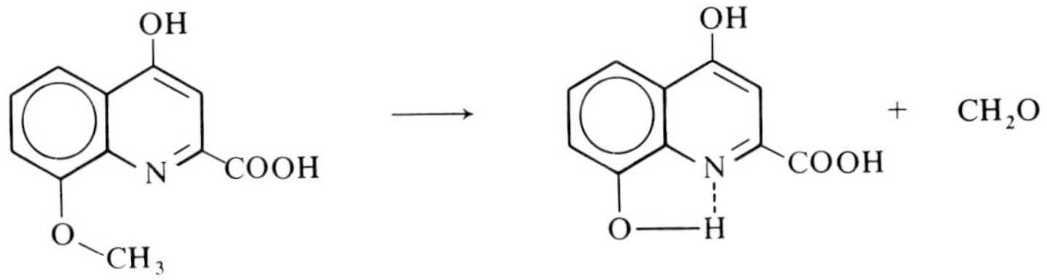

FIG. 75. O-Demethylation of 8-methoxykynurenic acid to xanthurenic acid and formaldehyde.

Enzymatic demethylation of this carcinogen was studied in male Donryu strain rats.[1210] An O-demethylating system was detected in the kidneys and according to the authors an O-demethylating enzyme system has not been detected previously in the kidney of any species. The demethylation of 8-methoxykynurenic acid appears to be regulated in part by the concentration ratio of intracellular pyridine nucleotides. Demethylase activity was determined in the liver and the kidney by estimating the amount of xanthurenic acid and formaldehyde formed. About 1 mole of formaldehyde was generated for each mole of xanthurenic acid formed. Formaldehyde was assayed with 2,4-pentanedione.[976] Ohira *et al.*[1210] suggest that the demethylation of 8-methoxykynurenic acid may be catalyzed by a different enzyme system from the sterol-demethylating system.[1217, 1218]

L-Dopa has proved useful in the treatment of patients with Parkinson's disease. It is believed L-dopa is converted to dopamine in the basal ganglia of patients receiving this drug. In the rat brain 3-O-methyldopa, XXIX, is demethylated to produce formaldehyde and dopa.[1219] It is postulated that 3-O-methyldopa accumulates in the striatum during L-dopa therapy and serves as a depot source for the formation of dopamine.[1220]

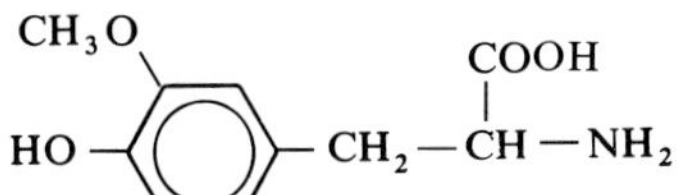

XXIX

Another reaction that must also involve formaldehyde formation is the microbial 3-O-demethylation of estratrienes.[1221] Estradiol-3-methylether, estrone-3-methylether, 17α-ethinyl-estradiol-3-methyl-ether and other 17α-substituted estratrienes were 3-O-demethylated to free 3-hydroxy compounds by fermentation with Corynebacterium sp. A hydroxy group in position 6α or 6β prevented the reaction. The opposite reaction, methylation of the 3-hydroxy group of estratrienes, was performed using *Mycobacterium smegmatis*.

The formaldehyde formation during these demethylations could prove useful in the analytical study of these reactions.

LV. S-METHYL COMPOUNDS

The formation of formaldehyde from methylthio compounds is a rarer phenomenon, one reason being the easier oxidation of the sulfur group.

Methionine reacted with ATP forms S-adenosylmethionine, XXX, a compound which contains an active methyl group. Formaldehyde can be readily formed in human blood from the activated methyl group of XXX.[1222, 1223]

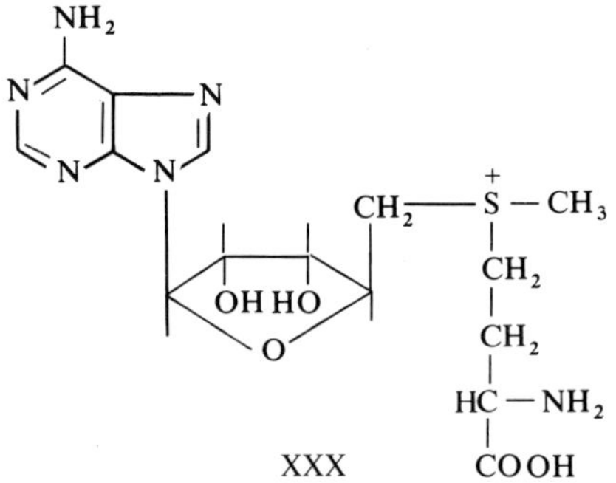

XXX

Some thioethers are oxidatively cleaved by mammalian liver enzymes, e.g. 6-methylthiopurine, XXXI,[991] and prometryne.[1224] Formaldehyde is the byproduct. Formaldehyde formation is enhanced by treatment of the microsomal enzyme system with polynuclear aromatic hydrocarbons.[1225]

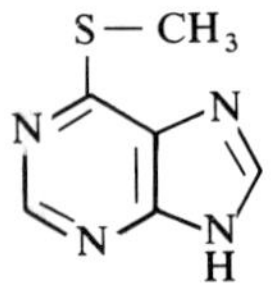

XXXI

LVI. METHYLENE, HYDROLYZABLE

A. Introduction

1. *Compound types.* The compounds belonging to this family contain the basic structure $—X—CH_2—Y—$, where X = Y = O, S, NR, SO_2, Table 4. A large number of these types of formaldehyde precursors are listed in Table 1. The hydrolysis, physical and physiological properties and analysis of these compounds through the derived formaldehyde will be discussed.

The simplest methylene compound which can be a formaldehyde precursor is methane. This compound has a residence time of about 4 to 7 years in the atmosphere as compared to about 16 years for Freon and 0·003 and 1 year for the carcinogens, methyl iodide and carbon tetrachloride, respectively. Since formaldehyde has a tritium content like that of atmospheric methane, it is postulated that the undiscovered tropospheric sink of methane is incomplete oxidation to formaldehyde.[1226]

2. *Formaldehyde standards.* Some of these methylene compounds can be purified readily and hydrolyzed quantitatively to formaldehyde and consequently can be used as a source of standard formaldehyde. Some of the materials which have been used at various times as sources of standard concentrations of formaldehyde include aqueous solutions of formaldehyde and liquid and solid polymers of formaldehyde.

The concentration of an aqueous formaldehyde solution can be determined with the help of dimedone through gravimetric analysis.[1227] See the Formaldehyde Section in Volume 1. Diethoxymethane, b.pt. 86–87°, could be purified through gas chromatography or through distillation. It could then be used as a source of standard formaldehyde, since it is readily hydrolyzed to ethanol and formaldehyde. Similarly, dimethoxymethane, b.pt. 41–42°, could be purified and used as a source of standard concentrations of formaldehyde. Paraformaldehyde, 1-hydroxymethyl-5,5-dimethylhydantoin, XXXII, and hexamethylenetetramine, XXXIII,[915] have been used to prepare standard solutions of formaldehyde.

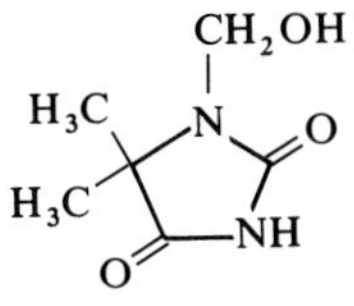

XXXII

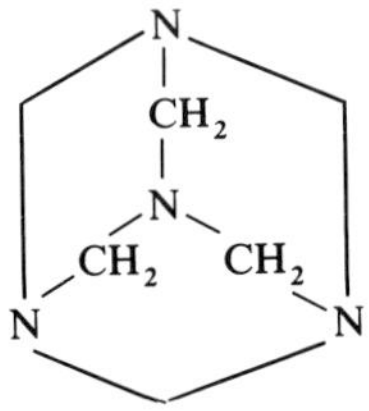

XXXIII

The most difficult problem is to prepare formaldehyde gas streams of known concentrations. An attractive and useful development was the preparation of formaldehyde permeation tubes.[1228] Attempts to utilize paraformaldehyde for this purpose were unsuccessful for the following reasons. The high amount of water (2 to 4% after desiccation) caused the polymerization of formaldehyde in the gas stream. Since the structure of paraformaldehyde is characterized by low molecular weight oligomers, the crystalline state of the molecule has amorphous random properties. The partial pressure of (i.e. the formation of formaldehyde from) paraformaldehyde and other polymeric forms of formaldehyde is due to "unzippering" depolymerization reactions at the hydroxyl end group. Thus, the rate of thermal decomposition to formaldehyde is dependent on the number and availability of end groups. Unfortunately, paraformaldehyde is characterized by widely varying chain lengths and ill-defined structures so that the attainment of equilibrium partial pressure is extremely difficult.

It was found that alpha-polyoxymethylene is a polymer which is characterized by uniform chain lengths, a high degree of polymerization, a well-defined crystalline structure, and a low water content.[421] The pure polymer is believed to have an average chain length of 200 and has a formaldehyde content of 99·7 to 99·9%. It was prepared in the following manner.

Preparation of pure alpha-polyoxymethylene.[421] Add potassium hydroxide pellets slowly to 40% methanol-free, aqueous formaldehyde solution until a molar ratio of 1:500 is obtained for $KOH:CH_2O$. Filter the precipitate by vacuum, wash with ethanol and ether, and store in a desiccator. Repeat the operation twice at $KOH:CH_2O$ molar ratios of 1:100 and 1:20.

The equilibrium pressures in the range 80 to 100° follow the equation:[1228]

$$\text{Log}\, P_e = 12{\cdot}02 - \frac{3{\cdot}75 \times 10^3}{T^\circ}$$

where P_e the partial pressure of formaldehyde and T° is temperature.

Preparation and use of the formaldehyde permeation tubes.[421] Construct permeation tubes with $\frac{1}{4}$ in. commercial Teflon tubing with sample chambers exactly 15 cm long and with a wall thickness of 0·03 inches. Fill with 1·2 g of alpha-polyoxymethylene. Close the ends with Teflon plugs and clamp shut. Conduct tests at temperatures of 100° and 130°. Use nitrogen as the gas stream at a flow rate of 1·0 SCFH. Calculate permeation rates via weight loss measurements every 24 hr at 100° and 130°. A gas stream containing 97 ppm CH_2O was obtained at 130° with a variance of ±2%. A 10 cm length of $\frac{1}{8}$ in. tubing at 100° yielded a gas stream containing 1·0 ppm CH_2O.

Marcote *et al.*[421] found that the variation in the permeation rate between different tubes was $\pm 7\%$ with respect to the arithmetical mean. This deviation is attributed to the physical properties of the tubing since the vapor pressure of alpha-polyoxymethylene is well characterized in this temperature region.

The permeation rate stability was determined individually for three tubes. The rates were found to be constant within $\pm 2\%$ for each of the tested tubes at both 100° and 130°C. This variation was decreased to $\pm 0{\cdot}5\%$ by placing a stainless steel coil in front of each of the tubes to minimize the effect of the carrier gas temperature. During this test it was determined that each permeation tube has a life expectancy of 6–8 months at 100°C.

By varying the dry nitrogen flow rate from 0·5 to 5 SCFH it was demonstrated that the permeation rate in this flow range is independent of the rate of removal of formaldehyde from the surface of the permeation tube. The formaldehyde permeation rate versus temperature was linear in all cases.

3. *Physiological activity*. Formaldehyde and other aldehydes are byproducts formed during the metabolism of most, if not all, genotoxicants, e.g. carcinogens, mutagens, etc. Explanations of the roles of the intracellularly released aldehydes in the genotoxic process have been completely unsatisfactory. It would appear that there is much more in this process than meets the eye. Aldehydes are very reactive molecules. One would expect that if they were released at a genetic "hot spot", they could cause a potent genotoxic effect. This certainly needs a much more thorough study. Let us summarize a few aspects of this problem, that indicate the need for such studies.

One indication of the potency of formaldehyde is its irritant and cogenotoxic effects on the mucociliary system. Ciliary activity ceased in the anesthetized tracheotomized rat on exposure to 0·5 ppm CH_2O for 150 sec or to 3 ppm for 50 sec.[1229] This would make the exposed cells much more susceptible to genotoxic chemicals.

What role formaldehyde formation plays in the carcinogenicity of safrole has never been explored. Of some interest is the presence of 2 formaldehyde precursor groups in the safrole molecule, e.g. the $-O-CH_2-O-$ and $C{=}CH_2$ groups.

What would be of interest is to utilize formaldehyde precursors in genotoxic studies which would release the formaldehyde in appropriate "hot spot" target areas. Of interest in this sense is the report that the formaldehyde precursor, hexamethylenetetramine causes sarcomas in rats following subcutaneous injection.[1230] It is postulated that the biological activity of $CH_2(OSO_2CH_3)_2$ is due to the specific intracellular release of formaldehyde.[1231] Cytogenetic studies in rat cells of the mutagenic activity of formaldehyde-containing resins have been followed through the chromosome aberrations at the ana-telophase stage. All the studied substances caused an increase in the number of chromosome aberrations and a cytostatic effect in animal bone marrow cells independent of the material fixation times.[1232]

Many theories of formaldehyde mutagenesis have been advanced.[527, 1233–1235] The possible mechanisms underlying the lethal and mutagenic actions of formaldehyde have been discussed.[1236]

One other toxic aspect of the highly reactive chemicals, such as formaldehyde, other aldehydes and alkylating agents, which should be mentioned is their acute toxic effects. These compounds, if volatile, can not only be genotoxic, but can also be lachrymators. We have discussed this previously but it needs further emphasis. Formaldehyde precursors, such as the styrenes, are more potent than the non-aromatic hydrocarbons in producing eye irritation following photooxidation.[867] Eye irritation in human beings in contact with photochemical smog is believed to be related to the total aldehyde concentration in polluted atmospheres. This direct relationship seemed to hold for formaldehyde at concentrations ranging from 0·01 to 0·1 ppm.[1237] The relationship was non-linear for acrolein, for which concentrations of 0·003 to 0·015 ppm resulted in more eye irritation than concentrations greater than 0·015 ppm. Synthetic atmospheres containing ethylene, propylene, isobutane, gasoline mixtures and auto exhaust have been irradiated. A high correlation was found between the eye irritation and the formaldehyde content of the irradiated mixture, which ranged from 0·1 to 4 ppm CH_2O.[1238] Other workers have confirmed that substantial eye irritation can be obtained from irradiated synthetic atmospheres containing NO_2, various hydrocarbons and the resulting formaldehyde.[1239] It was suggested that concentrations of formaldehyde as low as 12 μg/m^3 (0·01 ppm) could cause eye irritation.

4. *Atmospheric formaldehyde.* The problem with atmospheric pollutants is that everybody breathes them. So if genotoxicants are in a local atmosphere, every person in that high risk area is in a danger we do not understand and, as yet, do not know how to grapple with.

It is possible that in determining atmospheric formaldehyde, formaldehyde precursors hydrolyzable under analytical conditions are also determined. This is probably especially true near some chemical industries which pollute the air with a large variety of chemicals and in air polluted by auto traffic or Diesel exhaust[1240] and is most likely a problem in the chromotropic acid test for atmospheric "formaldehyde".[1241–1243]

A somewhat similar reagent with similar capabilities of determining formaldehyde precursors is 2-naphthol-6-sulphonic acid.[1244] The method permits the analysis of 0·2 μg of formaldehyde in 5 ml of solution. To detect the maximum allowable concentration of formaldehyde, 0·4–0·8 liters of air must be sampled.

Atmospheric formaldehyde has also been determined with pararosaniline[1245] or fuchsin.[1246]

2,4-Pentanedione or acetylacetone is another reagent which has been used to determine atmospheric formaldehyde under fairly mild conditions.[1247–1249] A continuous measuring instrument was used to determine atmospheric formaldehyde in Tokyo. Atmospheric formaldehyde is generated by incomplete combustion of automobile fuel, oxidation decomposition of lubricating oil, and photochemical reactions of pollutants. The pollutant stimulates the mucous membrane and reduces the field of vision. For formaldehyde measurements, a colorimetric apparatus that uses acetylacetone as the absorber and color former has been devised.[1247] Successive measurements have been performed with this apparatus since January 1968. The reagent is made by dissolving 150 g of ammonium acetate in water and adding 3 ml of glacial acetic acid and 2 ml of acetylacetone to the solution. After dilution with water to 2 liters, 0·2 g sodium chloride is then added to prevent the influence of sulfur dioxide. The absorbtion rate is greater than 95%. The stability of the solution is sufficiently high. On Iwaida Street in Kasumigasiki, Tokyo, as determined by the colorimetric apparatus, the traffic volume is about 40 000 cars a day. The minimum one hour formaldehyde value is 0·05 pphm and the maximum, 4·8 pphm. The minimum mean value in a day is 0·07 pphm and the maximum, 1·96 pphm. The minimum monthly mean value is 0·45 pphm and the maximum, 0·90 pphm. The yearly mean value is 0·76 pphm. Since formaldehyde is considered to be generated also by photochemical reactions, seasonal differences in its density are presumed to exist. However, the data obtained do not clearly indicate seasonal variations.

In another method[1248] measurements were made every hour with an automatic, continuous analyzer having a single-path colorimeter. The air sample was bubbled into a gas-absorbing bubbler containing 50 ml of acetylacetone for 54·5 min at a constant rate of 2 liter/min. The absorbance of formaldehyde-containing acetylacetone was measured at 420 nm. The concentration of formaldehyde was then plotted by a recorder, which was graduated from 0–20 pphm. This analyzer has been in use since 1967 at the continuous air monitoring station in the Kasumigasiki section of Tokyo.

Phenylhydrazine has also been used.[1250] The photometric determination of formaldehyde in air, based on its reaction with phenylhydrazine hydrochloride in the presence of potassium ferricyanide in alkaline medium, was found to have the sensitivity of 0·2 mg in the volume analyzed. Ammonia and phenol do not interfere with the determination, provided the concentration of the latter did not exceed that of formaldehyde by more than five times.

Atmospheric sulphur dioxide has been found to cause a negative interference in the MBTH determination of atmospheric formaldehyde and other aldehydes.[1251] Methods of measuring formaldehyde at concentrations in the ambient air were tested. The spectrophotometric method using 3-methyl-2-benzothiazolinone hydrazone was most sensitive and is very easy to use but sulfur dioxide in the air caused a negative interference with this method. Nitrogen dioxide gave no interference. To eliminate the SO_2 interference, it was necessary to remove the SO_2 without removing the formaldehyde before bubbling the sample air into the absorbing reagent. A $MnSO_4$ filter was most useful to eliminate SO_2 interference. The filter was prepared by dropping 10 ml of the aqueous solution of $MnSO_4$ at a concentration of 100 mg/ml over a 250 sq cm glass fiber filter. Then the filter was dried, cut to small pieces, and packed in a glass tube. This filter could thoroughly remove 0·35 ppm SO_2 in the air. The effective lifetime of the filter varied with the humidity of the air. At humidity above 88%, the removal efficiency over 95% was maintained for 3000 min for passing 0·35 ppm SO_2 at the flow rate of 1 liter/min. At low relative humidity of 15 to approximately 35%, the efficiency decreased gradually with the passage of SO_2. In ambient air with less SO_2 and humidity of 25 to approximately 75%, this filter was effective enough to remove SO_2. If the relative humidity is very low, it was desirable to use a new filter.

Investigations on the presence of formaldehyde in auto exhaust have been reported.[1252–1255] Here again formaldehyde precursors are probably measured with the formaldehyde.

5. *Methods.* We will briefly discuss some of the procedures which have been used for the determination of formaldehyde (discussed more fully in Volume 1). In some mixtures they also determine formaldehyde precursors since under the analytical conditions these precursors are hydrolyzed to formaldehyde. In addition these methods could also be used to determine formaldehyde precursors by appropriate changes in the analytical procedure. Thus, for example, the drug, dipyrone, releases formaldehyde under the analytical conditions of the chromotropic acid test and can be determined in the following manner.

Colorimetric determination of dipyrone (analginium or (antipyrinylmethyl-amino) methanesulphonic acid) in pharmaceutical preparations.[1256] To 1 ml of aqueous test solution add 1 ml of 2·5% chromotropic acid solution and 10 ml of concentrated H_2SO_4. Dilute to 25 ml and read the absorbance at 570 nm.

This type of procedure is an indication that, although formaldehyde can be determined in the atmosphere with chromotropic acid,[1257] there is a possibility that some formaldehyde precursors are being included in the determination.

Formaldehyde has also been determined in the presence of acrolein with the help of chromotropic acid by the following procedure.

Colorimetric determination of formaldehyde in the presence of acrolein.[1258] To the soln (5 ml) containing 0·025 to 0·05 mg of formaldehyde and up to a 5-fold excess of acrylaldehyde add 0·5 ml of chromotropic acid soln (0·9 g dissolved in 25 ml of H_2O; 50 mg of $SnCl_2 \cdot 2H_2O$ is added, and the mixture is filtered after a few hours) and 11·5 ml of H_2O; cool the soln in ice. Very slowly add 10 ml of H_2SO_4, then cool the soln in ice and dilute it to 50 ml with H_2SO_4. Heat this soln for 10 min at 80°, cool to 20° and measure the extinction at 410 and 575 nm against H_2SO_4. Factors obtained by regression analysis are used for calculation of the results.

Simplex optimization has been used to develop a method for the determination of formaldehyde with chromotropic acid which is believed to be superior to previous methods in sensitivity and usability.[1259] Simplex optimization is a statistically based empirical method for maximizing desired responses from a chemical system.[1260] It has also been used to optimize the J-acid[1261] and 2,4-pentanedione[1262] methods for formaldehyde. The simplex optimization method is generally much more rapid in optimizing an analytical method than purely manipulative empiricism. Unfortunately, the workers in this field do not report the apparent millimolar absorptivities of these modifications so that a readily perceived comparison of the sensitivities of previous methods with the simplex optimized methods would then become possible.

The pararosaniline method for formaldehyde is probably too complicated for most uses. The commercial dye is usually a mixture of dyes and is very difficult to purify. It has been recommended for measurement of formaldehyde.[1263] Since it is affected by a heating operation, a cooling operation and methods to preserve reagent solutions are presented. The effect of the heating operation on the degree of absorption is such that absorption increases with an increase in temperature, then decreases again, giving a peak. The maximum degree of absorption was recorded at 30° in 35 min. This method is not stable since fading occurs swiftly. Cooling operations were conducted by using water at different temperatures and by changing the cooling time. The degree of absorption decreased and reached a stable point in 5 min. The degree of absorption generally shows a low value in proportion to the cooling temperature.

The complexity of the aldehyde–parafuchsin reactions have been demonstrated in a TLC study.[1264] Dependent on the type of aldehyde used and on the age of the solution, up to 14 different spots were identified, some of them showing a bright fluorescence. With formaldehyde 7 components were isolated while with paraldehyde 10 components were found, with some of them fluorescent.

2,4-Pentanedione (or acetylacetone) has been a very popular reagent in the study of enzymatic demethylation reactions. It has also been used in the study of other types of mixtures, e.g. formaldehyde in maple syrup[382] and in sewage and sewage effluents.[1265]

During the collection of maple sap, paraformaldehyde pellets are commonly placed in the tap holes to prevent fermentation. Thus, the finished syrup could contain formaldehyde. The determination of formaldehyde usually involves distillation. Unfortunately, the distillation fails to separate the formaldehyde from unidentified syrup constituents that also react with the colour reagent. This interference is circumvented if the reaction is carried out in an aqueous dilution of the syrup and the chromogen is extracted into isobutanol for colorimetric measurement.

In the automated method for the determination of formaldehyde in sewage and sewage effluents an AutoAnalyzer procedure is used.[1265] The sample is dialyzed against H_2O to remove color and turbidity, then the diffusate is caused to react with acetylacetone and ammonium acetate; the colour of the resulting lutidine derivative is measured at 430 nm. This method is suitable for determining free formaldehyde in the range 0·1 to 25 mg l^{-1} without interference from 100 mg l^{-1} of acetone, methanol, formic acid, phenol, $CHCl_3$, NH_4^+, NO_3^- or NO_2^-; hexamine interferes, and correction for its presence is described. Recovery of added formaldehyde is at least 90% and the standard deviation (10 determinations), is 0·2 mg l^{-1} at the level of 20 mg l^{-1}. For combined (protein-bound) formaldehyde, a hydrolysis stage (3·25 M H_2SO_4 at 60°) is incorporated into the system. The method can also be used to automate the measurement stage of the determination of "distilled formaldehyde". Up to 100 samples can be analyzed per day.

Here again in the analysis of sewage samples one would expect the presence of formaldehyde precursors which could be hydrolyzed during the analytical procedure to formaldehyde, and thus would be determined as formaldehyde.

Another reagent which has been used for measuring formaldehyde release from resin-bonded boards is purpald or 4-amino-3-hydrazino-5-mercapto-1,2,4-triazole, XXXIV.[1266]

H_2N $NHNH_2$
HS

XXXIV

An alkaline soln, of purpald (4-amino-3-hydrazino-5-mercapto-1,2,4-triazole), which absorbs formaldehyde vapour from air and turns purple, is used in an air-flow test and in spot test to measure the release of formaldehyde from wood-particle board or fibreboard bonded with urea—formaldehyde or phenol—formaldehyde resins. Under standard conditions, the graph of extinction values (at 550 nm) for the purpald soln vs concn is rectilinear in the range 6 to 600 ng of formaldehyde per ml. These tests are simpler and more rapid than those previously reported and are suitable for use in board-manufacturing plants.

1,2-Diaminonaphthalene reacts selectively with aldehydes in acidic medium to give compounds which fluoresce intensely in an alkaline medium.[1267] With formaldehyde a fluorogen is obtained which fluoresces at *F*330/385. The determinable limit of formaldehyde is 60 ng/ml of test solution. The method can be used to determine all types of aldehydes but works best for aromatic aldehydes. The method is discussed more fully in the section on furfural precursors in Volume 6. The following procedure can be used.

1,2-*Diaminonaphthalene determination of formaldehyde.*[1267] *Reagent solution*, 1,2-*Diaminonaphthalene* (45 μg/ml)—dissolve 45 mg of the diamine in 15 ml concentrated H_2SO_4 and add approximately 900 ml of water, the solution being water-cooled to room temperature. Dilute with water to 1 liter. The solution is about 0·025 M in H_2SO_4 and keeps for 2 days at room temperature.

Procedure. To 2 ml of the diamine reagent solution add 1 ml of the test solution and heat at 100° for 20 min. Cool in ice-water and add 2·0 ml of 2·5 M aqueous sodium hydroxide solution. Read the fluorescence intensity at *F*330/385. The method can be used for some formaldehyde precursors. Some of these precursors would react under these analytical conditions. Other aldehydes interfere.

2,3-Dimethyl-2,3-bis(hydroxylamino)butane reacts with the simpler aliphatic aldehydes to form an anhydro product which can be converted to a stable free radical by the addition of sodium periodate or lead dioxide, Fig. 76.[1268] Spot test results in Table 22[1269] give some idea of the millimolar absorptivities one could expect with various aldehydes.

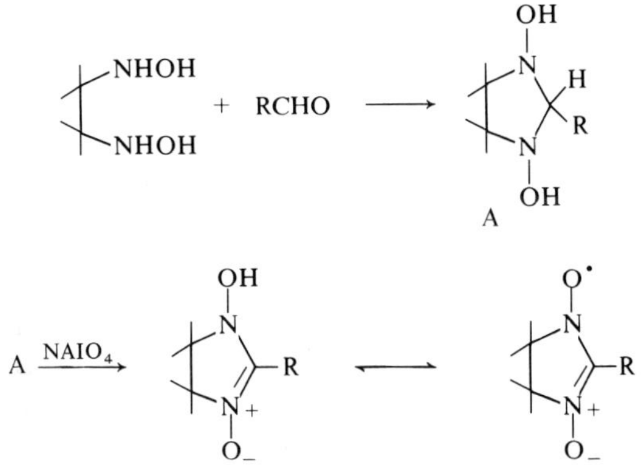

FIG. 76. Determination of aliphatic aldehydes with 2,3-dimethyl-2,3-bis(hydroxylamino)butane.

The formaldehyde stock solutions necessary for the method were analyzed by peroxide oxidation.[1270] 2,3-Dimethyl-2,3-bis(hydroxylamino)butane sulphate was recrystallized from a 2-propanol–water mixture to a melting point of 185° dec.

The following procedure was recommended.[1271] The reagent solution ($1{\cdot}32 \times 10^{-2}$ M in reagent) and the test solution containing formaldehyde were allowed to stand at 25° for 5 min, and 0·5 ml of this mixture was added to a cell containing 0·5 ml of 0·037 M sodium periodate and 2 ml of pyridine. Twenty-five minutes after the addition of the periodate the absorbance was measured at 515 nm.

Table 22. Spot test results with aldehydes[1269]

Compound	Minimum detectable quantity (μg)
Formaldehyde[a]	0·2[b]
Acetaldehyde	0·2
n-Propionaldehyde	2
n-Butraldehyde	1
Isobutyraldehyde	1
2-Ethylbutyraldehyde	10
n-Valeraldehyde	2
3,7-Dimethyl-2,6-octadienal[c]	200
3,7-Dimethyl-6-octenal[d]	20
Phenylacetaldehyde	50
Benzaldehyde	250
p-Hydroxybenzaldehyde	—
p-(Dimethylamino)benzaldehyde	—
5-Nitrosalicylaldehyde	—
trans-Cinnamaldehyde	—

[a] $m\varepsilon = 1{\cdot}16$.[1270]
[b] MDQ–Acetone, 80 μg; 2-butanone, 2000 μg.
[c] Citral.
[d] Citronellal.

The procedure needs much further study and needs to be optimized. Pyridine helped to stabilize the final chromogen. The color increased immediately after the addition of the sodium periodate and then stabilized after 15 min. During the next hour the absorbance decreased at a rate of 0·8%/hr. Formaldehyde reacted very rapidly in the procedure, the reaction being complete within 5 min at room temperature. Other aliphatic aldehydes reacted more slowly under the same conditions, Table 23.[1271] Acetone and benzaldehyde failed to react within 30 min under the same conditions and equivalent concentrations.

Chemiluminescent and bioluminescent methods have been shown to be of value in chemical analysis.[1272] The vigorous reaction between concentrated alkaline hydrogen peroxide, formaldehyde and pyrogallol results in a strong orange luminescence lasting several minutes.[1273] Since this method is not convenient for analytical purposes, diluted reagents and gallic acid have been applied for the determination of formaldehyde.[1274, 1275] The chemiluminescent base-catalyzed decomposition of the $CH_2O—H_2O_2$ system is postulated as a model for lipid peroxidation in biological membranes.[1276]

Table 23. Rate and extent of reaction of aldehydes and ketones with the bis-hydroxylamine[1271]

Compound	Wavelength of Maximum Absorbance, nm	t_{∞}[a], sec.	mε[b]
Formaldehyde	515	300	1·15
Acetaldehyde	538	600	1·9
Propionaldehyde	543	1000	1·6
Butyraldehyde	546	1000	1·45
Isobutyraldehyde	547	1400	1·35
Acetone		N.R.[c]	
Benzaldehyde		N.R.[c]	

[a] Time to reach maximum absorbance.
[b] Millimolar absorptivity calculated on basis of formaldehyde concentration in sample (and by inference the concentration of free radical).
[c] No reaction within 30 min.

The chemiluminescent method has been investigated for the determination of formaldehyde.[1275] Glyoxal and pyruvaldehyde could also be determined since they give chemiluminescent intensity comparable with that of formaldehyde. In a static method of chemiluminescent intensity measurement, the instant spike generated by glyoxal or pyruvaldehyde can be distinguished from the later spike of formaldehyde. In the flow method both aldehydes interfere with the formaldehyde determination. Other aldehydes give chemiluminescent intensities at least one order of magnitude lower than formaldehyde and are expected not to interfere with the determination. The relative standard deviation of the method is about 1 %. The minimal volume of the formaldehyde test solution is 2 ml, and the analysis time generally is less than 2 minutes. For the analytical assay of formaldehyde, the reaction was carried out in a flow system observing steady-state chemiluminescent intensity. The chemiluminescence produced by the oxidation of gallic acid and formaldehyde with H_2O_2 plus NaOH was followed in a continuous flow system.[1275] The linear range for formaldehyde analysis is 10^{-7} to 10^{-2} M. The limit of detection is 10^{-7} M CH_2O in a 2 ml sample at the signal-to-noise ratio 1:1. The actual detection limit is 2 ng of formaldehyde. Singlet oxygen is believed to be involved in the reaction of alkaline H_2O_2, CH_2O and pyrogallol.[1277–1279]

The reactions of indole derivatives with metaldehyde (or acetaldehyde) has been investigated.[1280] Formaldehyde could also react similarly. Thus, ergocristin (which is essentially a 3,4-disubstituted indole derivative) gave a blue color with metaldehyde while indole, N-methylindole, 2-methylindole and gramin gave red to violet chromogens and 3-methylindole and tryptophan gave yellow-brown and orange colors, respectively. 2,3-Disubstituted indoles do not react under the conditions of the reaction in 65% H_2SO_4.

On the other hand tryptophan in the presence of an oxidizing agent can react with formaldehyde to give a violet chromogen.[1006] We have discussed this reaction previously.

Formaldehyde and its precursors could also be determined enzymatically. Formaldehyde dehydrogenase (formaldehyde: NAD oxidoreductase, EC 1.2.1.1) is, with glyoxalase and maleylpyruvic acid isomerase, one of the few enzymes known to need glutathione as a specific cofactor. It catalyzes the following reaction in the presence of reduced glutathione.[1282,1283]

$$HCHO + NAD^+ + H_2O \longrightarrow HCOOH + NADH + H^+$$

On the basis of this reaction formaldehyde and its precursors could be measured through the derived NADH colorimetrically or fluorimetrically. The yeast enzyme cannot utilize NADP in the place of NAD. Glyoxal or pyruvaldehyde could probably also be determined by this reaction. Acetaldehyde, glycolaldehyde, glyoxylate, benzaldehyde or glucosone would not react. Glutathione, which is required in the reaction, cannot be replaced by cysteine, 2,3-dimercaptopropanol or thioglycolate.

6. *Formation of formaldehyde precursors.* To understand and utilize the formation of formaldehyde from its precursors in analytical methodology knowledge of the condensation of CH_2O with various chemicals is invaluable. In addition, these types of reactions could give clues as to what happens to the formaldehyde released in living tissue. Some of the reactions of formaldehyde are summarized in Table 24. In addition, many of these types of compounds formed from formaldehyde could be analyzed through the derived formaldehyde by many of the methods described in Volume 1 and in this volume.

Table 24. Reactions of formaldehyde emitted into or formed in the environment or cellular tissue[a]

Functional group	Reaction	Remarks
Acid, RCOOH	$2\,RCOOH + CH_2O \rightarrow CH_2(OOCR)_2 + H_2O$	Unimportant in aqueous solution
Acyl anhydride	$(R_2CO)_2O + CH_2O \rightarrow CH_2(OOCR)_2$	
Active hydrogen	$RH + CH_2O \rightleftharpoons RCH_2OH$	e.g. $CH_3NO_2 + CH_2O \rightarrow HOCH_2CH_2NO_2$
Active hydrogen (2 moles)	$2\,RH + CH_2O \rightarrow RCH_2R$	e.g. 2 H_3C H_3C O O + CH_2O ⟶ H_3C H_3C O O H_2C O O CH_3 CH_3
Amine (1 mole)	$RNH_2 + CH_2O \rightarrow RNH{-}CH_2OH \rightarrow RN{=}CH_2$	Schiff base formation
Amine (2 moles)	$2\,RNH_2 + CH_2O \rightarrow RNH{-}CH_2{-}NHR$	
Amine (+2 moles CH_2O)	$RNH_2 + 2\,CH_2O \rightarrow RN(CH_2OH)_2$	
Amine (+3 moles CH_2O)	$RNH_2 + 3\,CH_2O \rightarrow RN(CH_2O)_2{-}CH_2$	Ring formation
Amide (1 mole)	$RCONH_2 + CH_2O \rightarrow RCONHCH_2OH$	Methylolamides
Amide (2 moles)	$2\,RCONH_2 + CH_2O \rightarrow RCONH{-}CH_2{-}NHCO{-}R$	
Imine (1 mole)	$R_2NH + CH_2O \rightarrow R_2NCH_2OH$	
Imine (2 moles)	$2\,R_2NH + CH_2O \rightarrow R_2N{-}CH_2NR_2$	

Peptide (1 mole)	$CO{-}NH{-} + CH_2O \rightarrow CONH{-}CH_2OH$	
Peptide (2 moles)	$2CO{-}NH{=} + CH_2O \rightarrow {-}CONH{-}CH_2{-}NH{-}CO{-}$	
Polyamines $n = 0, 1$	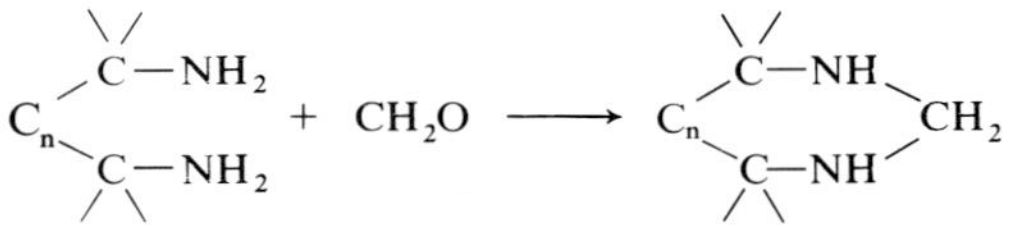 $C_n(C{-}NH_2)(C{-}NH_2) + CH_2O \longrightarrow C_n(C{-}NH)(C{-}NH)CH_2$	Ring formation
Mercaptoamines	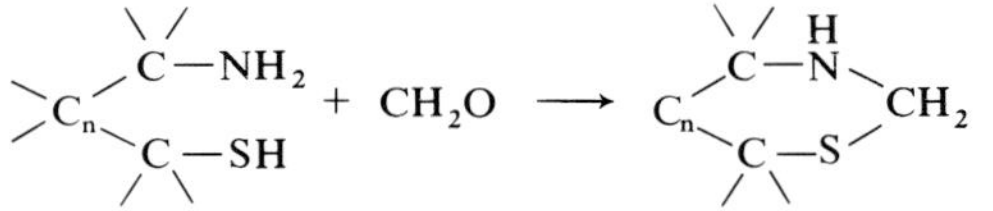 $C_n(C{-}NH_2)(C{-}SH) + CH_2O \longrightarrow C_n(C{-}NH)(C{-}S)CH_2$	e.g. Cysteine 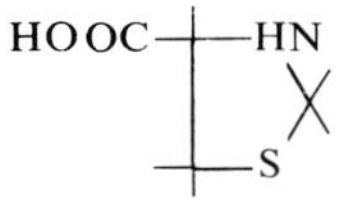 → HOOC–(ring)–HN, S
Amine + sulfite	$CH_2O + SO_2 + H_2O \rightarrow HOCH_2SO_3H$ $RNH_2 + HOCH_2SO_3H \rightarrow RNHCH_2SO_3H$	See PAS tests
Chloride (1 mole)	$HCl + HO(CH_2O)_nCH_2OH \rightarrow HO(CH_2O)_nCH_2Cl$	*m*-Trioxane and other polymers formed
Chloride (2 moles)	$2\,HCl + HO(CH_2O)_nCH_2OH \rightarrow Cl(CH_2O)_nCH_2Cl$	e.g. Bis-chloromethyl ether can thus be formed[b]
Hydroxyl (1 mole)	$ROH + CH_2O \rightarrow ROCH_2OH$	Hemiacetal formation
Hydroxyl (2 moles)	$2\,ROH + CH_2O \rightarrow ROCH_2OR$	Acetal formation
SH (1 mole)	$RSH + CH_2O \rightarrow RSCH_2OH$	Semithioacetal formed
SH (2 moles)	$2\,RSH + CH_2O \rightarrow RSCH_2SR$	Thioacetal formed
Aromatic hydrogen		
Phenols (1 mole)	$C_6H_5OH + CH_2OH \rightarrow HOCH_2C_6H_4OH$	
Phenols (2 moles)	$2\,C_6H_5OH + CH_2OH \rightarrow HOC_6H_4CH_2C_6H_4OH$	
$C_6H_5NR_2$ (1 mole)	$C_6H_5NR_2 + CH_2O \rightarrow HOCH_2C_6H_4NR_2$	
$C_6H_5NR_2$ (2 moles)	$2\,C_6H_5NR_2 + CH_2O \rightarrow R_2N{-}C_6H_4CH_2C_6H_4NR_2$	

Table 24—*continued*

Functional group	Reaction	Remarks
Azulenes (1 mole)	+ CH_2O ⟶ CH_2OH	
Azulenes (2 moles)	2 + CH_2O ⟶ CH_2	
PAH[c] (1 mole)	+ CH_2O ⟶ CH_2OH	
PAH (2 moles)	2 + CH_2O ⟶ ( )$_2$ CH_2	

Heterocyclic hydrogen
Pyrrols (1 mole)

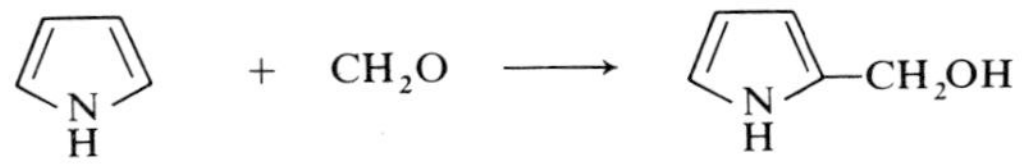

N H + CH_2O ⟶ N H $-CH_2OH$

Pyrrols (2 moles)

2 N H + CH_2O ⟶ N H $-CH_2-$ N H

Indoles (1 mole)

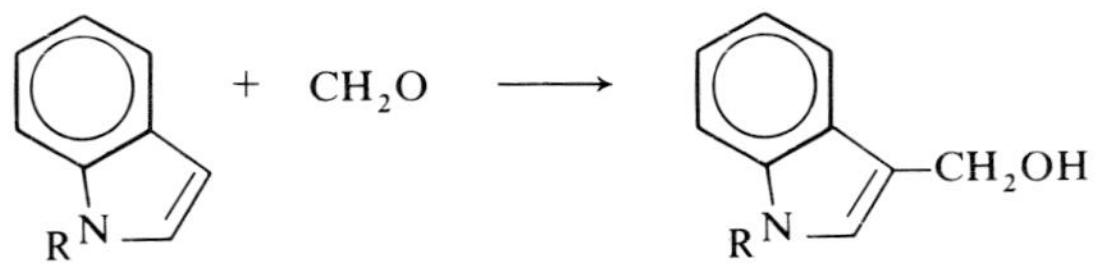

R N + CH_2O ⟶ R N $-CH_2OH$

Indoles (2 moles)

2 R N + CH_2O ⟶ R N $-CH_2-$ N R

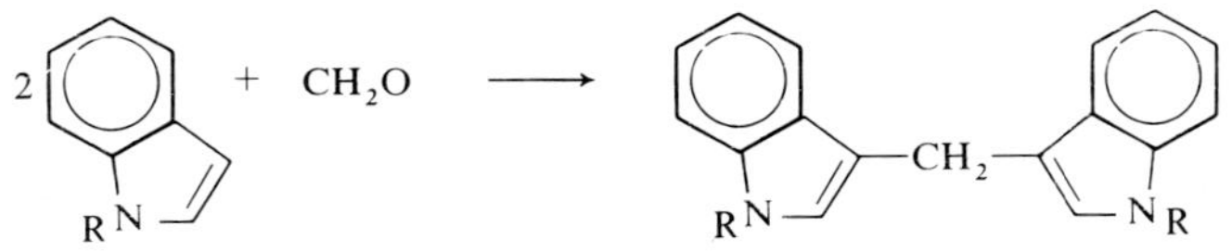

3-(2-Aminoethyl)-
indoles

N H $\overset{H_2}{C}$ CH—COOH | NH_2 + CH_2O ⟶ N H NH COOH

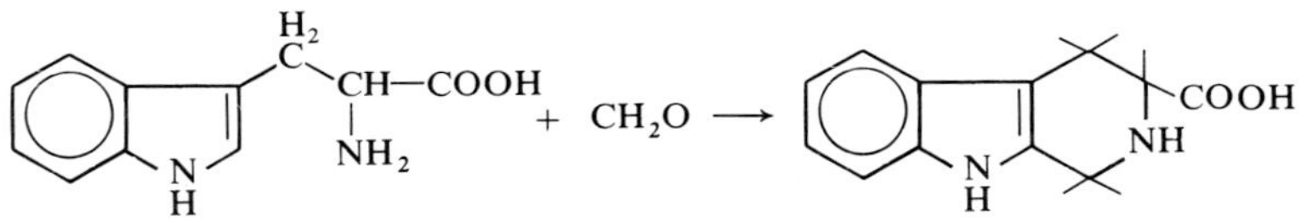

Table 24—*continued*

[a] Interaction between two different functional groups and formaldehyde has just been shown in a few reactions. For example, one reaction would be the reaction between the BaP formaldehyde addition product with adenine to give

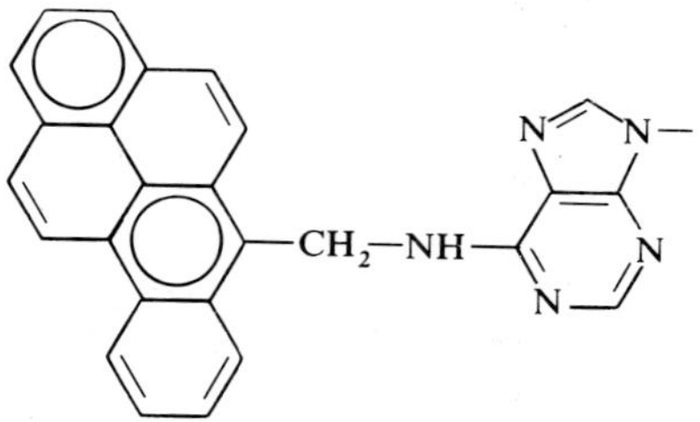

$-CH_2-NH-$

[b] With sulfate $-O-SO_2-O-CH_2-O-SO_2-O-$ or 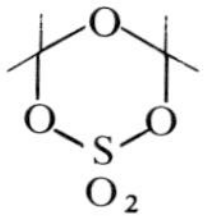 could be formed. Both of these would be powerful alkylating agents and possible carcinogens.

[c] PAH = polynuclear aromatic hydrocarbons.

In some cases when formaldehyde reacts with some chemical to form a heterocyclic derivative, this new compound can be remarkably stable towards both acid and alkali. This has been shown for reaction with cysteine where thiazolidine-4-carboxylic acid is formed.[1284] Cysteine is also an interference in the 2,4-pentanedione test for formaldehyde.[508]

The most interesting reactions of formaldehyde are with the biopolymers, e.g. the amino-containing polysaccharides, proteins, RNA and DNA. The one of intense interest is the reaction with DNA and its components. Formaldehyde can be used as a probe of DNA structure.[1285, 1286]

The ability of formaldehyde to react only with NH_2 and NH groups of purine and pyrimidine bases, which are free of hydrogen bonds, is used in studies of the secondary structure of nucleic acids.[1287–1289] Since formaldehyde reacts reversibly with nucleic acid monomers, it is capable of bringing about the reversible melting of DNA and is capable of being determined through appropriate microanalytical or histochemical methods.

Formaldehyde has been used for a determination of the defects in the secondary structure of nucleic acid which result through reaction with nitrous acid,[1290] ionizing and ultraviolet radiation,[1291, 1292] the presence in DNA of weak points enriched with A–T pairs,[1293] and changes in the secondary structure of DNA caused by RNAase–polymerase.[1294, 1295] These reactions of formaldehyde are probably of considerable importance from the genotoxic metabolite viewpoint since formaldehyde and other aldehydes are metabolic products of many genotoxic materials and radiations.

In this respect the studies of Simonov *et al.*[1295] are of considerable interest. These authors have established that the reaction of nucleotides with formaldehyde is accelerated in the presence of some amines, e.g. ethanolamine. Nucleotides are modified more efficiently by formaldehyde when amino acids are present, the effect on the reactivity of d-AMP decreasing in the order: glycine $>$ β-alanine $>$ leucine $>$ α-alanine $>$ valine. The degree of acceleration of the reaction depends on the concentration of these amines. Incubation of calf thymus DNA with formaldehyde in the presence of C^{14}-lysine results in the binding of the amino acid with the bases of DNA. This bound material is not eliminated by dialysis. DNA in a solution containing lysine and formaldehyde is despiralized far more rapidly than in the absence of lysine. Amino acids within a histone molecule retain the capacity to accelerate the denaturation of DNA by formaldehyde. Amino acids not only accelerate the modification of nucleotides by formaldehyde but also lead to the formation of a stable cross link between proteins and DNA. The reaction of formaldehyde with the exocyclic amino group of derivatives of adenine, cytosine and guanine is shown to be useful as a quantitative probe of DNA structure and results in the formation of a NH–CH_2OH group at low concentrations of formaldehyde and a dihydroxymethyl adduct at high formaldehyde concentrations. The following reactions are postulated.[1285]

$$RNH_2 + CH_2O \rightleftharpoons RNHCH_2OH + CH_2O \rightleftharpoons RN(CH_2OH)_2$$

These authors[1285] postulate that in more complicated systems such as ribosomes and chromatin, completely different types of reactions could occur and might even be expected to dominate. In this sense formaldehyde has been shown to lead to histone–DNA cross-links[1296] and to a considerable acceleration in reaction rate of formaldehyde with nucleotides in the presence of amino acids and histones[530] or morpholine.[1285] This latter compound is thought to be the nucleotide cross-linked to the amine through a methylene bridge.

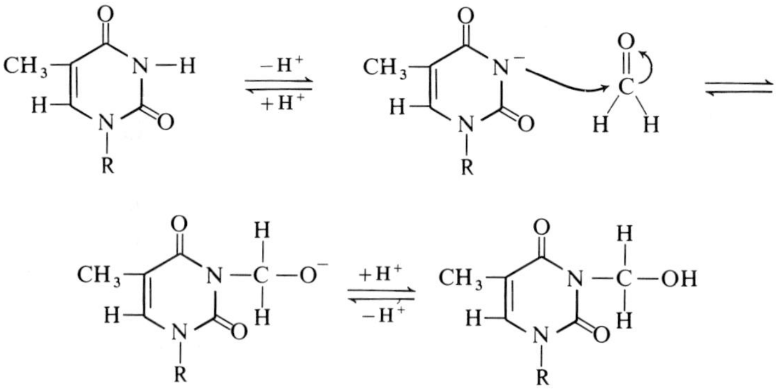

FIG. 77. Postulated reaction of thymidine with formaldehyde.[1286]

The formaldehyde reaction with the endocyclic imino groups of derivatives of thymine, uracil, halogenated uracils, poly(uridylic acid) and poly (inosinic acid) which lack an exocyclic amino group is postulated as follows (using thymidine as an example), Fig. 77.[1286]

Carcinogens can open up the DNA to reaction with formaldehyde. Such behavior has been shown with the carcinogen, N-acetoxy-N-2-acetylamino-fluorene.[1297] Native calf-thymus DNA was allowed to react with N-acetoxy-N-2-acetylaminofluorene and the dynamic behavior of this carcinogen-reacted DNA was studied by the formaldehyde unwinding technique. In all cases, the initial rate of formaldehyde attack was higher for DNA reacted with the carcinogen than for native DNA. In addition to the natural breathing of the DNA duplex it is demonstrated that each fixed F residue gives rise to weak points from which formaldehyde unwinding starts.

Thermal or acid-induced DNA denaturation is markedly influenced by formaldehyde in 2 different ways.[1298] If cells are exposed to CH_2O during heating, DNA denaturation is facilitated most likely by the direct action of CH_2O as a "passive" denaturing agent on DNA. If cells are pretreated with CH_2O which is then removed, DNA resistance to denaturation increases, presumably due to chromatin cross-linking.

B. —O—CH_2—O— compounds

Compounds with the methylenedioxyphenyl functional group are found widely distributed in alkaloids, essential oils, and other physiologically-active compounds of natural origin.[384] Compounds containing this moiety are extensively utilized in food additives, medicinals, perfumes, topical preparations, and as insecticides (primarily as synergists). Commercially, the most important pesticides are the methylenedioxyphenyl derivatives. Piperonyl butoxide is the most extensively used of the methylenedioxyphenyl synergists. Another compound of this type is methyl piperonylate which appears to be a component of cotton bracts and causes histamine release from chopped human lungs and thus could cause the acute reversible bronchoconstriction encountered following inhalation of cotton flax or hemp dust.[1299]

Table 25. Distribution of polyoxymethylene glycols in 50% formalin at 65°C, determined by NMR[1300]

	Relative % (m/m) of $HO(CH_2O)_nH$		
	$n = 1$	$n = 2$	$n \geq 3$
Pure solution	15·6	28·0	56·4
Formalin, diluted with an equal volume of DMSO-d_6	18·2	31·9	49·9

Dihydroxydimethyl peroxide, $HOCH_2-O-O-CH_2OH$ is a mutagen containing the $-O-CH_2OH$ structure.[770]

According to Walker[1064] formalin consists of free formaldehyde (<0·1%), methylene glycol, and polyoxymethylene glycols, and in very small concentrations methylal, methyl formate, trioxane, and acetals of the polyoxymethylene glycols. The relative amounts of the polyoxymethylene glycols in formalin have been determined, Table 25.[1300]

Formalin is usually stabilized with methanol. The addition of methanol causes the formation of hemiacetals, which are more soluble than the corresponding glycols, thus preventing precipitate formation.[1064] The following mechanism is postulated.[1300]

$$HO(CH_2O)_nH + CH_3OH \rightleftharpoons HO(CH_2O)_{n-1}H + CH_3OCH_2OH$$

$$HO(CH_2O)_{n-1}H + CH_3OCH_2OH \rightleftharpoons HO(CH_2O)_{n-2}H + CH_3O(CH_2O)_2H$$

Paraformaldehyde, HO—$(CH_2O)_n$—H, readily dissociates with heat to formaldehyde. It has been reported to be present in the Allende meteorite which fell in northern Mexico on 8 February, 1969.[420] The argument has been advanced that it is not an artifact. Based on the analytical data, it has been estimated that approximately 10^{14} g of formaldehyde have been deposited on the Earth by meteoric influx during the time from the origin of the Earth to the time when life is thought to have begun. It is theorized that the formaldehyde is a precursor of carbohydrates and with the organic nitrogen compounds, precursors of life itself.

Paraformaldehyde and formaldehyde concentrations in funeral home atmospheres have been determined with a chromotropic acid method.[1301] The sampling solution is 0·1% chromotropic acid in concentrated H_2SO_4. The sampling train consisted of sampling hose, impinger with 10 ml of sampling solution, trap impinger, rotameter, and pump. Air was drawn through 1·5 liters/min until a purple color was obtained. Both formaldehyde and paraformaldehyde react. The absorbance is read and the concentration of formaldehyde in the atmosphere is calculated. Alternative procedures are available (see Formaldehyde Sections and ref. 1302). The study of 187 samples from a variety of embalming room atmospheres gave a range of 0·09 to 5·26 ppm of formaldehyde. The paraformaldehyde particle geometrical mean particle size was found to be 1·6 μ, an optimum size for deposition and retention. These fine particles with the attached formaldehyde vapors can reach an area where they could cause serious lung damage. It is of some interest that 3 out of 7 of the embalmers suffered from asthma or sinus problems.

In the determination of low levels of formaldehyde in industrial chemical plant atmospheres it was found that methylal $(CH_2(OCH_3)_2)$ was a serious interference.[375] This was evident by comparing the millimolar absorptivities obtained with methylal (me 10.7) and formaldehyde (me 17·8) in the chromotropic acid method. Selective adsorption on Porapak Q could be utilized to collect the methylal and pass through the formaldehyde. Thus, methylal and formaldehyde could be determined separately.

The mechanism and catalysis for hydrolysis of acetals have been investigated.[1303] Acetals are weak oxygen bases, Table 26,[1304] whose hydrolysis involves acid catalysis. Values of pKa for the conjugate acids of the acetals yield invaluable information concerning the mechanism of hydrolysis of the various types of acetals. This has been discussed.[1303]

Table 26. Basicities of dioxymethanes[1304]
$R-O-CH_2-O-R$

R	pKa
2,2,2-Trifluoroethyl	−8·40
2,2,2-Trichloroethyl	−7·85
Phenyl	−6·53
2-Chloroethyl	−5·32
2-Fluoroethyl	−4·99
Methyl	−4·57
Ethyl	−4·13
Isopropyl	−3·70
tert-Butyl	−2·85

The methylenedioxy compounds can be assayed directly on a thin layer plate or in solution as shown in the following procedures, where chromotropic acid or gallic acid have been used.

Detection and estimation of methylenedioxy compounds on thin-layer plates.[433] Air dry and then evenly spray with reagent (Add 15 ml concentrated sulfuric acid to 1 g of disodium chromotropate in 15 ml water). Heat the plate at 110 to 120°C for 10 to 30 min dependent upon the compound. Scan the purple spots (transmission densitometry, white light) in a direction at right angles to that of solvent development.

Chromotropic acid location of formaldehyde precursors on thin-layer chromatograms.[48] Mix a 10% aqueous solution of chromotropic acid with a concentrated sulfuric acid–water solution (5:3, v/v) in a 1:5 (v/v) proportion and cool to room temperature. Spray with this fresh reagent. Heat 30 min at 105°C. Positive results are shown by a violet color, especially by compounds containing a methylenedioxy group, e.g. narcotine, hydrastine, sesamine, etc.

Chromotropic acid determination of piperine in pepper.[109] Extract 0·5 g of ground pepper with alcohol. To a 1 ml aliquot add 0·5 ml of 25% aqueous disodium chromotropate and 10 ml of concentrated sulfuric acid. Heat the mixture 30 min at 100°. Cool and dilute to 25 ml with sulfuric acid. Read at λ 580 nm.

Gallic acid determination of piperine in pepper.[435] To a 1 ml aliquot of an alcoholic extract of pepper add 9 ml of 0·05% gallic acid in 36 N sulfuric acid. Stopper the tubes loosely and heat at 100° for 60 min. Cool in an ice-water bath and allow to stand at room temperature (29 ± 1°C) for 30 min. Read absorbance at 660 nm.

1,3,5-Trioxane, XXXV, is entirely different from aqueous formaldehyde or the hydrated polymer, paraformaldehyde, since it is a colorless crystalline compound, m.p. 61–62°, and boils without decomposition at 115°.

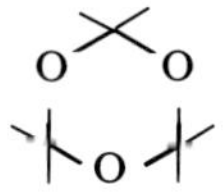

XXXV

Its properties have been summarized.[516] It has a pleasant chloroform-like odor, is readily soluble in organic solvents like acetic acid, benzene, toluene, trichloroethylene, alcohol or ether, and very soluble in water. It is a formaldehyde precursor and could be used as a formaldehyde standard. The rate of formation of formaldehyde can be controlled by choice of solvent and acid depolymerization catalyst. It is a completely anhydrous form of formaldehyde which is stable in neutral and alkaline solutions. Trioxane depolymerization at 35°, 45°, and 55° by 0·5 M toluenesulphonic acid has been studied.[1305] Formaldehyde solutions can be standardized titrimetrically by the iodometric method.[1064] Sulphuric acid was found to be a much more effective catalyst in acetic acid than in water; the reaction with a given acid concentration was of the order of a thousand times faster in the former solvent.[516] At 95° depolymerization proceeded at a measurable rate in glacial acetic acid with no added catalyst. The depolymerization of trioxane in aqueous solution was affected by sulphuric acid concentration and temperature, Table 27. Hydrochloric acid is more active than sulphuric acid of the same normality;

Table 27. Quantitative trioxane depolymerization at various temperatures with H_2SO_4 as catalyst[516]

H_2SO_4, N	20°	H_2SO_4, N	70°
8	22·5 days	0·5	7·5 days
12	40 hr	1	3·8 days
16	190 min	2	35 hr
20	14·5 min	4	7·5 hr
	40°	8	48 min
1	137 days	12	7·8 min
2	53 days		95°
4	11·8 days	0·1	500 days
8	23 hr	0·25	24 hr
12	160·0 min	0·5	12·8 hr
16	18·3 min	1	5·4 hr
		2	103 min
		4	33 min
		8	8·5 min

the difference becoming greater at higher acidities. Phosphoric is less active even when it is considered as a monobasic acid.

Compounds, such as the genotoxic material, methylene dimethanesulphonate, can be analyzed through the derived formaldehyde. The half-life of this compound at 37° in water is about 22 min.[1306] The following reaction takes place

$$CH_2(O_3SCH_3)_2 + H_2O \longrightarrow 2\,CH_3SO_3H + CH_2O$$

The acyloxymethyl derivatives of antibiotics such as the cephalosporins (e.g. pivaloyloxymethyl cephaloglycin) can be determined through the derived formaldehyde.[1307] Pivampicillin can be determined in a similar fashion with chromotropic acid.[1308] The following type of reaction takes place

$$RC-\overset{\overset{O}{\|}}{O}-CH_2-O-\overset{\overset{O}{\|}}{C}-C(CH_3)_3 \xrightarrow[H_2O]{H^+} RCOOH + CH_2O + (CH_3)_3CCOOH$$

In the following procedure a millimolar absorptivity of 16·1 was obtained. Maximum absorbance values are obtained in the procedure after heating for 30 min at 65°.[1309, 1310] Beer's law is followed from 0·1 to 1 μg of formaldehyde per ml. The formaldehyde can be standardized by methods described in Volume 1 or this volume or by the method described in the British Pharmacopoeia[1311] or in the British Pharmaceutical Codex.[1312]

Colorimetric determination of Pivampicillin with chromotropic acid.[1307] *Reagents.* Chromotropic acid, 4% in water. *Cupric chloride solution*—prepare by dissolving 50 g of copper oxide in 500 ml of 3 N HCl in the cold and dilute to 1 liter. *Aldehyde-free ethanol*—add 5 ml of 0·5% tetrazolium blue in ethanol and 5 ml of 1% tetramethylammonium hydroxide to 100 ml of alcohol. Let stand 2 hr and then distill the ethanol.

Procedure. Weigh a sample of 40–80 mg of pivampicillin hydrochloride accurately and introduce into a 250-ml two-necked flask. Add a 100-ml aliquot of 1·5 N hydrochloric acid and 20 ml of cupric chloride solution. This is not necessary for pivampicillin hydrolysis, but cupric chloride keeps (according to experimental evidence) colored products otherwise easily distillable in the steam current. Fit the flask with a condenser having an outlet adaptor dipping into a flask containing 6 ml of absolute ethanol and 15 ml of distilled water. Connect a second flask containing water in series with the first.

Heat the solution in the distillation flask to boiling (to effect the hydrolysis of the pivampicillin with consequent formation of formaldehyde) and slowly distill in a gentle current of nitrogen. Continue the distillation until about 10ml of liquid remains in the distillation flask. Combine the distillate and the water used to wash the condenser, the outlet adaptor, and the receivers and dilute to 200 ml in a graduated flask. The solution obtained in this way contains 0·3 μg/ml formaldehyde/mg pivampicillin hydrochloride in the weighed specimen.

Carry out the colorimetric reaction in accordance with the details given by Mathers and Pro.[1310] Take one or more 0·2–1-ml samples and transfer into an equal number of 25-ml graduated flasks. Dilute each sample to exactly 1 ml with 3% ethanol; add a 0·5-ml aliquot of 4% chromotropic acid and 7·5 ml of concentrated sulfuric acid to each flask with agitation in a bath of cold water. Heat at 65° for 30 min, cool immediately, and then dilute to 25 ml with distilled water.

Measure the absorbance of each solution at 570 nm against a blank prepared in the same way but with an identical volume of 3% ethanol instead of the sample. Determine the concentration of formaldehyde in the solution from the absorbance with the aid of a suitable calibration curve, and the concentration of pivampicillin by means of the factor 16·7

C. —O—CH_2—N< compounds

Free formaldehyde can result from the facile dissociation of the $HOCH_2$—N< compounds in textiles treated with formaldehyde resins (e.g. urea, melamine, and phenol; probably O—CH_2OH in the latter compound) to improve crease- and shrinkage-resistance. Contact dermatitis can result.[1313, 1314] The durable press resins used in clothes are usually N-methylol amides or N-methylol ureas; they can be as bad as the urea–formaldehyde resins which continually release formaldehyde or the N-methylol carbamates which contain the potentially genotoxic carbamate group.

The "free" formaldehyde content, both in ppm and in percentage, was determined by quantitative analysis on 112 fabrics obtained from American textile manufacturers and distributors.[1315] This fabric sampling included textiles of 39 different types of composition. All 112 samples, regardless of composition, contained some free formaldehyde. In an analysis of 15 samples of 100% polyester knit and 8 samples of 100% polyacrylonitrile (Orlon), it appears that these fabrics, in general, contain less free formaldehyde than those made up of 100% rayon, 100% cotton, 50% polyester and 50% cotton, and fabric composed of 65% synthetic polyester fiber (Dacron) and 35% cotton. Formaldehyde content in this study of clothing samples ranged from 0·1 to 3·517 ppm. The extraction by water of formaldehyde and other volatile substances from textile fabrics impregnated with compounds containing formaldehyde resins has been studied.[1316] Attempts to reduce formaldehyde emissions from the dryer of a plant manufacturing amino resin molding material, friction material, and polyurethane foam rubber are described.[1317]

Most of the gas is emitted from the amino resin molding process, about 1·5% from urea resin, and about 2% from melamine.

When a synthetic polymer material such as microporous rubber which contains phenol–formaldehyde resin, is used in the construction of buildings in hot climates, there is a potential hazard from the release of carbon monoxide, carbon dioxide and formaldehyde into the air in enclosed areas.[1318]

Probably the most obnoxious drawback to the more widespread use of particleboard and fiberboard in the construction of mobile homes, building and furniture is the gradual continuous release of small amounts of formaldehyde from some boards. The extent of formaldehyde discharge is believed to be directly proportional to the air temperature and the moisture content of the air and inversely proportional to the extent of air exchange and the time of board production.[1319] More formaldehyde is discharged from boards made from sawdust than from those made from wood shavings. The formaldehyde arises from free or loosely bound aldehyde (e.g. methylol, $O{-}CH_2OH$, groups in uncured residues of the phenol–formaldehyde or $-NH-CH_2OH$ groups in urea–formaldehyde resins used in the manufacture of the boards.[1320–1323] The buildup of formaldehyde fumes released from low quality boards or panels in confined or poorly ventilated areas, such as in cabinets, closets, drawers, mobile homes, self-panelled rooms, or trailers, can become so heavy that its pungent odor can become intolerable to humans.

The parameters which influence the rate and extent of formaldehyde release include the following: the amount of formaldehyde used in the resins, the formulation of the resin, the amount of resin applied, the curing conditions and curing times used, the plate pressures during forming, the duration of storage before use, and the environment and climate wherein the product is utilized.[1321,1322,1324,1325] Because of the complexity of these parameters board manufacturers are currently unable to predict the suitability of the boards in diverse building and manufacturing applications.

A method has been developed for measuring formaldehyde release from resin-bonded boards.[1266] An alkaline solution of purpald is used to absorb the formaldehyde quantitatively from air. The purple solution can then be assayed. An mε of 50 is obtained at λ 550 nm. The postulated reactions are shown in Fig. 78.

Determination of formaldehyde release from resin-bonded boards.[1266] *Reagent.* 1% of 4-amino-3-hydrazino-5-mercapto-1,2,4-triazole or purpald in 1 N aqueous sodium hydroxide solution.

Procedure. Obtain fines from a small test board by drilling holes with a sharp bit on a drill press. Place fines immediately in a tightly stoppered bottle. For each test a 5-gram sample of fines was transferred through a powder funnel into a Nalgene drying tube fitted with a cotton filter at the bulb end. Air was drawn at the rate of 1 milliliter per second (as measured with a flowmeter) through a gas wash bottle containing 20% aqueous sodium hydroxide (to absorb carbon dioxide and any formaldehyde from the air carrier gas), then through the Nalgene tube, the cotton wool, and finally through some capillary tubing through 30 ml of the reagent solution and out through a capillary tube into the open air. After 10 to 20 min of air flow, stopper the test tube tightly and measure the color visually or in a spectrophotometer at 550 nm.

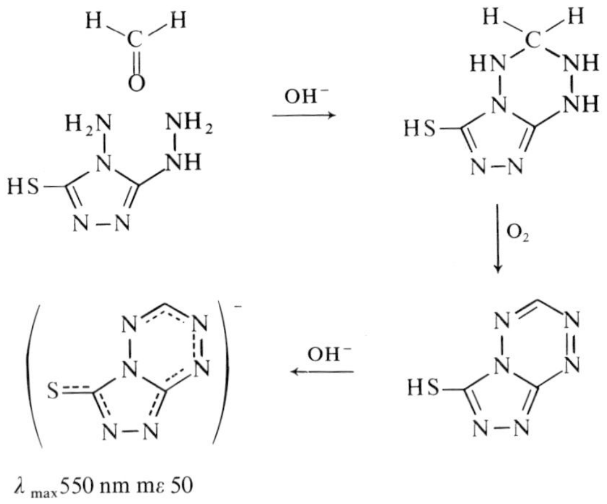

λ_{max} 550 nm mε 50

FIG. 78. Determination of formaldehyde with purpald in alkaline solution.[1266]

Beer's law is not followed, but formaldehyde can be determined from about 60 ($A = 0{\cdot}1$) to 600 ng/ml. The procedure can be considerably simplified, if necessary. Thus, powdered board or a standardized size of drilled board can be added to the reagent and then measured spectrally.

The airflow test can also be adapted to monitor air quality in board manufacturing plants.[1266] A measured volume of air can be drawn through a sintered glass filter into a measured volume of reagent in an air-sampling midget bubbler and the purple color can then be measured spectrophotometrically.

The color can fade in a few days due to absorption of CO_2 into the solution and with a decrease in alkalinity the formation of the less highly colored neutral compound. Other aldehydes would also be expected to react in these procedures. Methylol groups react slowly.

These methods need to be thoroughly evaluated and the procedures investigated for use with other aldehydes and their precursors. The procedures for formaldehyde and the other compounds would then need to be optimized for sensitivity, selectivity, simplicity and precision.

Cosmetic preservatives containing N—CH_2OH groups are readily hydrolyzed to formaldehyde which is then determined fluorimetrically with 2,4-pentanedione.[1326] The more recent type of preservatives are water soluble and have been incorporated into a wide variety of cosmetic formulations. These include 2-nitro-2-bromo-1,3-propanediol (to be discussed in a later section), methane bis[N,N′(5-ureido-2,4-diketo-tetrahydro-imidazole)N,N′-dimethylol], XXXVI, and 1-hydroxymethyl-5,5-dimethyl-hydantoin, XXXVII,

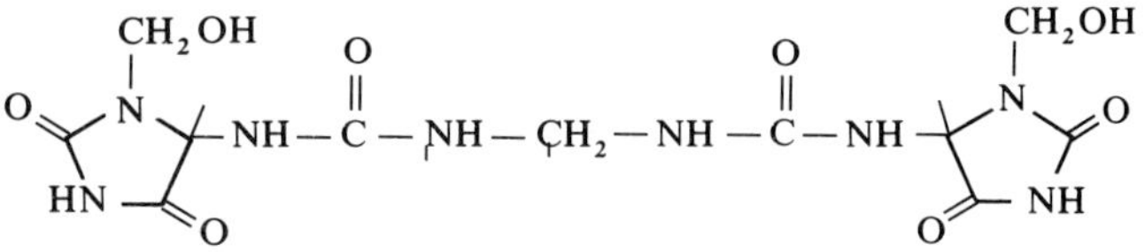

XXXVI

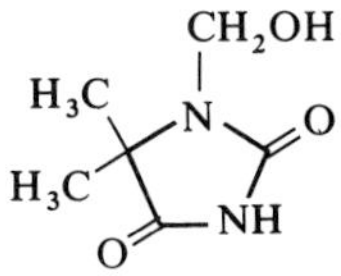

XXXVII

The reaction used for the preservatives has been applied to the determination of "free" formaldehyde in cosmetics.[61]

Fluorimetric determination of formaldehyde-releasing cosmetic preservatives.[1326] *Reagents. 2,4-Pentanedione solution*—prepare a 2 M aqueous ammonium acetate buffer solution daily, adjust pH to 6·0 with glacial acetic acid, and then make 0·02 M in acetylacetone (2,4-pentanedione, purified by distillation and the fraction boiling at 134–137° collected). The buffer solution (100 ml) contained 15·4 g of ammonium acetate, 0·30 ml of acetic acid and 0·2 ml of acetylacetone. Prepare formaldehyde standard solutions daily from 36·8% ACS formaldehyde in 10% methanol (0·1 to 0·8 μg of formaldehyde per ml).

Sample preparation. Weigh accurately approximately 100 mg of clear sample containing 0·02 to 1·0 mg of preservative, transfer to a volumetric flask with 10% methanol, and dilute to volume with 10% methanol. Dilute further, if necessary, to bring within the range of the formaldehyde standards.

Weigh accurately approximately 1 g of opaque sample containing 0·05 to 10 mg of preservative, transfer with a minimum of water to a separatory funnel, acidify with a few drops of concentrated HCl, and extract with two 30-ml portions of chloroform. Discard the chloroform. Make the aqueous phase 10% in methanol, transfer to a volumetric flask, and dilute to volume with 10% methanol. Dilute further, if necessary.

Procedure. To 2 ml of the reagent solution add 2 ml of the test solution. Heat at 60° for 1 hr, cool, and determine the fluorescence intensity at *F*410/510.

A comparative study has been made of the fixation on heterografts of aldehydes used for their tanning.[1327] Essentially, the fixation of formaldehyde, crotonaldehyde and glutaraldehyde on the aortic valves of the pig, elastin, and reconstituted collagen was studied with the help of the MBTH procedure. Acid-labile fixed formaldehyde is liberated, starting from tanned and washed samples, by distillation in the presence of acid.[1328] The formaldehyde in the distillate is then determined by the MBTH method.

Other methods for freeing fixed formaldehyde include submitting the tanned and carefully washed samples to the action of water at 100° or at 80° in a sealed tube for 5 hr or to the action of N H_2SO_4 for 24 hr at room temperature. The liberated aldehyde is then determined by the MBTH method. A blank involving non-tanned tissue should also be run.

In the MBTH procedures for formaldehyde, crotonaldehyde and glutaraldehyde Beer's law is followed from about 0·01 to 0·06 micromoles of aldehyde. The millimolar absorptivity obtained for formaldehyde at λ 623 nm was about twice that obtained for crotonaldehyde and glutaraldehyde.

In the fixation of the formaldehyde it is believed that it reacts with an amino group, e.g. the ε-amine and the lysine residue to form $—NHCH_2OH$. Further reaction with the diverse groupings in proteins forms acid-resistant groups, e.g. reaction with amide, guanidyl, indole (acid labile), imidazole and phenol.[1329, 1330] The formation of the phenolic acid-resistant bond has been demonstrated; it consists of the grouping lysine–methylene–tyrosine.[1331–1333] It has also been demonstrated that the ε-amino group of lysine (and of hydroxylysine in the case of collagen) is involved in the reaction between proteins and glutaraldehyde.[1334–1337]

Many of the combinations of formaldehyde with tissue proteins are reversible.[1338] Thus, these combinations of formaldehyde with the biopolymers should be capable of histochemical, cytochemical and *in vitro* microchemical analysis.

Colorimetric determination of aldehydes fixed on heterografts following tanning.[1327] *Reagent.* 1% aqueous MBTH·HCl. Ferric chloride solution, 0·2% aqueous.

Procedure. To 0·8 ml of test solution add 0·2 ml of reagent (pH of the mixture about 3 to 4). Heat at 100° for 3 min. (For glutaraldehyde, instead of heating let stand for 30 min at room temperature.) After cooling, add 2·5 ml of the ferric chloride solution. After standing 10 min at room temperature add 6·5 ml of acetone. Measure the absorbance at 623 nm.

Formaldehyde can also be derived from $—O—CH_2—N\langle$ compounds formed as intermediates. An example of this type of reaction is the addition of 1-methyl-1-*p*-nitrophenyl-2-methoxy azonium tetrafluoroborate to quinoline N-oxide to form a tricyclic compound which then fragments into 2-(*p*-nitrophenylazo)quinoline and formaldehyde, Fig. 79.[404]

2,4,6-Tri(hydroxymethylamino)-1,3,5-triazine (or TMM) is a compound (containing the very reactive $NHCH_2OH$ grouping) which is also anticarcinogenic and mutagenic.[770]

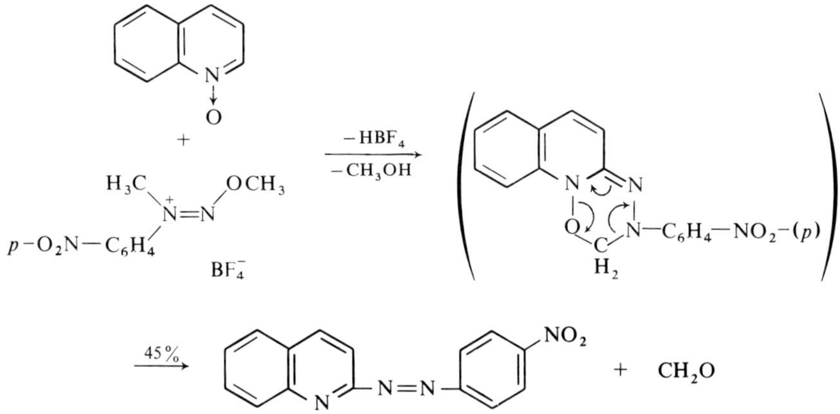

$$\text{quinoline N-oxide} + \underset{BF_4^-}{H_3C\overset{+}{N}(C_6H_4NO_2\text{-}p)=N\text{-}OCH_3} \xrightarrow[-CH_3OH]{-HBF_4} \left(\text{tricyclic intermediate: } N\text{-}C_6H_4\text{-}NO_2\text{-}(p)\right) \xrightarrow{45\%} \text{2-(}p\text{-nitrophenylazo)quinoline} + CH_2O$$

FIG. 79. Reaction of quinoline N-oxide with 1-methyl-1-*p*-nitrophenyl-2-methoxy azonium tetrafluoroborate.[404]

D. —O—CH_2—S— compounds

Formaldehyde reacts with various protein functions by addition.[1339] It has been shown that formaldehyde is bound to SH groups and can be demonstrated by the use of the mercuric chloride–Schiff mixture or to a lesser degree by Schiff's reagent alone.[383] This positive reaction of the bound formaldehyde (—S—CH_2OH probably) must be considered in the plasmal reaction when the sublimate–Schiff mixture is used as a detecting agent. This bound formaldehyde also interferes in the UV–Schiff reaction. There is no interference in the PAS reaction and in the plasmal reaction carried out with the separate application of sublimate and Schiff's reagent.[383] It is presumed that formaldehyde is released by the action of $HgCl_2$ and of SO_2 first from the SH groups in tissue sections and reacts with the Schiff reagent. The difference in intensity between the formaldehyde–Schiff and the formaldehyde–Schiff–$HgCl_2$ reactions has been postulated as due to a different capability of $HgCl_2$ and SO_2 to release formaldehyde from SH groups. Elleder and Lojda[383] state that the selective demonstration of SH-bound formaldehyde in keratin is an example of aldehyde–Schiff reactions in proteins. Contrary to the results obtained with formaldehyde which probably has to be released from the binding sites to react with S.r., acrolein gives a direct reaction with S.r. in places where it is bound.[1340] Also the glutaraldehyde–Schiff reaction demonstrates directly one free aldehyde group of glutaraldehyde bound to different protein functions. These results indicate that the reaction with bound formaldehyde is a positive interference in the plasmal reaction according to Hayes[1341] and in the UV–Schiff reaction.

E. $>N-CH_2-N<$ compounds

N-Methyl compounds have to be oxidized by vigorous methods in the test tube to form formaldehyde, as shown for the N-methylimidazole derivative, Fig. 80. This type of reaction has been discussed in the N-methyl section.

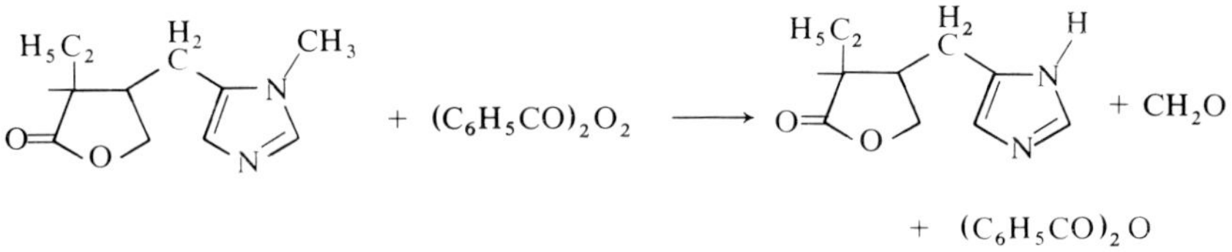

H_5C_2, H_2C, CH_3 + $(C_6H_5CO)_2O_2$ ⟶ H_5C_2, H_2C, H + CH_2O + $(C_6H_5CO)_2O$

FIG. 80. Oxidative N-demethylation of an N-methylimidazole derivative.

On the other hand the $-RN-CH_2-NR-$ moiety is readily hydrolyzed to formaldehyde in acid solution and thus compounds containing this group can be readily assayed. Sometimes this formaldehydogenic property can be troublesome. The administration of methenamine mandelate (Mandelamine) to normal individuals and patients with carcinoid results in a marked reduction in apparent urinary hydroxyindoleacetic acid as determined by the quantitative nitrosonaphthol method.[299] The interference is probably due to the reaction of the released formaldehyde with the indole molecule to give a derivative less readily attacked by nitrosonaphthol.

Hexamethylenetetramine (or methanamine or hexamine), a condensation product of formaldehyde and ammonia, liberates formaldehyde in the presence of acid and thus is antibacterial and possibly genotoxic. It is moderately effective against most urinary pathogens when the urine is brought below pH 5·5. It is usually utilized as the mandelate since the action of mandelic acid is primarily to contribute to the acidification of the urine.

A semi-automated colorimetric determination of hexamine and its mandelate in tablets is available.[289] Essentially the continuous flow-through system involves heating an ammoniacal solution of the tablets and colorimetrically determining the derived formaldehyde with chromotropic acid. Other methods can be used for the quantitation of this compound (see Formaldehyde Section in Volume 1 and data in this section), including those based on chromotropic acid reactions[386,915,1342] and the 2,4-pentanedione colorimetric[556] and fluorimetric[1343] reactions.

2-Hydrazinobenzothiazole has also been used to determine methenamine and its salts through the derived formaldehyde. The chromogen formed in the procedure is a formazan cation. Beer's law was followed from 0·005 to 0·04 μmole/5 ml of final solution. A fresh standard curve was prepared for each analysis.

Colorimetric determination of methenamine and its salts by 2-hydrazinobenzothiazole.[290] *Reagents. HBT*—0·5% 2-hydrazinobenzothiazole in 10% HCl. Stable for 2 weeks. *Ferricyanide*—1% aqueous potassium ferricyanide.

Procedure. Mix 0·5 ml of aqueous test solution with 0·5 ml of 10 N H_2SO_4. Heat at 100° for 15 min and allow to cool to room temperature before the addition of 0·5 ml of HBT. After 5 min, add 0·5 ml of ferricyanide. Let stand 20 min and add 3 ml of dimethylformamide. After 20 min, read the absorbance at 510 nm.

Addition of alkali to the solution forms the blue formazan anion, shifts the wavelength maximum to longer wavelength and increases the intensity of the color.

Hexamine is used as a methylene donor in the resorcinol bonding systems employed in tire vulcanizates.[291] These bonding systems enhance the adhesion necessary between the rubber and cord elements of a tire. A small amount of hexamine remains in the rubber after vulcanization.[1344] The hexamine can be removed from the cured rubber by extraction with chloroform. Water can then be used to selectively extract the highly water-soluble hexamine from a large number of other rubber additives by reextraction of the chloroform with water. Resotropin, a reaction product of hexamine and resorcinol is also used as a bonding agent in vulcanizates.[1345] Resotropin yields free hexamine upon extraction with chloroform. The following procedure was used to detect the hexamine and could be modified for quantitation.

Detection of hexamine in vulcanizates.[291] Extract 5 g of finely cut vulcanizate 2 hr in a soxhlet apparatus with chloroform. Concentrate the extract to less than 25 ml, transfer to a 25-ml volumetric flask and dilute to volume with chloroform. Add $\frac{1}{2}$ ml of water, stopper the flask, and shake vigorously several times during a 15- to 20-min period. Once the water layer has reformed, remove the clear portion for TLC analysis.

Analysis. Separate 5 µl of the clear aqueous solution on silica gel G with methanol–concentrated ammonium hydroxide (9:1). After development, the air-dried layer was sprayed with chromotropic acid solution and placed in an oven at 110° for 15 min. Hexamine was observed as a blue-violet spot on a light blue background at an R_f of 0·45. The chromotropic acid solution was prepared by mixing 5 parts of 60% H_2SO_4 with one part of aqueous 10% chromotropic acid. Five µl of a 0·25% aqueous solution of hexamine was used as a standard.

Atmospheric dimethylamine can react with formaldehyde to form bis(dimethylamino)methane (or N,N,N′,N′-tetramethylmethylenediamine). It registers as formaldehyde by the chromotropic acid or Schiff methods, 1% of it being equivalent to 0·29% formaldehyde. Methylamine can react with formaldehyde to form 1,3,5-trimethylcyclotrimethylenetriamine (or 1,3,5-trimethylhexahydro-s-triazine), XXXVIII. This compound registers as formaldehyde by the Schiff or chromotropic acid procedures.

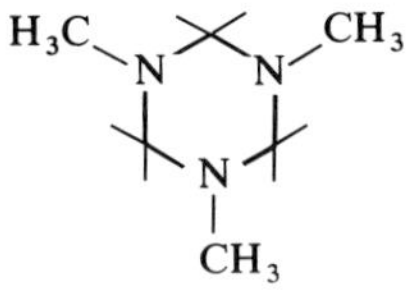

H_3C N N CH_3
N
CH_3

XXXVIII

When casein and formaldehyde were reacted together and washed for 12 days, the casein contained about 2% combined formaldehyde.[1346] Experiments in tanning of casein by formaldehyde at pH 6·0 showed that there was a condensation between the ε-amino groups of a lysine component and the peptide links of another chain. Much of the formaldehyde combined with protein could only be released by prolonged acid hydrolysis.

The irreversible binding of formaldehyde by proteins such as casein, gelatin and wool decreased sharply at a pH above 1.[1347] However, for carboxyhemoglobin this binding reached a maximum at pH 4. This was probably due to the formation of bridges involving the amide groups of asparagine which is present in large amounts in the hemoglobin.

Previously we have discussed briefly the carcinogenicity of the N-methyl 4-phenylazoaniline (or 4-methylaminoazobenzene) dyes. One of the metabolic pathways consists of the reactions

$$\phi\text{-N}{=}\text{N}-\phi\text{-NRCH}_3 \longrightarrow \phi\text{-N}{=}\text{N}-\phi\text{-NR}-\text{CH}_2\text{OH} \longrightarrow \phi\text{-N}{=}\text{N}-\phi\text{-NHR} + \text{CH}_2\text{O}$$

Much of our knowledge concerning these carcinogens has been derived from the research of the Millers.[967] These workers have found that the dye is bound through chemical linkages to certain of the liver proteins. They have suggested that these reactions may play a role in the carcinogenesis of the azo dyes.[1348, 1349] Their observations on the dye binding have been confirmed.[1350] The Millers suggest that the N-hydroxymethyl derivative of the dye is the metabolite that binds to the protein. N-Hydroxymethylgroups are highly reactive and form stable bonds with compounds which possess reactive hydrogens through elimination of water.[1330] N-Hydroxymethyl groups can thus crosslink an amino group with the activated carbon of a phenol, imidazole or indole compound or with the activated nitrogen of an exocyclic amino group or an endocyclic imino group. If such Mannich bases were formed with a protein grouping such as in tyrosine, histidine or tryptophan which has a reactive CH, the alkali stable $>N-CH_2-\overset{|}{\underset{|}{C}}-$ grouping would be formed. Otherwise, reaction would take place with reactive $>NH$ groups in the protein to yield the alkali labile $>N-CH_2-N<$ grouping. Since alkaline hydrolysis releases only small amounts of azo dye, it is believed that most of the dye is bound through the more stable $>N-CH_2-C\leqslant$ link. The Millers[967] have emphasized that protein dye could not be detected in the livers of guinea pigs, hamsters, rabbits, cotton rats, chipmunks, or chickens fed N,N-dimethyl 4-phenylazoaniline. No one has succeeded in producing liver tumours in these species by feeding this dye.

It is possible that some analytical use could be made of the formation of formaldehyde from the cross-linked compound (containing the group $>N-CH_2-N<$) formed from the reaction of a carcinogenic amine and a biopolymer.

Since the metabolism of so many carcinogens results in the by-product formation of formaldehyde (and other aldehydes), the genotoxic properties of formaldehyde and its many derivatives need thorough investigation. We have discussed the mutagenicity of formaldehyde. This property is usually related to the capacity of formaldehyde to form adenine dimers through methylene bridges.[843, 1351] N-Methylol groups are postulated as crosslinking agents responsible for mutagenicity of formaldehyde.[843] Considering the mechanism of formaldehyde reaction with DNA in cells, it should be remembered that formaldehyde interacts much more readily with the amino groups of proteins and amino acids to form $>N-CH_2OH$ compounds than with the nitrogen bases.[1352]

The $>N-CH_2-OH$ compounds are probably first formed when formaldehyde enters the cells. Monomethylol derivatives of amino acids can react with amide, guanidine, phenol, imidazole or indole groups(1329, 1330) to form cross-linked condensation products. Siomin *et al.*(530, 1353) have shown that the products of reaction between formaldehyde and amino acids actively react with nucleotides and nucleic acids. These authors have demonstrated that treatment of wild strains of *E. coli*, and strains deficient in excision repair, with a product of reaction between formaldehyde and amino acid produces an inactivation of the cells as well as single strand breaks in bacterial DNA.(1236) The breaks are successfully repaired in wild-type cells but remain unrepaired in bacteria deficient in DNA-polymerase I.

The report that a 3-fold excess of amino acid with respect to formaldehyde causes practically complete binding of aldehyde by amino acid is in accordance with values in the literature for the equilibrium constant for the formation of monomethylolglycine, $HOOC-CH_2-NH-CH_2OH$, during reaction of formaldehyde with glycine.(1352) The molecular weight of DNA sharply decreases under the action of the aminomethylol compounds and the acid-soluble products formed.(530, 1353) Since formaldehyde by itself is not capable of such action, Siomin *et al.* conclude that the single strand breaks in the DNA of bacterial cells may be due to the action of the N-methylol product or they may be caused by the excision repair only. From these various data it is postulated that the action of formaldehyde on bacterial DNA (leading to lethality and, possibly, to mutagenesis) is not exerted by formaldehyde itself but by the products of its reactions with amino containing compounds.

On this basis the microanalytical investigation of the formaldehyde precursor moieties in DNA could be of considerable importance in genotoxic studies.

F. $>N-CH_2-S-$ compounds

Formaldehyde can be derived from compounds containing this structure. Examples of pesticides which have this grouping are Imidan, O,O-dimethyl S-phthalimidomethyl phosphorodithioate, XXXIX, and Supracide, O,O-dimethyl phosphorodithioate S ester of 4-(mercaptomethyl)-2-methoxy Δ^2-1,3,4-thiazolidin-5-one, XL.

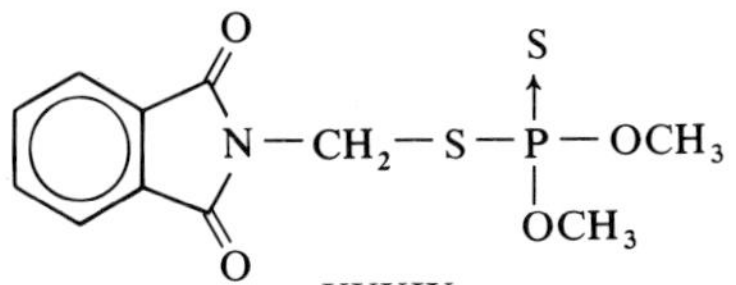

XXXIX

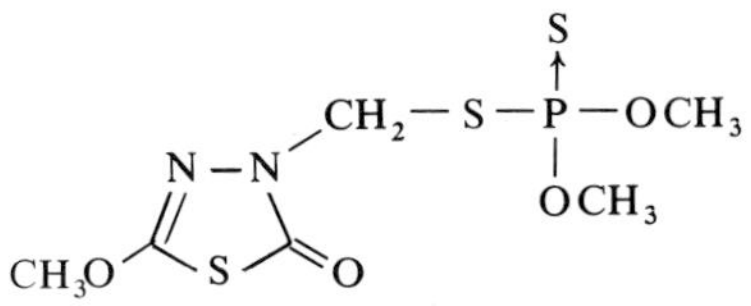

XL

Thiazolidine and many of its derivatives contain the $—S—CH_2NR—$ moiety. Formaldehyde can be obtained from the 2-unsubstituted compounds and assayed by the method shown for thiazolidine-4-carboxylic acid, Fig. 81. The following procedure was found to be useful.

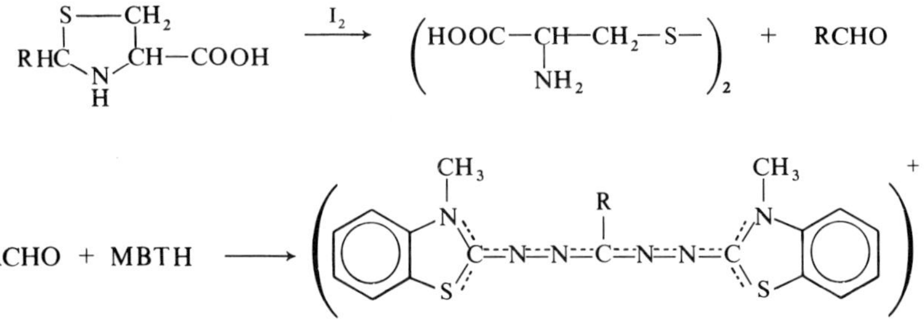

$$\text{RHC(S—CH}_2\text{)(NH)CH—COOH} \xrightarrow{I_2} \left(\text{HOOC—CH(NH}_2\text{)—CH}_2\text{—S—}\right)_2 + \text{RCHO}$$

$$\text{RCHO} + \text{MBTH} \longrightarrow \text{(3-methylbenzothiazolinylidene)—C=N—N=C(R)—N=N—C(benzothiazolium, CH}_3)^+$$

λ max 660–670, mε 42–58 (340)

FIG. 81. Determination of 2-alkyl-4-thiazolidinecarboxylic acid with MBTH following iodometric oxidation.

MBTH determination of thiazolidine-4-carboxylic acids.[484] To 0·5 ml of aqueous or 5% trichloroacetic acid test solution add 0·5 ml 0·4% aqueous MBTH·HCl. After 5 min at 25°C add 0·25 ml of iodine solution (127 mg of iodine dissolved in 100 ml of 5% aqueous potassium iodide). Shake vigorously for 0·5 min and let stand 15–180 min, depending on the thiazolidinecarboxylic acid under study. Add 2·5 ml of 0·2% fresh aqueous ferric chloride. After 10 min dilute to 10 ml with acetone. Read at λ 660–670 nm within 30 min after the addition of the acetone.

G. $—S—CH_2—S—$ compounds

Compounds containing the methylenedithio group can also be determined through released formaldehyde. Some of the pesticides have this functional group. An example is phorate which can be determined in the following fashion, Fig. 82. Many times it is worthwhile to oxidize some of these compounds to sulfones since a better yield of formaldehyde can be obtained on hydrolysis, e.g. as for ethion.[144,286] Perbenzoic acid can be used in the oxidation. The sulphones have to be hydrolyzed with alkali. The highest yield of formaldehyde is obtained from the sulphones. Even the sulphoxides are readily hydrolyzed. Thus, methyl methylthiomethyl sulphoxide, $CH_3SCH_2—SO—CH_3$ forms formaldehyde in acid solution.

Formaldehyde and other aldehydes can be formed by the following reaction.[1354]

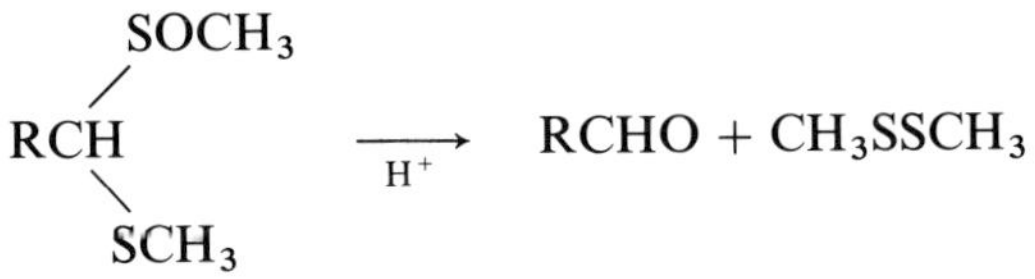

$$RCH(SOCH_3)(SCH_3) \xrightarrow{H^+} RCHO + CH_3SSCH_3$$

The hydrolysis is accomplished under mild conditions and the only byproduct is dimethyl disulphide.

Between pH 6 and 8 formaldehyde reacts with keratin without affecting the —S—S— bonds of cystine.[1355] In more alkaline solution it reduces —S—S— to 2 SH groups and reacts to form $—S—CH_2—S—$ in place of the original —S—S—. These groups could then be analytically determined.

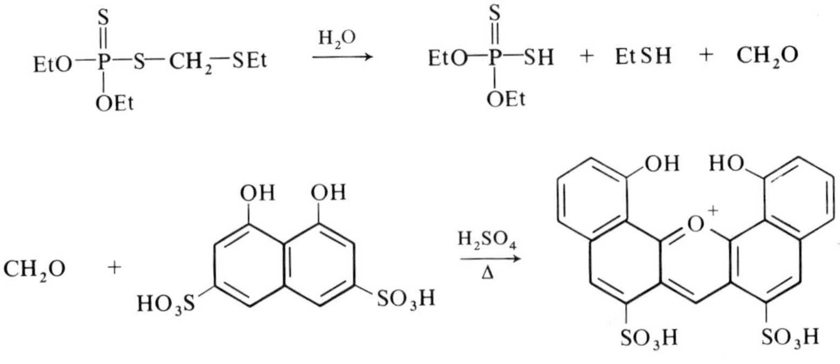

$$EtO—P(=S)(OEt)—S—CH_2—SEt \xrightarrow{H_2O} EtO—P(=S)(OEt)—SH + EtSH + CH_2O$$

$$CH_2O + \text{chromotropic acid (OH, OH; } HO_3S, SO_3H) \xrightarrow[\Delta]{H_2SO_4} \text{dye (OH, HO, } O^+; SO_3H, SO_3H)$$

λ max 578 me 15.7

FIG. 82. Determination of phorate with chromotropic acid.

Quite a few pesticides have the $—S—CH_2—S—$group and so could be considered as potential formaldehyde precursors. Examples of some of these pesticides are ethion or nialate, XLI, guthion, XLII, methyl trithion, XLIII, phenkapton, XLIV, phorate or thimet, XLV, and trithion or carbophenothion, XLVI.

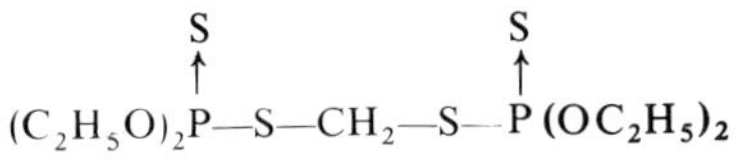

$$(C_2H_5O)_2P(\rightarrow S)—S—CH_2—S—P(\rightarrow S)(OC_2H_5)_2$$

XLI

$$(CH_3O)_2P(\rightarrow S)—S—CH_2—N(\text{benzotriazinone})$$

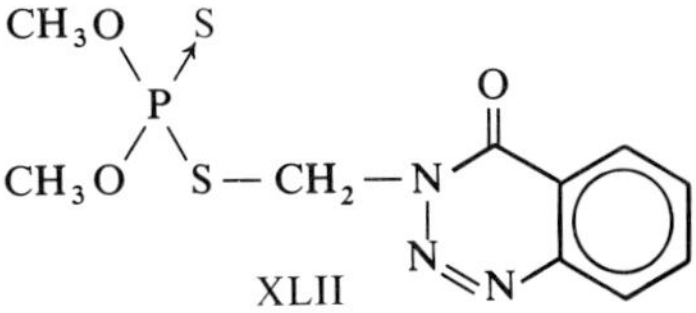

XLII

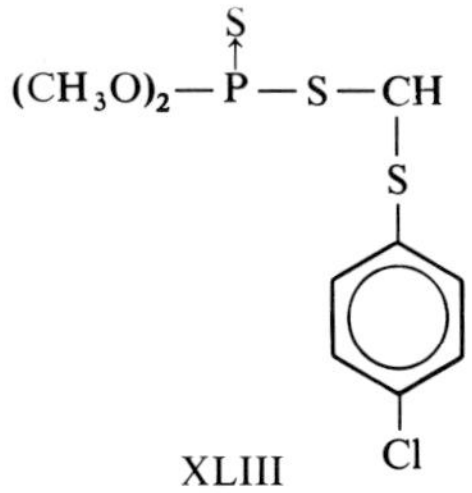

$(CH_3O)_2-\overset{S\uparrow}{P}-S-CH(-S-C_6H_4-Cl)$

XLIII

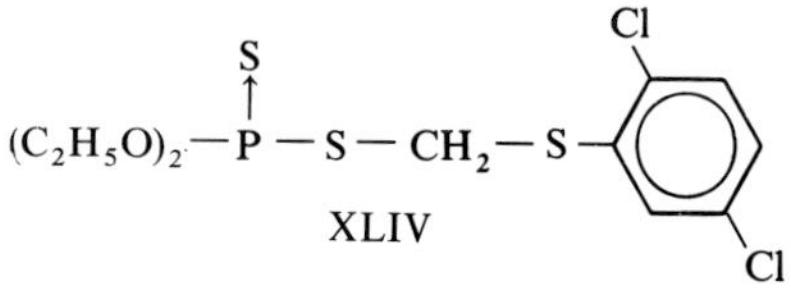

$(C_2H_5O)_2-\overset{S\uparrow}{P}-S-CH_2-S-C_6H_3Cl_2$

XLIV

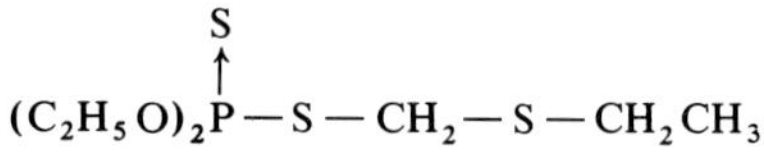

$(C_2H_5O)_2\overset{S\uparrow}{P}-S-CH_2-S-CH_2CH_3$

XLV

$(C_2H_5O)_2\overset{S\uparrow}{P}-S-CH_2-S-C_6H_4-Cl$

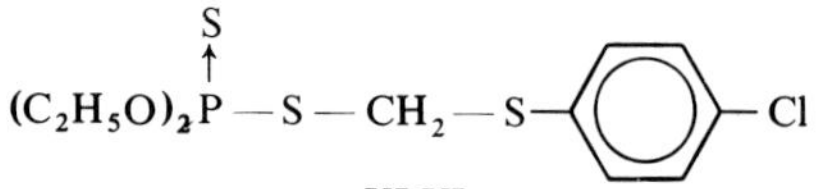

XLVI

H. $-O-CH_2-\overset{|}{\underset{|}{C}}-$ compounds

In compounds of this type a reactive carbon adjacent to the methylene group helps in the disproportionation. An example of this type of cleavage is shown by the formation of pyruvate and formaldehyde from 2-keto-4-hydroxypyruvate.[1356] Essentially the mammalian degradation of L-homoserine leads to formaldehyde by the following series of reactions, Fig. 83. The amino group of homoserine is removed via transamination. The 2- keto-4-hydroxybutyrate is a substrate for 2-keto-4-hydroxyglutarate aldolase. This aldolase has been prepared in pure form from extracts of rat liver,[1357,1358] bovine liver[1359] and *Escherichia coli.*[1360]

Formaldehyde participates in binding with the active-site lysyl residue of 2-keto-4-hydroxyglutarate aldolase.[1361] Cyanide irreversibly inactivates

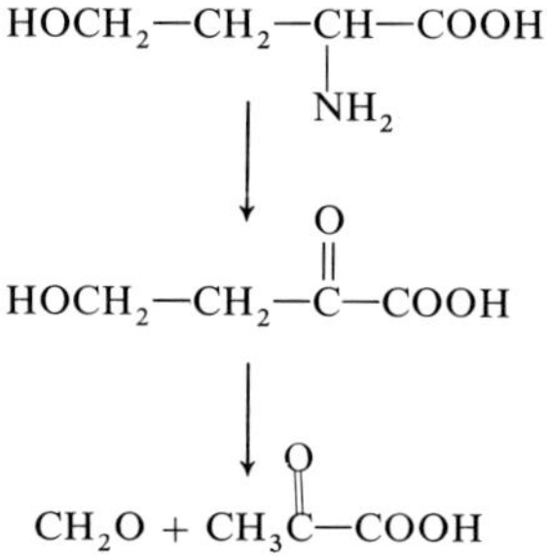

$$HOCH_2-CH_2-\underset{NH_2}{\underset{|}{CH}}-COOH$$

$$\downarrow$$

$$HOCH_2-CH_2-\overset{O}{\overset{\|}{C}}-COOH$$

$$\downarrow$$

$$CH_2O + CH_3\overset{O}{\overset{\|}{C}}-COOH$$

FIG. 83. Formaldehyde formation from homoserine.

this aldolase in the presence of formaldehyde due to aminonitrile formation, e.g., $CH_2{=}N{-}$enzyme + $CN^- \rightarrow NC{-}CH_2{-}NH{-}$enzyme. A somewhat similar direct interaction of formaldehyde with the ε-amino group of a lysyl residue is the serine transhydroxymethylase-catalyzed reaction where formaldehyde is probably transferred from serine to 5,10-methylene tetrahydrofolate through an intermediate enzyme-formaldehyde complex.[1362] These phenomena should be capable of utilization in microanalytical studies.

2-Nitro-1-alkanols (discussed previously) and oxetanes (to be discussed) are also examples of the $O{-}CH_2{-}\overset{|}{\underset{|}{C}}{-}$ type of molecule. An example of the analysis of such a compound through formaldehyde is 2-nitro-2-bromo-1,3-propanediol, XLVII.

$$HOCH_2{-}\overset{\displaystyle NO_2}{\underset{\displaystyle Br}{C}}{-}CH_2OH$$

XLVII

Theoretically 2 molecules of formaldehyde could be obtained from 1 molecule of XLVII. Actually, the yield of formaldehyde derived from XLVII is highly dependent on temperature and at 60° an average value of 28% of the theoretical was obtained.[1325] 2,4-Pentanedione was used to determine the compound fluorimetrically. Standards have to be run in the procedure. Stock solutions of formaldehyde can be standardized by precipitation with dimedone[1363] or by methods described previously.

I. Methylene halides

Methylene chloride is the main ingredient in most paint removers. The indoor use of paint removers containing methylene chloride results in the absorption of this solvent, which is then metabolized to carbon monoxide[1364–1366] and formaldehyde. Exposure for 2 to 3 hr can result in the elevation of carboxyhemoglobin to levels that stress the cardiovascular system.[1367] Since one-sixth of the 180 million kg of methylene chloride produced in the USA is being consumed in the rapidly expanding paint-remover market,[1368] the toxicity of methylene chloride and its metabolic products needs more thorough study.

From reports available on the metabolism of the halogenomethanes by tissue suspensions it appears probable that several quite distinct routes of metabolism are operative.

An alternative route is formaldehyde formation.[1369] Rat liver slices, homogenates and extracts dehalogenate a number of bromo- and chloroalkanes. The products were halide ion and formaldehyde.

$$CH_2BrCl + H_2O \rightarrow Br^- + Cl^- + 2H^+ + CH_2O$$

The system was activated by cyanide, glutathione or cysteine, and dehalogenation was more active under nitrogen than in the presence of oxygen. The process was decreased by heating the tissue sample for 5 min at 60°C and liver activity was increased by repeated exposure of the rat to bromochloromethane; kidney slices were more active than liver cells. This particular enzymic system appears to be non-oxidative in its mechanism, the properties indicating that it is *unrelated* to the metabolic degradation of carbon tetrachloride that is important for the development of necrosis.

J. Formaldehyde and the genotoxic methylene group

Mammals can convert many types of formaldehyde (or aldehyde) precursors to ultimate mutagens. These mutagens can be detected with a quantitative microsomal mutagenesis assay method utilizing an appropriate micro-organism.[1370,1371] The active mutagen metabolites may be short-lived, electrophilic agents which can combine with many common chemical groups containing atoms with an open electron pair, e.g. $—\ddot{N}H_2$, so that the active mutagen metabolite has to be in close spatial and temporal contact with the appropriate genic groupings in the organism to imitate a mutagenic pathway or it is used up in non-genic reactions. For example, in the case of dimethylnitrosamine the ultimate mutagen or carcinogen is usually postulated to be the alkylating CH_3^+ group, or some appropriate precursor of this group. However, formaldehyde is a byproduct in the metabolism of dimethyl-nitrosamine, is also a metabolic byproduct from many other types of carcinogens, and is readily formed from many types of activated methylene compounds.

Formation of formaldehyde at genic hot spots could initiate a mutagenic pathway also. Recent work indicates this as a possibility to consider. For example, let us examine the relationship between formaldehyde formation (or N-demethylation) and mutagenicity in the liver microsomes of C_3H and $C_{57}BL6$ mice.[1370] *Salmonella typhimurium* was used as the test organism. It was found that microsomes from C_3H mice activated dimethylnitrosamine to a mutagen more than did those of $C_{57}BL6$ mice. Pretreatment with 3-methylcholanthrene which induces activity of mixed function oxidase microsomal enzymes increased activation of dimethylnitrosamine by both strains but to different degrees. Formaldehyde production was measured[1372] and correlated well with the induction of histidine reversions, Table 28.[1370] With other conditions possibly being equal, formaldehyde formation from dimethylnitrosamine is proportional to its activation to a mutagen. Formaldehyde formation (or the initial demethylation of dimethylnitrosamine) is probably the rate controlling step in the activation of dimethylnitrosamine to

Table 28. Effect of 3-methylcholanthrene enzyme induction of livers of C_3H and $C_{57}BL/6$ mice in terms of dimethylnitrosamine demethylase activity and metabolism by liver microsomes of dimethylnitrosamine to a mutagen[1370]

Mouse strain	Inducer	Demethylase activity[a]	Revertants per 10^6 survivors
$C_{57}BL/6$	3-MeChol.	4·66	42·9
C_3H	3-MeChol.	3·88	31·8
C_3H	Control	3·53	20·3
$C_{57}BL/6$	Control	2·66	13·8

[a] nmoles CH_2O/mg protein/min.

a mutagen.[1370] The authors emphasize that alternative pathways proportional to demethylation (actually formaldehyde formation) are also possible.

In line with this type of phenomenon is the report that formaldehyde, a widely distributed compound in man's environment,[770,1373] is mutagenic for a wide variety of organisms, including viruses, procaryotes and eucaryotes.[770,1374] The mutagenic activity of formaldehyde has been discussed in our volumes previously. Formaldehyde also produces crossing-over in *Drosophila melanogaster*.[1375–1377] It induces intra- and intergenic recombination in the eucaryotic organism, *Saccharomyces cerevisiae*.[1378] On the basis of the higher sensitivity to formaldehyde of *Escherichia coli* mutants defective in the excision-repair of UV-induced lesions[1351] or in DNA polymerase I[1236] as compared to the corresponding wild type, Chanet *et al.*[1378] postulated that the damage induced by CH_2O and by UV irradiation may be repaired by similar processes.

Because of its high order of activity formaldehyde reacts with a wide variety of organic cellular compounds, e.g. proteins, nucleic acids, etc.[1374] Consequently, in living tissue formaldehyde can take many pathways. The mutagenic pathway of formaldehyde has been postulated as involving the formation of methylene-bis compounds of purines in DNA[843,1374,1379] with the 6-amino group of adenylic acid (or adenosine) mediating the mutagenic activity of formaldehyde. Formation of the methylene-bis compounds of purines is extremely slow as compared with the reactions of formaldehyde with amino groups of amino acids to form aminomethylol compounds.[1374] In line with this reasoning is the postulate that N-methylols are cross-linking agents responsible for the mutagenicity of formaldehyde.[843] The N-methylol compounds formed in the cells exposed to formaldehyde react efficiently with nucleotides leading to DNA breaks which may be subject to repair in *Escherichia coli*.[1236] In *Drosophila* and *Escherichia coli*, formaldehyde-induced mutations appear to occur at the replication point.[1380,1381] Whatever the mechanism of action of formaldehyde, this widespread agent has not only a mutagenic effect[527,1379] but also a recombinogenic effect on higher[1375,1382] and lower eucaryotic forms. These effects would probably also be caused by many of the biological precursors of formaldehyde and especially the activated methylene compounds which can be hydrolyzed to formaldehyde.

LVII. 5-METHYLTETRAHYDROFOLIC ACID

The capacity to form formaldehyde from 5-methyltetrahydrofolic acid is shared by "formaldehyde forming enzyme"[1383] and methylene reductase (5-methyltetrahydrofolate–NAD–oxidoreductase, EC 1.1.1.68),[1384] an enzyme present in rat brain tissue, other organs[1385] and blood platelet preparations.[1386] In this sense a highly positive correlation has been found in the rank order of 12 rat brain regions with respect to methylene reductase activity[1387] and the activity reported for the "formaldehyde forming enzyme" in the same areas.[1388] Other evidence has also been presented indicating these 2 enzymes are identical.[1384] The following reactions are postulated, Fig. 84.[1384,1389] These reactions could be studied and the enzyme could be assayed with the help of one of the many methods described in these volumes for the determination of formaldehyde. FAD, XLVIII, is required as a cofactor as well as an electron acceptor[1389] in the reaction, with a pH optimum of about 6·5.[1389,1390] $FADH_2$ is the 1,10-dihydro derivative of FAD.

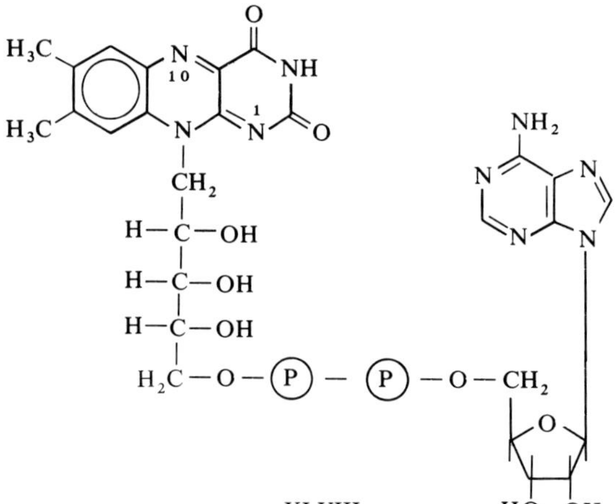

XLVIII

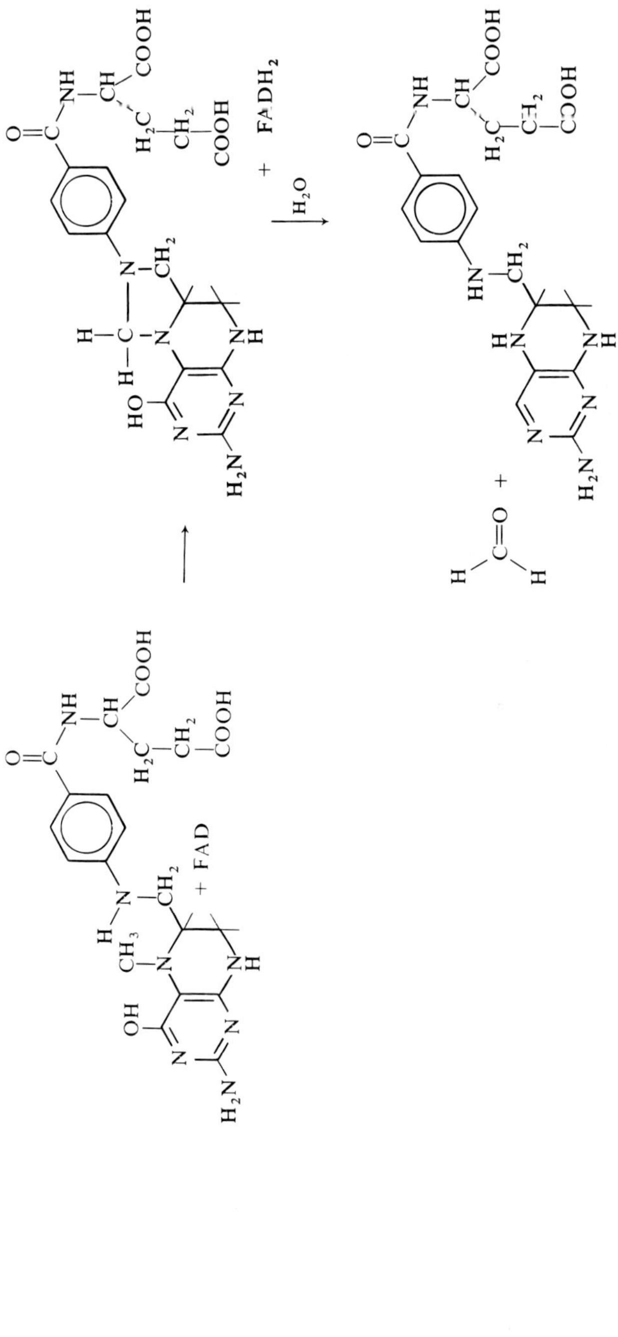

FIG. 84. Formaldehyde formation from 5-methyltetrahydrofolic acid.[1384]

It has been shown that the 5-methyltetrahydrofolate–N-methyl–transferase activity believed to be present in various animal tissues[1391–1393] is actually due to the enzymatic formation of formaldehyde which subsequently condenses non-enzymatically with the phenethylamines added to the reaction mixtures.[1386, 1390, 1394, 1395] This formation of a heterocyclic compound could also be utilized to follow the reaction and determine the formaldehyde.

Formaldehyde formation from 5-methyltetrahydrofolic acid,[1396] as well as its further reaction with β-phenylethanolamine to form a tetrahydroisoquinoline derivative, is activated in rat crude tissue extracts by the addition of menadione, XLIX,

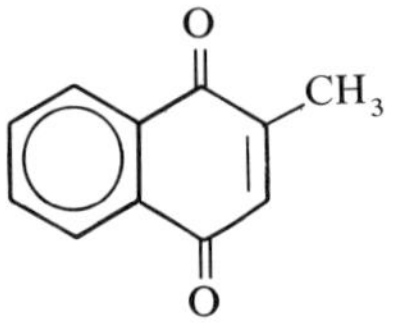

XLIX

to the incubation mixtures.[1384] FAD also stimulates the process only in the presence of oxygen. Together, FAD and menadione maximally stimulate the formation of formaldehyde, or of the amine condensation product, under anaerobic conditions.

It appears that nonenzymatic breakdown of 5-methyltetrahydrofolic acid to formaldehyde can occur.[1386,1391] Ordinarily methylene reductase is needed to catalyze the dehydrogenation of 5-methyltetrahydrofolic acid to 5,10-methylene tetrahydrofolic acid, Fig. 84. The latter compound has the $>N-CH_2-N<$ group so that it can be non-enzymatically hydrolyzed to formaldehyde and tetrahydrofolic acid.[1387]

Under defined conditions S-adenosylmethionine, L, can generate formaldehyde in human blood by an enzymatic process.[1386] Ordinarily L is the chief donor of methyl groups in living tissue.

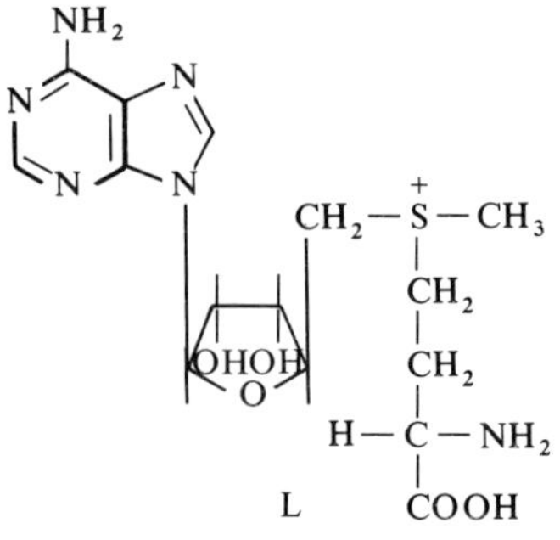

L

Evidence has been presented that incubation of amines such as dopamine with S-adenosylmethionine or 5-methyltetrahydrofolic acid results in the enzymatic formation of formaldehyde which condenses nonenzymatically with dopamine to form 6,7-dihydroxytetrahydroisoquinoline,[1386,1397] Fig. 85.

In the absence of the amine formaldehyde can be identified by the dimedone precipitation method[451, 1398] or by one of the photometric methods described in this volume or Volume 1.

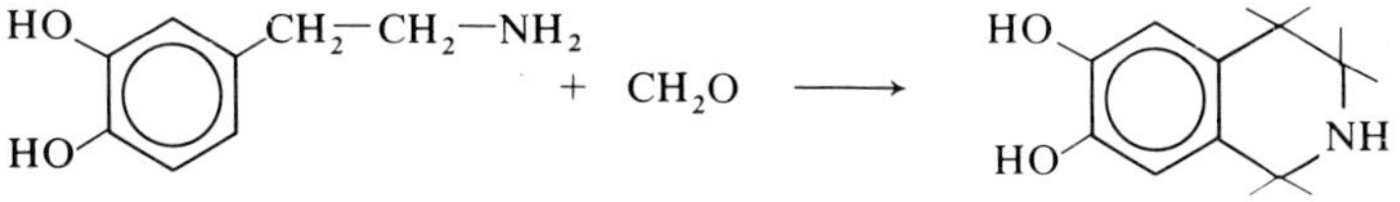

FIG. 85. Reaction of dopamine with formaldehyde.

Since alkaloids such as tetrahydroisoquinoline and tetrahydropapaveroline were detected in urine of parkinsonian patients treated with L-Dopa,[1399] formaldehyde precursors could play a role in the formation of these heterocyclic compounds. It has even been postulated that because of formaldehyde's high affinity for biogenic amines and probably also for certain macromolecules, formaldehyde production could play some role in the aging process,[1394] as does malonaldehyde. (See Malonaldehyde Precursors Section in forthcoming volume.)

It has been demonstrated that formaldehyde is enzymatically formed from 5-methyltetrahydrofolic acid,[1394, 1400] but one metabolic pathway can involve reaction of formaldehyde with a biological phenethylamine. For example, incubating epinine with a pig brain enzyme resulted in the formation of 2-methyl-6,7-dihydroxy-1,2,3,4-tetrahydroisoquinoline, Fig. 86, as identified by thin-layer chromatography.[1400] 5-Methyltetrahydrofolic acid was the source of the formaldehyde. Similar results were obtained with other amines, like tryptamine, N-methyltryptamine and 4-methoxy-3-hydroxyphenylethylamine using 5 different solvent systems. This is in line with the postulate that alkaloids of the tetrahydroisoquinoline or the β-carboline group could be formed *in vivo* by the Pictet–Spengler condensation of corresponding amines with formaldehyde.[1399] Similarly, dopamine can condense with acetaldehyde

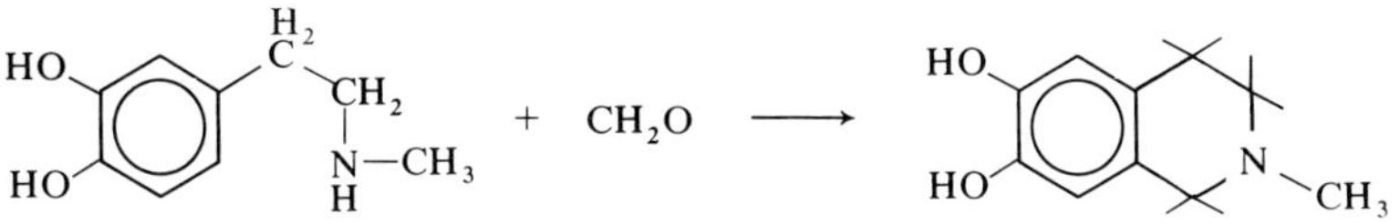

FIG. 86. Reaction of epinine with formaldehyde to form 2-methyl-6,7-dihydroxy-1,2,3,4-tetrahydroisoquinoline.

yielding salsolinol, LI, as has been demonstrated *in vivo* in the urine of parkinsonian patients being treated with L-Dopa[1399]

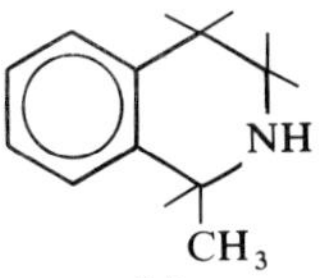

NH
CH_3
LI

Among the alkaloids derived from the 3-indolylethylamines, harmaline, LII, and harmine, LIII, are potent hallucinogens while certain tetrahydroisoquinolines derived from dopamine are reported to be inhibitors of neuramine metabolism[1401] and of catecholamine uptake,[1402] both at high concentrations only. On the basis of these types of *in vitro* reactions it has been hypothesized that these alkaloids might be produced in abnormal amounts in pathological conditions, and perhaps even in schizophrenic disorders[1400]

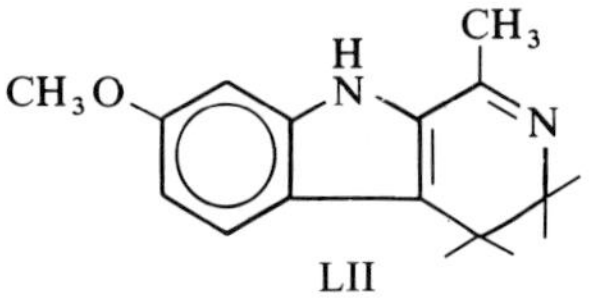

CH_3O H N CH_3 N
LII

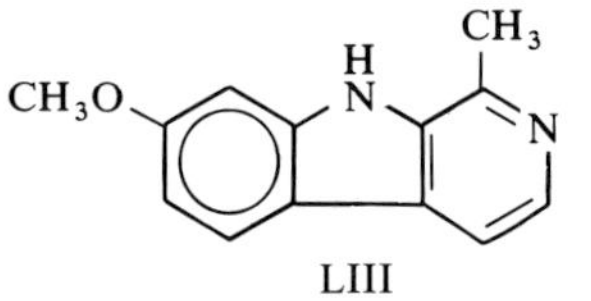

CH_3O H N CH_3 N
LIII

In line with this work is the report that an enzymatic preparation from human brain converts tryptamine, N-methyltryptamine and 5-hydroxytryptamine to tryptoline (9H-1,2,3,4-tetrahydropyrido[3,4-b]indole), LIV, and its 1-methyl- and 5-hydroxy-derivatives, respectively, in the presence of the formaldehyde precursor, 5-methyltetrahydrofolic acid.[1403]

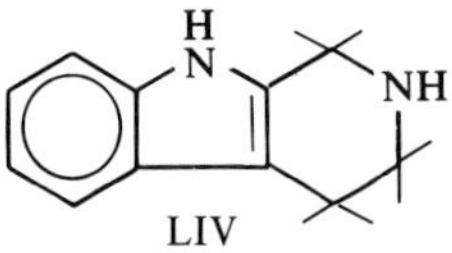

H N NH
LIV

S-Adenosylmethionine can also be a formaldehyde precursor in the formation of a tryptoline. It is postulated that methanol is formed first and this is oxidized enzymatically to formaldehyde which then condenses to form the tryptoline.[1223] It is felt that 5-methyltetrahydrofolic acid is enzymatically converted to a formaldehyde precursor which releases formaldehyde for reaction with an indolyl or phenyl ethylamine.[1404] Similar enzymatic products have been obtained from tryptamines[1395] and catecholamines.[1386] In this respect tryptolines have been isolated from the reaction between tryptamines and 5-methyltetrahydrofolate in human platelets.[1405]

One other series of related factors wherein natural formaldehyde precursors may play an important role is in the relationships between schizophrenia, epilepsy and cancer. Schizophrenia and epilepsy are believed to be biologically antagonistic diseases.[1406] Reynolds[1407] has postulated that the biochemical disturbances in untreated epileptics render them less prone to schizophrenia and vice versa. However, when epilepsy is treated with anticonvulsants, schizophrenic-like psychoses are seen;[1408] this result is attributed to the effect of anticonvulsants on folate metabolism.[1409,1410] Anticonvulsants, especially diphenylhydantoin, LIV, lower the serum folate in significant numbers of epileptic patients[1411]

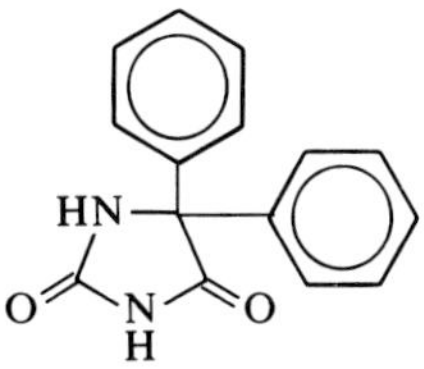

HN
O N O
H

LIV

Since the anticonvulsants are amides, they could react with the formaldehyde formed from the biological formaldehyde precursors to form reactive $N-CH_2OH$ and thus interfere with folate metabolism.

The mentally ill, especially those with schizophrenia, rarely get cancer.[1411a–d] It has been suggested that the total amount of labile methyl groups available may be a determinant in the initiation of acute schizophrenia as well as the different chronic clinical forms.[1411a]

Some of the natural formaldehyde precursors appear to be involved in these various processes, e.g. 5-methyltetrahydrofolic acid, methylenetetrahydrofolic acid and S-adenosylmethionine.[1411a]

The formaldehyde precursors could be assayed through the derived formaldehyde by some of the fluorimetric and colorimetric methods described in Volume 1 and this volume. The formaldehyde could probably also be determined fluorimetrically through the fluorogens formed on reaction with an appropriate phenethylamine or 3-indolylethylamine. These types of reactions have been used in the determination of natural phenethylamines and tryptamines with formaldehyde. The formaldehyde fluorescence reaction with catecholamines and tryptamines is a two-step reaction involving a ring-closure, followed by a dehydrogenation step resulting in the formation of strongly fluorescent 3,4-dihydroisoquinolines and 3,4-dihydro-β-carbolines, respectively.[1412,1413] The 3,4-dihydroisoquinolines can exist in 2 forms which are in a pH-dependent equilibrium, and it is the tautomeric quinoidal form which gives rise to the strong fluorescence at *F*410/480 nm.[1412] It has been stated that the acid-catalyzed histochemical formaldehyde reaction exhibits a good

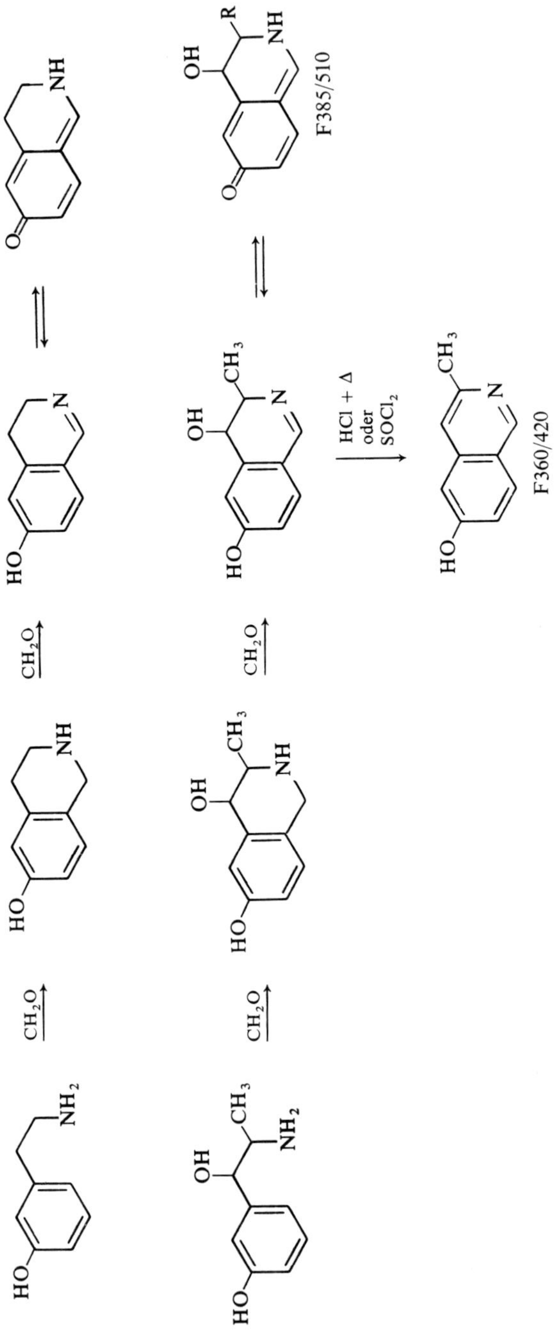

FIG. 87. Histochemical determination of *m*-tyramine and metaraminol.[1416]

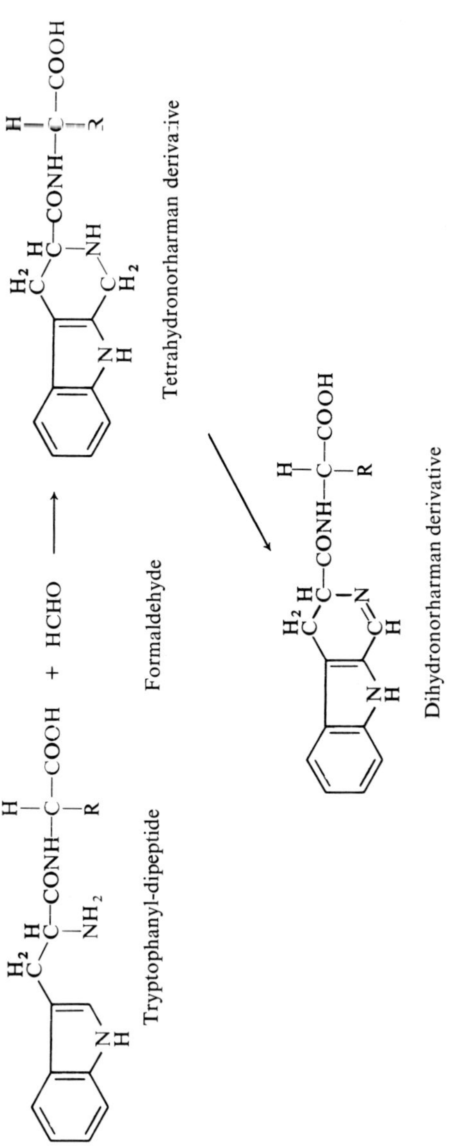

Tryptophanyl-dipeptide + HCHO (Formaldehyde) ⟶ Tetrahydronorharman derivative ⟶ Dihydronorharman derivative

FIG. 88. Reaction of tryptophanyl dipeptide with formaldehyde.[1419, 1420]

specificity for indolylethylamines and 3-hydroxylated or 3-methoxylated phenylethylamines.[1414] Corrected excitation and emission peaks of formaldehyde-induced fluorescence have been summarized for metaraminol, *F*385/415; dopamine, *F*410/500; adrenaline, *F*410/480; tryptamine, *F*370/500; 5-hydroxytryptophan, *F*410/540; 5-methoxytryptamine, *F*400/560; L-tryptophanyl-L-glycine, *F*375/500; and many others.[1415] These results have been applied to the microspectrophotofluorimetry of tissues, single cells and even organelles. The reactions of some of the phenylethylamines with formaldehyde are shown in Fig. 87.[1416]

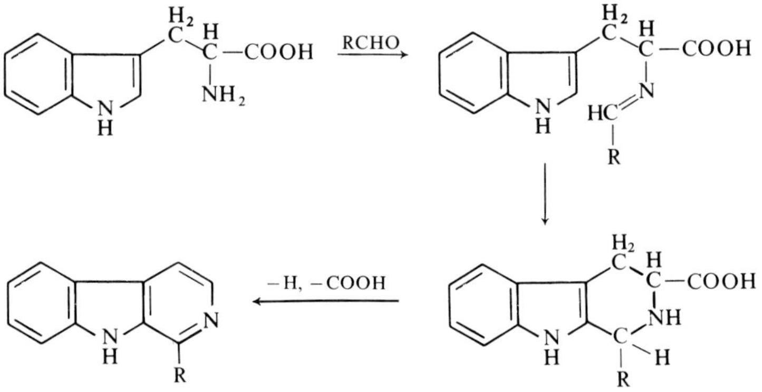

A, R = H; B, R = CH_3

FIG. 89. The formation of harman (A) and norharman (B) from tryptophan.[1421]

Tryptamine, which is found in steer, dog and human brains,[1417] can fluoresce at *F*370/490[1418] or *F*350/440.[1419] N-Terminal tryptophan-peptides have been determined fluorimetrically after formaldehyde condensation, Fig. 88[1419, 1420] as has tyrptophan in human plasma, liver and urine, Fig. 89.[1421]

The 1,2,3,4-tetrahydro-β-carbolines and the 3,4-dihydro-β-carbolines formed on reaction between the 3-indolylethylamines and formaldehyde could probably also be determined with MBTH. In a somewhat similar fashion it has been shown that indole and 9-methylcarbazole react with oxidized MBTH to give bands at λ 508 nm, mε 32 and λ 598 nm, mε 31, respectively.[120, 121]

LVIII. NITRITES, ALKYL

Various types of alkyl nitrites and benzyl nitrites have been photolyzed at low concentrations in a photochemical reaction cell containing air or oxygen.[1422]

In all cases some formaldehyde was formed. The products obtained in the photolysis of methyl nitrite are shown in Fig. 90. At 60 min irradiation in oxygen, methyl nitrite, ethyl nitrite, propyl nitrite and benzyl nitrite give 36%, 7%, 2% and a trace of formaldehyde, respectively. From ethyl nitrite, propyl nitrite and benzyl nitrite, 15% acetaldehyde, 33% propionaldehyde and approximately 30% benzaldehyde are obtained, respectively. For these latter compounds the main product was a peroxyacyl nitrate, approximately 60% being formed.

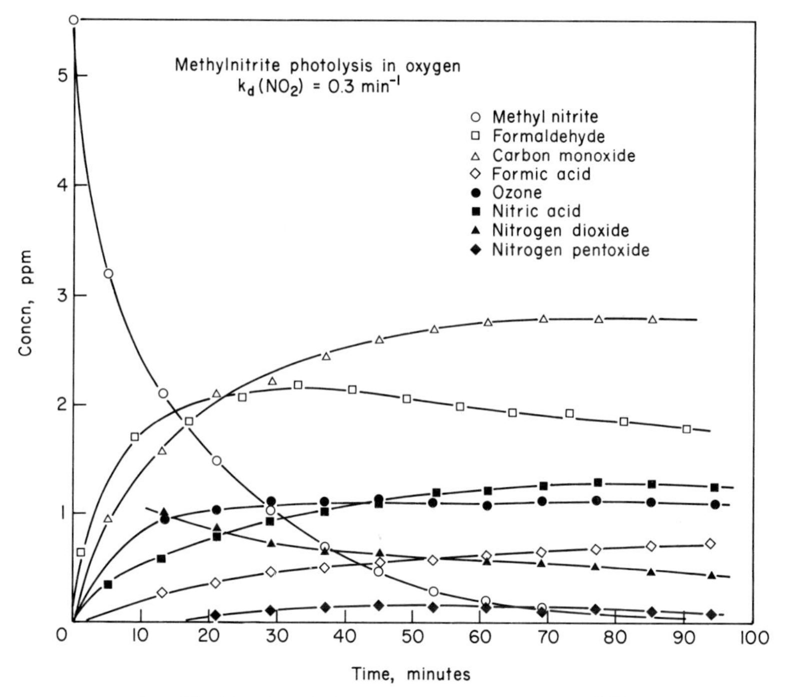

FIG. 90. The photolysis of methyl nitrite in oxygen.[1422]

LIX. OLEFINS

$R_2C{=}CH_2$ compounds (Section II) and ethylene (Section XIX) have been discussed previously. Thus, the photooxidation of ethylene,[1423, 1424] propylene,[1425] and toluene and *m*-xylene, the 2 most abundant atmospheric alkylbenzenes,[1426] results in the formation of formaldehyde as one of the products.

LX. OXALIC ACID

This acid can be determined by reduction to glycolic acid, oxidation of this acid to formaldehyde, and then reaction with chromotropic acid.[413,414] Oxalic acid can be directly determined by heating with indole in 50% sulfuric acid.[1427] A red chromogen is formed which absorbs at 525 nm. Formic acid, which can be considered to be a hydroxyformaldehyde, also reacts to give the same product that formaldehyde or oxalic acid give. Thus, the mechanism of reaction is probably identical to that described for the indole test for total aldehydes.[1428]

Table 29. Effect of various compounds on recovery of oxalic acid[414]

Addition to 100 ml sample		Recovery of oxalic acid %
1. None	—	99
2. Glucose	3 g	135
3. Glucose	400 mg	99
4. Glycolic acid	8 mg	109
5. Oxaloacetic acid	1 mg	109
6. Uric acid	58 mg	105
7. Tryptophan	3 mg	102
8. Oxaluric acid	2 mg	101
9. Succinic acid	1 mg	101
10. Formic acid	3 mg	100
11. Lactic acid	50 mg	98
12. Citric acid	102 mg	98
13. Malonic acid	1 mg	98
14. L-Glyceric acid	33 mg	98
15. Glyoxylic acid	17 mg	97
16. Sodium pyruvate	10 mg	97
17. Ribose	400 mg	96
18. Glycine	100 mg	96

In the reductive method oxalic acid can be determined in 0·5 ml of urine.[414] Beer's law is followed from 10 to 100 μg/100 ml. With chromotropic acid as the reagent an $m\varepsilon$ of 18 is obtained at 570 nm. The coefficient of variation of replicate determinations on aqueous solutions taken through the entire procedure was 2·0%. For urine, the mean value was 5·2%. The recovery of 25 μg of oxalic acid added to different samples of urine was within the range 75–98%, with a mean value of 85·6%. Recoveries approaching 100% can be obtained when urine is diluted 10 or 20 times instead of 4 times but this is at the lower limit of the method. Twelve or more urine analyses can be carried out conveniently in a normal working day. Of the compounds tested, Table 29, only 3 caused significant interference. A high concentration of glucose, (3%) corresponding to that in diabetic urine, gave a recovery of 135% but 0·4% glucose had no detectable effect. Glycolic acid in a concentration corresponding to that observed maximally in normal urine or in urine from patients with hyperoxaluria after diluting four times, yielded a recovery of 109%. Oxaloacetic acid also caused a high result but this compound is not usually detectable in normal or pathological urines.

Chromotropic acid determination of oxalic acid in urine.[(414)] *Reagents. Electrolytic zinc*—electrolytic zinc wire, diam. 3 mm (London Zinc Mills, Enfield, Middlesex) is cut into short lengths measuring approximately 5 mm and weighing approximately 250 mg. Immediately before use the zinc is cleaned by immersing briefly in freshly prepared 10 N HNO_3 (two volumes of concentrated HNO_3 to one volume of water). After washing thoroughly in distilled water the zinc is ready for use. *Chromotropic acid solution*—dissolve 1 g of 4,5-dihydroxynaphthalene 2,7-disulphonic acid, disodium salt "for formaldehyde determinations" in 100 ml of water. Store at 4° and prepare freshly once a week. *Oxalic acid standard*—Dissolve 1·0231 g of potassium oxalate monohydrate in 100 ml of water. Store at 4° and prepare freshly once a month. This solution contains 5 mg of anhydrous oxalic acid per ml.

Method. Acidify the urine sample by adding concentrated HCl (1 ml per 100 ml of urine) to ensure the solution of any crystals of calcium oxalate which may be present. Measure 0·5 ml of urine into a 25 ml graduated stoppered centrifuge tube, followed by 1·5 ml of water and one drop of 0·04% bromothymol blue indicator solution. Adjust the solution to pH 7·0 (green) by the addition of dilute NaOH or dilute acetic acid solutions. Add 2 ml of a saturated aqueous solution of calcium sulphate, followed by 14 ml of ethanol, mix gently and allow the solution to stand at room temperature for at least 3 hr or preferably overnight.

Centrifuge at 2000 rev/min for 10 min, carefully decant the supernatant fluid and allow the tube to drain for a few minutes on a filter paper. Wipe the mouth of the tube with clean tissue and dissolve the precipitate in 2 ml of 2 N H_2SO_4 solution. Add a piece of freshly cleaned zinc and heat in a boiling water bath for 30 min. (The tube is left unstoppered to allow evaporation to occur and the final volume should be less than 0·5 ml to ensure full color development.)

Remove the zinc with a bent glass rod. Wash the zinc with 0·5 ml of 1% chromotropic acid solution, adding the washings to the tube. This operation is most conveniently carried out by fixing the tube almost horizontally in a retort clamp and sliding the piece of zinc to the mouth of the tube where it can be washed with the chromotropic acid solution before final removal.

Add 5 ml of concentrated H_2SO_4 slowly, with mixing, and heat in a boiling water bath for 30 min. Cool, dilute to 20 ml with 10 N H_2SO_4 and determine the absorbance at 570 nm. The color is stable for several hours.

LXI. OXETANES

In the gas phase the heterocyclic compounds decompose under ultraviolet radiation, as shown for oxetane, Fig. 91.[416,417] In the case of 2,2-dimethyloxetane two pathways are possible under photolysis: acetone and ethylene and formaldehyde and isobutylene are formed.[416]

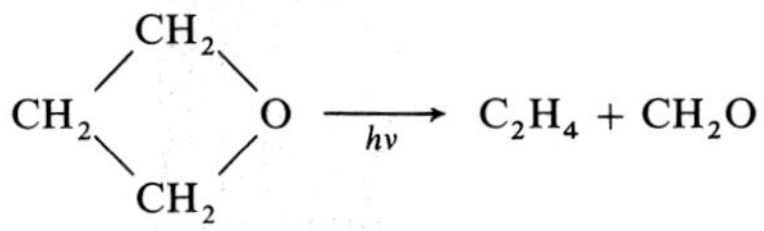

$$\text{(CH}_2)_3\text{O} \xrightarrow{h\nu} C_2H_4 + CH_2O$$

FIG. 91. Photodecomposition of trimethylene oxide to ethylene and formaldehyde.

This type of cleavage has been made use of in the search for biodegradable but active derivatives of the pesticides. Thus, the dimethyloxetane analog of DDD (2,2-bis-(*p*-chlorophenyl)-1,1-dichloroethane) is active but shows a high degree of degradability, forming formaldehyde and inactive 1,1-di-*p*-chlorophenyl-2,2-dimethylethylene. The reaction is shown for the *p*-ethoxy analog in Fig. 92.[418]

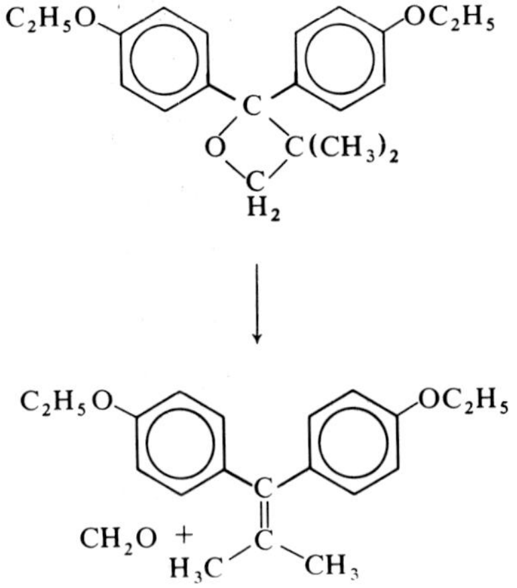

FIG. 92. Degradation of a dimethyloxethane to 1,1-di(4-ethoxyphenyl)-2,2-dimethylethylene.[418]

The thermal decomposition of dioxetanes has been investigated experimentally for several different dioxetanes under a variety of conditions.[1429–1436] 1,2-Dioxetane decomposes thermally to two formaldehyde molecules, e.g.

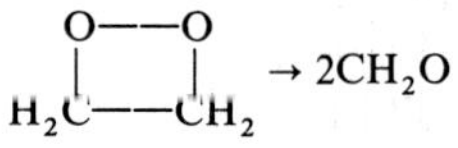

$$\begin{array}{cc} O & \!\!-\!\!-\!\! & O \\ | & & | \\ H_2C & \!\!-\!\!-\!\! & CH_2 \end{array} \rightarrow 2CH_2O$$

LXII. OZONIDES

The formation of ozonides (1,2,4-trioxolanes) from alkenes and ozone consists of a succession of three [2 plus 3] cycloadditions or cycloreversions involving primary ozonides (1,2,3-trioxolanes) and aldehyde or ketone oxides as decisive intermediates, all of which have finite lifetimes.[1437] The "ozonides" have been shown to be 1,2,4-trioxolanes,[1438] some of which are stable distillable liquids or crystalline solids. Criegee[1437] postulates the following reactions, Fig. 93. In the first step a primary ozonide is formed. This decomposed into a carbonyl compound (e.g. formaldehyde in the case of ethylene), which can be isolated, and a "carbonyl oxide". In the third step the "carbonyl oxide" adds to the carbonyl compound to form the 1,2,4-trioxolane. This latter ozonide could be a source of formaldehyde and other aldehydes depending on the substituents in the molecule.

The primary ozonides are thermally more labile and are stronger oxidizing agents than ozone because they lack resonance stabilization. Thus formaldehyde (or other aldehydes) could be formed from the oxidation of tetracyanoethylene by a primary ozonide at −78°, Fig. 94.[1437, 1439] Tetracyanoethylene is stable to ozone at this temperature. Isolation of the primary ozonide is not necessary, since ozonization of the olefin in the presence of one equivalent of tetracyanoethylene will form the carbonyl compounds.

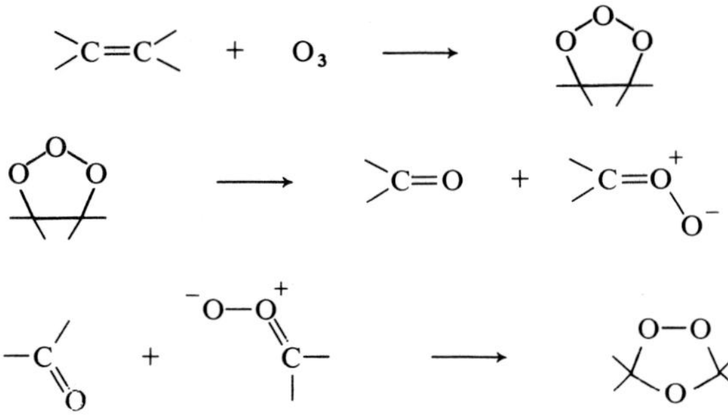

$$>C=C< \; + \; O_3 \longrightarrow \text{primary ozonide (1,2,3-trioxolane)}$$

$$\text{primary ozonide} \longrightarrow\; >C=O \; + \; >C=\overset{+}{O}-O^-$$

$$-C(=O)- \; + \; {}^-O-\overset{+}{O}=C< \longrightarrow \text{1,2,4-trioxolane}$$

FIG. 93. Postulated mechanism of the ozonolysis of an alkene.[1437]

FIG. 94. Reaction of an ozonide with tetracyanoethylene.[1437, 1439]

LXIII. PILOCARPINE

This drug, LIX, causes a direct cholinergic receptor stimulation and is used principally in the treatment of chronic, simple glaucoma. Its assay is based on oxidative cleavage by benzoyl peroxide to formaldehyde.[432]

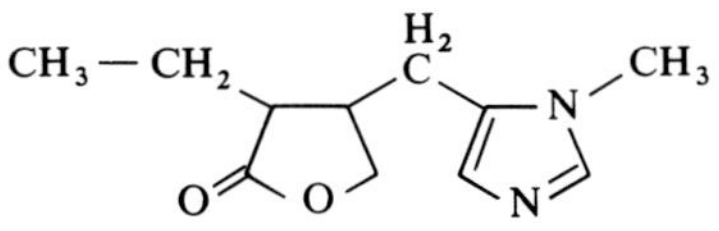

LIX

Chromotropic acid is the reagent and the color reaches its maximum intensity after 10 min and remains stable for about 1 hr. Beer's law is followed from 100 to 500 μg pilocarpine nitrate per 5 ml of final solution with a precision of ±2·2%. Interfering precursors containing $C—OCH_3$ or $N—CH_3$ should be separated from the pilocarpine prior to analysis. Eserine which contains 3 $N—CH_3$ groups will interfere so it is eliminated by a TLC step, which separates pilocarpine (R_f 0·0) and eserine (R_f 0·38).

Chromotropic acid determination of pilocarpine.[432] *Reagents. Chromotropic acid*—1% in concentrated H_2SO_4. *Pilocarpine nitrate standard solution*—0·2% (w/v) in water.

Determination of pilocarpine in standards. Transfer 2 ml accurately measured pilocarpine nitrate standard solution (4 mg) to 250 ml distillation flask containing 20 ml distilled water. Weigh ca. 0·25 g benzoyl peroxide, 5 g NaCl and 5 g purified sand, and pulverize mixture thoroughly; add mixture to distillation flask and start distillation by rapidly heating to boiling. Collect 20 ml distillate under ca 2 ml water and filter into 25 ml volumetric flask. Wash container and filter with 2–3 ml water and adjust to 25 ml with water. Measure exactly 2 ml solution and transfer to test tube. Cautiously add 3 ml chromotropic acid reagent by letting reagent run down sides of tube and mix by gently swirling tube. Let tube stand 10 min; then measure absorbance of color at 545 nm vs. blank analyzed simultaneously.

Determination of pilocarpine in formulations. (a) *Pilocarpine eye drops.*—Contains 2% (w/v) pilocarpine nitrate in isotonic solution. Apply chromotropic acid method described above to suitable aliquot.

(b) *Pilocarpine eye ointment.*—Contains 1% (w/v) pilocarpine nitrate in fatty base. Weigh 1 g ointment in beaker, quantitatively transfer to separatory funnel with $CHCl_3$–water (1 + 1) rinses, and shake. Let layers separate and drain $CHCl_3$ into beaker. Extract aqueous layer in separatory funnel with successive aliquots of $CHCl_3$ until fatty base is completely removed. Wash combined $CHCl_3$ extracts (ca. 40 ml) with 5 ml water and add aqueous washes to original aqueous liquid. Make alkaline with NH_4OH solution and extract with four 10 ml portions of $CHCl_3$ or until pilocarpine is completely extracted. Wash combined $CHCl_3$ extracts with 1 ml water and then evaporate $CHCl_3$ to dryness. Dissolve residue in acidified water, dilute to 10 ml, and apply chromotropic acid method to suitable aliquot of solution.

(c) *Pilocarpine hair tonic.*—Contains pilocarpine nitrate 1·5 g, salicylic acid 1 g, benzoic acid 1 g, resorcinol 1 g, 5% panthenol 30 ml, castor oil 5 ml, oil of lavender q.s., and 50% ethanol to 100 ml. Transfer 1 ml liquid (equivalent to 15 mg pilocarpine nitrate) to small separatory funnel containing 20 ml water acidified with HCl. Extract with three 10 ml portions of $CHCl_3$. Wash combined $CHCl_3$ extracts with 5 ml water and discard $CHCl_3$ layer. Add aqueous washing to original aqueous liquid, make solution alkaline with NH_4OH and proceed as in (b).

(d) *Dimiotic eye drops.*—Contains 1% pilocarpine nitrate and 0·5% eserine salicylate. Transfer 2 ml eye drops (equivalent to 20 mg pilocarpine nitrate) and extract base as described in (c). Chromatograph sample containing ca 5 mg pilocarpine (divide in 10 spots) on silica gel G plate activated at 105°C, using cyclohexane–diethylamine (9 + 2) as developer. Quantitatively scrape spot areas of pilocarpine (R_f 0) and extract with ethanol. Apply chromotropic acid method after evaporation of ethanol.

(e) *Piloserine eye ointment.*—Contains 2% pilocarpine nitrate and 0·2% eserine salicylate. Extract base as described in (b) and proceed with thin layer chromatographic technique in (d).

(f) *Pilocarpine-phenylephrine drops.*—Contains 1% pilocarpine nitrate and 1% phenylephrine HCl in isotonic solution. Transfer 2 ml drops to separatory funnel, extract pilocarpine with $CHCl_3$ after making solution alkaline with NaOH, and continue as described above.

LXIV. POLYMETHYLENE OXIDES

Members of this family include the epoxides (or oxiranes), LV, the oxetanes, LVI, the tetrahydrofurans, LVII, the tetrahydropyrans, LVIII, etc.

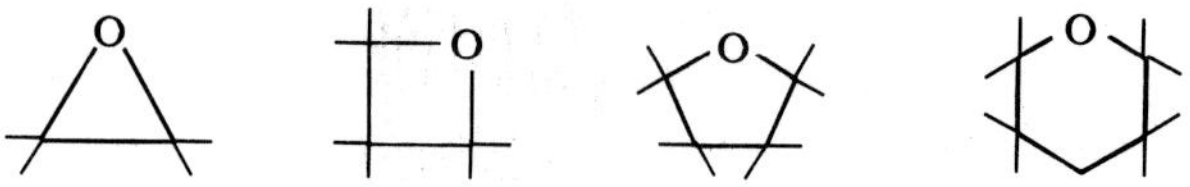

LV LVI LVII LVIII

Although some of these are formaldehyde precursors and have already been discussed, some comparative reactions will be given. Thus, epoxides can form formaldehyde as previously shown. In addition, rearrangement reactions can take place,[1440] e.g.

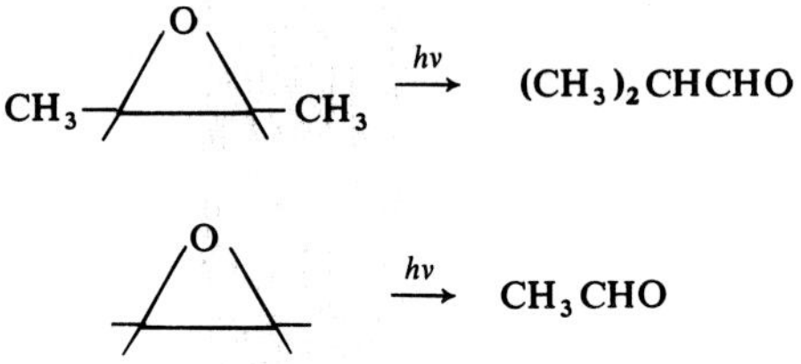

$$CH_3\text{-}(\text{epoxide})\text{-}CH_3 \xrightarrow{h\nu} (CH_3)_2CHCHO$$

$$(\text{epoxide}) \xrightarrow{h\nu} CH_3CHO$$

Or a ring rearrangement can take place as shown for limonene, Fig. 95.[1441] The photorearrangement of oxetane to formaldehyde and ethylene[416, 1442] has been discussed. The photorearrangement of tetrahydrofuran to ethylene and acetaldehyde is a slower reaction.[1442] Formaldehyde can also be obtained by heating 2-phenyltetrahydropyran with titanium dioxide and alumina, Fig. 96.[1443]

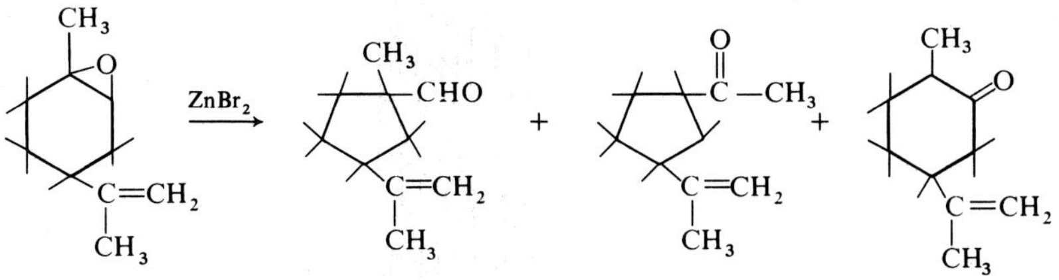

CH_3, O, $ZnBr_2$, $C{=}CH_2$, CH_3 → CH_3, CHO, $C{=}CH_2$, CH_3 + O, C—CH_3, $C{=}CH_2$, CH_3 + CH_3, O, $C{=}CH_2$, CH_3

FIG. 95. Catalyzed rearrangement of an epoxide.[1441]

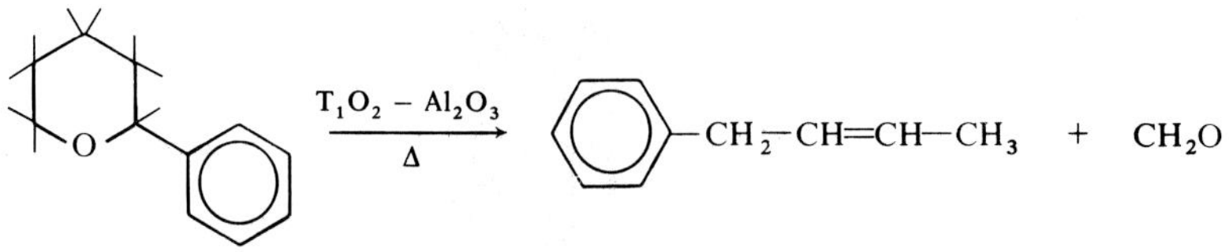

$$\xrightarrow[\Delta]{T_1O_2\ -\ Al_2O_3} C_6H_5\text{—}CH_2\text{—}CH{=}CH\text{—}CH_3 + CH_2O$$

FIG. 96. Catalyzed ring-opening of 2-phenyltetrahydropyran to formaldehyde and 1-phenyl-2-butene.[1443]

LXV. POLYOLS

Much of this material has been covered in other parts of this section, especially see 1,2-glycols. The following reaction scheme has been proposed for the periodate method.[116, 1444, 1445]

$$CH_2OH—(CHOH)_n—CH_2OH + (n + 1)HIO_4 \longrightarrow 2CH_2O + nHCOOH + H_2O + (n + 1)HIO_3$$

LXVI. β-PROPIOLACTONE

This powerful alkylating agent is prepared from the reaction of ketene and formaldehyde.[760,1446]

$$CH_2{=}CO + CH_2O \longrightarrow$$ (β-propiolactone ring structure with O and =O)

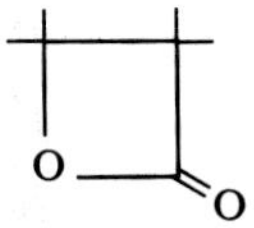

It is used fairly extensively as a sterilizing agent and a vapor phase decontaminant. It has proven useful in the sterilization of grafts, plasma and vaccines. Once absorbed on surfaces, it may require a relatively long time to be desorbed. It is a highly reactive molecule with a half-life of about 3 hr in water at 25°, the half-life being much less in the presence of nucleophiles at body temperature[1447] as for example cysteine[1448] and albumin.[1448] Consequently, only a small fraction of the dose reaches the genic biopolymers.

β-Propiolactone is mutagenic[1450] in *Vicia faba*[1451, 1452] and in *E. coli* and *Serratia macescens*.[1453, 1454] It, like the carcinogenic nitrogen mustard, hydrazine and diethyl sulphate, can revert the histidine-requiring mutant of *Salmonella typhimurium*, strain TA1530.[724] The compound can initiate skin tumour formation in the mouse with the help of croton oil as the promoting agent.[1455]

On subcutaneous injection in mice or rats[600, 1456] β-propiolactone causes tumours. Fed to rats it is carcinogenic.[600] Solutions of it in acetone or corn oil painted on mouse skin result in a highly significant number of skin tumours.[1457]

It causes chromosomal aberrations in *Allium*,[1458] *Neurospora*[1449] and *Vicia faba*.[1449–1451, 1458] It binds to DNA, RNA and protein.[1447] A study of the binding of β-propiolactone to mouse skin RNA and DNA has indicated that binding of the β-propiolactone to DNA, and not to RNA or protein, correlates with tumour-initiating potency.[1459]

β-Propiolactone is an alkylating agent and a formaldehyde precursor so that it can be determined through either of these 2 properties. Thus, β-propiolactone can be determined as an alkylating agent by 4-(4′-nitrobenzyl)pyridine[602, 1460–1463] and by 4-pyridinaldehyde 4-nitrophenylhydrazone type reagents.[602, 1464]

The intriguing bit of information about β-propiolactone is that it can form formaldehyde by 2 different methods of oxidation and can then be determined colorimetrically or even fluorimetrically. In one method the propiolactone is oxidized by periodate and the resultant aldehyde is reacted with chromotropic acid[556] in concentrated sulphuric acid to a violet chromogen which is extracted with chloroform and then determined.

An alternative analytical method is to oxidize the β-propiolactone with strong H_2SO_4 during the analytical reaction.[1464] This method is stated to be more sensitive than the periodate method. The color is stable for more than 1 hr. Beer's law is followed from about 5 to 40 μg of β-propiolactone per milliliter. The following procedure is used.

2,7-*Dihydroxynaphthalene determination of* β-*propiolactone.*[1464] *Reagent.* Prepare before use a solution containing 1·6 g of 2,7-dihydroxynaphthalene in 10 ml of concentrated sulphuric acid.

Procedure. Carefully add 1 ml of the reagent to 1 ml of the aqueous test solution containing about 10 μg of β-propiolactone. Read the absorbance at the λ_{max} 540 nm.

Both the chromotropic acid and 2,7-dihydroxynaphthalene methods need to be studied more thoroughly and optimized for best results. Other previously described colorimetric and fluorimetric reagents could also be used.

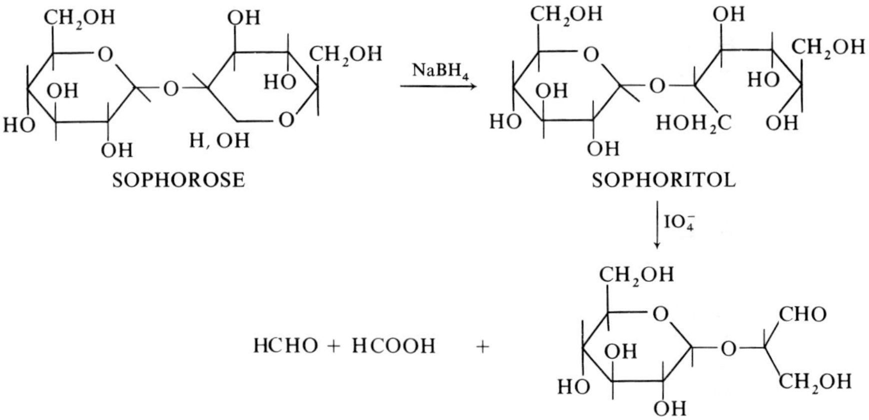

SOPHOROSE → ($NaBH_4$) SOPHORITOL → (IO_4^-) HCHO + HCOOH + carbohydrate aldehyde

FIG. 97. Reduction of sophorose to sophoritol which is then oxidized to formaldehyde and a carbohydrate aldehyde.

LXVII. SOPHOROSE AND SOPHORITOL

When the reducing disaccharide, sophorose, is oxidized by periodate, it produces about 0·3 mole of formaldehyde.[459] However, when sophorose is reduced and then oxidized, it produces 1 mole of formaldehyde, Fig. 97.

The periodic acid oxidation of similar polyhydroxy compounds have been discussed and enumerated in the older literature. Examples of some of these formaldehyde precursors (which are listed with other aldehyde precursors in Jackson's review) are given in Table 30.[1465]

Table 30. Polyhydroxy compounds oxidizable by periodate to formaldehyde[1465]

Compound	% Yield of CH_2O
Carbohydrates	
2,3-Dimethylglucose	100
Fructose	86
Galactose	100
Glucoheptose	—
Glucose	100
Glucose phenylosazone	93
Mannose	100
Mannose phenylhydrazone	35
2-Methylglucose	81
3-Methylglucose	—
Ribose-3-phosphoric acid	61
Sorbose	90
2,3,4-Trimethylglucose	5
Xylose	100
α-Ketols	
Dihydroxyacetone	90
Hydroxyamino compounds	
Hydroxylysine	100
3-Methylglucosamic acid	100
Serine	98
Polyhydroxy acids	
Gluconic acid	89
Polyhydroxy alcohols	
Adonitol	—
Erythritol	100
Ethylene glycol	100
Glycerol	100
α-Glycerophosphoric acid	84
Mannitol	100
Sorbitol	100

Table 6—*continued*

Compound	% yield of CH_2O
Steroids	
Allopregnane-3,17,20,21-tetrol	—
Allopregnane-3,11,17,21-tetrol-20-one	100
Allopregnane-3,17,-21-triol-11,20-dione	84
Corticosterone	—
Dehydrocorticosterone	100
4,5-Dihydrocorticosterone	60
Pregnane-3,21-diol-11,20-dione	—
Δ^4-Pregnene-17,21-diol-3,11,20-trione	83
Δ^4-Pregnene-11,17,21-triol-3,20-dione	80
Miscellaneous	
Dihydroxydihydrobetulin	57
Lactoflavin (riboflavin)	60
Leucodrin methyl ether	—

LXVIII. THIOL-BOUND FORMALDEHYDE

This type of grouping has been discussed previously in the methylene section in the part concerned with $—S—CH_2—O$ compounds. We will briefly mention some additional points here.

Formaldehyde fixation is recommended in lipid histochemistry even in the case of the plasmal reaction.[1466] However, differences have been observed between various fixation procedures.[1341, 1467] In sections of human scalp and rat tongue formaldehyde is bound to SH groups and can be demonstrated by the use of the sublimate–Schiff mixture[383, 1467] and to a lesser degree by Schiff's reagent alone. Thus, the reaction with bound formaldehyde must be considered in the plasmal reaction when sublimate–Schiff mixture is used as a detecting agent. The bound formaldehyde also interferes in the UV–Schiff reaction.[1339] There is no interference in the periodic acid–Schiff reaction and in the plasmal reaction carried out with the separate application of sublimate and Schiff's reagent.[1467] When sublimate is applied alone the formaldehyde escapes into the solution and subsequent treatment of the sections with the Schiff reagent cannot produce any staining of the keratin.[383] However, if Schiff reagent is present with the sublimate, formaldehyde is released and instantly gives a purple color with the Schiff reagent and this chromogen being soluble, diffuses and stains many structures including those stained with the Schiff–mercuric chloride reaction.

LXIX. TRIGLYCERIDES

A. Introduction

Measurements of triglyceride levels in serum or plasma are useful in the diagnosis and classification of inherited and acquired abnormalities of lipoprotein metabolism, and are useful for following patients with athereosclerosis who commonly possess elevated levels of cholesterol and triglyceride.[488] Since hyperlipidemia is frequently accompanied by coronary heart disease, triglyceride methods are needed to screen large populations. Both manual and automated methods have been developed for the colorimetric or fluorimetric analysis of total triglycerides through the derived formaldehyde. With the help of appropriate clean-up techniques and gas–liquid chromatography individual molecular species of triglycerides as well as simplified groups of triglyceride molecules can now be distinguished.[1468]

The determination of triglycerides through the derived formaldehyde has been extensively studied. Some of the reagents which have been used to determine the triglyceride-derived formaldehyde include chromotropic acid,[193–205, 207–215, 488, 493, 498, 1469, 1470] 2,4-pentanedione,[192, 216–221, 273, 485, 486, 489–492, 494–506, 1468, 1471–1480] phenylhydrazine,[222–226, 1481] and MBTH.[227, 487]

Table 31. Popular triglyceride methods in clinical laboratories[1482]

Method	Percentage of use in clinical laboratories
2,4-Pentanedione—colorimetric—manual	31
Enzymatic—UV—manual	22
2,4-Pentanedione—fluorimetric—manual	17
Enzymatic—UV—automated	13
2,4-Pentanedione—fluorimetric—automated	11
Chromotropic acid—manual	5
$FeCl_3$—NH_2OH—manual	2

Analysis of two proficiency testing surveys has revealed the methods selected by clinical chemistry laboratories for 13 common assays; the data for the triglyceride method is given in Table 31.[1482] The pentanedione and chromotropic acid methods are formaldehyde-precursor methods. The study concludes that it appears inevitable that the future will see a greater role for automation and instrumental analysis in the clinical chemistry laboratory.

Normal concentrations of serum triglycerides for fasting individuals are reported to be 30–135 mg/dl (1483).

On the basis of theoretical considerations and studies on the *in vitro* hydrolysis of triglycerides Rautela *et al.*[492] have suggested that blank determinations are unnecessary in routine measurements.

B. Chromotropic acid

An example of a method where interfering phospholipids are removed by adsorption onto silicic acid is the following.

Chromotropic acid determination of plasma triglycerides.[203] Mix 500 mg of silicic acid, 6 ml of freshly prepared isopropyl ether–ethanol (19:1) and 0·1 ml plasma in a test tube. For the blank, replace the plasma by water; for standards, evaporate 0·1 ml of a solution of tripalmitin in chloroform (2 mg per ml) in a tube, then add the silicic acid, the solvent mixture and 0·1 ml of water. Shake for 30 sec, set aside for 10 min, then shake again and centrifuge. Transfer 4 ml of supernatant fluid to a clean tube, add 0·4 ml freshly prepared ethanolic KOH (6 M KOH–ethanol (1:19)), stopper the tubes, and heat for 15 min at 55 to 60°. Cool, add 1 ml 0·3 M sulfuric acid, shake and centrifuge. Add 0·3 ml of the lower layer to 0·1 ml 0·02 M potassium periodate and, after 10 min, add 0·1 ml of 0·2 M of sodium arsenite. After 5 to 10 min add 3 ml of CA reagent (1 g of disodium salt of CA dissolved in 100 ml of water and mixed with H_2SO_4–H_2O (300:150)). Heat at 100° for 30 min, cool and read at 570 nm.

Since cholesterol and fatty acids interfere in the colorimetric determination of triglycerides, they are removed by solvent partition with petroleum ether prior to chromotropic acid analysis.[488] This method is a modification and improvement over a previous method.[194]

Chromotropic acid determination of serum and plasma triglycerides.[488] *Reagents. Alcoholic KOH* (4%)—freshly prepared weekly and kept in a refrigerated brown bottle. 0·1 M *sodium periodate*—10·7 g $NaIO_4$ dissolved in 500 ml water. *1* M *sodium arsenite solution*—22·5 g of NaOH and 50 g of As_2O_3 dissolved in 500 ml of water. *Chromotropic acid*—dissolve 1 g of the disodium salt in 100 ml water. Dilute 350 ml of concentrated H_2SO_4 with 175 ml water, cool and bring chromotropic acid solution to 500 ml. Store in refrigerated brown bottle.

Procedure. Measure 1 ml of each serum or plasma sample with an Eppendorf Micropipett (accuracy ±1%) into a glass-stoppered test tube containing 25 ml of acetone–ethanol (1:1, v/v) and mix on the rotary extractor for 10 min. Remove the precipitated protein by centrifugation at 2000 rev/min. for 15 min. Withdraw 6 ml aliquots from the supernatants, dry in test tubes under a stream of air in a 55° water bath, and take up in 3 ml of petroleum ether–diethyl ether (95:5, v/v). Pack long-stemmed funnels (stem dimensions, 300 mm × 4 mm O.D.) plugged with surgical cotton previously cleaned with chloroform with

700–800 mg of dry Florisil to a height of 125 mm. Condition each column with 10 ml of chloroform–acetone–water (99:99:2, v/v/v) prior to the addition of sample.

Transfer samples dissolved in the petroleum ether–diethyl ether solution to the columns, quantitative transfer being effected by washing the sample tube twice with 2 ml of the solvent each time. Add 8 ml of chloroform–acetone–water to each column and collect the effluents in glass tubes calibrated at 25 ml. Do not charge 2 columns in the group with sample; collect petroleum ether–diethyl ether and chloroform–acetone–water effluents from one of these columns for use as the reagent blank, and use the other column for the tripalmitin standard. Charge this column with 600 μg of tripalmitin dissolved in 3 ml petroleum ether–diethyl ether and elute with the solvents listed above. Use spectrophotometric readings obtained on the tripalmitin standards to calculate serum triglyceride concentrations.

Dry column effluents under a stream of air at 55°, add 0·25 ml of alcoholic KOH to each tube and wash down with 2·5 ml of 95% ethanol. Mix the contents and saponify for 30 min in a 55° water bath. Add 0·25 ml of 10 N sulfuric acid and wash down with 2·5 ml of water. Remove cholesterol and fatty acids from the saponified mixtures at this point by mixing with 15–20 ml of petroleum ether (boiling point 60–75°) for 10 min on the rotary extractor. Allow the mixtures to stand for approximately 5–10 min for the petroleum ether phase to become separated from the aqueous phase. Remove the upper phases containing cholesterol and fatty acids by aspiration, and drive off trace amounts of petroleum ether by placing the tubes in a 70° block for 30 min. Allow the tubes to cool to room temperature before proceeding to the next step.

Add 1 ml of 0·1 M sodium periodate solution followed by 2 ml of 1 M sodium arsenite solution exactly 5 min later. Shortly thereafter, liberation of iodine produced a light brown color which faded rapidly. Bring the mixtures to a final volume of 25 ml with water. Place 2 ml aliquots of each sample in 20 mm × 50 mm glass test tubes and mix with 10 ml of chromotropic acid solution. Place the tubes in a 100° bath for 30 min, cool to room temperature in a basin of cool tap water, and read at 570 nm against the reagent blank.

Use the following formula to translate absorbance readings into serum triglyceride concentrations:

$$\frac{26}{6} \times \frac{\mu\text{g tripalmitin standard}}{A \text{ of standard}} \times A \text{ of unknown} \times \frac{100}{1000} = \text{mg triglyceride/100 ml serum.}$$

In an alternative procedure Beer's law is obeyed from 2 to 25 μg of glycerol per ml. Results on human control serum are reproducible to approximately ±9% (±2SD). Recovery ranged around 100 ± 15%.

Chromotropic acid estimation of serum triglyceride.[498] *Reagents. Sulfuric acid* (4 N)—add 107 ml of conc. H_2SO_4 analytical reagent, to approx. 600 ml water. Make up to 1 liter. Stable for months. Dilute fourfold to make 1 N H_2SO_4. *Silicic acid*—heat silicic acid (Mallinckrodt, 100 mesh powder, $SiO_2 \cdot H_2O$) in a flat dish in a 100°C oven for 2 hours. Allow to cool in a vacuum desiccator and store in a tightly stoppered bottle. Keep in a desiccator over sulfuric acid. *Potassium hydroxide–barium hydroxide solution*—mix equal volumes of (a) and (b) just before use. (a) Potassium hydroxide (4 N): Dissolve 22·4 g of KOH, analytical reagent in water and make up to 100 ml. Stable if kept in a tightly closed polyethylene bottle. (b) Barium hydroxide, saturated: Shake approximately 10 g of $Ba(OH)_2 \cdot 8H_2O$ analytical reagent with 100 ml of water in a polyethylene bottle. Stable if prevented from absorbing CO_2. *Sodium periodate* (0·02 M)—dissolve 428 mg of $NaIO_4$ (anhydrous) analytical reagent, in water and make up to 100 ml. Stable in the refrigerator for at least 1 month. *Sodium arsenite* (0·2 M)—dissolve 2·6 g of sodium meta-arsenite ($NaAsO_2$), analytical reagent, and make up to 100 ml with water. Stable at room temperature for at least 2 months. *Chromotropic acid reagent*—add 300 ml of conc. H_2SO_4, analytical reagent, to 150 ml of water, and cool to room temperature. Dissolve 1 g of chromotropic acid (4,5-dihydroxy-naphthalene-2,7-disulfonic acid, disodium salt) in 100 ml of water and filter. Add the diluted sulfuric acid, slowly, to the chromotropic acid solution. Keep refrigerated in a glass stoppered, dark colored bottle. Stable for 2 months. *Tristearin standard* (2·5 mg/100 ml)—dissolve 250·0 mg of tristearin in chloroform and make up to 100 ml. Dilute 1 ml of this stock standard to 100 ml with chloroform (working standard). The solution is stable if evaporation of chloroform is prevented. *Glycerol standard* (0·5 mg/100 ml)—dissolve 526 mg of glycerol in water, and make up to 1 liter. Dilute 1 ml of the stock solution with 1 N sulfuric acid to 100 ml for the working standard. Stable for 2 weeks in the refrigerator. This standard is not the primary standard. It is used daily to check the efficacy of the reagents for oxidation and development of color.

Method. To 100 μl of serum in a 15-ml centrifuge tube, fitted with a ground glass stopper, add 4 ml of 1 N H_2SO_4 and 4 ml of chloroform. Stopper tightly and shake on the shaking machine for 10 minutes. Centrifuge and aspirate the aqueous top layer and the protein button (middle layer). Filter the chloroform layer through a filter paper (4-cm diameter, Whatman no. 40) into a second centrifuge tube that contains 200 mg of dried silicic acid. Shake for 10 min on the shaking machine. Centrifuge for 3 min at 2000 rpm in an International no. 1 centrifuge.

Evaporate a 2-ml aliquot of the chloroform solution to dryness in a calibrated centrifuge tube. This is done by placing a rack of test tubes in a stainless steel pan containing lukewarm water, in the hood, and raising the temperature gradually to 75–80°C with a Bunsen burner. When the chloroform is almost all evaporated, the temperature is raised to boiling. In another calibrated centrifuge tube place 2 ml of a 2·5 mg/100 ml tristearin solution in chloroform and evaporate to dryness. This is the standard.

Dissolve the residue in 0·4 ml of ethanol, add 0·1 ml of KOH–$Ba(OH)_2$ solution and mix well. Cap the tubes with loose Teflon caps. Incubate for 30 min in a 75–80°C water bath. Set up a blank using 0·4 ml of ethanol and 0·1 ml of the KOH – $Ba(OH)_2$ solution. Treat as for the unknown. This is the blank for the unknown and the tristearin. Cool the tubes to room temperature and make up to the 1-ml mark with 4 N H_2SO_4. Mix well, allow to stand for 10 min, and centrifuge.

To 0·5-ml aliquots of the supernatants, to 0·5 ml of the 0·5 mg/100 ml glycerol standard, and to 0·5 ml of 4 N H_2SO_4 for the glycerol standard blank, add 0·1 ml of sodium periodate solution. Mix, and let stand for 10 min. Add 0·1 ml of arsenite solution to all tubes and mix. Add 3 ml of chromotropic acid reagent to each tube. Mix well, and heat in a heating block, set to 105–110°C, for 30 min. Cool to room temperature and read the color at 575 nm.

Calculations. With the glycerol standard:

$$\frac{\text{Abs. unknown} - \text{Abs.blank}}{\text{Abs. glycerol} - \text{Abs. glycerol blank}} \times 95 = \text{mg of triclycerides/100 ml.}$$

With the tristearin standard:

$$\frac{\text{Abs. unknown} - \text{Abs. blank}}{\text{Abs. tristearin} - \text{Abs. blank}} \times 98 = \text{mg of triglycerides/100 ml.}$$

A TLC–chromotropic acid procedure has been developed which enables the determination of triglycerides in the presence of a large excess of paraffin,[493] Beer's law is followed in the following procedure.

TLC determination of triglycerides.[493] Separate palm-kernel oil, which contains a large number of triglycerides (C_8 to C_{18} saturated fatty acids and $C_{18:1}$ and $C_{18:2}$ unsaturated fatty acids) by TLC on plates impregnated with silver nitrate followed by reversed-phase chromatography on Kieselguhr G. The impregnation medium is liquid paraffin and the solvent system is acetone–acetonitrile (8:2) (saturated to 80% with liquid paraffin). Following separation visualize the triglycerides in UV light (360 nm) with 0·01% aqueous fluorescein. Free the triglyceride fractions from fluorescein by passage through Kieselgel, with elution by peroxide-free ethyl ether. Then evaporate 5 to 10 ml of the eluate under nitrogen and, after the addition of 0·5 ml of 0·4% alcoholic KOH, saponify at 70° for 80 min. Acidify the product with 1 ml of 1% H_2SO_4. Remove the liquid paraffin by the addition of 2·5 ml of peroxide-free ether. Heat 1 ml of the aqueous phase to remove the ethanol. Oxidize the glycerol product with 0·05 ml of 0·5% $NaIO_4$ solution. After 20 min, reduce excess periodate with 0·05 ml of 5% $NaHSO_3$ solution. After 15 to 20 min add 5 ml of 0·2% sodium chromotropate in 60% H_2SO_4 and heat the sample for 80 min at 105° in a drying oven. After cooling, read the absorbance at 570 nm.

C. MBTH

MBTH has also been used in an automated colorimetric method for the determination of triglycerides in plasma or serum.[(487)] The samples are extracted with 2-propanol containing activated alumina. The following equations have been postulated for the formation of aldehydes from the triglycerides

$$\text{Triglycerides} \xrightarrow{\text{KOH}} \text{glycerol} + \text{fatty acid salts}$$

$$\text{Glycerol} + HIO_4 \xrightarrow{-H_2O} CH_2O + HOCH_2CHO + HIO_3$$

The glycolaldehyde can also react with MBTH or it can be further oxidized to formaldehyde (and formic acid), so that there is more formaldehyde for reaction with MBTH. Beer's law is obeyed up to at least 400 mg of triglycerides per 100 ml, with a sampling rate of 60 per hr. Recoveries were measured with each run and over a typical 6 day period varied from 96% to 104%. In the 2,4-pentanedione colorimetric method an mε of 8·0 is obtained at λ_{max} 412 nm under optimum conditions; in the MBTH method an mε of 65 is obtained at λ_{max} 670 nm. The results obtained with this method correlate well with a fluorimetric[(489)] and a totally enzymatic[(487)] method.

D. 2,4-Pentanedione

Colorimetric and fluorimetric methods with 2,4-pentanedione for the determination of the formaldehyde derived from triglycerides are available in the manual and automated modes.

1. *Colorimetric manual.* A variety of manual methods utilizing 2,4-pentanedione are available.[(216, 219–221, 273, 494, 496, 499, 501, 505, 506, 1468, 1472, 1473, 1478–1480)] The main advantages of one of these methods are that (a) a single glycerol-containing lipid sample (50 to 200 μmoles) could be used after mild hydrolysis for analysis of fatty acids, glycerol and phosphorus; (b) low blank values due to use of highly purified, freshly distilled reagents and of purified glycerol color reagent through treatment with charcoal; (c) application to the analysis of small quantities of lipids from tissue culture cells, plants and bacteria; and (d) at lower concentrations method can be used fluorimetrically.[(505)]

In an alternative method triglycerides are extracted from serum (made ~0·05 N in H_2SO_4) with isopropanol–heptane (7:4), phospholipids remaining in the aqueous phase.[501] Treatment of the organic extract with 0·1 M sodium methoxide in anhydrous isopropanol frees the glycerol which is then oxidized with sodium metaperiodate to formaldehyde, which is condensed with pentanedione and ammonium acetate to give yellow 3,5-diacetyl-1,4-dihydrolutidine, the absorbance of which is measured at 410 nm.

In another procedure stable reagents have been described.[485] The procedure is similar to that described by Fletcher[220] except that alumina replaces a zeolite mixture in the preparation of triglyceride extracts of serum. The alumina adsorbs interfering substances. The extracts are made as follows.

Preparation of triglyceride extracts of serum.[485] To labeled 13 × 100 mm screw-capped glass tubes, add 0·1 ml of water, serum or standard. Add 4 ml isopropanol and mix well. Add 0·4 g of washed alumina (Woelm "neutral", activity grade I for chromatography) to all tubes and place them on a mechanical rotator for 15 min. Centrifuge and transfer 2·0 ml of the supernatant fluid to appropriately labeled 13 × 100 mm cuvets.

Saponification and analysis was then performed according to Fletcher's procedure.[220] The method is rapid—30 to 40 specimens can be processed in 3 hr.

The striking features of this method are the stable reagents. The more critically unstable reagent contains acetylacetone, and invariably it has been prepared by adding acetylacetone to an ammonium buffer. Ammonia is a reactant in the Hantzsch condensation, so it was postulated that the ammonia present in the reagent must be reacting with the acetylacetone component, because this mixture always becomes yellow. When the aceylacetone reagent was prepared in water–isopropanol (4:1, v/v) containing no ammonia, the acetylacetone neither became yellow nor deteriorated. The ammonia needed for the Hantzsch condensation is easily added to the periodate reagent, which is not affected adversely by including ammonium acetate in its formulation and remains stable so long as no alcohol is included in its formulation.

Essentially, the composition of these reagents are as follows.

Saponification reagent: Dissolve 5·0 g of KOH in 60 ml of distilled water and add 40 ml of isopropanol. Stable for six months at room temperature.

Sodium metaperiodate reagent: Dissolve 77 g of anhydrous ammonium acetate in about 700 ml of distilled water. Add 60 ml of glacial acetic acid and 650 mg of sodium metaperiodate. Dissolve and dilute to 1 liter with distilled water. Stable for at least six months at room temperature.

Acetylacetone reagent: Add 0·75 ml of 2,4-pentanedione to 20 ml of isopropanol and mix well. Add 80 ml of distilled water and mix. Stable for at least six months at room temperature.

Standard: Stock standard consists of 1·0 g of triolein per 100 ml of isopropanol (Sigma Chemical Co.). A working standard of 400 mg/dl is prepared by diluting 4·0 ml of the stock solution to volume with isopropanol in a 10-ml volumetric flask. Stable for at least six months at 4°C in a tightly sealed container.

An alternative method has been described that foregoes the adsorption step, requires less than 0·2 ml of serum, requires simple inexpensive equipment, can assay 5 to 10 samples per hour, has an average recovery of 99·9% (97·4 to 101·8%) triolein, and has an average precision of ±6·5% with a 95% confidence level.

Manual micromethod for determining serum triglyceride.[499] *Reagents. Extraction solvent*—n-Nonane: isopropanol, 2:3·5, by volume. *Sulfuric acid*—0·04 M/liter, in distilled water. *Sulfuric acid*—0·8 mol/liter, in distilled water. *Sodium ethoxide*—0·1 M/liter, in isopropanol. The solution is allowed to stand overnight to allow the fine sediment to settle. The clear supernatant liquid is stable for three to four weeks. *Chloroform. Sodium meta-periodate*—0·02 M, in distilled water. $NaIO_4$, 4·280 g/liter. *Sodium arsenite*—0·2 M, in distilled water. $NaAsO_2$, 25·800 g/liter. *Ammonium acetate*—3 M, adjusted to pH 6·00 ± 0·05 with glacial acetic acid. *Acetylacetone reagent*—One to five drops (0·05 to 0·25 ml) of 2,4-pentanedione are added to 25 ml of the 3 M/liter ammonium acetate solution, and mixed thoroughly. Use within one hour. *Standard triolein solution*—100 mg/100 ml, in extraction solvent. Pure triolein is kept refrigerated under nitrogen before weighing and dissolving; the standard solution is kept in small, tightly-closed bottles.

Procedure. Add to a stoppered tube: 2·5 ml of extraction solvent, 0·5 ml of 0·04 M sulfuric acid solution, and 0·15 ml of serum. Into a similar tube add 2·35 ml of extraction solvent, 0·15 ml of triolein standard solution, 0·5 ml of 0·04 M sulfuric acid solution, and 0·15 ml of distilled water. To prepare a blank, add to a similar tube: 2·5 ml of extraction solvent, 0·5 ml of 0·04 M sulfuric acid solution and 0·15 ml of distilled water. Stopper all tubes and mix with a Vortex mixer for 15 sec. Allow to stand for 1–2 min so the phases may separate. Centrifugation is not usually necessary. From each tube pipet 0·50 ml of the upper phase into a similar tube. Add 0·5 ml of sodium ethoxide solution to each tube. Stopper and mix, then incubate for 15 min in a water bath at 60°C. After transesterification, add 1·0 ml of 0·8 M sulfuric acid solution and 2 ml of chloroform to each tube. Stopper tubes and mix with a Vortex mixer for 15 sec, Allow to stand for 1–2 min to again separate the phases. Centrifugation is not necessary. From each tube, pipet 0·80 ml of the upper phase into a disposable test tube. At this time, prepare the acetylacetone reagent (at least 15 min before use). Add 0·10 ml of sodium meta-periodate solution to each tube. Mix by swirling. After 10 min, add 0·10 ml of sodium arsenite solution to each tube. Mix by swirling. After 5 min, add 2·0 ml of freshly-made acetylacetone reagent to each tube. Mix with a Vortex mixer for 2–3 sec and incubate 10 min at 60°C. Read absorbance at 415 nm against blank set at zero absorbance.

For *fluorometric* reading, 0·10 ml of each final solution is diluted with 2·9 ml of distilled water.

A recent manual method has been described which is based on an heptane extraction procedure(496) with stable saponification, oxidation, and color development reagents.(494) Beer's law is followed to 3 g/liter. Comparison with the automated Block and Jarrett procedure(1484) showed no significant difference. The coefficient of variation (47 duplicate samples) for the method was 6·3%. Utilizing the following procedure serum triglyceride concentrations after fasting ranged from 41–266 mg/dl, with a mean of 115 mg/dl, for 16 lean or obese men, age 18–22 years.(1485) Similar results have been reported for children 1 to 19 years of age(1486) where an upper 95% triglyceride level of 140 mg/dl was also found.

Manual procedure for determination of serum triglycerides with 2,4-*pentanedione.*(1485) *Reagents. Saponification reagent*—10 g KOH in 75 ml water and 25 ml of isopropanol. *Periodate*—to 77 g of anhydrous ammonium acetate in 700 ml of water add 60 ml of glacial acetic acid and 650 mg of sodium metaperiodate. Dilute to 1 liter with water and mix thoroughly. 2,4-*Pentanedione reagent*—0·4 ml of the dione in 100 ml of isopropanol. All three of these reagents are stable for more than 2 months at room temperature in a brown glass bottle. *Triolein stock standard solution*—1 g/dl of isopropanol. Working standards of 100, 200, and 300 mg/dl of isopropanol.

Procedure. To appropriately labeled 13 × 100 mm screw-capped tubes (blanks, standards, unknowns) add 0·5 ml of water, standard, or serum, respectively. Add an additional 0·5 ml of water to all standards. Add 2·0 ml heptane to all tubes, followed by 3·5 ml of isopropanol to the blank and unknown tubes and 3·0 ml isopropanol to the standards. Add 1·0 ml of 0·04 M sulfuric acid to all tubes followed by a 30-sec mixing. In all steps, mixing was with a vortex-type mixer. Allow the phases to separate without centrifugation. To 0·2 ml of the heptane (upper) layer from the extraction procedure, add 2·0 ml of isopropanol and 0·60 ml of saponification reagent. Mix and allow to stand at room temperature for at least 5 min. Add 1·5 ml of metaperiodate reagent and mix. Add 1·5 ml of acetylacetone reagent and mix. Cap each tube and place all tubes in a 65–70°C water bath for at least 15 min. Remove the tubes and allow them to cool to room temperature. Read the absorbance within 45 min at 415 nm.

A somewhat similar manual colorimetric procedure for triglycerides in serum has been described.(1480) Heptane–isopropanol has been used in the extraction of the triglycerides and sodium methoxide has been used in the transesterification of the triglycerides as postulated in Fig. 98. The chromogen, 3,5-diacetyl-1,4-dihydrolutidine, absorbing at λ_{max} 410 nm, mε 7·0[1473] is formed.

With the availability of antihyperlipidaemic drugs it is becoming desirable to classify cases of hyperlipidaemia according to the World Health Organization scheme.[1487] Thus, with the help of lipoprotein electrophoresis and the estimations of triglycerides and cholesterol, patients can be typed.

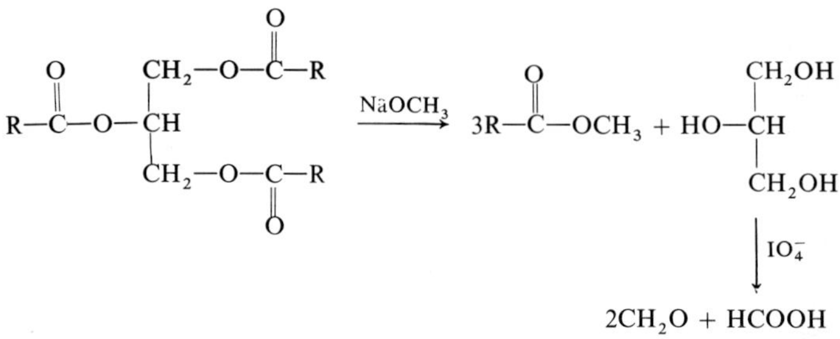

$$\begin{array}{c} CH_2{-}O{-}\overset{O}{\overset{\|}{C}}{-}R \\ | \\ R{-}\overset{O}{\overset{\|}{C}}{-}O{-}CH \\ | \\ CH_2{-}O{-}\underset{O}{\underset{\|}{C}}{-}R \end{array} \xrightarrow{NaOCH_3} 3R{-}\overset{O}{\overset{\|}{C}}{-}OCH_3 + \begin{array}{c} CH_2OH \\ | \\ HO{-}CH \\ | \\ CH_2OH \end{array} \xrightarrow{IO_4^-} 2CH_2O + HCOOH$$

FIG. 98. Transesterification of a triglyceride to glycerol which is then oxidized to CH_2O and HCOOH.

Many methods have been used to isolate triglycerides in a mixture amenable to analysis. Because of the interference of the phospholipids in the determination of the triglycerides the first stage of triglyceride analysis involves the separation of the phospholipids and extraction of the triglycerides. Some of the solvents which have been used include chloroform,[211] chloroform–methanol,[195] and isopropanol.[489, 494, 495, 1476, 1478] However, with these solvents phospholipids are also extracted and have to be removed by adsorption on to Zeolite,[211, 491] activated alumina, silicic acid,[1488] florisil, Kieselguhr, Lloyd's reagent or calcium hydroxide. A much more efficient method is to extract the test mixture with nonane–isopropanol[218, 499, 504, 506, 1472, 1474, 1479] or heptane–isopropanol[496, 1477, 1480, 1485] which differentially extracts triglycerides.

2. *Colorimetric automated.* Several automated colorimetric methods utilizing 2,4-pentanedione are available for the analysis of plasma or serum triglycerides.[490,492,495,502–504,1471,1472,1474,1476] In one of these methods an all-liquid procedure for extracting triglycerides from serum is used.[502] The acidified serum is extracted with nonane–isopropanol (2:3·5, v/v). Saponification is with sodium methoxide solution. Figure 99 shows the flow diagram for the procedure. Samples are analyzed at the rate of 40 per hr with 1:1 wash ratio. Standard curves were linear at least to 300 mg%. A typical equation for the standard line calculated by the method of least squares is:

$$C = 548A - 23{\cdot}8$$

where C = concentration in mg%; A = absorbance. The reproducibility of the method was good; the mean recovery was 103% (range 99 to 109%).

The colorimetric adaptation of a modified fluorimetric automated method[489, 1475] resulted in a procedure wherein 30 unknowns and standards or 40 blanks could be run each hour.[1471] Reagents were as described by Noble and Campbell[1475] except silicic acid was used instead of zeolite. Serum or plasma (0·5 ml) was treated with 5 ml isopropanol and 0·5 g of silicic acid, shaken vigorously for 10 min, centrifuged and the extract analyzed automatically.[1471]

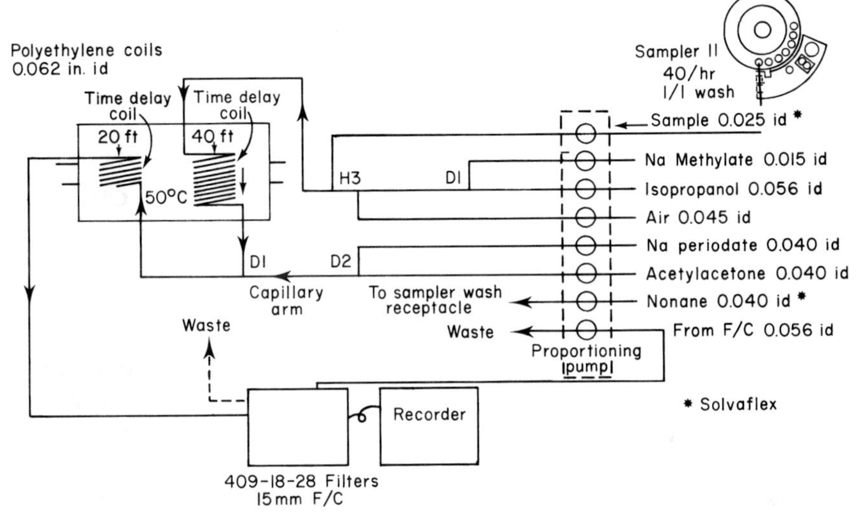

FIG. 99. Automated determination of triglycerides with 2,4-pentanedione (acetylacetone).[502]

In another colorimetric method even the extraction is automated so that a fully automated 2,4-pentanedione procedure is available.[504] Beer's law is obeyed from 0·4 to 8 m mole triglycerides per liter. The overall coefficient of variation was 0·047. Comparison with the fluorimetric procedure of Kessler and Lederer[489] gave a correlation coefficient (r) of 0·97, $P = 0{\cdot}01$, $n = 20$. Other semiautomated procedures have been described which utilize 2,4-pentanedione in a colorimetric modification of the Kessler and Lederer procedure.[489,495,1476] Since it is believed that a combined determination of cholesterol and triglycerides in serum is the simplest and most reliable way to detect hyperlipidemia in man,[1489,1490] some of the automated methods have been set up to determine glycerol and triglycerides simultaneously with a dual-channel AutoAnalyzer. One recent method uses 2,4-pentanedione to determine

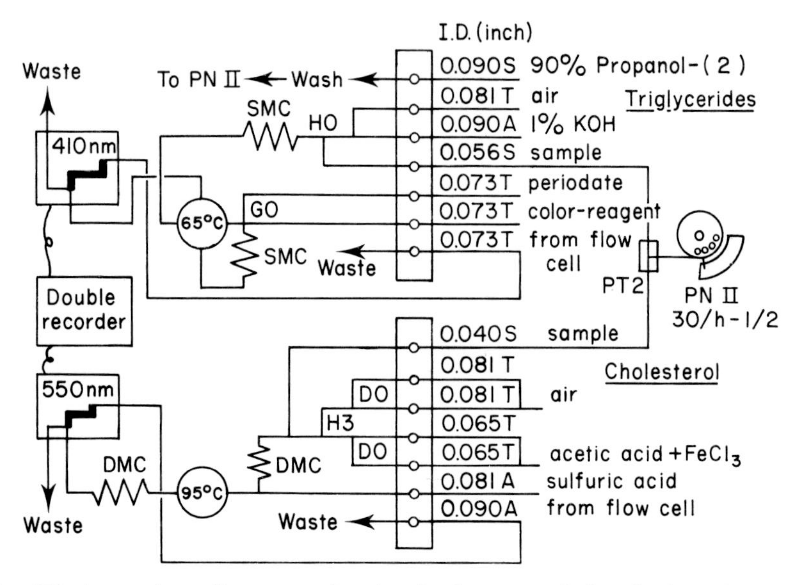

FIG. 100. Autoanalyzer flow-system for the simultaneous colorimetric determination of cholesterol and triglycerides in an adsorbent treated isopropanol extract of serum. Different tube qualities were used: T = normal tygon tube, A = Acidflex tube, S = solvaflex tube.[495]

the formaldehyde derived from the triglycerides, Fig. 100.[495] In this procedure zeolite was added to the serum extract in order to eliminate the interfering phospholipids, glycerol, glucose and other substances.

3. *Fluorimetric manual.* Many of the colorimetric methods for triglycerides utilizing 2,4-pentanedione can be modified slightly into fluorimetric methods. In addition, manual,[192, 217, 218, 485, 486, 491, 505, 1478] semi-automated[486, 489, 490, 497] and fully automated[500, 504, 1475] fluorimetric methods have been described.

An example of a manual method is the following.

Fluorimetric determination of serum triglycerides.[485] Serum (10 µl) is mixed with $CHCl_3$ (4 ml) and Kieselgel (Merck, activated for 4 hr at 120°) (50 mg) and shaken mechanically for 15 min, then centrifuged for 3 min at 5000 g. For the reagent blank, H_2O is used in place of serum, and as standard a soln of tristearin in $CHCl_3$ (4 ml) (prepared by diluting a stock 0·025% soln 1 in 100) is mixed with H_2O (10 µl) and Kieselgel (50 mg). The $CHCl_3$ extract (1 ml) is evaporated to dryness in a stream of N at 40°, and the residue is saponified by heating for 60 min at 60° with isopropyl alcohol (0·1 ml) and 0·05 N NaOH (20 µl). After cooling, 0·4 N HCl (100 µl) and 0·005 M $NaIO_4$ (20 µl) are added and after 10 min at room temperature unconsumed $NaIO_4$ is treated with 0·05 M KI (20 μl) and the liberated iodine is reduced with 0·05 M $Na_2S_2O_3$ (20 μl). A 0·1% (v/v) soln of 2,4-pentanedione in 2 M ammonium acetate (0·2 ml) is added and the soln is kept at 60° for 30 min, then cooled and isopropyl alcohol (0·5 ml) is added. The fluorescence is then read at *F*405/485 and the concentration of triglycerides is calculated from the test and standard readings corrected for blank fluorescence, the standard representing 100 mg of triglycerides per 100 ml.

A method involving the simultaneous determination of serum cholesterol and triglycerides after preliminary alumina column chromatography has been described.[1478] Transesterification with sodium methoxide was used, the glycerol was oxidized to formaldehyde with metaperiodate and the formaldehyde was determined at F405/485 with appropriate filters and a filter fluorimeter. Cholesterol and triglycerides are determined from a single serum extract. Furthermore, this procedure is simple while remaining quite specific. It involves a single extraction of the serum to acquire the fraction containing cholesterol and triglycerides. This fraction is then eluted through a commercial ("Lipo-Frax") chromatographic column which retains phospholipids, glucose, glycerol and bilirubin. Samples are taken of the eluent and analyzed spectrophotometrically for cholesterol and fluorometrically for triglycerides.

The extraction of 0·3 ml of serum is accomplished with 5·7 ml of isopropanol. The whole mixture is poured into the chromatographic column. The authors used prefilled columns, 10 mm by 40 mm, containing 2·5 g of specially treated activated alumina.

4. *Fluorimetric semiautomated.* An example of a semiautomated fluorimetric method is one[486] that utilizes the better features of two previous methods.[489, 1475] In this method plasma lipids are extracted in isopropanol containing a mixture of Zeolite, Lloyd's reagent and Van Slykes copper lime mixture. Each of these components removes phospholipids, bilirubin, chromogens and glucose, respectively. The sample stream and alcoholic KOH are mixed in the AutoAnalyzer, and heated to 50° to saponify the glycerides. The glycerol is then oxidized by periodate to formaldehyde which forms the fluorogen with 2,4-pentanedione and ammonia.

If human adipose tissue is to be assayed by such a procedure it can be extracted with petroleum ether.[1491]

5. *Fluorimetric automated.* The difficulty with these methods of analysis is that manual extraction of the triglycerides is used prior to analysis.[218, 1492] A fully automated fluorimetric procedure and equipment for extracting and quantitating serum or plasma triglycerides has been developed.[500] Fifty samples or standards can be analyzed per hour with a relative standard deviation of 3% at a triglyceride concentration of 125 mg%. The lowest

detectable concentration of triglyceride for which the 95% confidence limit did not include zero, was 7 mg%.

Any compound capable of being oxidized by periodate to formaldehyde could produce falsely elevated triglyceride values in the formaldehydogenic methods.[489, 1493, 1494] To avoid such interference these compounds are removed by extraction or adsorption before analysis[489] as discussed previously. For example, propylene glycol is an interference which can be corrected for by using a serum-containing blank[497] or extraction by the Folch procedure.[1495]

LXX. URIC ACID

This compound is not a formaldehyde precursor in several methods of analysis wherein its determination depends on the formation of formaldehyde from methanol with the help of a uricase–catalase system.[1496, 1497] Uric acid in serum or urine can be determined. The principle of the method is as follows:

$$\text{Uric acid} + 2H_2O + O_2 \xrightarrow{\text{Uricase}} \text{Allantoin} + CO_2 + H_2O_2$$

$$H_2O_2 + CH_3OH \xrightarrow{\text{Catalase}} HCHO + 2\,H_2O$$

The formaldehyde is then determined with 2,4-pentanedione at λ410 nm through the chromogen, 3,5-diacetyl-1,4-dihydrolutidine. This type of analysis is an example of the use of a reagent as the formaldehyde precursor to determine a test substance. Uricase added to serum gave an average recovery of 100·6% (range 97·5 to 103·3%). Beer's law was followed from about 4 to 20 mg uric acid/100 ml. The coefficient of variation over a month was $\pm$1·2%.[1497]

An analogous alternative method for uric acid has been described which utilizes MBTH for the formaldehyde derived from the methanol.[367] Beer's law is followed from about 20 to 150 mg of uric acid/liter. The following procedure was recommended.

Enzymatic micromethod for the determination of uric acid in biological fluids.[367] Mix 0·05 ml of test solution with 0·25 ml of enzyme solution (mix and dilute 20 ml of 0·2 M Tris HCl of pH 8, 0·4 ml of methanol, 90 000 i.μ. of catalase and 2·5 i.μ. of urate oxidase to 25 ml) and let stand for 20 min at room temperature or 8 min at 37°. Add 0·4 ml of 0·2% MBTH in glycine—0·5 M HCl buffer, pH 3 $\pm$ 0·05, followed after 5 min by 0·4 ml of 0·8% $FeCl_3 \cdot 6H_2O$ in 0·01 M HCl. After a further 10 min, dilute the mixture with 2 ml of water and measure the absorbance at 630 nm.

LXXI. VINYL CHLORIDE

A. Introduction

This compound can also be a formaldehyde precursor. Formaldehyde is one of the metabolic products of vinyl chloride. Formaldehyde can also be formed from vinyl chloride oxidatively for analytical purposes.

We have discussed and will discuss many halogenated organic compounds which are aldehyde precursors and have genotoxic properties. The halogenated compounds are of considerable interest because of their many varied uses and their toxic properties. More than five billion kilograms of halogenated organic compounds were produced in the United States in 1974.[1498] The following production figures have been reported for the USA in 1974: vinyl chloride, $2{\cdot}5 \times 19^9$ kg, trichloroethylene, $2{\cdot}0 \times 10^8$ kg, and tetrachloroethylene, $3{\cdot}3 \times 10^8$ kg. It is estimated that approximately 6% of the monomer (approximately $1{\cdot}5 \times 10^8$ kg) was lost to the air during processing in 1974.

Vinyl chloride has a 30- to 40-year history of industrial use. It was fairly recently thought of as an inert material and was even considered for use as a surgical anesthetic. It is a colorless gas with a faint sweet odor, boiling at $-13{\cdot}9°$. It was used as a propellant for deodorants, pesticides, furniture polishes, and hair sprays. Its main use is as a monomer (VCM) in a series of thermoplastic resin homopolymers and copolymers under the name polyvinylchloride (PVC). The resin is an incredibly useful material necessary to the production of many types of plastic containers, electric insulation, pipes, bottles, food wrapping films, raincoats, upholstery, and other products. Exposures to vinyl chloride have been greatest in occupational situations and wherever vinyl chloride has been used as a propellant in home, beauty parlor and workplace.[1499] Polyvinyl chloride floor coverings present another area for human exposure to vinyl chloride. The continuous liberation of vinyl chloride in enclosed living areas could be hazardous to human health. Migration of vinyl chloride from polyvinyl chloride bottles and films into food or beverages is believed to be only a slight hazard especially if the vinyl chloride content of the bottles or films is less than 1 ppm.

B. Atmospheric and metabolic reactions

1. *Atmospheric oxidation.* Since large amounts of vinyl chloride are emitted into the atmosphere, the fate of this genotoxic material is of some concern. Of primary interest are the rates of photooxidation and the identity of photooxidation intermediates and final products. A study of the ultraviolet irradiation of vinyl chloride in photochemical chambers has yielded some insight into a number of aspects of the photooxidation processes for vinyl chloride and other halogenated ethylenes.[545] The concentration–time plots

for the irradiation of vinyl chloride in the presence of NO_2 has shown that at 160 min of irradiation time approximately 40% of the vinyl chloride had reacted. The major products observed were formic acid, hydrochloric acid, carbon monoxide, formaldehyde, and ozone, Fig. 101.[545] Trace amounts of formyl chloride and nitric acid were also detected.

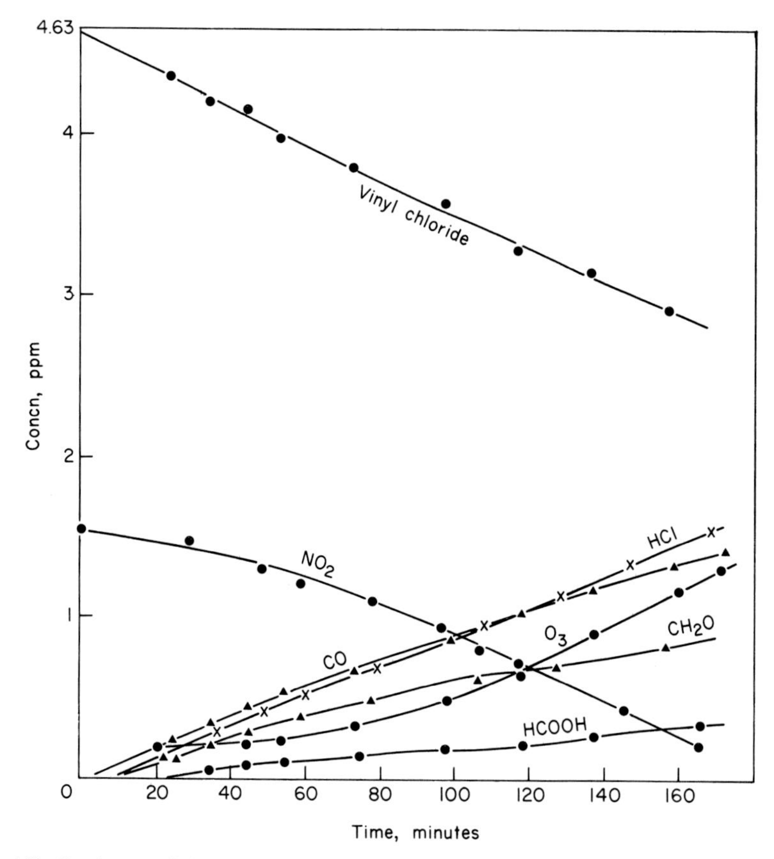

FIG. 101. Products and their variation with time from the irradiation with ultraviolet light of a mixture of 4·63 ppm vinyl chloride with 1·5 ppm of NO_2.[545]

Vinyl chloride can also react with ozone; the postulated products are formaldehyde, carbon dioxide, hydrochloric acid, formyl chloride, carbon monoxide and water. The photo products formed in a chamber reaction are shown in Fig. 102.[545] The rate constant for the reaction, assuming a second-order mechanism, is $0{\cdot}34 \times 10^{-3}$ ppm^{-1} min^{-1}.

The reaction of vinyl chloride with NO has also been studied in a 335 ft^3

chamber.[545] After 5 hr of irradiation 39% of the vinyl chloride (1·8 ppm) reacted, with the NO_2 maximum of 0·82 ppm at 130 min and the formaldehyde maximum of 0·32 ppm at 140 min. The concentration of formaldehyde leveled off at this value for the rest of the run. The authors suggest that the production rate of formaldehyde is approximately equal to the rate of destruction after 140 min The ozone observed after 330 min was 0·44 ppm. In all cases the

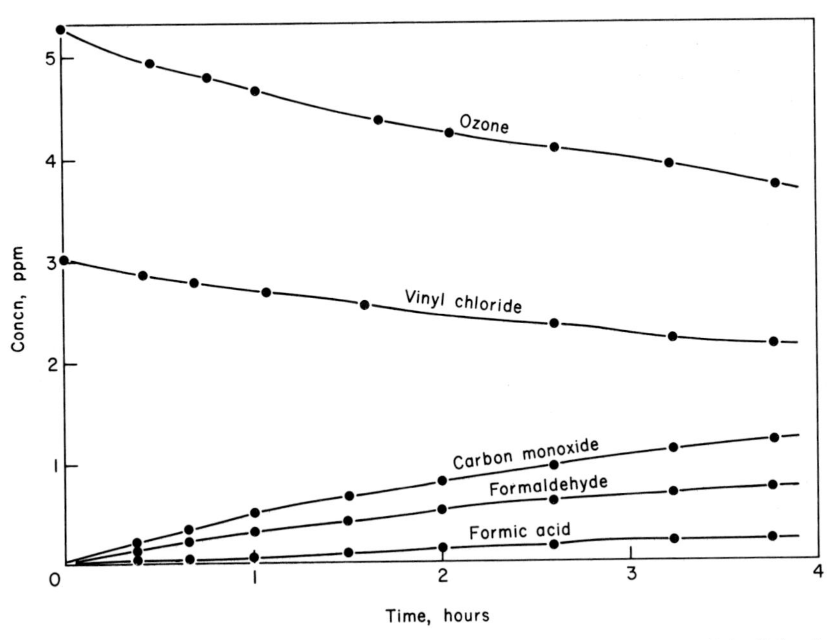

FIG. 102. Products and their variation with time from the irradiation with ultraviolet light of a mixture of 3 ppm vinyl chloride and 5·3 ppm ozone.[545]

formaldehyde was measured with the chromotropic acid method.[1500] These various reactions give an idea of what products are formed from the photooxidation of vinyl chloride in the atmosphere.

2. *Metabolism.* Vinyl chloride is metabolized in rats and it appears that some of the metabolites may be ultimutagens and/or ulticarcinogens.[1501–1506] However, there is some evidence that vinyl chloride may be (also?) an ulticarcinogen and ultimutagen without metabolic activation. Thus, the induction of angiosarcomas in rats by intraperitoneal injection of vinyl chloride[1507] implies that vinyl chloride, and not a product of its metabolism, is the causative agent in this case.[1508] In addition, the compound also may be a

direct mutagen since it does not require activation by liver microsomal enzymes.[1501, 1509]

The main eliminative route for [^{14}C]vinyl chloride after oral, i.v. or i.p. administration to rats is pulmonary; both unchanged vinyl chloride and vinyl chloride-related CO_2 are excreted by that route and the other [^{14}C] metabolites via the kidneys.[1508] This agrees with Schaumann's observations[1510] that in mammals the pulmonary excretion of unchanged vinyl chloride follows its inhalational administration.

Biotransformation of vinyl chloride into S-(2-chloroethyl) cysteine and N-acetyl-2-(2-chloroethyl) cysteine occurs through addition of cysteine, and biotransformation into: (i) chloroacetic acid, thiodiglycollic acid and glutamic acid, and (ii) into formaldehyde (methionine, serine), CO_2 and urea is explicable in terms of an associative reaction with molecular O_2 involving a singlet oxygen bonded transition state in dynamic equilibrium with cyclic peroxide ground state.[1508] No evidence for chloroethylene oxide formation could be found.

Green and Hathway[1508] believe that oxidative biotransformation of vinyl chloride might involve an associative reaction with molecular oxygen to form a singlet oxygen bonded form in dynamic equilibrium with a cyclic peroxide form. Chloroacetic acid would be formed from the singlet oxygen bonded transition state. Formaldehyde, and through it CO_2 and urea, would be formed by dismutation of the cyclic peroxide transition state. Evidence for formaldehyde formation rests on the production of CO_2 and of urea, and on the detection of the trace (vinyl chloride) metabolites, methionine and serine, themselves established metabolic products of formaldehyde.[1511] Formation of formaldehyde and CO_2 by this metabolic pathway finds confirmation in their production *in vitro* amongst the reaction products of a cyclic peroxide of vinyl chloride.[1512, 1513]

Exposure of a cell-free liver microsomal preparation to vinyl chloride results in the formation of chloroacetaldehyde.[1504] Chloroacetic acid, probably derived from chloroacetaldehyde, has been found in vinyl chloride workers.[1514]

Vinyl chloride can also effect the N-demethylation (formaldehyde formation) of the N-methyl drugs. This has been demonstrated when liver microsomal activities were measured 24 hr after a 6 hr exposure to 5% vinyl chloride in male rats pretreated with phenobarbital, Aroclor 1254 or the control vehicle.[1515] These authors reported that liver injury as indicated by serum transaminase elevations after vinyl chloride exposure, was found only in phenobarbital and Aroclor 1254 treated animals while cytochrome P-450 contents and oxidative N-demethylation of aminoantipyrine and ethylmorphine were markedly decreased in all groups. Oxidative N-demethylation of dimethylaminoantipyrine and ethylmorphine was assayed by the methods, respectively, of Orrenius[1516] and Gram *et al.*[1517]

C. Physiological activity

1. *Toxicity.* A symposium on the toxicity of vinyl chloride–polyvinyl chloride has been published.[1518] A notable feature of this symposium was a bibliography of 389 references on the toxicology of vinyl chloride and polyvinyl chloride.[1519] The industrial toxicology of vinyl chloride has been reviewed.[1520–1523]

2. *Clastogenicity.*[1524] Examination of lymphocyte cultures from 11 vinyl chloride polymerization workers and 10 controls revealed a significantly higher incidence of aberrations in the exposed population. Most of the excess damage was of the unstable variety and involved the grossest kinds of changes such as fragments or rearrangements. When these complex changes were regarded as the product of 2 breaks, the incidence of all breaking events was also significantly increased in the workers. The presence of chromosome damage is indicated in vinyl chloride exposed workers.

3. *Mutagenicity.* Some data has been given previously on the direct mutagenicity of vinyl chloride. Human, rat and mouse liver-mediated mutagenicity of vinyl chloride in *Salmonella typhimurium* strains has been investigated.[1501] Exposure of strains TA 1530, TA 1535 and G-46 to the carcinogen vinyl chloride increased the number of His^+ revertants/plate 16, 12 or 5 times over the spontaneous mutation rate. After 6 hr of exposure to vinyl chloride, the mutagenic response for TA 1530 strain was enhanced 7-, 4- or 5-fold when fortified postmitochondrial liver fractions from humans, rats or mice were added. The enzyme-mediated vinyl chloride mutagenicity was dependent on an NADPH generating system and the enzyme activity was localized in a liver microsomal fraction. Phenobarbitone pretreatment of rats and mice increased the mutagenic response by 15–40% as compared to untreated controls. The relative mutagenic activities of vinyl chloride, taking the value from mouse liver as 100, for TA 1530 strain mediated by tissue fractions were: rat liver, 80; mouse and rat kidney, 20 and 16; mouse and rat lung, less than 7; human liver (from 4 biopsy specimens), 170, 64, 70 and 46.

Experiments with Salmonella bacteria have shown that the carcinogenic hazard of vinyl chloride could have been predicted with a mutagenicity test.[1503] The data in these experiments indicate that vinyl chloride is not mutagenic per se but becomes mutagenic after a metabolic activation in the liver. On the other hand, vinyl chloride is mutagenic to the TA 100 strain of *Salmonella typhimurium* while its possible metabolite, chloroacetaldehyde, is hundreds of times more potent than vinyl chloride.[1509]

4. *Carcinogenicity.* The development of hepatic angiosarcomas in workers exposed to vinyl chloride gas in the manufacture of polyvinyl chloride has been

well documented.[1521, 1525–1527] Humans exposed to high concentrations of vinyl chloride can develop angiosarcoma of the liver. However, a recent epidemiological study indicated that vinyl chloride and polyvinyl chloride workers may develop cancers at multiple sites.[1528] An increase in deaths due to malignant neoplasms is found in employees in the highest exposure category.[1527] The individuals in this category experienced repeated excursions of several thousand parts per million vinyl chloride or were exposed to time-weighted average concentrations of 200+ ppm over time spans from one month to 18 years. The authors suggest that exposure to vinyl chloride may increase susceptibility to malignant neoplasms among individuals who are at risk due to other factors. Vinyl chloride produces tumours in rats, mice, hamsters and humans.[1529] Liver angiosarcomas are observed in all these animal species. In the small animals there appears to be a transplacental effect of vinyl chloride.

5. *Teratogenicity*. Parental exposure to vinyl chloride may be responsible for certain types of birth defects.[1530] The wives of men who work with vinyl chloride are twice as likely to have miscarriages or stillbirths.[1531]

D. Analysis

Many superior methods are available for the analysis of vinyl chloride. For vinyl chloride by itself, gas chromatography is a method of choice; for air polluted with vinyl chloride, GC–MS–COMP is the method of choice for the members of this mixture. However, photometric methods can be used as screening or small laboratory procedures if the analyst understands the shortcomings of the method. Thus, atmospheric vinyl chloride has been analyzed colorimetrically after collection on activated charcoal, extraction, oxidation to formaldehyde and reaction with chromotropic acid.[1531] Ethylene, methanol and other $R{-}CH{=}CH_2$ compounds would interfere. Other reagents could be used to determine formaldehyde.

LXXII. VINYL COMPOUNDS

Compounds containing the $R_2C{=}CH_2$ group have been discussed in Section II of this book. Other compounds of this type which have been discussed are ethylene, olefins and of course, vinyl chloride. As shown in this section even acrolein can be a formaldehyde precursor.

There are many types of vinyl compounds in the industrial atmosphere. Some of them are halogenated, possibly genotoxic, and in some cases could reach levels of 1 000 000 ng/m^3 even in the ambient atmosphere outside an industrial plant. These concentrations would be very high and of considerable concern, especially when you consider the fact that average urban levels of the carcinogen, benzo[a]pyrene, range around 2 ng/m^3. If necessary, general screening tests for such compounds could be developed based on formaldehyde determination.

A. Vinyl butyl ether and acrolein

Industrial atmospheres have been analyzed for these compounds.[518] Air at a rate of 20 liter/hr is passed through an oxidizing-absorbing mixture, 4 ml in each of 2 impingers. For acrolein the absorbent contained 100 ml of 2% ammonium acetate, 4 ml of 1·5% HIO_4 in 5% $KHSO_4$ and 4 ml of 2% $KMnO_4$, for vinyl butyl ether 100 ml of 2% ammonium acetate, 8 ml of aqueous 1·5% HIO_4 and 4 ml 2% $KMnO_4$. Both of these compounds are oxidized at the double bonds to formaldehyde. Oxidation is stopped with the addition of sulfite and the formaldehyde is determined with chromotropic acid. Formaldehyde and other formaldehyde precursors would be serious interferences in this method.

B. 2-Vinylpyridine and 5-vinyl-2-picoline

In a somewhat similar fashion these compounds could be determined in or near industrial atmospheres except that they would be collected in 0·01 N H_2SO_4.[520] They would be oxidized by periodate–permanganate to formaldehyde and the latter determined with chromotropic acid. Pyridine does not interfere; formaldehyde and its precursors would.

LXXIII. XYLONIC, CELLOBIONIC AND GLUCONIC ACIDS

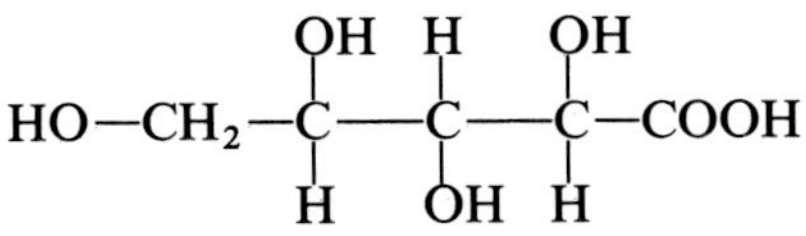

$$\mathrm{HO{-}CH_2{-}\underset{H}{\overset{OH}{C}}{-}\underset{OH}{\overset{H}{C}}{-}\underset{H}{\overset{OH}{C}}{-}COOH}$$

This acid can be readily separated from its oligomeric aldonic acids by anion-exchange chromatography and can be distinguished from its oligomers (e.g. xylobionic, xylotrionic, etc., acids) by the fact that it can form formaldehyde on periodate oxidation while they cannot. A three-channel analyzer was used as the analysis system.[1523, 1533] The analysis systems utilized the carbazole, chromic acid and periodate–formaldehyde methods. An example of a separation is given in Fig. 103.[1532] This chromatogram demonstrates the usefulness of the application of the three-channel analyzer for identification purposes. All oligomeric species gave a response in the carbazole channel, whereas no reaction occurred with xylonic and gluconic acids. However, these latter two acids gave a strong response in the periodate–formaldehyde channel, as did the cellobionic and cellotrionic acids. In contrast the oligomeric acids belonging to the xylonic acid series, which are lacking a primary hydroxyl group vicinal to a free hydroxyl, did not give rise to formaldehyde.

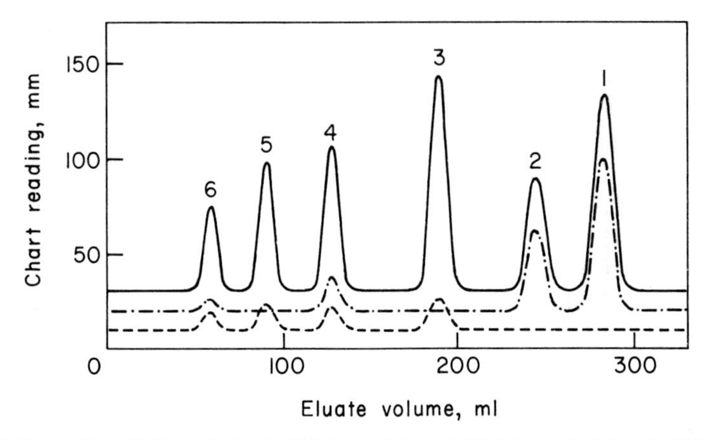

FIG. 103. Separation of 1·0 mg of xylonic (1), 0·6 mg of gluconic (2), 1·0 mg of xylobionic (3), 0·6 mg of cellobionic (4), 0·5 mg of xylotrionic (5) and 0·3 mg of cellotrionic acid (6). Channels: ———, chromic acid; —·—·—·, periodate–formaldehyde; — — — —, carbazole. Eluent: 0·02 M sodium acetate, pH 5·9. Nominal flow-rate: 8·5 cm. min^{-1}. Resin bed: 4 × 670 mm, Dowex 1 × 8, 13–18 μm.[1529]

LXXIV. BASIC PRECURSOR STRUCTURES

The wide variety of functional groups from which formaldehyde can be derived is shown in Table 4. Examples are given to show the types of molecules that are involved in this useful phenomenon.

One basic structure that does not fit in this Table is carbon monoxide, but these reactions have been studied from the prebiotic viewpoint, i.e. in the beginning there were simple organic chemicals and their reactions were the first blind movements toward life. Thus, the irradiation of carbon monoxide, ammonia and water adsorbed on highly effective substrata such as volcanic ash slate or clay minerals gave formaldehyde, formamide and urea.[1534] There is a photocatalytic production of organic compounds from carbon monoxide and water in a simulated Martian atmosphere.[1535] Essentially, the ultraviolet irradiation of siliceous materials or alumina in atmospheres containing carbon monoxide and water vapor diluted in a large volume of carbon dioxide or nitrogen results in the synthesis of large amounts of formic acid and smaller amounts of formaldehyde, acetaldehyde and glycolic acid. The photocatalytic formation of formaldehyde and acetaldehyde from the reaction of carbon monoxide and water on surfaces was also demonstrated by Hubbard *et al.*[1536] Even ultrasonic vibrations can produce compounds such as formaldehyde and hydrocyanic acid in water saturated with nitrogen, carbon monoxide (or methane) and hydrogen. The reactions occur in cavitation bubbles which on collapsing, generate intense hydraulic shocks accompanied by brief but very high pressure and temperature pulses.[1537]

LXXV. REAGENTS AND REACTIONS IN FORMALDEHYDE FORMATION

Some of the reactions useful in the direct formation of formaldehyde are shown in Table 32. Examples of these various types are given in Table 1.

Hot sulfuric acid is usually used in the decarboxylation reactions. This is the way formaldehyde is formed from glyoxylic acid[78] and from tartronate.[482,483] In the determination of glycine and other amino acids with ninhydrin, decarboxylation is one of the reactions taking place.[183,243–245]

Table 32. Some reactions useful in formaldehyde formation

Acetylation	Lössen rearrangement
Adduct formation	Oxidation
Decarboxylation	Oxidative dealkylation
Dehydration	Oxidative deamination
Diazotization	Reduction
Disproportionation	Tautomerism
Hydrolysis	

Glycolic acid and its precursors can be made to disproportionate in hot sulfuric acid.

The formation of formaldehyde through hydrolysis is usually done in acid solution (see Table 33). In some cases alkaline conditions are necessary before the formaldehyde-formation step, e.g. Alar, methyl methacrylate, and raceophenidol, Table 1. Methylene chloride and phenylephrine can form formaldehyde in alkaline solution.

Nitromethanol can be determined by hydrolysis, with aqueous sodium hydroxide, to formaldehyde, which is then reacted with chromotropic acid. Depending on the interferences present, some of the previously described reagents could also be used here in the final step.

Table 33. Analysis for formaldehyde precursors in test mixtures

Mixture	Precursor	Procedure[a]	Ref.
Apples	Alar	H → (O) → HBT → λ 582	10
Glycoproteins	Aminohexoses	H → (O) → 2,4-PD → YG fluor. on paper	33
Urine	Arabitol	CC → (O) → CA → λ 570	45
Fruit, alfalfa, clover, corn	Aramite	Extn (benzene) → (O) → Distill → PH → λ 520	46
Air particulates	Carbohydrates	Extn → alkali → MBTH	2, 66
Air	Chlorethanol	(O) → 2,4-PD → λ 412	81
Air	1-Chloro-2,3-epoxypropane	(O) → CA → λ 580	82
Water	Chlorophenoxyacetic acids	→ $CHCl_3$ → PC → elute → CA → λ 580	83
Impure chlorphenesin-1-carbamate	Chlorphenesin	(O) → CA → λ 570	86
Grain and seed	2,4-Dichlorophenoxyacetic acid[b]	DPr → Extn → CC → CA → λ 565	102
Milk	2,4-Dichlorophenoxyacetic acid, 2,4-D	DPr → Extn → CC → CA → λ 565	103
Surface waters	2,4-D[c]	Extn → H → CC → CA → λ 565	104
Shellfish and fish	2-Butoxyethyl 2,4-D[d]	H → L-L(benzene → buffer → CCl_4) → CC → CA → λ 570	106
Water	2,4-D[e]	→ $CHCl_3$ → PC → elute → CA → λ 580	107
Air	Diethanolamine	Collection in water → CA → λ 570	108
Polyethylene terephthalate	Diethylene glycol	(O) → CH_2O (not determined)	112
Cherries	Dimethoate	Extn → H → (O) → CA	117
Air	Ethylene	Coll. in H_2SO_4 → (O) → CA → λ 570	150
Air	Ethylene glycol	Coll. in H_2O → (O) → $Pb(OAc)_2$ → Cent → CA	152

Body fluids	Ethylene glycol	DPr → (O) → p-rosaniline	154
Body fluids and tissues	Ethylene glycol	DPr → (O) → 2,4-PD	156
Sterilized pharmaceuticals	Ethylene oxide	Distill → CC → (O) → CA → λ 580	159, 160
Air	Ethylene oxide	H → (O) → CA → λ 570	161, 166
Fumigated spices	Ethylene oxide[f]	H → (O) → CA → λ 570	162
Gases	Ethylene oxide	Extract into H_2O → H → (O) → CA	163
Liver tissue	Ethylmorphine	DPr → enz.DM → 2,4-PD	166
Wines	Formic acid	Steam Distn → (O) → CA → λ 570	171
Body fluids	Formic acid	(H) → CA → λ 570	174
Fruit juice	Formic acid	Steam Distn → (H) → CA → λ 570	178
Natural waters	Formic acid	L-L(H_2O → ether → H_2O) → (H) → CA → λ 570	175
Blood, plasma + serum	Glucose	DPr → CA → λ 570	12, 186, 187
Human serum	Glucose	DPr → MBTH → λ 620	188
Human urine	Glyceric acid	CC → PC → (O) → CA → λ 570	191
Rat adrenal glands	Glycerides	Sap. → (O) → CA → λ 570	193
Serum or plasma	Glycerides	Extn → Sap. → (O) → CA	194–196, 198, 200–203, 205[g], 207–209, 211–214
Lipids	Glycerides	Sap → PC → (O) → CA	197, 199[h]
Seed fat	(tri)glycerides	TLC → CC → (O) → CA	204
Ice cream	(1-mono)glycerides	Extn(ether) → (O) → CA	206
Mosquito tissue	(tri)glycerides	Silica acid → Extn($CHCl_3$) → Sap → (O) → CA	210
Serum or plasma[i]	(tri)glycerides	Extn → Sap → (O) → 2,4-PD	216–221
Serum or plasma	(tri)glycerides	Extn → Sap → (O) → PH	222–226
Fermentation solutions	Glycerol	(O) → CA	139, 236
Phospholipid hydrolyzate	Glycerol	H → PC → (O) → CA	142
Rat plasma	Glycerol	PC → TLC → (O) → 2,4-PD	238
Wines and liquors	Glycerol	PC → (O) → CA	232
Blood	Glycerol	DPr → (O) → CA	233
Phosphatides	Glycerol	H → into $CHCl_3$ → (O) → CA	237

Table 33—*continued*

Mixture	Precursor	Procedure[a]	Ref.
Blood and urine	Glycine	DPr → (DC + DA + (O)) → CA	243–244
Cat spinal cord and roots	Glycine	DA → DP → CA	248
Carboxymethylcellulose	Glycolic acid (free)	Extn(80% ethanol) → CA	261
Periodate oxidn products	Glycolic acid (free)	(H) → CA	263
Resin breakdown products	Glycolic acid (free)	Extn → CA	264
Used antifreeze solutions	Glycolic acid (free)	DP → 2,7-dihydroxynaphthalene	268
Irradiated oxalates	Glycolic acid (free)		269
Air particulates	1,2-Glycols	Collection → Extn(H_2O) → (O) → MBTH	2, 66
Tissue	1,2-Glycols	Sectioning → (O) → *p*-Rosaniline	274
Cottonseed and its oils	Guthion	Extn($CHCl_3$) → CC(Al_2O_3) → Distill-(HCl) → CA	286
Milk	Imidan	Extn → CC(Al_2O_3) → CA	319
Pharmaceutical prepns	Inositol	Extn(H_2O) → (O) → 2,7-naphthalenediol	320
Plasma + urine	Mannitol	DPr[j] → (O) → CA	328
Serum	Mannitol	DPr → (O) → CA	330
Essential oils	Methanol	(O) → CA	333, 335
Alcohol beverages	Methanol	(O) → CA	337, 352
Biological fluids	Methanol	(O) → CA	338, 339, 342, 344, 354, 358, 360
Combustion effluents	Methanol	(O) → CA	340
Fermentation products	Methanol	(O) → CA	347
Impure lactic acid	Methanol	(O) → CA	350
Wines	Methanol	(O) → CA	351
Soil	Methanol	(O) → CA	353
Blood	Methanol	(O) → 2,7-Dihydroxynaphthalene	357
Forensic materials	Methanol	(O) → *p*-Rosaniline	359

Air	2-Methylaminoethanol	Collect in water → (O) → CA	108
Urine	Oxalic acid	DPr → Pptn as CaC_2O_4 → (H) → CA	413
Plant tissue	Phorate	Extn → (O) → CC → CA	429, 430
Plant tissue	Phorate	Extn → (O) → Extn → TLC → CA	431
Pepper	Piperine	Extn(alc) → H → CA	109, 434
Pepper	Piperine	Extn → H → Gallic acid	435
Impure proresid	Podophyllotoxin	TLC → elute ($CHCl_3$) → CA	439
Urine	Polyols	CC(ion exchange, borate buffer) → (O) → CA	45
Impure tetranitrotetrazine	RDX	PC → elution → 2,7-dihydroxynaphthalene	447
Tablets	Riboflavin	Extn → (O) → CA	448
Soft drinks	Serine	CC → (O) → Dist → CA	454
Blood, urine, feces	Sorbitol	DPr → Reflux(alkali) → CC → CA	460, 461
Blood plasma	Trypsin	R[k] → (O) → 2,4-PD	401
Air	Vinyl chloride	Coll. (active C) → elution → (O) → CA	519

[a] Ca = chromotropic acid, CC = column chromatography, Cent = centrifuge, Coll = collection, DA = deamination, DC = decarboxylation, DM = demethylation, DP = disproportionation, DPr = deproteination, Extn = extraction, H = hydrolysis, HBT = 2-hydrazinobenzothiazole, L—L = liquid liquid extraction, MBTH = 3-methyl-2-benzothiazolinone hydrazone, (O) = oxidant, PC = paper chromatography, 2,4-PD = 2,4-pentanedione, PH = phenylhydrazine, Pptn = precipitation, Sap = saponification, TLC = thin-layer chromatography.
5 μg can be determined in 200 g sample.
[c] Sensitivity = 7 μg/liter of water.
[d] Recoveries of 63·9 to 75·2%.
[e] Beer's law obeyed from 5–150 μg.
[f] 1 ppm for 20 g sample.
[g] Semiautomated procedure.
[h] TLC separation of 3 types of glycerides.
[i] Also tissues.(221)
[j] For plasma.
[k] R = *p*-tosyl arginine methyl ester.

Glyoxylic acid in sulfuric acid at 100° is hydrolyzed to formaldehyde and can then be determined.

Dihydroxyacetone can also be determined by heating the sulfuric acid solution of chromotropic acid at a higher temperature.

Compounds containing the structures $O{-}CH_2{-}O$, $S{-}CH_2{-}O$, $N{-}CH_2S$, and $N{-}CH_2{-}N$ can also be hydrolyzed to formaldehyde with hot sulfuric acid and then determined with chromotropic acid, or one of the other reagents. Examples of some compounds of this type analyzed in this fashion are urea formaldehyde condensates, piperine, piperonyl butoxide, safrole, S-trithiane, piperonal, piperonylic acid, and hexamethylenetetramine. For better yields of formaldehyde, some of these compounds need to be heated to 150° or 180° in sulfuric acid. In the trace analysis of some mixtures, however, heating at 180° for a few minutes can lead to extensive charring. Even methanol under these conditions is oxidized to formaldehyde and thus gives a positive test. Consequently caution must be exercised in the use of hot sulfuric acid for trace analysis of mixtures.

Some of the oxidizing agents which have been used in the oxidation step to formaldehyde include:

Bismuthate
Chloramine T
Chromium trioxide
Cobalt salts
Enzymatic methods
Ferric chloride + MBTH
Halogen
Lead tetraacetate
Mercuric sulfate
Ninhydrin
Nitric acid
Nitrogen dioxide
Ozone
Perbenzoate
Periodate
Permanganate
Permanganate–periodate
Persulfate
Ruthenium dioxide + periodate
Selenium dioxide
Sulfuric acid

Of these, the periodate oxidizing agents are the most popular.

Many compounds have been analyzed indirectly by the following steps: (a) oxidation with periodate, permanganate, persulfate, or some other oxidizing agent, (b) destruction of the excess oxidizing agent with bisulfite, or occasionally arsenite, and (c) reaction with chromotropic acid. Examples of this type are methanol, ethylene oxide, 1,2-glycols, glycerol, other polyhydroxy compounds, compounds with a terminal methylene group, and serine. Methoxy compounds have been determined by hydrolysis to methanol, followed by oxidation and reaction with chromotropic acid. Instead of the chromotropic acid method, the phenylhydrazine procedure has been used to determine free methanol and also combined methanol in the pectin methylesterase determination. After oxidation, polyhydroxy compounds have also been determined by the phenylhydrazine and Schiff test methods.

Glycine has also been determined by oxidation with ninhydrin at pH 5 and then determination of the formaldehyde with chromotropic acid.

In many of the procedures some of the new methods could be used for the determination. In some cases this might be advantageous.

Sometimes in periodate oxidation it is useful to trap the formaldehyde with semicarbazide so as to avert its loss by side reactions without affecting its estimation.[1099,1100]

Reducing agents have seen small use probably because those in use are not too efficient or convenient and because they have not been investigated too thoroughly. Magnesium or zinc in acid solution is most frequently used.

LXXVI. CHROMOGEN-FORMING REAGENTS

These reagents have been discussed in the formaldehyde section of the volume devoted to the aldehyde functional group. These reagents and the spectral data of the final chromogens are given in Table 6. Of these reagents the most highly selective are chromotropic acid, J-acid, phenyl J-acid and especially 2,4-pentanedione. 3-Methyl-2-benzothiazolinone hydrazone is one of the more sensitive reagents of this family but other aliphatic aldehydes also react with it.

LXXVII. FLUOROGEN-FORMING REAGENTS

These reagents are also discussed in the formaldehyde section and enumerated in Table 6. 2,4-Pentanedione is the most highly selective reagent of this group. J-Acid is also a selective reagent for formaldehyde but under the conditions of the procedure there are quite a few miscellaneous precursors that can form formaldehyde.

LXXVIII. APPLICATION

The large number of methods for the photometric determination of formaldehyde can be used in the determination of the wide variety of organic compounds found in as wide a variety of complex mixtures. A representative number of these mixtures, their analyzed components and the analytical procedures are summarized briefly in Table 33. Thus, formaldehyde precursors of all types and kinds can be determined in our internal environment and in the environments around us. Many more examples are given in the body of this section.

REFERENCES

1. L. Hough and M. I. Taha, *J. Chem. Soc.* 2042 (1956).
2. L. Hough and M. I. Taha, *J. Chem. Soc.* 3994 (1957.
3. J. F. O'Dea and R. A. Gibbons, *Biochem. J.* **55**, 580 (1953).
4. E. Sawicki, R. Schumacher and C. R. Engel, *Microchem. J.* **12**, 377 (1967).
5. J. Mes and L. Kamm, *J. Chromatog.* **38**, 120 (1968).
6. G. M. Brearley and J. B. Weiss, *Biochem. J.* **110**, 413 (1968).
7. E. Sawicki, C. R. Engel and M. Guyer, *Anal. Chim. Acta*, **39**, 505 (1967).
8. E. Sawicki and R. A. Carnes, *Mikrochim. Acta, 1968*, 602.
9. R. U. Lemieux and E. von Rudloff, *Can. J. Chem.* **33**, 1710 (1955).
10. V. P. Lynch, *J. Sci. Food Agric.* **20**, 13 (1969).
11. H. C. Tun, J. F. Kennedy, M. Stacey and R. R. Woodbury, *Carbohyd. Res.* **11**, 225 (1969).
12. B. Klein and M. Weissman, *Anal. Chem.* **25**, 771 (1953).
13. L. Hough, D. B. Powell and B. M. Woods, *J. Chem. Soc.* 4799 (1956).
14. W. Godicke and U. Gerike, *Steroids*, **17**, 59 (1971).
15. P. Fleury, J. Courtois and M. Grandchamp, *Bull. Soc. Chim. France*, 88 (1949).
16. B. Smith, R. Ohlson and A. Olson, *Acta Chem. Scand.*, **16**, 1463 (1962).
17. Y. R. Naves, *Helv. Chim. Acta*, **32**, 1151 (1949).
18. B. J. Finlayson, J. N. Pitts, Jr. and H. Akimoto, *Chem. Phys. Lett.* **12**, 495 (1972).
19. A. P. Altshuller, *Air Water Pollution Int. J.* **10**, 713 (1966).
20. A. Levy, S. E. Miller and F. Scofield, "Proceedings of the Second International Clean Air Congress", Academic Press, New York, 1971, p. 305.
21. R. F. Gould, Ed., "Photochemical Smog and Ozone Reactions, Advances in Chemistry", 113 (1972).
22. A. C. Stern, Ed., "Air Pollution", Vol. 1, Academic Press, New York, 1968.
23. H. E. O'Neal and C. Blumstein, *Int. J. Chem. Kinet.* **5**, 397 (1973).
24. A. Polgar and J. L. Jungnickel, *Organic Analysis*, **3**, 203 (1956).
25. M. L. Karnovsky and W. S. Rapson, *J. Soc. Chem. Ind.* **65**, 138 (1946).
26. M. Pesez and J. Bartos, *Talanta*, **14**, 1097 (1967).
27. M. Pesez and J. Bartos, *Analusis*, **1**, 257 (1972).
28. B. H. Nicolet and L. A. Shinn, *J. Am. Chem. Soc.* **61**, 1615 (1939).
29. M. W. Rees, *Biochem. J.* **68**, 118 (1958).
30. E. Sawicki and C. R. Engel, *Chemist-Analyst*, **56**, 7 (1967).
31. J. M. McKibbin, *Methods of Biochemical Analysis*, **7**, 111 (1959).
32. B. Zaar and A. Gironwall, *Scand. J. Clin. Lab. Invest.* **13**, 588 (1961).
33. J. B. Weiss and I. Smith, *Nature*, **215**, 638 (1967).
34. A. Poland and E. Glover, *Mol. Pharmacol.* **10**, 349 (1974).
35. A. Poland and D. Nebert, *J. Pharmacol. Exp. Ther.* **184**, 269 (1973).
36. J. T. Wilson and J. R. Fouts, *Biochem. Pharmacol.* **16**, 215 (1967).
37. C. Dalton and D. Di Salvo, *Technicon Quarterly*, **4**, 20 (1972).
38. M. Kitada, T. Kamataka and H. Kitagawa, *Chem. Pharm. Bull.* **22**, 752 (1974).
39. C. F. Huebner, R. Lohmar, R. J. Dimler, S. Moore and K. P. Link, *J. Biol. Chem.*, **159**, 503 (1945).
40. J. L. Bose, A. B. Foster and R. Stephens, *J. Chem. Soc.* 3314 (1959).
41. E. Sawicki, T. R. Hauser and S. McPherson, *Anal. Chem.* **34**, 1460 (1962).
42. M. Beroza, *Anal. Chem.* **26**, 1970 (1954).
43. J. E. Forrest, R. Richard and R. A. Heacock, *J. Chromatog.* **65,** 439 (1972).
44. S. L. Tompsett and D. C. Smith, *Analyst*, **78**, 209 (1953).

45. N. Spencer, *J. Chromat.* **30**, 566 (1967).
46. M. E. Brokke, U. Kiigemagi and L. C. Terriere, *J. Agric. Food Chem.* **6**, 26, 471 (1958).
47. "Official Methods of Analysis of the AOAC", 10th ed, 24.095– 24.099 (1965).
48. M. Beroza, *Agr. Food Chem.* **11**, 51 (1963).
49. R. Preussmann, H. Hengy, D. Lubbe and A. Von Hodenberg, *Anal. Chim. Acta*, **41**, 497 (1968).
50. H. Druckrey, *Xenbiotica*, **3**, 271 (1973).
51. H. S. Shieh, *Nature*, **212**, 1608 (1966).
52. H. S. Shieh, *Can. J. Microbiol.* **10**, 837 (1964); **11**, 375 (1965).
53. J. Fog and E. Jellum, *Nature*, **195**, 590 (1962).
54. J. T. G. Overbeek, C. L. J. Vink and H. Deenstra, *Rec. Trav. Chim.* **74**, 85 (1955).
55. D. W. Hutchinson, B. Johnson and A. J. Knell, *Biochem. J.* **127**, 907 (1972).
56. M. Beljean and M. Pays, *Clin. Chim. Acta*, **44**, 119 (1973).
57. J. Fog and E. Jellum, *Nature*, **195**, 490 (1962).
58. K. Lundquist and L. Ericsson, *Acta Chem. Scand.* **24**, 3681 (1970).
59. E. Sawicki and C. R. Sawicki, *Annals N. Y. Acad. Sci.* **163**, 895 (1969).
60. B. L. Van Duuren, C. Katz, B. M. Goldschmidt, K. Frenkel and A. Sivak, *J. Nat. Cancer Inst.* **48**, 1431 (1972).
61. C. H. Wilson, *J. Soc. Cosmet. Chem.* **25**, 67 (1974).
62. J. C. Speck, Jr., *Anal. Chem.* **20**, 647 (1948).
63. M. Pesez and J. Bartos, *Ann. Pharm. Franc.* **22**, 609 (1964).
64. M. Pesez, P. Poirier and J. Bartos, "Practice of Colorimetric Organic Analysis", Masson + Cie, Paris, 1966.
65. R. Belcher, G. Dryhurst and A. M. G. MacDonald, *J. Chem. Soc.* 3964 (1965).
66. K. Fujie, *J. Japan Soc. Air Pollution*, **3**, 153 (1969).
67. H. J. H. Fenton and H. A. Sisson, *Proc. Cambridge Philosoph. Soc.* **14**, 385 (1908).
68. R. W. Eyler, E. D. Klug and F. Diephus, *Anal. Chem.* **19**, 24 (1947).
69. C. R. Szalkowski and W. J. Mader, *J. Am. Pharm. Assoc.* **44**, 533 (1955).
70. H. D. Graham, *J. Dairy Sci.* **55**, 42 (1972).
71. H. D. Graham, *J. Fd. Sci.* **36**, 1052 (1971).
72. J. P. Viccaro and E. L. Ambye, *Microchem. J.* **17**, 710 (1972).
73. F. S. H. Head and G. Hughes, *J. Chem. Soc.* 603 (1954).
74. E. Paart and O. Samuelson, *Carbohyd. Res.* **15**, 111 (1970).
75. D. E. Kramm and C. L. Kolb, *Anal. Chem.* **27**, 1076 (1955).
76. S. Ramanathan, J. Rivlin, O. A. Stamm and H. Zollinger, *Textile Res. J.* **38**, 63 (1968).
77. F. Feigl and R. Moscovici, *Analyst*, **80**, 803 (1955).
78. T. W. Stanley and E. Sawicki, *Anal. Chem.* **37**, 938 (1965).
79. R. C. Bonino, *Rev. Asoc. Bioquim. Argentina*, **28**, 115 (1963).
80. E. M. Barilari and M. Katz, *Arch. Bioquim., Quim. Farm., Tucuman*, **9**, 75 (1961); through *Chem. Abstr.* **57**, 12814 (1962).
81. J. W. Daniel and J. C. Gage, *Analyst*, **81**, 594 (1956).
82. W. Jaraczewska and W. Kaszper, *Medycyna Praca*, **18**, 168 (1967).
83. K. Erne, *Acta Chem. Scand.* **17**, 1663 (1963).
84. O. M. Aly and S. D. Faust, *Anal. Chem.* **36**, 2201 (1964).
85. F. S. Tanaka, H. R. Swanson and D. S. Frear, *Phytochem.* **11**, 2701 (1972).
86. A. A. Forist and R. W. Judy, *J. Pharm. Sci.* **53**, 1244 (1964).
87. D. P. Schwartz, *Anal. Chem.* **30**, 1855 (1958).
88. S. C. Pan, *J. Chromatog.* **9**, 81 (1962).
89. B. P. Lisboa, *J. Chromatog.* **16**, 136 (1964).

90. B. P. Lisboa, *Steroids*, **7**, 41 (1966).
91. P. Desgrez, J. Haas and S. H. Weinmann, *Ann. Biol. Clin.* (*Paris*), **15**, 289 (1957); through *Chem. Abstr.* **51**, 16643 (1957).
92. B. E. Lowenstein, A. C. Corcoran and I. H. Page, *Endocrinology*, **39**, 82 (1946).
93. D. C. Smith and S. L. Tompsett, *Analyst*, **79**, 53 (1954).
94. J. Rabinovitch, J. Decombe and A. Freedman, *Lancet*, **261**, 1201 (1951).
95. C. J. W. Brooks and J. K. Norymberski, *Biochem. J.* **55**, 371 (1953).
96. R. W. H. Edwards and A. E. Kellie, *Biochem. J.* **56**, 207 (1954).
97. R. B. Clark and R. T. Rubin, *Anal. Biochem.* **29**, 31 (1969).
98. M. Spatz, "Toxic and Carcinogenic Alkylating Agents from Cycads", presented at Conference on Biological Effects of Alkylating Agents, New York Academy of Sciences, New York City, 16–18 September, 1968.
99. H. Matsumoto and F. M. Strong, *Arch. Biochem.* **101**, 299 (1963).
100. D. K. Dastur and R. S. Palekar, *Nature*, **210**, 841 (1966).
101. E. von Rudloff, *Can. J. Chem.* **43**, 2660 (1965).
102. R. P. Marquardt and E. N. Luce, *J. Agr. Food Chem.* **3**, 51 (1955).
103. R. P. Marquardt and E. N. Luce, *Anal. Chem.* **23**, 1484 (1951).
104. O. M. Aly and S. D. Faust, *J. Am. Water Works Assoc.* **55**, 639 (1963).
105. V. H. Freed, *Science*, **107**, 98 (1948).
106. J. E. Coakley, J. E. Campbell and E. F. McFarren, *J. Agr. Food Chem.* **12**, 262 (1964).
107. L. C. Erickson and B. C. Brannaman, *Hilgardia*, **23**, 175 (1954).
108. F. A. Miller, *Am. Ind. Hyg. Assoc. J.* **30**, 411 (1968).
109. L. A. Lee, *Anal. Chem.* **28**, 1621 (1956).
110. A. J. Harrison and J. S. Lake, *J. Phys. Chem.* **63**, 1489 (1959).
111. H. E. Carter, F. J. Glick, W. P. Norris and G. E. Philips, *J. Biol. Chem.* **170**, 285 (1947).
112. J. R. Kirby, A. J. Baldwin and R. H. Heidner, *Anal. Chem.* **37**, 1306 (1965).
113. A. Colon, G. E. Herpich, R. G. Johl, J. D. Neuss and H. A. Frediani, *J. Am. Pharm. Assoc.* **39**, 335 (1950).
114. E. A. Garlock, Jr. and D. C. Grove, *J. Clin. Invest.* **28**, 843 (1949).
115. W. A. Vail and C. E. Bricker, *Anal. Chem.* **24**, 975 (1952).
116. P. F. Fleury and J. Lange, *Compt. Rend.* **195**, 1395 (1932).
117. R. Santi and B. Bazzi, *Chimica*, **12**, 325 (1956).
118. C. H. Mitchell and D. M. Ziegler, *Anal. Biochem.* **28**, 261 (1969).
119. A. T. Boltini and R. E. Olsen, *J. Org. Chem.* **27**, 452 (1962).
120. E. Sawicki, T. W. Stanley, T. R. Hauser, W. Elbert and J. L. Noe, *Anal. Chem.* **33**, 722 (1961).
121. E. Sawicki, T. R. Hauser, T. W. Stanley, W. Elbert and F. T. Fox, *Anal. Chem.* **33**, 1574 (1961).
122. D. M. Ziegler and F. H. Pettit, *Biochemistry*, **5**, 2932 (1966).
123. J. M. Machinist, W. H. Orme-Johnson and D. M. Ziegler, *Biochemistry*, **5**, 2939 (1966).
124. F. V. Neumann and C. W. Gould, *Anal. Chem.* **25**, 751 (1953).
125. A. Hawks, P. F. Swann and P. N. Magee, *Biochem. Pharmacol.* **21**, 432 (1972).
126. N. V. Sutton, *Anal. Chem.* **36**, 2120 (1964).
127. R. Preussmann, H. Hengy and A. Von Hodenberg, *Anal. Chim. Acta*, **42**, 95 (1968).
128. R. Kato, H. Shoji and A. Takanaka, *Gann.* **58**, 467 (1967).
129. J. A. Brouwers and P. Emmelot, *Exptl. Cell Res.* **19**, 467 (1960).
130. K. Morihara, *Bull. Chem. Soc. Japan*, **37**, 1787 (1964).

131. V. M. Craddock, *Chem-Biol. Interactions*, **4**, 149 (1972).
132. A. R. Archibald and J. G. Buchanan, *Carbohydrate Res.* **11**, 558 (1969).
133. R. G. Wilkins, *Quart. Rev.*, (*London*), **16**, 316 (1962).
134. R. E. Hamm and K. Schroeder, *Inorg. Chem.* **3**, 391 (1964).
135. G. Kakabadse and H. J. Wilson, *Analyst*, **86**, 402 (1961).
136. G. Schwarzenbach, *Helv. Chim. Acta*, **32**, 839 (1949).
137. A. Krynska, *Ochrana Pracy*, (*Warsaw*), **18**, 93 (1968).
138. V. Ulbrich and J. Makeš, *J. Chrom.* **15**, 371 (1964).
139. M. Lambert and A. C. Neish, *Can. J. Res.* **28B**, 83 (1950).
140. F. C. Charalampous, *Methods in Enzymology*, **5**, 283 (1962).
141. F. C. Charalampous, *J. Biol. Chem.* **211**, 249 (1954).
142. L. W. Wheeldon, M. Brinley and D. A. Turner, *Anal. Biochem.* **4**, 433 (1962).
143. G. S. Salyamon and M. V. Popelkovskayu, *Gig. Sanit.* **36**, 60 (1971).
144. P. A. Giang and M. S. Schechter, *J. Agr. Food Chem.* **8**, 51 (1960).
145. E. S. Gronsberg, *Trudy Khim. khim. Teknol.* 186 (1970).
146. D. M. Soignet, R. J. Berni, J. I. Wadsworth and R. R. Benerito, *Anal. Lttrs.* **5**, 395 (1972).
147. R. U. Lemieux and E. von Rudloff, *Can. J. Chem.* **33**, 1701 (1955).
148. E. S. Gronsberg, *Khim. Prom.* 775 (1970).
149. S. W. Nicksic, J. Harkins and B. A. Fries, *J. Air Poll. Control Assoc.* **14**, 224 (1964).
150. N. A. Krylov, *Gigiena i Sanit.* **26**, 48 (1961); through *Chem. Abstr.* **56**, 5076 (1952).
151. T. Beyrich, *Pharm. Zentralhalle*, **108**, 837 (1969).
152. G. Mandric, *Igiena* (*Bucharest*), **10**, 271 (1961).
153. P. Desnuelle and M. Naudet, *Bull. Soc. Chim. France*, **12**, 871 (1945); *Industries Corps Gras*, **1**, 113 (1945).
154. R. N. Harger and R. B. Forney, *J. Forensic Sci.* **4**, 136 (1959).
155. H. B. S. Conacher and D. I. Rees, *Analyst*, **91**, (1966).
156. J. C. Russell, E. W. McChesney and L. Goldberg, *Food Cosmet. Toxicol.* (*London*), **7**, 107 (1969).
157. W. H. Evans and A. Dennis, *Analyst*, **98**, 782 (1973).
158. D. A. White, D. S. Miyada and R. M. Nakamura, *Clin. Chem.* **20**, 645 (1974).
159. N. Adler, *J. Pharm. Sci.* **54**, 735 (1965).
160. C. E. Bricker and J. K. Lee, *J. Am. Pharm. Assoc., Sci. Ed.* **41**, 346 (1952).
161. N. E. Bolton and N. H. Ketcham, *Arch. Environ. Health*, **8**, 711 (1964).
162. F. E. Critchfield and J. B. Johnson, *Anal. Chem.* **29**, 797 (1957).
163. M. Jaworski, A. Zielasko and K. Gasior, *Chem. Anal., Warsaw*, 6, 1005 (1961).
164. J. C. Gage, *Analyst*, **82**, 587 (1957).
165. G. S. Salyamon and M. V. Popelkovskaya, *Gig. Sanit.* **37**, 117 (1972).
166. R. E. Stitzel, F. E. Greene, R. Furner and H. Conway, *Biochem. Pharmacol.* **15**, 1001 (1966).
167. S. L. Sachdev, J. P. Lodge, Jr. and P. W. West, *Anal. Chim. Acta*, **58**, 141 (1972).
168. J. L. Bowling, J. A. Dean and W. D. Shults, *Anal. Letters*, **6**, 933 (1973).
169. L. Legradi, E. Pungor and O. Szabadka, *Acta Chim. Hung.* **42**, 89 (1964).
170. S. M. Bose, K. T. Joseph and B. M. Das, *Tanner* (*India*), **4**, No. 3, 23; No. 4, 16 (1956).
171. W. Diemair and C. Gundermann, *Z. Lebensmitt. Untersuch.* **110**, 261 (1959).
172. R. Fabre, R. Truhaut and A. Singerman, *Ann. Pharm. Franc.* **12**, 409 (1954).
173. W. M. Grant, *Anal. Chem.* **20**, 267 (1948).
174. N. Rietbrock and W. D. Hinrichs, *Klin. Wochschr.* **42**, 981 (1964).
175. A. G. Stradomskaya and I. A. Goncharova, *Gidrokhim. Mater.* **43**, 57 (1967).
176. E. Eegriwe, *Z. Anal. Chem.* **110**, 22 (1937).

177. E. Sawicki, *in* "Analytical Chemistry", Elsevier Publishing Company, Netherlands, 1962, pp. 62–69.
178. K. Gierschner, *Ind. Obst. Gemüseverwert.* **53**, 355 (1968).
179. S. Rolski and G. M. Maciak, *Microchem. J.* **18**, 181 (1973).
180. M. Ishidate, M. Matsui and M. Okada, *Anal. Biochem.* **11**, 176 (1965).
181. M. Matsui, M. Okada and M. Ishidate, *Anal. Biochem.* **12**, 143 (1965).
182. J. C. Speck, Jr. and A. A. Forist, *Anal. Chem.* **26**, 1942 (1954).
183. S. L. Tompsett, *Anal. Chim. Acta*, **19**, 360 (1958).
184. Y. Houminer and S. Hoz, *Israel J. Chem.* **8**, 97 (1970).
185. M. J. Houle and R. L. Powell, *Anal. Biochem.* **13**, 562 (1965).
186. F. W. Sunderman, Jr. and F. W. Sunderman, *Tech. Bull. Reg. Med. Technologists*, **31**, 93 (1961).
187. F. W. Sunderman, Jr. and F. W. Sunderman, *Am. J. Clin. Pathol.* **36**, 75 (1961).
188. A. Tsuji, T. Kinoshita, A. Sakai and M. Hoshino, *Chem. Pharm. Bull.* **17**, 1304 (1969).
189. E. Chargoff and B. Magasanik, *J. Am. Chem. Soc.* **69**, 1459 (1947).
190. P. F. Fleury and Y. Fievet-Guinard, *Ann. Pharm. Franc.* **5**, 504 (1947).
191. H. E. Williams and L. H. Smith, *J. Lab. Clin. Med.* **71**, 495 (1968).
192. W. Godicke and U. Gerike, *Mikrochim. Acta*, 603 (1972).
193. R. Angelico, G. Cavina, A. D'Antona and G. Giocoli, *J. Chromatog.* **18**, 57 (1965).
194. D. H. Blankenhorn, G. Rouser and T. I. Wiemer, *J. Lipid Res.* **2**, 281 (1961).
195. L. A. Carlson and L. B. Wadstrom, *Clin. Chim. Acta*, **4**, 197 (1959).
196. A. Christophe and F. Mathijs, *Bull. Soc. Chim. Belg.* **73**, 592 (1964).
197. A. S. W. DeFreitas and F. Depocas, *Can. J. Biochem.* **42**, 195 (1964).
198. M. Eggstein, *Klin. Wochenschr.* **44**, 267 (1966).
199. C. Franzke, K. O. Heims and I. Vollgraf, *Nahrung*, **11**, 515 (1967).
200. S. N. Jagannathan, *Can. J. Biochem.* **42**, 566 (1964).
201. A. Kaplan and V. F. Lee, *Proc. Soc. Exptl. Biol. Med.* **118**, 296 (1965).
202. M. Kawade, *Mie. Med. J.* **11**, 399 (1962).
203. S. Laurell, *Scand. J. Clin. Lab. Invest.* **18**, 668 (1966).
204. C. Litchfield, M. Farquhar and R. Reiser, *J. Am. Oil Chem. Soc.* **41,** 588 (1964).
205. H. B. Lofland, Jr., *Anal. Biochem.* **9**, 393 (1964).
206. H. E. Schmidt, *Fette, Seif., Anstrichmitt.* **65**, 488 (1963).
207. R. D. Stewart, *Can. J. Biochem. Physiol.* **32**, 679 (1954).
208. M. Suehiro, *Sekagaku*, **33**, 387 (1961); through *Chem. Abstr.* **55**, 23644 (1961).
209. E. Van Handel, *Clin. Chem.* **7**, 249 (1961).
210. E. van Handel, *Anal. Biochem.* **11**, 266 (1965).
211. E. Van Handel and D. B. Zilversmit, *J. Lab. Clin. Med.* **50**, 152 (1957).
212. G. Vanzetti and E. Denegri, *Ital. J. Biochem.* **13**, 391 (1964).
213. W. M. Butler, Jr., H. M. Maleng, M. G. Horning and B. B. Brodie, *J. Lipid Res.* **2**, 95 (1961).
214. H. Ignatowska and A. Michajlik, *Polski Tygodnik Lekarshi*, **13**, 1037 (1958); through *Chem. Abstr.* **54**, 3575 (1960).
215. M. Kraml and L. Cosyns, *Clin. Biochem.* **2**, 373 (1969).
216. J. R. Claude, F. Corre and C. Levallois, *Clin. Chim. Acta*, **19**, 231 (1968).
217. D. G. Cramp and G. Robertson, *Anal. Biochem.* **25**, 246 (1968).
218. M. E. Royer and H. Ko, *Anal. Biochem.* **29**, 405 (1969).
219. V. F. Dunsbach, *Z. Clin. Chem.* **4**, 262 (1965).
220. M. J. Fletcher, *Clin. Chim. Acta*, **22**, 393 (1968).
221. V. M. Sardesai and J. A. Manning, *Clin. Chem.* **14**, 156 (1968).

222. G. C. Buckley, J. M. Cutler and J. A. Little, *Can. Med. Assoc. J.* **94**, 886 (1966).
223. F. Galletti, *Clin. Chim. Acta* **15**, 184 (1967).
224. A. Jover, *J. Lipid Res.* **4**, 228 (1963).
225. A. Randrup, *Scand. J. Clin. Lab. Invest.* **12**, 1 (1960).
226. M. R. Lloyd and R. B. Goldrick, *Med. J. Aust.* **2**, 493 (1968).
227. M. Pays, P. Malangeau and R. Bourdon, *Ann. Pharm. Franc.* **25**, 29 (1967).
228. C. Szonyi and K. Sparrow, *J. Am. Oil Chemists Soc.* **41**, 535 (1964).
229. B. Baehler, *Pharm. Acta Helv.* **40**, 226 (1965).
230. J. W. B. Erskine, C. R. N. Strouts, G. Walley and W. Lazarus, *Analyst*, **78**, 630 (1953).
231. G. Jurriens, B. de Vries and L. Schouten, *J. Lipid Res.* **5**, 267 (1964).
232. K. G. Bergner and H. Meyer, *Deut. Lebensm. Rundschau*, **56**, 49 (1960).
233. D. Biesold and E. Strack, *Z. Physiol. Chem.* **311**, 115 (1958).
234. W. R. Frisell, L. A. Meech and C. G. Mackenzie, *J. Biol. Chem.* **207**, 709 (1954).
235. H. M. Kalckar, *J. Biol. Chem.* **167**, 461 (1947).
236. A. Mizsei, M. Iglóy and G. Veress, *Magy. Kém Lap.* **19**, 503 (1964).
237. O. Renkonen, *Biochim. Biophys. Acta*, **56**, 367 (1962).
238. A. S. W. de Freitas, *Can. J. Biochem.* **45**, 1041 (1967).
239. F. L. Pyman and H. A. Stevenson, *J. Chem. Soc.* 448 (1934).
240. W. Weigel, *Z. Physiol. Chem.* **353**, 113 (1972).
241. K. E. Bharucha and F. D. Ganstone, *J. Sci. Food Agr.* **6**, 378 (1955).
242. G. Maerker and E. T. Haeberer, *J. Amer. Oil Chemists Soc.* **43**, 97 (1966).
243. B. Alexander, G. Landwehr and A. Seligman, *J. Biol. Chem.* **160**, 51 (1945).
244. E. Orlando and L. Ferrari, *Giorn. bioquim.* **3**, 147 (1954).
245. S. Ruheman, *J. Chem. Soc.* **99**, 792 (1911).
246. P. D. Frazier and G. K. Summer, *Anal. Biochem.* **44**, 66 (1971).
247. H. N. Christensen, T. H. Riggs and N. E. Ray, *Anal. Chem.* **23**, 1521 (1951).
248. M. H. Aprison and R. Werman, *Life Sciences*, **4**, 2075 (1965).
249. J. Giroux and A. Puech, *Ann. Pharm. Franc.* **21**, 469 (1963).
250. A. Puech, Glycine and Industrial Gamma Globulin, Thesis, University of Montpellier, France, 1962.
251. V. M. Sardesai and H. S. Provido, *Clin. Chim. Acta*, **29**, 67 (1970).
252. J. F. Goodwin and S. Stampwala, *Clin. Chem.* **19**, 1010 (1973).
253. F. Feigl, V. Gentil and C. Starkmayer, *Mikrochim. Acta*, 348 (1957).
254. E. L. Coe, *Anal. Biochem.* **10**, 236 (1965).
255. L. Josimovic, *Anal. Chim. Acta*, **62**, 210 (1972).
256. K. Takahashi, *J. Biochem.* **71**, 563 (1972).
257. F. Feigl, "Spot Tests in Organic Analysis", 7th edn. Elsevier, New York, 1966, p. 465.
258. E. Eegriwe, *Z. Anal. Chem.* **89**, 123 (1932).
259. G. Denigès, *Bull. Trav. Soc. Pharm. Bordeaux*, **49**, 193 (1909).
260. S. Dagley and A. Rodgers, *Biochim. Biophys. Acta*, **12**, 591 (1953).
261. M. Easterwood, *Anal. Chem.* **29**, 981 (1957).
262. P. Fleury, J. Courtois and R. Perles, *Mikrochem. Acta*, **36/37**, 863 (1951).
263. P. R. Fleury, R. Perles and L. LeDizet, *Ann. Pharm. Franc.* **11**, 581, (1953).
264. S. J. Lyle and A. R. Sani, *Anal. Chim. Acta*, **33**, 619 (1965).
265. R. H. Still, K. Wilson and B. W. J. Lynch, *Analyst*, **93**, 805 (1968).
266. V. P. Calkins, *Ind. Eng. Chem., Anal. Ed.* **15**, 762 (1943).
267. R. A. Gibbons, *Analyst*, **87**, 178 (1962).
268. H. Green, *Analyst*, **82**, 107 (1957).
269. L. Josimovic and O. Gal, *Anal. Chim. Acta*, **36**, 12 (1966).

270. S. R. Sarfati and P. Szabo, *Carbohydrate Research*, **12**, 290 (1970).
271. L. Maros and S. Szebeni, *Acta Chim. Sci. Hung.* **48**, 11 (1966).
272. N. Shaw, *Biochem. Biophys. Acta*, **164**, 435 (1968).
273. V. E. Vaskovsky and S. V. Isay, *Anal. Biochem.* **30**, 25 (1969).
274. J. F. A. McManus, Periodate oxidation techniques, in J. F. Danielli, Ed., "General Cytochemical Methods", Vol. 2, Academic Press, New York, 1961, pp. 171–201.
275. J. F. A. McManus and C. Hoch-Ligeti, *Lab. Invest.* **1**, 19 (1952).
276. J. M. Bobbitt, Periodate oxidation of carbohydrates, in M. L. Wolfrom and R. S. Tipson, Eds, "Advances in Carbohydrate Chemistry", Vol. 11, Academic Press, New York, 1956, pp. 1–41.
277. U. Zeidler, H. Lepper and W. Stein, *Fette Seifen Anstrichm.* **76**, 260 (1974).
278. G. de Vries and A. Schors, *Tetrahedron Letters*, 5689 (1968).
279. E. Klenk and G. Uhlenbruck, *Z. physiol. Chem.* **307**, 266 (1956).
280. E. Martensson, A. Raal and L. Svennerholm, *Biochim. Biophys. Acta* **30**, 124 (1958).
281. J. Uriel and P. Grabar, *Anal. Biochem.* **2**, 80 (1961).
282. G. D. Vogels and C. V. D. Drift, *Anal. Biochem.* **33**, 143 (1970).
283. M. F. Ishak and T. Painter, *Acta Chem. Scand.* **27**, 1268 (1973).
284. R. D. Guthrie, *Advances Carbohydrate Chem.* **16**, 105 (1961).
285. T. J. Painter and B. Larsen, *Acta Chem. Scand.* **24**, 813 (1970).
286. P. A. Giang and M. S. Schechter, *J. Agr. Food Chem.* **6**, 845 (1958).
287. N. E. Vernazza, *Acad. Sci. Torino, Classe Sci. Fis. Mat. Nat.* **70**, 404 (1935).
288. A. Vercillo and M. C. Grossi, *Boll. Lab. Chim. Provinciali* (Bologna), **17**, 413 (1966).
289. M. L. Dow, *J. Assoc. Off. Anal. Chemists*, **56**, 647 (1973).
290. J. K. Lim and C. CC. Chen, *J. Pharm. Sci.* **62**, 1503 (1973).
291. J. G. Kreiner, **88**, 423 (1974).
292. V. Stankovič, *Chem. Zvesti* **17**, 274 (1963).
293. R. Engst, L. Prahl and E. Jarmatz, *Nahrung*, **13**, 417 (1969).
294. D. A. Cronin and J. Gilbert, *J. Chromatog.* **87**, 387 (1973).
295. S. A. Barker, M. J. How, P. V. Peplow and P. J. Somers, *Anal. Biochem.* **26**, 219 (1968).
296. J. M. Heuss, G. N. Nebel and B. A. D'Alleva, *Environ. Sci. Technol.* **8**, 641 (1974).
297. H. Kala, *Pharmazie*, **20**, 82 (1965).
298. R. Fried, *Mitt. Deut. Pharm. Ges.* **34**, 189 (1962).
299. M. E. Shils, *Clin. Chem.* **13**, 397 (1967).
300. J. Bartos, *Ann. Pharm. Franc.* **20**, 650 (1962).
301. F. A. L. Anet and L. Marion, *Can. J. Chem.* **33**, 849 (1955).
302. D. D. Van Slyke, A. Hiller and D. A. MacFadyen, *J. Biol. Chem.* **141**, 681 (1941).
303. P. B. Hamilton and R. A. Anderson, *J. Biol. Chem.* **213**, 249 (1955).
304. P. Desnuelle and S. Antonin, *Compt. Rend.* **216**, 206 (1945).
305. R. B. Aronson, F. M. Sinex, C. Franzblau and D. D. Van Slyke, *J. Biol. Chem.* **242**, 809 (1967).
306. F. M. Sinex and D. D. Van Slyke, *Federation Proc.* **16**, 250 (1957).
307. O. O. Blumenfeld, M. A. Paz, P. M. Gallop and S. Seifter, *J. Biol. Chem.* **238**, 3835. (1963).
308. W. T. Butler and L. W. Cunningham, *J. Biol. Chem.* **240**, PC 3449 (1965).
309. H. Hörmann and G. Fries, *Z. Physiol. Chem.* **311**, 19 (1958).
310. R. J. Schlueter and A. Veis, *Biochemistry*, **3**, 1657 (1964).
311. H. Zahn and L. Zurn, *Z. Naturforsch.* **12b**, 788 (1957).
312. P. Desnuelle and S. Antonin, *Biochem. Biophys. Acta*, **1**, 50 (1947).

313. N. Blumenkrantz and D. J. Prockop, *Anal. Biochem.* **30**, 377 (1969).
314. G. A. Crowe, Jr. and C. C. Lynch, *J. Am. Chem. Soc.* **71**, 3731 (1949).
315. G. A. Crowe, Jr. and C. C. Lynch, *J. Am. Chem. Soc.* **70**, 3795 (1948).
316. H. Hift and H. R. Mahler, *J. Biol. Chem.* **198**, 901 (1952).
317. H. E. Carter and H. E. Neville, *J. Biol. Chem.* **170**, 301 (1947).
318. B. H. Nicolet and L. A. Shinn, *J. Biol. Chem.* **139**, 687 (1941).
319. M. C. Bowman and M. Beroza, *J. Assoc. Off. Agr. Chemists*, **48**, 922 (1965).
320. P. C. Bose, R. K. Singh and G. K. Ray, *Ind. J. Pharm.* **25**, 419 (1963).
321. M. L. Wolfrom, A. Thompson, A. N. O'Neill and T. T. Galkowski, *J. Am. Chem. Soc.* **74**, 1062 (1952).
322. B. T. White, *J. Dairy Sci.* **44**, 1791 (1961).
323. J. F. O'Dea, *Chem. Ind.* 1338 (1953).
324. M. Guernet, C. Ciuro and P. Malangeau, *Bull. Soc. Chim. Biol.* **47**, 1777 (1965).
325. F. B. Anderson, E. L. Hirst and D. J. Manners, *Chem. Ind.* 1178 (1957).
326. F. B. Anderson, E. L. Hirst, D. J. Manners and A. G. Ross, *J. Chem. Soc.* 3233 (1958).
327. G. Dryhurst, "Periodate Oxidation of Diol and Other Functional Groups", Pergamon Press, London, 1970, pp. 89–91.
328. A. C. Corcoran and J. H. Page, *J. Biol. Chem.* **170**, 165 (1947).
329. S. L. Kanter, *Clin. Chim. Acta* **16**, 177 (1967).
330. G. Tibbling, *Scan. J. Clin. Invest.* **22**, 7 (1968).
331. M. Pays and M. Beljean, *Ann. Pharm. Franc.* **28**, 241 (1970).
332. E. A. Ovsoyan and A. B. Nalbandian, *Armyanskii Khim. Z.* **22**, 1057 (1969).
333. T. Beyrich and R. Pohloudek-Fabini, *Ernahrungsforschung*, **5**, 441 (1960); through *Chem. Abstr.* **55**, 3930 (1961).
334. R. N. Boos, *Anal. Chem.* **20**, 964 (1948).
335. E. Bremanis, *Z. Lebensm.–Untersuch. Forsch.* **93**, 1 (1951); through *Chem. Abstr.* **45**, 8941 (1951).
336. W. Deckenbrock and M. Sprick, *Branntweinwirtschaft*, **75**, 1 (1953).
337. E. Elvove, *Ind. Eng. Chem.* **9**, 295 (1917).
338. M. Feldstein and N. C. Klendshoj, *Can. J. Med. Technol.* **16**, 135 (1954).
339. M. Feldstein and N. C. Klendshoj, *Anal. Chem.* **26**, 932 (1954).
340. M. Feldstein, "Symposium on Air Pollution Control", Special Technical Publication No. 281, p. 23 (1959).
341. F. R. Georgia and R. Morales, *Ind. Eng. Chem.* **18**, 305 (1926).
342. J. Hindberg and J. O. Wieth, *J. Lab. Clin. Med.* **61**, 355 (1963).
343. M. Holden, *Biochem. J.* **39**, 172 (1945).
344. C. D. Hough, *Analyst*, **85**, 921 (1960).
345. M. Langejan, *Pharm. Weekblod*, **92**, 667 (1957).
346. A. Kamibayashi, M. Miki and H. Ono, *Hakko Kyokaishi*, **16**, 165 (1958).
347. E. F. Mariño, J. A. S. Puerta, T. H. Zaballos, E. R. Matía and T. H. Díez, *Boln. Inst. nac. Invest. agron., Madr.* **24**, 453 (1964).
348. A. P. Mathers, *J. Assoc. Off. Agr. Chemists*, **38**, 753 (1955).
349. M. J. Maurice and B. Veen, *Anal. Chem.* **163**, 13 (1958).
350. R. L. Maute, R. H. Benson and E. Martelli, *Anal. Chem.* **40**, 1380 (1968).
351. H. Rebelein, *Dt. LebensmittRdsch.* **61**, 211 (1965).
352. M. Ritz, M. Šunić and M. Filajdić, *Kem. u Ind., Zagreb*, **13**, 267 (1964).
353. G. P. Sedova, *Gigiena i Sanitariya*, 56, (1963).
354. F. Stratton, *Am. J. Clin. Pathol.* **25**, 179 (1955).
355. L. A. Williams, R. A. Linn and B. Zak, *J. Forensic Sci.* **6**, 119 (1961).
356. S. Vamamura and T. Matsuoka, *J. Soc. Brewing, Japan*, **49**, 111 (1954).

357. O. E. Skraug, *Scand. J. Clin. Lab. Invest.* **8**, 338 (1956).
358. O. T. Khachidze, *Sadovodstvo, Vinogradarstvo i Vinodeli Moldevi*, **12**, 32 (1957).
359. I. Odler, *Soudni Lekarstvi*, **2**, 49 (1957); through *Chem. Abstr.* **54**, 24980 (1960).
360. V. M. Sardesai and H. S. Provido, *J. Lab. Clin. Med.* **64**, 977 (1964).
361. R. F. Milton and W. A. Walters, "Methods of Quantitative Microanalyses", Arnold, London, 1949, p. 317.
362. B. M. Pogell, J. M. Mosely, C. J. Likes and D. F. Koenig, *J. Agr. Food Chem.* **5**, 301 (1957).
363. L. O. Wright, *Ind. Eng. Chem.* **19**, 750 (1927).
364. C. C. Leong and T. C. Yam, *Analyst*, **96**, 367 (1971).
365. R. H. Dyer, *J. Assoc. Off. Anal. Chemists*, **54**, 785 (1971).
366. P. J. Wood and I. R. Siddiqui, *Anal. Biochem.* **39**, 418 (1971).
367. S. Lartillot and C. Vogel, *Biochimie*, **55**, 829 (1973).
368. S. B. Schryver and C. C. Wood, *Analyst*, **45**, 165 (1920).
369. K. Morihara, *Bull. Chem. Soc. Japan*, **37**, 1785 (1964).
370. T. L. Lunder, *Anal. Biochem.* **49**, 585 (1972).
371. R. B. Tupeeva, *Hyg. Sanit.* **32**, 72 (1967).
372. A. P. Mathers and M. A. Pro. *Anal. Chem.* **27**, 1662 (1955).
373. T. Pavolini and A. Malatesta, *Ann. Chim. Appl.* **37**, 495 (1947).
374. F. H. Bernhardt, H. Staudinger and V. Ullrich, *Z. Physiol. Chem.* **351**, 467 (1970).
375. L. S. Frankel, P. R. Madsen, R. R. Siebert and K. L. Wallisch, *Anal. Chem.* **44**, 2401 (1972).
376. L. Horner, E. Winkelmann, K. H. Knapp and W. Ludwig, *Ber.* **92**, 288 (1959).
377. H. Uehleke, *Xenobiotica*, **1**, 327 (1971).
378. H. Uehleke *in* E. Jucker, Ed., "Progress in Drug Research", Vol. 15, Basle, Birkhauser, 1971, p. 147.
379. "IARC Monographs on the Evaluation of Carcinogenic Risk of Chemicals to Man", Vol. I, International Agency for Research on Cancer, Lyon, France, 1972, p. 164.
380. W. Lijinsky, L. Keefer, E. Conrad and R. Van de Bogart, *J. Nat. Cancer Inst.* **49**, 1239 (1972).
381. E. S. Gronsberg, *Trudy Khim. Khim Technol* (*Gorkii*), 131 (1964), through *Anal. Abstr.* **12**, 4879 (1965).
382. A. B. Karasz, F. De Cocco, J. J. Maxstadt and A. Curthoys, *JAOAC*, **57**, 541 (1974).
383. M. Elleder and Z. Lojday, *Histochemie*, **30**, 325 (1972).
384. L. Fishbein and H. L. Falk, *Environ. Res.* **2**, 297 (1969).
385. E. Bovalini and A. Casini, *Ann. Chim. Roma*, **49**, 1059 (1959).
386. C. E. Bricker and H. R. Johnson, *Ind. Eng. Chem., Anal. Ed.* **17**, 400 (1945).
387. F. Feigl and L. Hainberger, *Mikrochim. Acta*, 806 (1955).
388. S. W. Gunner and T. B. Hand, *J. Chromatog.* **37**, 357 (1968).
389. O. R. Hansen, *Acta Chem. Scand.* **7**, 1125 (1953).
390. T. Kaniewska and B. Borkowski, *Dissnes pharm., Warsz.* **20**, 111 (1968).
391. M. Largejan, *Pharm. Weekblad.* **92**, 693 (1957).
392. M. S. Blum, *J. Agr. Food Chem.* **3**, 122 (1955).
393. A. Labat, *Bull. Soc. Chim.* **5**, 1745 (1909).
394. A. Pictet and G. A. Kramers, *Ber.* **43**, 1334 (1910).
395. P. T. Allen, H. F. Beckman and J. F. Fudge, *J. Agr. Food Chem.* **10**, 248 (1962).
396. L. Fishbein and J. Fawkes, *J. Chromatog.* **20**, 521 (1965).
397. J. J. Velenovsky, *J. Assoc. Offic. Agr. Chemists*, **43**, 350 (1960).
398. E. Späth and H. Quietensky, *Ber.* **60**, 1883 (1927).

399. S. D. Bruck, *J. Res. Nat. Bur. Stand., A*, **66**, 251 (1962).
400. S. Granick, *J. Biol. Chem.* **236**, 1168 (1961).
401. V. M. Sardesai and H. S. Provido, *J. Lab. Clin. Med.* **65**, 1023 (1965).
402. A. M. Siegelman, A. S. Carlson and T. Robertson, *Arch. Biochem. Biophys.* **97**, 159 (1962).
403. N. G. Polezhaev and L. V. Kuznetsova, *Gig. Truda prof. Zabol.* 60 (1965).
404 T. Eicher, S. Hünig and P. Nikolaus, *Chem. Ber.* **102**, 3176 (1969).
405. R. E. Reeves, *J. Am. Chem. Soc.* **63**, 1476 (1941).
406. F. Feigl and E. Silva, *Drug Standards*, **23**, 113 (1955).
407. I. Lewandowska and I. Sokolowska, *Dissnes pharm., Warsz.* **20**, 81 (1968).
408. Y. Yoichi and A. Sano, *Japan Analyst*, **16**, 708 (1967).
409. L. R. Jones and J. A. Riddick, *Anal. Chem.* **28**, 254 (1956).
410. F. Feigl and D. Goldstein, *Anal. Chem.* **29**, 1521 (1957).
411. R. S. Pereira, *Mikrochem. Mikrochim. Acta, 36/37*, 398 (1951).
412. C. M. Vega, *Rev. Farm.* **93**, 22 (1951).
413. A. Hodgkinson and P. Zarembski, *Analyst*, **86**, 16 (1961).
414. A. Hodgkinson and A. Williams, *Clin. Chim. Acta*, **36**, 127 (1972).
415. E. N. McIntosh, M. Purko and W. A. Wood, *J. Biol. Chem.* **228**, 499 (1957).
416. J. D. Margerum, J. N. Pitts, Jr., J. G. Rutgers and S. Searles, *J. Am. Chem. Soc.* **81**, 1549 (1959).
417. R. Gomer and W. A. Noyes, Jr., *J. Am. Chem. Soc.* **72**, 101 (1950).
418. G. Holan, *Nature*, **232**, 644 (1971).
419. F. Feigl and V. Anger, "Spot Tests in Organic Analysis", 7th edn, Elsevier, Amsterdam, 1966, p. 324.
420. I. A. Breger, P. Zubovic, J. C. Chandler and R. S. Clarke, Jr., *Nature*, **236**, 155 (1972).
421. R. V. Marcote, T. H. Johnston and R. Chand, Presented at the 19th Annual ISA Analysis Instrumentation Symposium, St. Louis, Missouri, 24–26 April, 1973, pp. 31–44.
422. G. Parker, R. J. Cox and D. Richards, *J. Pharm. Pharmacol.* **7**, 683 (1955).
423. J. H. Morris, L. Berrens and E. Young, *Clin. Chim. Acta*, **12**, 407 (1965).
424. K. Ida, T. Onga, K. Kinoshita and S. Kawaji, *J. Antibiotics* (Japan) *Ser. B.*, **10**, 37 (1957); through *Chem. Abstr.* **53**, 17224 (1959).
425. K. Erne, *Acta Vet. Scand.* **7**, 77 (1966).
426. R. C. d'A. de Carnevale Bonino, *Revta Asoc. bioquím argent.* **30**, 115 (1965).
427. R. Preussmann, A. Von Hodenberg and H. Hengy, *Biochem. Pharmacol.* **18**, 1 (1969).
428. J. R. Jones and W. A. Waters, *J. Chem. Soc.* 2772 (1960).
429. American Cyanamide Co., "Colorimetric Method for Thimet Phorate Residue in Potatoes", June, 1960.
430. A. C. Waldron, N. R. Pasarela, M. H. Woolford and J. H. Ware, *J. Agr. Food Chem.* **11**, 241 (1963).
431. R. C. Blinn, *J. Am. Oil Chemists' Soc.* **46**, 952 (1963).
432. M. S. Karawya, *J. Assoc. Off. Agr. Chemists*, **55**, 1180 (1972).
433. S. W. Gunner, *J. Chromatog.* **40**, 85 (1969).
434. C. Genest, D. M. Smith and D. G. Chapman, *J. Agric. Food Chem.* **11**, 508 (1963).
435. H. D. Graham, *Nature*, **207**, 526 (1965).
436. H. A. Jones, H. J. Ackermann and M. E. Webster, *J. Assoc. Offic. Agr. Chem.* **35**, 771 (1952).
437. W. H. Munday, *J. Assoc. Offic. Agric. Chemists*, **43**, 707 (1960).

438. M. Beroza and W. A. Jones, *Anal. Chem.* **34**, 1029 (1962).
439. S. Klosowski, *Dissnes pharm., Warsz.* **20**, 335 (1968).
440. A. M. Unrau and F. Smith, *Chem. Ind.* (*London*), 330 (1957).
441. N. Argant, *Bull. Soc. Chim. Biol.* **31**, 485 (1949).
442. C. G. Huggins and O. N. Miller, *J. Biol. Chem.* **221**, 377 (1956).
443. G. S. Salyamon, *Hyg. Sanit.* **34**, 202 (1969); through *APCA Abstracts*, **16**, No. 13140 (1970).
444. M. Tomasz and R. W. Chambers, *Biochem. Biophys. Acta*, **108**, 510 (1965).
445. M. Tomasz, Y. Sanno and R. W. Chambers, *Biochemistry*, **4**, 1710 (1965).
446. E. L. Pratt and M. E. Auerbach, *J. Assoc. Off. Agr. Chemists*, **49**, 329 (1966).
447. S. K. Yasuda and R. N. Rogers, *Anal. Chem.* **32**, 910 (1960).
448. E. V. Rao and M. N. Narayanan, *Indian J. Pharm.* **30**, 70 (1968).
449. C. Genest, D. M. Smith and D. G. Chapman, *J. Ass. Off. Agric. Chem.* **44**, 631 (1961).
450. M. J. Boyd and M. A. Logan, *J. Biol. Chem.* **146**, 279 (1942).
451. W. R. Frisell and C. G. Mackenzie, The determination of formaldehyde and serine in biological systems, in D. Glick, Ed., "Methods of Biochemical Analysis", Vol. VI, Interscience, New York and London, 1958, p. 63.
452. M. Hayashi, Y. Nakijima, K. Inowe and K. Miyaki, *Chem. Pharm. Bull., Tokyo*, **11**, 1200 (1963).
453. M. Hayashi, K. Miyaki and T. Unemoto, *Chem. Pharm. Bull., Tokyo*, **8**, 904 (1960).
454. R. H. Morgan, *J. Assoc. Public Analysts*, **4**, 73 (1966).
455. B. A. Neidig and W. C. Hess, *Anal. Chem.* **24**, 1627 (1952).
456. M. W. Rees, *Biochem. J.* **40**, 632 (1946).
457. H. L. Haller, F. B. LaForge and W. N. Sullivan, *J. Org. Chem.* **7**, 185 (1942).
458. G. Durand, J. Feger, M. Coignoux, J. Agneray and M. Pays, *Anal. Biochem.* **61**, 232 (1974).
459. M. J. Clancy, *J. Chem. Soc.* 4213 (1960).
460. J. M. Bailey, *J. Lab. Clin. Med.* **54**, 158 (1959).
461. L. H. Adcock, *Analyst*, **82**, 427 (1957).
462. D. Exley, S. C. Ingall, J. K. Norymberski and G. F. Woods, *Biochem. J.* **81**, 428 (1961).
463. D. Exley and J. K. Norymberski, *J. Endocrin.* **29**, 303 (1964).
464. G. T. Bassil and A. M. Hain, *Nature*, **165**, 525 (1950).
465. A. M. Bongiovanni and W. R. Eberlein, *J. Lab. Clin. Med.* **48**, 320 (1956).
466. R. Borth, *Acta Endocrinol.* **22**, 125 (1956).
467. A. C. Corcoran and I. H. Page, *J. Lab. Clin. Med.* **33**, 1326 (1948).
468. W. H. Daughaday, H. Jaffe and R. H. Williams, *J. Clin. Endocrinol.* **8**, 166 (1948).
469. L. L. Engel, P. Carter and L. L. Fielding, *J. Biol. Chem.* **213**, 99 (1955).
470. V. P. Hollander, S. Di Mauro and O. H. Pearson, *Endocrinology*, **49**, 617 (1951).
471. G. F. Marrian, *Recent Progr. in Hormone Research*, **9**, 303 (1954).
472. H. L. Mason, *Recent Progr. in Hormone Research*, **9**, 267 (1954).
473. J. Y. F. Paterson and G. F. Marrian, *Acta Endocrinol.* **14**, 259 (1953).
474. L. P. Romanoff, J. Plager and G. Pincus, *Endocrinology*, **45**, 10 (1949).
475. S. A. Simpson, J. F. Tait, A. Wettstein, R. Neher, J. von Euw and T. Reichstein, *Experimentia*, **9**, 333 (1953).
476. E. Venning, *Recent Progr. in Hormone Research*, **9**, 300 (1954).
477. J. Wheeler, S. Freeman and C. Chen, *J. Lab. Clin. Med.* **42**, 758 (1953).
478. H. Wilson, *Recent Progr. in Hormone Research*, **9**, 310 (1954).
479. H. Wilson, *J. Clin. Endocrinol. and Metabolism*, **13**, 1465 (1953).

480. J. R. Turvey, *Advances Carbohydrate Chem.* **20**, 194 (1965).
481. Y. Kato, K. Ohashi, H. Koike and Y. Kinoshita, *Yakugaku Zasshi*, **88**, 393 (1968).
482. W. K. Rieben and A. B. Hastings, *Helv. Physiol. Pharmacol. Acta*, **4**, C52 (1946).
483. L. G. Wesson, *Anal. Chem.* **30**, 1080 (1958).
484. G. G. Guidotti, A. F. Borghetti and L. Loreti, *Anal. Biochem.* **17**, 513 (1966).
485. W. Godicke and U. Gerike, *Clin. Chim. Acta*, **30,** 727 (1970).
486. E. A. Pachtman, *Fluor. News*, **6**, 4 (1971).
487. W. E. Neeley, G. E. Goldman and C. A. Cupas, *Clin. Chem.* **18**, 1350 (1972).
488. H. P. Chin, S. S. Abd El-Meguid and D. H. Blankenhorn, *Clin. Chim. Acta*, **31**, 381 (1971).
489. G. Kessler and H. Lederer, "Automation in Analytical Chemistry, Technicon Symposia", Mediad, New York, 1966, pp. 341–344.
490. S. Tytko, C. E. Willis and J. W. King, "Automation in Analytical Chemistry, Technicon Symposia", Mediad, New York, 1968, pp. 29–33.
491. L. B. Foster and R. T. Dunn, *Clin. Chem.* **19**, 338 (1973).
492. G. S. Rautela, S. Slater and D. A. Arvan, *Clin. Chem.* **19**, 1193 (1973).
493. H. Wessels, *Fette Seifen Anstrichmittel*, **75**, 478 (1973).
494. B. P. Neri and C. S. Frings, *Clin. Chem.* **19**, 1201 (1973).
495. B. Kohring and R. Kattermann, *Z. Klin. Chem. Klin. Biochem.* **12**, 282 (1974).
496. S. P. Gottfried and B. Rosenberg, *Clin. Chem.* **19**, 1077 (1973).
497. P. H. Lenz, D. I. Cargill and A. I. Fleischman, *Clin. Chem.* **19**, 1071 (1973).
498. P. Haux and S. Natelson, *Microchem. J.* **16**, 68 (1971).
499. F. G. Soloni, *Clin. Chem.* **17**, 529 (1971).
500. H. Ko and M. E. Royer, *Biochem. Med.* **6**, 144 (1972).
501. S. Lartillot and C. Vogel, *Feuill. Biol.* **11**, 39 (1970).
502. A. L. Levy and C. Keyloun, "Automation in Analytical Chemistry, Technicon Symposia", Mediad, New York, 1970, pp. 497–502.
503. F. Bjorksten, *Clin. Chim. Acta*, **40**, 143 (1972).
504. W. R. Holub, *Clin. Chem.* **19**, 1391 (1973).
505. D. Townsend, B. Livermore and H. Jenkin, *Microchem. J.* **16**, 456 (1971).
506. M. L. Rojkin and J. R. Repetto, *Revta Assoc. Bioquim. Argent.* **37**, 177 (1972).
507. O. C. Sundsvold, B. Uppstad, G. W. Ferguson, D. Feeley and T. McLachlan, *J. Assoc. Public Analysts*, **9**, 86 (1971).
508. C. H. Castell and B. Smith, *J. Fish Res. Board Can.* **30**, 91 (1973).
509. K. Amano and H. Tozawa, Irradiation cleavage of trimethylamine oxide in fish muslce, *in* R. Kreuzer, Ed., "Freezing and Irradiation of Fish", Fishing News Ltd, London, 1969, pp. 467–471.
510. K. Amano and K. Yamada, *Bull. Jap. Soc. Sci. Fish.* **30**, 430 (1964).
511. C. H. Castell, *J. Am. Oil Chem. Soc.* **48**, 645 (1971).
512. T. Tokunaga, *Bull. Hokkaido Reg. Fish. Res.* **29**, 108 (1964); **30**, 90 (1965).
513. K. Yamada and K. Amano, *Bull. Jap. Soc. Sci. Fish.* **31**, 60 (1965).
514. F. Soudan, The natural formal of fishery products, in E. Heen and R. Kreuzer, Eds, "Fish in Nutrition", Fishing News Ltd., London, 1962, p. 78.
515. H. Z. Lecher and W. B. Hardy, *J. Am. Chem. Soc.* **70**, 3789 (1948).
516. J. F. Walker and A. F. Chadwick, *Ind. Eng. Chem.* **39**, 3 (1947).
517. D. Kutter, *Pharm. Acta Helv.* **48**, 371 (1973).
518. E. Gronsberg, *Gigiena Truda Prof. Zabolevaniia*, **12**, 54 (1968).
519. E. S. Gronsberg, *Khim. Prom.*, (*Moscow*), **7**, 30, 510 (1966).
520. E. S. Gronsberg, *Khim. Prom.*, 866 (1971).
521. J. R. Dyer, *Methods of Biochemical Analysis*, **3**, 111 (1956).

522. E. Sawicki in N. Cheronis, Ed., "Microchemical Techniques", Interscience, New York, 1962, pp. 59–106.
523. T. Matsunaga, T. Soejima, Y. Iwata and F. Watanabe, *Gann.* **45**, 451 (1954).
524. F. H. Sobels, *Am. Naturalist*, **88**, 109 (1954).
525. T. Alderson, *Nature*, **207**, 164 (1965).
526. C. Auerbach, *Nature*, **210**, 104 (1966).
527. K. A. Jensen, I. Kirk, G. Kolmark and M. Westergaard, *Cold Spring Harbor Symp. Quant. Biol.* **16**, 245 (1951).
528. M. Demerec, G. Bertani and J. Flint, *Am. Naturalist*, **85**, 119 (1951).
529. C. Auerbach, *Hereditas*, **37**, 1 (1951).
530. Y. A. Siomin, V. V. Simonov and A. M. Poverenny, *Biochim. Biophys. Acta*, **331**, 27 (1973).
531. H. I. Li, *Biopolymers*, **11**, 835 (1972).
532. J. W. Flesher and K. L. Sydnor, *Int. J. Cancer*, **11**, 433 (1973).
533. N. H. Sloane and M. Heinemann, *Biochim. Biophys. Acta*, **201**, 384 (1970).
534. T. Watabe and N. Yamada, *Biochem. Pharmacol.* **24**, 1051 (1975).
535. T. Watabe and K. Akamatsu, *Chem. Pharm. Bull. Tokyo*, **22**, 2155 (1974).
536. B. L. Van Duuren, L. Langseth, B. M. Goldschmidt and L. Orris., *J. Nat. Cancer Inst.* **39**, 1217 (1967).
537. R. Stewart in K. B. Wiberg, Ed., "Oxidation in Organic Chemistry", Academic Press, New York, 1966, p. 42.
538. E. Sawicki, T. R. Hauser, T. W. Stanley and W. C. Elbert, *Anal. Chem.* **33**, 93 (1961).
539. J. A. Kerr, J. G. Calvert and K. L. Demerjian, *Chemistry in Great Britain*, **8**, 252 (1972).
540. B. Dimitriades, *Env. Sci. Technol.* **6**, 253 (1972).
541. J. N. Pitts, Jr. and B. J. Finlayson, *Angew Chem., Internat. Edit.* **14**, 1 (1975).
542. P. A. Leighton, "Photochemistry of Air Pollution", Academic Press, New York, N.Y., 1961.
543. C. K. K. Yeung and C. R. Phillips, *Env. Sci. Technol.* **9**, 732 (1975).
544. W. E. Scott, E. R. Stephens, P. L. Hanst and R. C. Doerr, *Proc. Am. Petrol. Inst.* **37**, 171 (1957).
545. B. W. Gay, Jr., P. L. Hanst, J. J. Bufalini and R. C. Noonan, *Env. Sci. Technol.* **10**, 58 (1976).
546. E. A. Schuck and H. J. Doyle, "Photooxidation of Hydrocarbons in Mixtures Containing Oxides of Nitrogen and Sulfur Trioxide", Report No. 29, San Marino, California: Air Pollution Foundation, 1959.
547. E. Sawicki, S. P. McPherson, T. W. Stanley, J. Meeker and W. C. Elbert, *Int. J. Air Water Poll.* **9**, 515 (1965).
548. S. P. McPherson, E. Sawicki and F. T. Fox, *J. Gas Chromatog.*, 156 (1966).
549. Y. A. Peregud and Y. V. Gernet, "Chemical Analysis of the Air of Industrial Enterprises", Khimya Press, Leningrad, 1973.
550. Y. S. Gronsberg, *Khim. Prom.* (*Moscow*), No. **7**, 513 (1967).
551. N. A. Yakovleva and M. E. Son'kin, *Gig. Sanit.* **4**, 60 (1974).
552. M. Friedman and C. W. Sigel, *Biochemistry*, **5**, 478 (1966).
553. D. J. McCaldin, *Chem. Revs.* **60**, 39 (1960).
554. R. F. Scherberger, F. A. Miller and D. W. Fassett, *Am. Ind. Hyg. Assoc. J.* **21**, 471 (1960).
555. J. Axelrod, *J. Pharmacol. Exp. Therap.* **117**, 322 (1956).
556. T. Nash, *Biochem. J.* **55**, 416 (1953).
557. S. Ivankovic, D. Schmahl and W. J. Zeller, *Z. Krebsforsch.* **81**, 269 (1974).

558. T. Kamataki and H. Kitagawa, *Biochem. Pharmacol.* **22**, 3199 (1973).
559. B. S. Cohen and R. W. Estabrook, *Arch. Biochem. Biophys*, **143**, 54 (1971).
560. R. W. Estabrook and B. S. Cohen in J. R. Gillette, A. H. Conney, G. J. Cosmides, R. W. Estabrook, J. R. Fouts and G. J. Mannering, Eds, "Microsomes and Drug Oxidations", Academic Press, New York, 1969, p. 95.
561. T. Mack, K. J. Freundt and D. Henschler, *Biochem. Pharmacol.* **23**, 607 (1974).
562. K. J. Freundt and W. Dreher, *Naunyn-Schmiedebergs Arch. Pharmak.* **263**, 208 (1969).
563. R. Preussmann, H. Druckrey, S. Ivankovic and A. von Hodenberg, *Ann. N. Y. Acad. Sci.* **163**, 697 (1969).
564. L. Tomatis and C. M. Goodall, *Int. J. Cancer*, **4**, 219 (1969).
565. T. Tanaka, IARC Scientific Publication No. 4, pp. 100–111 (1973).
566. "IARC Monographs on the Evaluation of Carcinogenic Risk of Chemicals to Man", Vol. I, International Agency for Research on Cancer, Lyon, 1972, pp. 157–163.
567. L. Fishbein, W. G. Flamm and H. L. Falk, "Chemical Mutagens", Academic Press, New York, 1970, pp. 172–176.
568. R. C. Shank and P. N. Magee, *Biochem,. J.* **105**, 521 (1967).
569. J. A. Miller, *Federation Proc.* **23**, 1361 (1964).
570. L. Fishbein, "Chromatography of Environmental Hazards, Vol. I: Carcinogens, Mutagens and Teratogens", Elsevier, Amsterdam, 1972, pp. 394–398.
571. M. Spatz, *Ann. N. Y. Acad. Sci.* **163**, 848 (1969).
572. G. L. Laqueur and H. Matsumoto, *J. Natl. Cancer Inst.* **37**, 217 (1966).
573. M. Spatz and G. L. Laqueur, *J. Natl. Cancer Inst.* **38**, 233 (1967).
574. M. Spatz, W. J. Dougherty and D. W. E. Smith, *Proc. Soc. Exptl. Biol. Med.* **124**, 476 (1967).
575. H. J. Teas and J. G. Dyson, *Proc. Soc. Exptl. Biol. Med.* **125**, 988 (1967).
576. M. G. Gabridge, A. Denunzio and M. S. Legator, *Science*, **163**, 689 (1969).
577. M. Spatz and G. L. Laqueur, *Proc. Soc. Exptl. Biol. Med.* **127**, 281 (1968).
578. J. M. Cooper, *Proc. Roy. Soc. N. S. W.* **74**, 450 (1941).
579. N. V. Riggs, *Aust. J. Chem.* **7**, 123 (1954).
580. M. S. Zedeck, S. S. Sternberg, R. W. Poynter and J. McGowan, *Cancer Res.* **30**, 801 (1970).
581. Y. Nagata and H. Matsumoto, *Proc. Soc. Exp. Biol.* **132**, 383 (1969).
582. Anonymous, *Chem. Eng. News*, 5 (17 September 1973).
583. B. K. Leong, H. N. MacFarland and W. H. Reese, Jr., *Arch. Env. Health*, **22**, 663 (1971).
584. S. Laskin, M. Kuschner, R. T. Drew, V. P. Cappiello and N. Nelson, *Arch. Env. Health*, **23**, 135 (1971).
585. Anonymous, *J. Env. Health*, **34**, 632 (1972).
586. B. L. Van Duuren, S. Laskin and N. Nelson, *Chem. Eng. News*, p. 62, March 27, 1972.
587. B. L. Van Duuren, S. Laskin and B. M. Goldschmidt, *Env. Sci. Technol.* **7**, 744 (1973).
588. J. F. Walker, "Formaldehyde", 3rd ed, Reinhold, New York, 1964, pp. 37, 180, 255.
589. J. C. Tou and G. J. Kallos, *Am. Ind. Hyg. Assoc. J.* **35**, 419 (1974).
590. L. S. Frankel, K. S. McCallum and L. Collier, *Environ. Sci. Technol.* **8**, 356 (1974).
591. L. I. Larsson and O. Samuelson, *Microchim. Acta*, **2**, 21 (1967).
592. O. Samuelson and H. Stromberg, *Carbohydrate Res.* **3**, 89 (1966).
593. B. Sahagian and V. E. Levine, *Clin. Chem.* **10**, 116 (1964).
594. P. Roschlau, E. Bernt and W. Gruber, *Z. Klin. Chem. Klin. Biochem.* **12**, 403 (1974).

595. L. L. Abell, B. B. Levy, B. B. Brodie and F. E. Kendall, in D. Seligson, Ed., "Standard Methods of Clinical Chemistry", Vol. 2, Academic Press, N.Y., 1958, p. 26.
596. H. Greim, G. Bonse, Z. Radwan, D. Reichert and O. Henschler, *Biochem. Pharmacol.* **24**, 2013 (1975).
597. Chem. Eng. News, 8 (20 January, 1975).
598. R. J. Jaeger, *Annals N. Y. Acad. Sci.*, **246**, 150 (1975).
599. P. C. Koller, *Mutation Res.* **8**, 199 (1969).
600. B. L. Van Duuren, Ed., Biological effects of alkylating agents, *Annals N.Y. Acad. Sci.* **163**, 589 (1969).
601. R. P. Bratzel, A survey of alkylating agents, *Cancer Chemotherapy Rpts*, No. 26, 1963.
602. E. Sawicki, D. F. Bender, T. R. Hauser, R. M. Wilson and J. Meeker, *Anal. Chem.* **35**, 1479 (1963).
603. W. J. Serfontein and J. H. Smit, *Nature*, **214**, 169 (1967).
604. G. Neurath, *Experentia*, **23**, 400 (1967).
605. D. E. Johnson, J. D. Miller and J. W. Rhoades, National Cancer Institute Monograph No. **28**, 181 (1968).
606. D. H. Fine, D. P. Rounbehler, N. M. Blecher and S. S. Epstein, N-Nitroso compounds in air and water. International Conference on Environmental Sensing and Assessment, Las Vegas, October, 1975.
607. D. H. Fine, D. P. Rounbehler, E. Sawicki, K. Krost and G. A. De Marrais, *Anal. Letters*, **9**, 595 (1976).
608. D. H. Fine, D. P. Rounbehler, E. D. Pellizari, J. E. Bunch, R. W. Berkley, J. McCrae, J. T. Bursey, E. Sawicki, K. Krost and G. A. De Marrais, *Bull. Env. Contam. Toxicol.*, **15**, 739 (1976).
609. K. Bretschneider and J. Matz, *Arch. Geschwulstforsch.* **42**, 36 (1973).
610. H. Druckrey, R. Preussmann, S. Ivankovic and D. Schmähl, *Z. Krebsforsch.* **69**, 103 (1967).
611. R. L. Tate III and M. Alexander, *J. Nat. Cancer Inst.* **54**, 327 (1975).
612. J. Dressel, *Landwirt Forsch. Sonderh.* **28**, 273 (1973).
613. D. T. Burns and G. V. Alliston, *J. Food Technol.* **6**, 433 (1971).
614. T. G. Alliston, G. B. Cox and R. S. Kirk, *Analyst*, **97**, 915 (1972).
615. T. Panalaks, J. R. Iyengar and N. P. Sen, *J. Assoc. Off. Anal. Chemists*, **56**, 621 (1973).
616. M. Castegnaro, B. Pignatelli and E. A. Walker, *Analyst*, **99**, 156 (1974).
617. T. Fazio, R. H. White, L. R. DuSold and J. W. Howard, *J. Assoc. Off. Anal. Chemists*, **56**, 919 (1973).
618. N. P. Sen, J. R. Iyengar, B. Donaldson and T. Panalaks, *J. Agr. Food Chem.* **22**, 540 (1974).
619. F. Ender, G. M. Havre, R. Madsen, L. Ceh and A. Helgebostad, *Z. Therphysiol. Tiere Nahry-Futtermittelkde*, **22**, 181 (1967).
620. Y. Y. Fong and W. C. Chan, *Food Cosmet. Toxicol.* **11**, 841 (1973).
621. S. S. Mirvish, *J. Nat. Cancer Inst.* **44**, 633 (1970).
622. G. E. Hein, *J. Chem. Educ.* **40**, 181 (1963).
623. F. Schweinsberg and J. Sander, *Z. Physiol. Chem.* **353**, 1671 (1972).
624. N. P. Sen in I. E. Liener, Ed., "Toxic Constituents in Animal Foodstuffs", Academic Press, New York, N.Y., 1974, p. 131.
625. S. S. Hecht, R. M. Ornaf and D. Hoffmann, *Anal. Chem.* **47**, 2046 (1975).
626. D. Hoffmann, S. S. Hecht, R. M. Ornaf and E. L. Wynder, *Science*, **186**, 265 (1974).
627. A. Ayanaba, W. Verstraete and M. Alexander, *J. Nat. Cancer Inst.* **50**, 811 (1973).
628. A. Ayanaba, W. Verstraete and M. Alexander, *Proc. Am. Soil Sci. Soc.* **37**, 565 (1973).

629. A. Ayanaba and M. Alexander, *J. Env. Qual.* **3**, 83 (1974).
630. S. Miyahra, *Kagaku Zasshi*, **81**, 19 (1966).
631. E. Preusser, *Biol. Zentr.* **85**, 19 (1966).
632. D. H. Fine and D. P. Rounbehler, N-Nitroso compounds in water, Presented at the First Chemical Congress of the North American Continent, Mexico City, December, 1975.
633. D. H. Fine, D. P. Rounbehler, N. M. Belcher and S. S. Epstein, *Science*, **192**, 1328 (1976).
634. B. S. Alam, J. B. Saporoschetz and S. S. Epstein, *Nature*, **232**, 116 (1971).
635. T. S. Mysliwy, E. L. Wick, M. C. Archer, R. C. Shank and P. M. Newberne, *Brit. J. Cancer*, **30**, 279 (1974).
636. J. Sander, F. Schweinsberg and H. P. Menz, *Z. Physiol. Chem.* **349**, 1691 (1968).
637. M. Greenblatt, S. Mirvish and B. T. So, *J. Nat. Cancer Inst.* **46**, 1029 (1971).
638. M. Greenblatt, S. Mirvish and B. T. So, *J. Nat. Cancer Inst.* **50**, 119 (1973).
639. J. Sander, *Arzneimittel-Forsch.* **21**, 1572, 1707, 2034 (1971).
640. E. Sawicki, The chemical composition and potential genotoxic aspects of polluted atmospheres, presented at the Workshop for the "Investigations on the Carcinogenic Burden by Air Pollution in Man" in Hanover, Germany on 22–24 October, 1975.
641. M. A. Robertson, J. S. Harington and E. Bradshaw, *Brit. J. Cancer*, **25**, 377, 385 (1971).
642. J. S. Harington, J. R. Nunn and L. Irwig, *Nature*, **241**, 49 (1973).
643. B. C. Challis and C. D. Bartlett, *Nature*, **254**, 532 (1973).
644. E. Boyland, E. Nice and K. Williams, *Food Cosmet. Toxicol.* **9**, 639 (1971).
645. W. Lijinsky, E. Conrad and R. Van de Bogart, *Nature*, **239**, 165 (1972).
646. J. H. Ridd, *Quart. Rev.* **15**, 418 (1961).
647. L. F. Keefer and P. P. Roller, *Science*, **181**, 1245 (1973).
648. P. C. Caldwell, *Int. Rev. Cytol.* **5**, 229 (1956).
649. G. S. Rao and G. Krishna, *J. Pharm. Sci.* **64**, 1579 (1975).
650. W. Lijinsky, E. Conrad and R. Van de Bogart, Formation of carcinogenic nitrosamines by interaction of drugs with nitrite, in P. Bogovski, R. Preussman and E. A. Walker, Eds, "N-nitroso Compounds: Analysis and Formation", IARC Scientific Publication No. 3, International Agency for Research on Cancer, Lyon, 1972, pp. 130–133.
651. W. Lijinsky, H. W. Taylor, C. Snyder and P. Nettesheim, *Nature*, **244**, 176, (1973).
652. W. Lijinsky and M. Greenblatt, *Nature New Biol.* **236**, 177 (1972).
653. E. Boyland and S. A. Walker, *Arznemittel. Forsch.* **24**, 1181 (1974).
654. W. Lijinsky, *Cancer Res.* **34**, 255 (1974).
655. B. R. Simoneit and A. L. Burlingame, *Nature*, **234**, 210 (1971).
656. B. C. Challis and M. R. Osborne, *Chem. Commun.* 518 (1972).
657. "IARC Monographs on the Evaluation of Carcinogenic Risk of Chemicals to Man", Vol. I, International Agency for Research on Cancer, Lyon, 1972, pp. 95–140.
658. P. Bogovski, R. Preussman and E. A. Walker, "N-Nitroso Compounds: Analysis and Formation", IARC Scientific Publication No. 3, International Agency for Research on Cancer, Lyon, 1972.
659. H. Druckrey, R. Preussmann and S. Ivankovic, *Annals of N.Y. Acad. Sci.* **163**, 676 (1969).
660. H. Druckrey, R. Preussmann, D. Schmahl and M. Muller, *Naturwissenschaften*, **49**, 19 (1962)

661. R. Schoenthal and P. N. Magee, *Brit. J. Cancer*, **16**, 92 (1962).
662. H. Druckrey, R. Preussmann, D. Schmahl and M. Muller, *Naturwissenschaften*, **48**, 165 (1961).
663. P. N. Magee and J. M. Barnes, *J. Path. Bact.* **84**, 19 (1962).
664. F. G. Zak, J. H. Holzner, E. J. Singer and H. Popper, *Cancer Res.* **20**, 96 (1960).
665. P. N. Magee and J. M. Barnes, *Brit. J. Cancer*, **10**, 114 (1956).
666. J. M. Barnes and P. N. Magee, *Brit. J. Ind. Med.* **11**, 167 (1954).
667. H. Druckrey, R. Preussman, D. Schmahl and M. Muller, *Naturwissenschaften*, **48**, 134 (1961).
668. H. Druckrey and D. Steinhoff, *Naturwissenschaften*, **49**, 487 (1961).
669. H. Druckrey and R. Preussmann, *Naturwissenschaften*, **49**, 111 (1962).
670. H. Druckrey, D. Steinhoff, R. Preussmann and S. Ivankovic, *Z. Krebsforsch.* **66**, 1 (1964).
671. H. Druckrey, R. Preussmann, G. Blum and S. Ivankovic, *Naturwissenschaften*, **50**, 100 (1963).
672. H. Druckrey, *Angew. Chem., Int. Ed.* **9**, 742 (1970).
673. H. Druckrey, *Z. Krebsforsch.* **71**, 135 (1968).
674. H. Brune, *Z. Krebsforsch.* **69**, 307 (1967).
675. P. N. Magee, *Annals N.Y. Acad. Sci.* **163**, 717 (1969).
676. R. Preussmann, On the significance of N-nitroso compounds as carcinogens and on problems related to their chemical analysis, in P. Bogovski, R. Preussmann and E. A. Walker, Eds., "N-Nitroso Compounds: Analysis and Formation", IARC Scientific Publication No. 3, International Agency for Research on Cancer, Lyon, 1972, pp. 6–9.
677. L. den Engelse, P. A. J. Bentvelzen and P. Emmelot, *Chem.-Biol. Interactions*, **1**, 395 (1969/1970).
678. P. N. Magee and J. M. Barnes, *Acta Unio Intern. Contra Cancrum*, **15**, 187 (1959).
679. G. B. Toth, *Cancer Res.* **28**, 727 (1968).
680. G. Della Porta and B. Terracini, *Progr. Exp. Tumor Res.* **11**, 334 (1969).
681. G. Jasmin and J. L. Riopelle, *Revue Can. Biol.* **23**, 129 (1964).
682. B. Terracini, P. N. Magee and J. M. Barnes, *Brit. J. Cancer*, **21**, 559 (1967).
683. R. Terracini, G. Palestro, M. R. Gigliardi and R. Montesano, *Brit. J. Cancer*, **20**, 871 (1966).
684. H. Druckrey, *Arzneiforsch.* **13**, 320 (1963).
685. A. W. Pound, T. A. Lawson and L. Horn, *Brit. J. Cancer*, **27**, 451 (1973).
686. P. Nettesheim, D. A. Creasia and T. J. Mitchell, *J. Nat. Cancer Inst.* **55**, 159 (1975).
687. N. Venkatesan, J. C. Arcos and M. F. Argus, *Life Sci.* **7**, 1111 (1968).
688. C. Hoch-Ligeti, M. F. Argus and J. C. Arcos, *J. Natl. Cancer Inst.* **40**, 535 (1968).
689. W. Kunz, G. Schaude and C. Thomas, *Z. Krebsforsch.* **72**, 291 (1969).
690. H. V. Malling, *Mutation Res.* **3**, 537 (1966).
691. D. F. Heath, *Biochem. J.* **85**, 72 (1962).
692. V. M. Craddock, *Biochim. Biophys. Acta*, **312**, 202 (1973).
693. P. F. Swann and A. E. M. McLean, *Biochem. J.* **124**, 283 (1971).
694. H. V. Malling, *Mutation Res.* **13**, 425 (1971).
695. C. N. Frantz and H. V. Malling, *Cancer Research*, **35**, 2307 (1975).
696. N. Venkatesan, J. C. Arcos and M. F. Argus, *Cancer Res.* **30**, 2556 (1970).
697. D. Hadjiolov, *Z. Krebsforsch.* **76**, 91 (1971).
698. A. E. M. McLean and P. N. Magee, *Brit. J. Exptl. Pathol.* **51**, 587 (1970).
699. L. den Engelse and P. Emmelot, *Chem.-Biol. Interactions*, **4**, 321 (1971).
700. J. D. Judah, A. E. M. McLean and E. K. McLean, *Am. J. Med.* **49**, 609 (1970).

701. A. E. M. McLean and E. K. McLean, *Brit. Med. Bull.* **25**, 278 (1969).
702. A. E. M. McLean and H. G. Verschuuren, *Brit. J. Exp. Path.* **50**, 22 (1969).
703. D. Schmall, F. W. Kruger, S. Ivankovic and P. Preissler, *Arzneimittel-Forsch.* **21**, 1560 (1971).
704. R. C. Garner and A. E. M. McLean, *Biochem. Pharmacol.* **18**, 645 (1969).
705. B. Stripp, M. E. Hamrick and J. R. Gillette, *Biochem. Pharmacol.* **21**, 745 (1972).
706. A. W. Pound and T. A. Lawson, *Brit. J. Exp. Pathol.* **55**, 203 (1974).
707. A. W. Pound, L. Horn and T. A. Lawson, *Pathology*, **5**, 233 (1973).
708. G. Ugazio, R. R. Koch and R. O. Rechnagel, *Exp. Mol. Pathol.* **18**, 281 (1973).
709. J. V. Dingell and M. Heinberg, *Biochem. Pharmacol.* **17**, 1269 (1968).
710. E. A. Glende, *Biochem.* **21**, 1697 (1972).
711. T. F. Slater, *Nature*, **209**, 36 (1966).
712. M. F. Argus, G. M. Bryant, K. M. Pastor and J. C. Arcos, *Cancer Res.* **35**, 1574 (1975).
713. J. C. Arcos, G. M. Bryant, N. Venkatesan and M. F. Argus, *Biochem. Pharmacol.* **24**, 1544 (1975).
714. A. Somogyi, A. H. Conney, R. Kuntzman and B. Solymoss, *Nature New Biol.* **237**, 61 (1972).
715. L. Fiume, G. Campadelli-Fiume, P. N. Magee and J. Holsman, *Biochem. J.* **120**, 601 (1970).
716. J. C. Arcos, M. F. Argus and N. P. Buu-Hoi, *Fed. Proc.* **32**, 702 (1973).
717. D. Hadjiolov and D. Mundt, *J. Nat. Cancer Inst.* **52**, 753 (1974).
718. D. Hadjiolov and D. Markov, *J. Nat. Cancer Inst.* **50**, 979 (1973).
719. B. Stripp, G. Sipes and H. M. Maling, *Drug. Metab. Disposition*, **2**, 464 (1974).
720. H. M. Maling, B. Highman, M. A. Williams, W. Saul, W. Butler and B. B. Brodie, *Toxicol. Appl. Pharmacol.* **27**, 380 (1974).
721. E. K. Weisburger, J. M. Ward and C. A. Brown, *Toxicol. Appl. Pharmacol.* **28**, 477 (1974).
722. P. Czygan, H. Greim, A. J. Garro, F. Hutlerer, F. Schaffner, H. Popper, O. Rosenthal and D. Y. Cooper, *Cancer Res.* **33**, 2983 (1973).
723. F. J. De Serres and H. V. Malling, in A. Hollaender, Ed., "Environmental Mutagenesis: Principles and Methods for Their Detection", Vol. 2, Plenum, New York, 1970, p. 311.
724. B. N. Ames and C. Yanofsky in A. Hollaender, Ed., "Chemical Mutagens", Vol. 1, Plenum, New York, 1971, pp. 267–282.
725. P. N. Magee and J. M. Barnes, *Advances Cancer Res.* **10**, 163 (1967).
726. M. G. Gabridge and M. S. Legator, *Proc. Soc. Exp. Biol. Med.* **130**, 831 (1969).
727. V. Wunderlich, M. Schutt, M. Bottger and A. Graffi, *Biochem. J.* **118**, 99 (1970).
728. M. A. Friedman and D. R. Sawyer, *Fed. Proc.* **32**, 833 (1973).
729. D. B. Couch and M. A. Friedman, *Mutation Res.* **26**, 371 (1974).
730. L. D. Engelse, *Chem. Biol. Interactions*, **8**, 329 (1974).
731. S. Magour and J. G. Nievel, *Biochem. J.* **123**, 8P (1971).
732. S. Orrenius, J. L. E. Ericsson and L. Ernster, *J. Cell Biol.* **25**, 627 (1965).
733. P. Czygan, H. Greim, A. Garro, F. Schaffner and H. Popper, *Cancer Res.* **34**, 119 (1974).
734. N. Venkatesan, J. C. Arcos and M. F. Argus, *Cancer Res.* **30**, 2563 (1970).
735. E. Zeiger, *Env. Health Perspectives*, **6**, 101 (1973).
736. M. W. Anders and G. J. Mannering, *Mol. Pharmacol.* **2**, 319 (1966).
737. R. Montesano and P. N. Magee, *Proc. Am. Assoc. Cancer Res.* **12**, 14 (1971).
738. W. E. Heston, *Cancer Res.* **25**, 1320 (1965).

739. V. Y. Weekes, *J. Nat. Cancer Inst.* **55**, 1199 (1975).
740. N. K. Clapp, A. W. Craig and R. E. Toya, *Int. J. Cancer*, **5**, 119 (1970).
741. S. Takayama and K. Oota, *Gann.* **56**, 189 (1965).
742. S. Takayama and K. Oota, *Gann.* **54**, 465 (1963).
743. N. K. Clapp and R. E. Toya, Sr., *J. Nat. Cancer Inst.* **45**, 495 (1970).
744. V. Weeks and D. Brusick, *Mutation Res.* **31**, 175 (1975).
745. American Conference of Governmental Industrial Hygienists. Documentation of Threshold Limit Values, Third Edition. ACGIH, Cincinnati, Ohio (1971).
746. R. Montensano and P. N. Magee, *Nature*, **228**, 173 (1970).
747. K. Y. Lee and K. Spencer, *J. Nat. Cancer Inst.* **33**, 957 (1964).
748. P. N. Magee, Mechanisms of transplacental carcinogenesis by nitroso compounds, in "Transplacental Carcinogenesis", IARC Scientific Publication, 143–148 (1972).
749. H. C. Hodge and S. H. Sterner, *Am. Ind. Hyg. Assoc. Quart.* **10**, 93 (1949).
750. H. A. Freund, *Ann. Intern. Med.* **10**, 1144 (1937).
751. N-Nitroso Compounds News Letter, No. 6 (April, 1975).
752. H. Druckrey, A. Schildbach, D. Schmahl, R. Preussmann and S. Ivankovic, *Arzneimittel-Forsch.* **13**, 841 (1963).
753. I. J. Mizrahi and P. Emmelot, *Cancer Res.* **22**, 339 (1962).
754. H. Druckrey, R. Preussmann and S. Ivankovic, *Annals N.Y. Acad. Sci.* **163**, 589 (1969).
755. B. N. LaDu, H. G. Mandel and E. L. Way, "Fundamentals of Drug Disposition", Williams and Wilkins Co., Baltimore, 1971.
756. P. N. Magee and E. Farber, *Biochem. J.* **83**, 114 (1962).
757. G. A. Olah, D. J. Donovan and L. K. Keefer, *J. Nat. Cancer Inst.* **54**, 465 (1975).
758. Y. L. Chow, *Can. J. Chem.* **45**, 53 (1967).
759. L. Fishbein, "Chromatography of Environmental Hazards", Vol. I, Elsevier, Amsterdam, 1972, pp. 76–87.
760. L. Fishbein, *Annals N.Y. Acad. Sci.* **163**, 869 (1969).
761. B. Bain and L. Lowenstein, Effect of type of culture tubes to the mixed leucocyte reaction. Rept. to Med. Res. Council of Canada (Grant MBT 1664).
762. B. L. Van Duuren, C. Katz, B. M. Goldschmidt, K. Frenkel and A. Sivak, *J. Nat. Cancer Inst.* **48**, 1431 (1972).
763. F. Oesch, *Xenobiotica*, **3**, 305 (1973).
764. F. Mukai and W. Troll, *Annals N.Y. Acad. Sci.* **163**, 828 (1969).
765. I. A. Rapoport, *Dokl. Akad. Nauk. SSSR*, **60**, 469 (1948).
766. G. Kolmark and M. Westergaard, *Hereditas*, **39**, 209 (1953).
767. M. J. Bird, *J. Genet.* **50**, 480 (1952).
768. Z. Hartman, *Carnegie Inst. Wash. Publ.* **612**, 107 (1956).
769. J. Nemenzo and C. H. Hine, *Toxicol. Appl. Pharmacol.* **14**, 653 (1969).
770. L. Fishbein, W. G. Flamm and H. L. Falk, Chemical Mutagens, Academic Press, New York, 1970.
771. I. A. Rapoport, *Dokl. Akad. Nauk.* **54**, 65 (1946); **56**, 537 (1947).
772. W. D. Kaplan, *Science*, **108**, 43 (1948).
773. E. B. Freese, J. Gerson, H. Taber, H. J. Rhaese and E. Freese, *Mutation Res.* **4**, 517 (1967).
774. F. H. Sobels, *Radiation Res., Suppl.* 3, 171 (1963).
775. B. L. Van Duuren, *Intern. J. Environ. Anal. Chem.* **1**, 233 (1972).
776. B. L. Van Duuren, *Annals, N.Y. Acad. Sci.* **163**, 633 (1969).

777. B. L. Van Duuren, N. Nelson, L. Orris, E. D. Palmes and F. L. Schmitt, *J. Nat. Cancer Inst.* **31**, 41 (1963).
778. B. L. Van Duuren, L. Orris and N. Nelson, *J. Nat. Cancer Inst.* **35**, 707 (1965).
779. B. L. Van Duuren, L. Langseth, L. Orris, G. Teebor, N. Nelson and M. Kuschner, *J. Nat. Cancer Inst.* **37**, 825 (1966).
780. B. L. Van Duuren, L. Langseth, L. Orris, M. Baden and M. Kuschner, *J. Nat. Cancer Inst.* **39**, 1213 (1967).
781. C. S. Weil, N. Condra, C. Haun and J. A. Striegel, *Am. Ind. Hyg. Assoc. J.* 305 (1963).
782. A. L. Walpole, *Ann. N.Y. Acad. Sci.* **68**, 750 (1958).
783. P. D. Lawley and M. Jarman, *Biochem. J.* **126**, 893 (1972).
784. P. L. Grover, A. Hewer and P. Sims, *Biochem. Pharmacol.* **21**, 2713 (1972).
785. M. Lacomme, G. LeMoan and M. Chaigneau, *Ann. Pharm. Fr.* **32**, 411 (1974).
786. P. Lepsi, *Pracovni Lekar* (*Prague*), **25**, 330 (1973).
787. G. J. Buist and C. A. Bunton, *J. Chem. Soc.* 1406 (1954).
788. C. C. Price and H. Kroll, *J. Am. Chem. Soc.* **60**, 2726 (1938).
789. C. C. Price and M. Knell, *J. Am. Chem. Soc.* **64**, 552 (1942).
790. C. A. Pons and R. P. Custer, *Am. J. Med. Sci.* **211**, 544 (1946).
791. D. I. Peterson, J. E. Peterson and M. G. Hardinge, *J. Am. Med. Assoc.* **186**, 955 (1963).
792. R. J. Haggerty, *New Engl. J. Med.* **261**, 1296 (1959).
793. G. Rajagopal and S. Ramakrishnan, *Anal. Biochem.* **65**, 132 (1975).
794. D. D. Van Slyke, *J. Biol. Chem.* **32**, 455 (1917).
795. A. P. Altshuller, I. R. Cohen, M. E. Meyer and A. F. Wartburg, Jr., *Anal. Chim. Acta*, **25**, 101 (1961).
796. K. J. Hughes and R. W. Hurn, A preliminary survey of hydrocarbon-derived oxygenated material in automobile exhaust gases", presented at the 53rd Annual Meeting of the Air Pollution Control Assoc., Cincinnati, Ohio, 22–26 May, 1960.
797. B. D. Tebbens and J. D. Torrey, *Science*, **120**, 662 (1954).
798. P. P. Mader, G. Cann and L. Palmer, *Plant Physiol.* **30**, 318 (1955).
799. E. D. Barber, F. T. Fox, J. P. Lodge and L. M. Marshall, *J. Chromatog.* **2**, 615 (1959).
800. E. R. Stephens, P. L. Hanst, R. C. Doerr and W. E. Scott, *J. Air Poll. Control Assoc.* **8**, 333 (1959).
801. H. W. Gibson, *Chem. Revs.* **69**, 673 (1969).
802. H. Droller, *Z. Physiol. Chem.* **211**, 57 (1932).
803. Y. Houminer and S. Patal, *Israel J. Chem.* **7**, 513, 525, 535 (1969).
804. G. Kessler and H. Lederer, "Technicon Symposium: Automation in Analytical Chemistry", Mediad, New York, 1965, p. 863.
805. A. M. Eastham and G. A. Latremouille, *Can. J. Res.* **28B**, 264 (1950).
806. F. D. Gunstone, *J. Chem. Soc.* 1611 (1954).
807. F. Feigl, Spot Tests in Organic Analysis, Elsevier, Amsterdam—New York, 1966, p. 501.
808. M. L. Efron and M. G. Ampola, *Pediat. Clin. N. Amer.* **14**, 881 (1967).
809. J. R. Clamp and L. Hough, *Biochem. J.* **101**, 120 (1966).
810. T. H. Plummer, Jr., *J. Biol. Chem.* **246**, 2930 (1971).
811. K. Takahashi, W. H. Stein and S. Moore, *J. Biol. Chem.* **242**, 4682 (1967).
812. I. A. Kapustin and V. I. Gergalov, *Izv. Akad. Nauk. Bieloruss SSR*, Ser. Khim. Nauk, 113 (1973).
813. G. D. Christian and J. R. Moody, *Anal. Chim. Acta*, **41**, 269 (1968).

814. L. Berrens in P. Kallos and B. H. Waksman, Eds, "Progress in Allergy", Vol. 14, Karger, Basel, 1970, pp. 259–339.
815. J. R. Spies, *J. Agr. Food Chem.* **22**, 30 (1974).
816. L. Berrens in P. Kallos, M. Hasek, T. M. Inderbritzen, P. A. Miescher, and B. H. Waksman, Eds., "Monographs in Allergy", Vol. 7, Karger, Basel, 1971, pp. 1–298.
817. T. P. King, P. S. Norman and L. M. Lichtenstein, *Biochemistry*, **6**, 1992 (1967).
818. B. W. Griffiths, *J. Chromatogr.* **69**, 391 (1972).
819. T. P. King, P. S. Norman and L. M. Lichtenstein, *Ann. Allergy*, **25**, 541 (1967).
820. L. Berrens, *Immunochemistry*, **4**, 81 (1967).
821. L. Berrens, *Immunochemistry*, **5**, 585 (1968).
822. S. Kalyanaraman, T. V. Ramakrishna, M. Srinivasan and S. Thiagarajan, *Z. Anal. Chem.* **276**, 73 (1975).
823. W. A. Krotoski and H. E. Weimer, *Arch. Biochem. Biophys.* **115**, 337 (1966).
824. J. Feger, R. Cacan, R. Durand and J. Agneray, J. Colloque International sur les Glycoconjugués organisé par le C.N.R.S. Villeneuve d'Ascq, 21–27 juin 1973.
825. M. Pays and M. Beljean, *Ann. Pharm. Franc.* **28**, 153 (1970).
826. G. Kohlhaw, B. Deus and H. Holzer, *J. Biol. Chem.* **240**, 2135 (1965).
827. R. J. S. Duncan and K. F. Tipton, *Eur. J. Biochem.* **11**, 58 (1969).
828. W. Frank and W. De Boer, *Z. Physiol. Chem.* **314**, 70 (1959).
829. W. A. Warren, *J. Biol. Chem.* **245**, 1675 (1970).
830. E. Galas, *Nesz. Polytech. Lodz, Chem. S pzyw.* **No. 14**, 17 (1968); through *Chem. Abstr.* **68**, 112390n (1968).
831. N. Hasan, N. Nassif and I. F. Durr, *Int. J. Biochem.* **3**, 607 (1972).
832. J. Stig and R. Tabova, *Biochem Med.* **11**, 1 (1974).
833. J. Havlicek and O. Samuelson, *Chromatographia*, **7**, 361 (1974).
834. J. Havlicek and O. Samuelson, *Carbohyd. Res.* **22**, 307 (1972).
835. B. Arwidi and O. Samuelson, *Svensk Papperstid.* **68**, 330 (1965).
836. O. Samuelson, "Ion Exchange Separations in Analytical Chemistry", Almqvist and Wiksell, Stockholm; Wiley, New York, 1963.
837. J. Havlicek, G. Petersson and O. Samuelson, *Acta Chem. Scand.* **26**, 2205 (1972).
838. B. C. V. Mitchley, S. A. Clarke, T. A. Connors and A. M. Neville, *Cancer Res.* **35**, 1099 (1975).
839. T. A. Connors, *FEBS Lett.* **57**, 223 (1975).
840. D. T. North, *Mutation Res.* **4**, 225 (1967).
841. G. Rohrborn, *Z. Vererbungslehre*, **93**, 1 (1962).
842. H. Schmidt-Elmendorff, W. Schmidt and K. H. Schreyer, *Med. Klin.* (*Munich*), **50**, 2189 (1955).
843. T. Alderson, *Nature New Biol.* **244**, 3 (1973).
844. J. F. Kennedy and W. R. Butt, *Biochem, J.* **115**, 225 (1969).
845. J. F. Kennedy, M. F. Chaplin and M. Stacey, *Carbohydrate Res.* **36**, 369 (1974).
846. B. Toth, *Europ. J. Cancer*, **8**, 341 (1972).
847. R. Preussmann, S. Ivankovic, C. Landschutz, J. Gimmy, E. Flohr and V. Griesbach, *Z. Krebsforsch-Klin. Onkol.* **81**, 285 (1974).
848. J. H. Weisburger, *Cancer*, **28**, 60 (1971).
849. H. Druckrey, R. Preussmann, F. Matzkus and S. Ivankovic, *Naturwissenschaften*, **54**, 285 (1967).
850. B. S. Reddy, J. H. Weisburger, T. Narisawa and E. L. Wynder, *Cancer Res.* **34**, 2368 (1974).
851. H. K. Jain and R. N. Raut, *Nature*, **211**, 652 (1966).
852. F. J. C. Roe, G. A. Grant and S. M. Millican, *Nature*, **216**, 375 (1967).

853. E. Freese, E. Bautz and E. B. Freese, *Proc. Nat. Acad. Sci. U.S.* **47**, 845 (1961).
854. B. Toth and H. Shimizu, *Cancer Res.* **33**, 2744 (1973).
855. R. A. Prough, J. A. Wittkop and D. J. Reed, *Arch. Biochem. Biophys.* **140**, 450 (1970).
856. W. Kreis, *Cancer Res.* **30**, 82 (1970).
857. M. L. Murphy, *Clin. Proc. Childrens Hosp.* **18**, 307 (1962).
858. H. Druckrey, S. Ivankovic and R. Preussmann, *Experentia*, **23**, 1042 (1967).
859. R. Fahrig, *Mutation Res.* **26**, 19 (1974).
860. T. Ong and F. J. de Serres, *Mutation Res.* **13**, 276 (1971).
861. E. Vogel, *Mutation Res.* **11**, 397 (1971).
862. E. Vogel, R. Fahrig and G. Obe, *Mutation Res.* **21**, 123 (1973).
863. R. Fahrig, *Mutation Res.* **13**, 436 (1971).
864. R. C. S. Audette, T. A. Connors, H. G. Mandel, K. Merai and W. C. J. Ross, *Biochem. Pharmacol.* **22**, 1855 (1973).
865. J. L. Skibba, D. D. Beal, G. Ramirez and G. T. Bryan, *Cancer Res.* **30**, 147 (1970).
866. J. L. Skibba, G. Ramirez, D. D. Beal and G. T. Bryan, *Biochem. Pharmacol.* **19**, 2043 (1970).
867. J. M. Heuss and W. A. Glasson, *Env. Sci. Technol.* **2**, 1109 (1968).
868. A. C. Lloyd, W. P. Carter and J. L. Sprung, *California Air Environment*, **4**, 1 (Spring, 1974).
869. E. Adler, *Angew. Chem.*, **69**, 272 (1957).
870. E. Adler, K. Holmberg and L. Ryrfors, *Acta Chem. Scand.* **883**, (1974).
871. C. A. Bunton and V. J. Shiner, Jr., *J. Chem. Soc.* 1593 (1960).
872. W. Rigby, *J. Chem. Soc.* 1907 (1950).
873. K. Lundquist and G. E. Miksche, *Tetrahedron Lett.* 2131 (1965).
874. K. Lundquist, G. E. Miksche, L. Ericsson and L. Berndtson, *Tetrahedron Lett.* 4587 (1967).
875. S. Larsson and G. E. Miksche, *Acta Chem. Scand.* **23**, 917, 3337 (1969).
876. J. C. Pew and W. J. Connors, *J. Org. Chem.* **34**, 580 (1969).
877. E. Adler and K. Lundquist, *Acta Chem. Scand.* **15**, 223 (1961).
878. E. McNelis, *J. Am. Chem. Soc.* **88**, 1074 (1966).
879. P. Claus, P. Schilling, J. S. Gratzl and K. Kratzl, *Monatsh. Chem.* **103**, 1178 (1972).
880. E. Ziegler and K. Gartler, *Monatsh, Chem.* **79**, 637 (1948); **80**, 759 (1949).
881. J. Gierer, *Acta Chem. Scand.* **8**, 1319 (1954).
882. D. D. Van Slyke, A. Hiller, D. A. MacFadyen, A. B. Hasting and F. W. Klemperer, *J. Biol. Chem.* **133**, 287 (1940).
883. M. R. Stetten and R. Schoenheimer, *J. Biol. Chem.* **153**, 113 (1944).
884. J. V. Taggart and R. E. Krakaur, *J. Biol. Chem.* **177**, 641 (1949).
885. M. R. Stetten, *J. Biol. Chem.* **189**, 499 (1951).
886. H. J. Vogel and B. D. Davis, *J. Am. Chem. Soc.* **74**, 109 (1952).
887. H. J. Strecker, *J. Biol. Chem.* **235**, 3218 (1960).
888. H. J. Strecker and A. B. Johnson, *J. Biol. Chem.* **237**, 1876 (1962).
889. G. Y. Wu and S. Seifter, *Anal. Biochem.* **67**, 413 (1975).
890. E. Sawicki, T. R. Hauser and S. McPherson, *Chemist-Analyst*, **50**, 68 (1961).
891. G. Brunow and G. Miksche, *Acta Chem. Scand.* **B29**, 349 (1975).
892. G. Brunow and G. E. Miksche, *Acta Chem. Scand.* **23**, 1444 (1969).
893. H. Aminoff, G. Brunow, K. Falck and G. E. Miksche, *Acta Chem. Scand.* **B28**, 373 (1974).
894. J. Gierer, I. Peterson, L. A. Smedman and I. Wennberg, *Acta Chem. Scand.* **27**, 2083 (1973).

895. E. Adler and S. Häggroth, *Acta Chem. Scand.* **3**, 86 (1949).
896. T. Ashorn, *Soc. Sci. Fennica Commentationes Phys. Math.* 25 (1961).
897. G. Dryhurst, "Periodate Oxidation of Diol and Other Functional Groups", Pergamon Press, Oxford, 1970, pp. 90, 91.
898. Organic Chemistry Section Summary of Activities, July, 1970 to June, 1971 NBS Technical Note 587, p. 44 (1972).
899. S. Igarashi, *Japan Analyst*, **22**, 444 (1973).
900. E. E. Wigg, *Science*, **186**, 785 (1974).
901. R. G. Thurman, H. G. Ley and R. Scholz, *Eur. J. Biochem.* **25**, 420 (1972).
902. K. Takeshige and S. Minakami, *J. Biochem.* **76**, 1151 (1974).
903. G. L. Conney and D. W. Levine, *Adv. Appl. Microbiology*, **15**, 337 (1972).
904. K. Ogata and N. Kato, *Petroleum and Microorganisms* (*Japan*), No. 8, 4 (1972).
905. K. Sakaguchi, R. Kurane and M. Murata, *Agr. Biol. Chem.* **39**, 1695 (1975).
906. N. Kato, K. Tsuji, Y. Tani and K. Ogata, *J. Ferment. Technol.* **52**, 917 (1974).
907. B. N. Hemsworth, *J. Reprod. Fert.* **18**, 15 (1969).
908. M. Partington and A. J. Bateman, *Heredity*, **19**, 191 (1964).
909. J. W. Conklin, A. C. Upton, K. W. Christenberry and T. P. McDonald, *Radiation Res.* **19**, 156 (1963).
910. A. C. Upton, *Nat. Cancer Inst. Monograph*, **22**, 329 (1966).
911. N. K. Clapp, A. W. Craig and R. E. Toya, Sr., *Science*, **161**, 913 (1968).
912. B. N. Hemsworth, *J. Reprod. Fert.* **17**, 325 (1968).
913. D. J. Pillinger, B. W. Fox and A. W. Craig, in L. J. Roth, Ed., "Isotopes in Experimental Pharmacology", Univ. Chicago Press, Chicago, 1965, pp. 415–432.
914. A. A. Harrington and R. E. Kallio, *Can. J. Microbiol.* **6**, 1 (1960).
915. D. A. MacFadyen, *J. Biol. Chem.* **158**, 107 (1945).
916. J. C. Fruchart, B. B. Tsumbu, J. J. Jaillard and G. Sezille, *Clin. Chim. Acta*, **63**, 399 (1975).
917. L. L. Abell, B. B. Levy, B. B. Brodie and F. E. Kendall, *J. Biol. Chem.* **195**, 357 (1952).
918. D. A. Holaday, S. Rudofsky and P. S. Treuhaft, *Anesthesiology*, **33**, 579 (1970).
919. R. A. Van Dyke and C. L. Wood, *Anesthesiology*, **39**, 614, (1973).
920. M. H. Karger and Y. Mazur, *J. Am. Chem. Soc.* **91**, 5663 (1969).
921. R. F. Kaiko, N. Chatterjie and C. E. Inturrisi, *J. Chromatog.* **109**, 247 (1975).
922. R. F. Kaiko and C. E. Inturrisi, *J. Chrom.* **82**, 315 (1973).
923. E. Ackermann, *Biochem. Pharmacol.* **19**, 1955 (1970).
924. R. V. Smith and P. W. Erhardt, *Anal. Chem.* **47**, 2462 (1975).
925. L. Ernster and S. Orrenius, *Fed. Proc. Fed. Am. Soc. Exp. Biol.* **24**, 1190 (1965).
926. H. Remmer, J. B. Schenkman, R. W. Estabrook, H. Sesame, J. R. Gillette, S. Narasimhulu, D. Y. Cooper and O. Rosenthal, *Mol. Pharmacol.* **2**, 187 (1966).
927. Y. Imai and R. Sato, *Biochem. Biophys. Res. Commun.* **22**, 620 (1966).
928. S. Orrenius, M. Das and Y. Gnosspelius, Overall biochemical effects of drug induction on liver microsomes, in J. R. Gillette, A. H. Conney, G. J. Cosmides, R. W. Estabrook, J. R. Fouts and G. J. Mannering, Eds, "Microsomes and Drug Oxidation", Academic Press, 1969, pp. 251–270.
929. G. Dallner, P. Siekevitz and G. Palade, *Biochem. Biophys. Res. Commun.* **20**, 135 (1965).
930. T. F. Slater, *Biochem. J.* **106**, 155 (1968).
931. G. Powis, *Biochem. J.* **148**, 269 (1975).
932. A. H. Conney, C. Davison, R. Gastel and J. J. Burns, *J. Pharmacol. Exp. Therap.* **130**, 1 (1960).
933. A. H. Conney, *Pharmacol. Rev.* **19**, 317 (1967).

934. I. B. Tsyrlov and V. V. Lyachovich, *Biochem. Pharmacol.* **21**, 2540 (1972).
935. T. Tabei, K. Fukushima and W. L. Heinrichs, *Endocrinology*, **96**, 815 (1975).
936. E. D. Wills, Lipid peroxide formation and drug hydroxylation by microsomes, in D. Shugar, Ed., "Biochemical Aspects of Antimetabolites and of Drug Hydroxylation", Academic Press, New York, 1969, pp. 273–278.
937. A. K. Gayathri, M. R. S. Rao and G. Padmanaban, *Indian J. Biochem. Biophys.* **10**, 31 (1973).
938. H. Remmer, J. B. Schenkman and H. Greim, Spectral investigations on cytochrome P-450, in J. R. Gillette, A. H. Conney, G. J. Cosmides, R. W. Estabrook, J. R. Fouts and G. J. Mannering, Eds, "Microsomes and Drug Oxidation", Academic Press, 1969, pp. 189–197.
939. A. Y. H. Lu and W. Levin, *Biochim. Biophys. Acta*, **344**, 205 (1974).
940. H. Denk, J. B. Schenkman, P. G. Bacchin, F. Hutterer, F. Schaffner and H. Popper, *Exp. Mol. Pathol.* **14**, 263 (1971).
941. L. Gabriel, R. A. Canuto, E. Gravela, R. Garcea, and F. Feo, *Life Sci.* **15**, 2119 (1974).
942. F. Sperling, H. K. U. Evenike and T. Farber, *Environ. Res.* **5**, 164 (1972).
943. J. T. Wilson, *Biochem. Pharmacol.* **17**, 1449 (1968).
944. J. B. Schenkman, J. A. Ball and R. W. Estabrook, *Biochem. Pharmacol.* **16**, 1071 (1967).
945. W. Chen, P. A. Vrindten, P. G. Dayton and J. J. Burns, *Life Sci.* **1**, 35 (1962).
946. M. E. Dean and B. H. Stock, *Drug Metab. Disposition*, **3**, 325 (1975).
947. M. E. Dean and B. H. Stock, *Toxicol. App. Pharmacol.* **28**, 44 (1974).
948. A. S. M. Giasuddin, C. P. J. Caygill, A. T. Diplock and E. H. Jeffery, *Biochem. J.* **146**, 339 (1975).
949. G. E. R. Hook, J. K. Haseman and G. W. Lucier, *Chem.-Biol. Interactions*, **10**, 199 (1975).
950. G. W. Lucier, O. S. McDaniel, G. E. R. Hook, B. A. Fowler, B. R. Sonaware, and E. Faeder, *Env. Health Perspectives*, **5**, 199 (1973).
951. G. L. Sparschu, F. L. Dunn and V. K. Rowe, *Toxicol. Appl. Pharmacol.* **17**, 317 (1970).
952. G. L. Sparschu, F. L. Dunn and V. K. Rowe, *Food Cosmet. Toxicol.* **9**, 405 (1971).
953. D. J. Clegg, *Food Cosmet. Toxicol.* **9**, 195 (1971).
954. M. P. Carpenter, *Fed. Proc. Fed. Am. Soc. Exp. Biol.* **26**, 475 (1967); **27**, 677 (1968).
955. M. P. Carpenter, *Ann. N.Y. Acad. Sci.* **203**, 81 (1972).
956. A. K. Cho and G. T. Miwa, *Drug Metab. Disposition*, **2**, 477 (1973).
957. J. C. Arcos, A. H. Conney and N. P. Buu-Hoi, *J. Biol. Chem.* **236**, 1291 (1961).
958. A. H. Conney, E. C. Miller and J. A. Miller, *Cancer Res.* **16**, 450 (1956).
959. K. Takamiya, S-H. Chen and H. Kitagawa, *Gann.* **64**, 363 (1973).
960. T. Saito and T. Sugimura, *Gann.* **64**, 373 (1973).
961. M. Ishidate, M. Watanabe and S. Odashima, *Gann.* **58**, 267 (1967).
961. E. C. Miller, J. A. Miller, R. R. Brown and J. C. MacDonald, *Cancer Res.* **18**, 469 (1958).
962. G. C. Mueller and J. A. Miller, *Cancer Res.* **11**, 271 (1951).
963. A. P. Poland, E. Glover, J. R. Robinson and D. W. Nebert, *J. Biol. Chem.* **249**, 5599 (1974).
964. G. J. Mannering, N. E. Sladek, C. J. Parli and D. W. Shoeman, Formation of a new P-450 hemoprotein after treatment of rats with polycyclic hydrocarbons, in J. R. Gillette, A. H. Conney, G. J. Cosmides, R. W. Estabrook, J. R. Fouts and G. J. Mannering, Eds, "Microsomes and Drug Oxidation", Academic Press, 1969, pp. 303–330.

965. N. E. Sladek and G. J. Mannering, *Mol. Pharmacol.* **5**, 174 (1969).
966. A. H. Conney and J. J. Burns, *Nature*, **184**, 363 (1959).
967. J. A. Miller and E. C. Miller, *Advances in Cancer Research*, **1**, 339 (1953).
968. S. Wu and E. A. Smuckler, A comparison of NADP dependent reductive cleavage and N-demethylation of aminoazobenzene (AB), monomethyl AB + Dimethyl AB by cell free preparations from rat liver, in J. R. Gillette, A. H. Conney, G. J. Cosmides, R. W. Estabrook, J. R. Fouts and G. J. Mannering, Eds, "Microsomes and Drug Oxidation", Academic Press, 1969, pp. 189–197.
969. D. S. Hewick and J. R. Fouts, *Biochem. Pharmacol.* **19**, 457 (1970).
970. R. M. Kaschnitz and M. J. Coon, *Biochem. Pharmacol.* **214**, 295 (1975).
971. D. Ryan, A. Y. H. Lu, J. Kawalek, S. West, and W. Levin, *Biochem. Biophys. Res. Commun.* **64**, 1134 (1975).
972. A. R. Hansen and J. R. Fouts, *Biochem. Pharmacol.* **20**, 3125 (1971).
973. C. Lin, R. Chang, C. Casmer, and S. Symchowicz, *Drug. Metab.* **1**, 611 (1973).
974. F. DeMatteis, A. J. Donnelly and W. J. Runge, *Cancer Res.* **26**, 721 (1966).
975. E. W. Hurst and G. E. Paget, *Brit. J. Dermatol.* **75**, 105 (1963).
976. J. Cochin and J. Axelrod, *J. Pharmacol. Exp. Ther.* **125**, 105 (1959).
977. A. H. Conney, A. Y. H. Lu, W. Levin, A. Somogyi, S. West, M. Jacobson, D. Ryan and R. Kuntzman, *Drug Metab. Dispos.* **1**, 199 (1973).
978. A. Y. H. Lu, A. Somogyi, S. West, R. Kuntzman and A. H. Conney, *Arch. Biochem. Biophys.* **152**, 457 (1972).
979. D. L. Bull and D. A. Linquist, *J. Agr. Food Chem.* **12**, 310 (1964).
980. R. J. Kuhr and J. E. Casida, *J. Agr. Food Chem.* **15**, 814 (1967).
981. G. M. Cohen and G. J. Mannering, *Mol. Pharmacol.* **9**, 383 (1973).
982. N. Narasimhachari and R.-L. Lin, *Res. Commun. Chem. Path. Pharmacol.* **8**, 341 (1974).
983. D. E. Green and I. S. Forrest, *Canad. Psych. Assoc. J.* **11**, 299 (1966).
984. S. H. Curry and J. H. Marshall, *Life Sci.* **7**, 9 (1968).
985. A. H. Beckett, M. A. Beaven and A. E. Robinson, *Biochem. Pharmacol.* **12**, 779 (1963).
986. H. S. Posner, R. Cuplan and J. Levine, *J. Pharmacol. Exp. Therap.* **141**, 377 (1963).
987. H. M. Berman and M. A. Spirtes, *Biochem Pharmacol.* **20**, 2275 (1971).
988. P. F. Coccia and W. W. Westerfeld, *J. Pharmacol. Exp. Therap.* **157**, 446 (1967).
989. H. P. Harke, D. Schuller, B. Frahm and A. Mauch, *Res. Commun. Chem. Path. Pharmacol.* **9**, 595 (1974).
990. D. M. Ziegler, C. H. Mitchell and D. Jollow, The properties of a purified hepatic microsomal mixed function amine oxidase, in J. R. Gillette, A. H. Conney, G. J. Cosmides, R. W. Estabrook, J. R. Fouts and G. J. Mannering, Eds, "Microsomes and Drug Oxidation", Academic Press, 1969, pp. 173–188.
991. P. Mazel, J. F. Henderson and J. Axelrod, *J. Pharmacol. Exp. Therap.* **143**, 1 (1964).
992. P. Mazel, A. Kerza-Kwiatecki and J. Simanis, *Biochim. Biophys. Acta*, **114**, 72 (1966).
993. P. L. Berquist and R. E. F. Matthews, *Biochim. Biophys. Acta*, **34**, 567 (1959).
994. D. B. Dunn, *Arch. Environ. Health*, **10**, 842 (1965).
995. T. R. Devereux and J. R. Fouts, *Chem.-Biol. Interactions*, **8**, 91 (1974).
996. A. I. Archakov, I. I. Karuzina, A. I. Bokhon'ko, T. A. Alexandrova and L. F. Panchenco, *Biochem. Pharmacol.* **21**, 1595 (1972).
997. M. Knecht, *Z. Naturforsch.* **21**, 799 (1966).
998. D. Hadjiolov and D. Markov, *J. Nat. Cancer Inst.* **50**, 979 (1973).
999. A. E. M. McLean and P. A. Day, *Biochem. Pharmacol.* **23**, 1173 (1974).

1000. A. W. Pound and T. A. Lawson, *Brit. J. Exp. Pathol.* **56**, 77 (1975).
1001. H. V. Malling and C. N. Frantz, *Mutation Res.* **25**, 179 (1974).
1002. E. Englesberg, *J. Bacteriol.* **63**, 1 (1952).
1003. A. J. Lemin, *J. Agr. Food Chem.* **14**, 409 (1966).
1004. J. L. Holtzman, V. Rothman and S. Margolis, *Biochem. Pharmacol.* **21**, 581 (1972).
1005. R. S. Chhabra, R. J. Pohl and J. R. Fouts, *Drug Metab. Disposition*, **2**, 443 (1974).
1006. J. Chrastil and J. T. Wilson, *Anal. Biochem.* **63**, 202 (1975).
1007. R. M. Welch, J. Cavallito, and D. D. Gillespie, *Drug Metab. Dispos.* **1**, 211 (1973).
1008. A. P. Alvares, D. R. Bickers, and A. Kappas, *Proc. Nat. Acad. Sci. USA*, **70**, 1321 (1973).
1009. H. A. Price and R. L. Welch, *Env. Health Perspect.* **1**, 73 (1972).
1010. A. R. Yobs, *Env. Health Perspect.* **1**, 78 (1972).
1011. F. J. Biros, A. C. Walker and A. Medbery, *Bull. Env. Contam. Toxicol.* **5**, 317 (1970).
1012. L. Acker and E. Schulte, *Naturwissenschaften*, **57**, 497 (1970).
1013. J. R. Allen and D. H. Norback, *Science*, **179**, 498 (1973).
1014. T. Kamataki and H. Kitagawa, *Biochem. Pharmacol.* **23**, 1915 (1974).
1015. W. F. Bousquet, B. D. Rupe and T. S. Miya, *J. Pharmacol. Exp. Ther.* **147**, 376 (1965).
1016. R. E. Stizel and R. L. Furner, *Biochem. Pharmacol.* **16**, 1489 (1967).
1017. B. Stripp, R. H. Menard, N. G. Zampaglione, M. E. Hamrick, and J. R. Gillette, *Drug Metab. Dispos.* **1**, 216 (1973).
1018. E. Jeffery and G. J. Mannering, *Mol. Pharmacol.* **10**, 1004 (1974).
1019. N. E. Sladek and G. J. Mannering, *Mol. Pharmacol.* **5**, 186 (1969).
1020. A. K. Cho, B. K. Hodshon and B. B. Brodie, *Biochem. Pharmacol.* **19**, 1817 (1970).
1021. B. Solymoss, S. Varga and H. G. Classen, *Eur. J. Pharmacol.* **10**, 127 (1970).
1022. B. Solymoss. H. G. Classen and S. Varga, *Proc. Soc. Exp. Biol. Med.* **132**, 940 (1969).
1023. C. L. Litterst, E. G. Mimnaugh, R. L. Reagan and T. E. Gram, *Biochem. Pharmacol.* **23**, 2391 (1974).
1024. J. R. Cooper and B. B. Brodie, *J. Pharmacol. Exp. Therap.* **114**, 409 (1955).
1025. A. H. Conney and J. J. Burns, *Advances Pharmacol.* **1**, 31 (1962).
1026. A. H. Conney, J. R. Gillette, J. K. Inscoe, E. G. Trams and H. S. Posner, *Science*, **130**, 1478 (1959).
1027. N. C. Jain, N. J. Leung, R. D. Budd and T. C. Sneath, *J. Chromatog.* **103**, 85 (1975).
1028. J. A. McIntyre, A. E. Armandi, L. P. Rosei, W. Ling and G. C. Haberfelde, *Clin. Chem.* **21**, 109 (1975).
1029. J. W. Smith and T. J. Sheets, *J. Agr. Food Chem.* **15**, 577 (1967).
1030. J. Raaflaub and D. E. Schwartz, *Experentia*, **21**, 44 (1965).
1031. G. Weitzel, F. Schneider and A. M. Fretzdorff, *Experentia*, **20**, 38 (1964).
1032. G. Weitzel, F. Schneider, W. Hirschmann, D. Durst, R. Thauer, H. Ochs and D. Kummer, *Z. Physiol. Chem.* **348**, 443 (1967).
1033. K. W. Brunner and C. W. Young, *Ann. Intern. Med.* **63**, 69 (1965).
1034. M. L. Samuels, W. V. Leary, R. Alexanian, C. D. Howe and E. Frei, *Cancer*, **20**, 1187 (1967).
1035. M. G. Kelly, K. Gadekar, S. T. Yancey and V. T. Oliverio, *Cancer Chemother. Rep.* **39**, 77 (1964).
1036. J. C. Heuson and R. Heimann, *Eur. J. Cancer*, **2**, 385 (1966).
1037. S. Chaube and M. L. Murphy, *Teratology*, **2**, 23 (1969).
1038. A. M. Abdel-Wahab, R. J. Kuhr and J. E. Casida, *J. Agr. Food Chem.* **14**, 290 (1966).
1039. T. E. Gram, Enzymatic N-, O-, and S-dealkylation of foreign compounds by hepatic microsomes, in B. B. Brodie and J. R. Gillette, Eds, "Concepts of

Biochemical Pharmacology", Vol. 28, Part 2, in "Handbook of Experimental Pharmacology", Springer-Verlag, 1971, pp. 334–348.
1040. R. E. McMahon, *J. Pharm. Sci.* **55**, 457 (1966).
1041. E. M. Mrak, Report of the Secretary's Commission on Pesticides and their Relationship to Environmental Health, Washington, D.C., USDHEW, Parts 1 and 2, 677 pp.
1042. D. G. Crosby, *Ann. Rev. Plant Physiol.* **24**, 467 (1973).
1043. R. L. Blakeley, Interrelationship between folate and cobalamin metabolism, in S. Neuberger and E. L. Tatum, Eds, "The Biochemistry of Folic Acid and Related Compounds", John Wiley and Sons, New York, 1969, pp. 389–438.
1044. E. Borek and P. R. Srinivasan, *Ann. Rev. Biochem.* **35**, 275 (1966).
1045. R. Jackson and T. Hunter, *Nature*, **227**, 672 (1970).
1046. P. Grant, *Develop. Biol.* **2**, 197 (1960).
1047. I. Parsa, W. H. Marsh and P. J. Fitzgerald, *Am. J. Pathol.* **59**, 1 (1970).
1048. M. C. Poirier, L. A. Poirier and R. Lepage, *Cancer Res.* **32**, 1104 (1972).
1049. P. M. Newberne, Carcinogenicity of aflatoxin-contaminated peanut meal, in G. N. Wogan, Ed, "Mycotoxins in Foodstuffs", MIT Press, Cambridge, 1965, pp. 187–208.
1050. E. Farber, *Advan. Cancer Res.* **7**, 383 (1963).
1051. R. W. Engle and D. H. Copeland, *Cancer Res.* **12**, 905 (1952).
1052. E. LeBreton, *Compt. Rend*, **238**, 2446 (1954).
1053. E. J. Kensler, E. Bierman and G. Condouris, *J. Natl. Cancer Inst.* **15**, 1569 (1955).
1054. G. Shklar, E. Cataldo and A. L. Fitzgerald, *Cancer Res.* **26**, 2218 (1966).
1055. S. Bulba and J. A. Stekol, *Fed. Proc.* **26**, 696 (1967).
1056. V. M. Doctor, J. P. Chang and M. K. Richards, *Cancer Res.* **27**, 546 (1967).
1057. H. Kroger, P. Kahle and H. Kessel, *Z. Physiol. Chem.* **349**, 1725 (1968).
1058. J. C. Reid, O. S. Temmer and M. D. Bacon, *J. Natl. Cancer Inst.* **17**, 189 (1956).
1059. P. Mazel in B. N. LaDu, H. G. Mandel and E. L. Way, Eds, "Fundamentals of Drug Metabolism and Drug Disposition", Williams and Wilkins, Baltimore, Md., 1971.
1060. T. Nash, *Nature*, **170**, 976 (1952).
1061. C. Huygen, *Anal. Chim. Acta*, **28**, 349 (1963).
1062. B. N. LaDu, L. Gaudette, N. Trousof and B. B. Brodie, *J. Biol. Chem.* **214**, 741 (1955).
1063. W. R. Frisell and C. G. Mackenzie, in D. Glick, Ed., "Methods of Biochemical Analysis", Vol. VI, Interscience, New York, N.Y., 1958.
1064. J. F. Walker, "Formaldehyde", 3rd edn, Reinhold, New York, 1964.
1065. J. A. Zapp, Jr., *Science*, **190**, 422 (1975).
1066. A. R. Jones and H. Jackson, *Biochem. Pharmacol.* **17**, 2247 (1968).
1067. H. F. Fraser and H. Isbell, *J. Pharmacol. Exp. Ther.* **105**, 458 (1952).
1068. J. H. Jaffe and E. C. Senay, *J. Am. Med. Assoc.* **216**, 1303 (1971).
1069. A. Zaks, M. Fink and A. M. Freedman, *J. Am. Med. Assoc.* **220**, 811 (1972).
1070. J. H. Jaffe, E. C. Senay, C. R. Schuster, P. R. Renault, B. Smith and S. DiMenza, *J. Am. Med Assoc.* **222**, 437 (1972).
1071. R. Levine, A. Zaks, M. Fink and A. M. Freedman, *J. Am. Med. Assoc.* **226**, 316 (1973).
1072. C. Y. Sung and E. L. Way, *J. Pharmacol. Exp. Ther.* **110**, 260 (1954).
1073. R. M. Veatch, T. K. Adler and E. L. Way, *J. Pharmacol. Exp. Ther.* **145**, 11 (1964).
1074. R. E. McMahon, H. W. Culp and F. J. Marshall, *J. Pharmacol. Exp. Ther.* **149**, 436 (1965).
1075. R. F. Kaiko and C. E. Inturrisi, *Fed. Proc., Fed. Am. Soc. Exp. Biol.* **33**, 473 (1974).

1076. R. E. Billings, R. E. McMahon and D. A. Blake, *Fed. Proc., Fed. Am. Soc. Exp. Biol.* **33**, 473 (1974).
1077. R. E. Billings, R. Booher, S. E. Smits, A. Pohland and R. E. McMahon, *J. Med. Chem.* **16**, 305 (1973).
1078. D. S. Hewick and J. R. Fouts, *Biochem. J.* **117**, 833 (1970).
1079. M. S. Dar, *General Pharmacol.* **6**, 275 (1975).
1080. T. A. Smith, *Phytochemistry*, **14**, 865 (1975).
1081. P. J. Large, R. R. Eady and D. J. Murden, *Anal. Biochem.* **32**, 402 (1969).
1082. R. R. Eady and P. J. Large, *Biochem. J.* **111**, 37P (1969).
1083. E. Schmidt and M. Fischer, *Ber.* **53**, 1537 (1920).
1084. E. Schmidt and R. Schumacher, *Ber.* **54**, 1414 (1921).
1085. P. A. Smith and R. N. Loeppky, *J. Am. Chem. Soc.* **89**, 1147 (1967).
1086. D. C. Malins, W. T. Roubal and P. A. Robisch, *J. Agric. Food Chem.* **18**, 740 (1970).
1087. P. N. Magee, *Eur. Occup. Hyg.* **15**, 19 (1972).
1088. J. E. Saxton, Ed., "The Alkaloids", Vol. I, etc., The Chemical Society, Burlington House, London, 1971, etc.
1089. T. Fujiwara-Arasaki and N. Mino, "Proc. 7th Int. Seaweed Symp.", p. 506, (1972).
1090. T. Bernard and G. Goas, *Compt. Rend.*, (*Ser. D*), **267**, 622 (1968).
1091. P. J. Large and H. McDougall, *Anal. Biochem.* **64**, 304 (1975).
1092. K. Amano and K. Yamada, *Bull. Jap. Soc. Sci. Fish*, **30**, 639 (1964).
1093. K. Norpoth, U. Witting, M. Springorum and C. Witting, *Int. Arch. Arbeitsmed.* **33**, 315 (1974).
1094. H. W. Leber, E. Degkwitz and H. Staudinger, *Z. Physiol. Chem.* **350**, 439 (1969).
1095. A. G. Hildebrandt, M. Speck and I. Roots, *Arch. Pharmacol.* **281**, 371 (1974).
1096. W. H. Orme-Johnson and D. M. Ziegler, *Biochem. Biophys. Res. Commun.* **21**, 78 (1965).
1097. C. S. Lieber and L. M. de Carli, *J. Biol. Chem.* **245**, 2505 (1970).
1098. E. Feytmans and F. Leighton, *Biochem. Pharmacol.* **22**, 249 (1973).
1099. W. R. Frisell and C. G. Mackenzie, *J. Biol. Chem.* **217**, 275 (1955).
1100. J. M. Johnston and C. G. Mackenzie, *J. Biol. Chem.* **221**, 301 (1956).
1101. S. Orrenius and J. L. E. Ericsson, *J. Cell Biol.* **28**, 181 (1966).
1102. M. P. Carpenter, *Fed. Proc. Fed. Am. Soc. Exp. Biol.* **26**, 475 (1967).
1103. A. Haddow, R. J. Harris, G. A. R. Kon and E. M. F. Roe, *Trans. Roy. Soc.* (*London*), **A241**, 147 (1948).
1104. S. S. Navran and R. T. Louis-Ferdinand, *Res. Commun. Chem. Pathol. Pharmacol.* **12**, 685 (1975).
1105. R. R. Brown, J. A. Miller and E. C. Miller, *Federation Proc.* **11**, 192 (1952).
1106. J. A. Miller, R. R. Brown, E. C. Miller and G. C. Mueller, *Cancer Res.* **11**, 269 (1951).
1107. A. E. Takemori and G. J. Mannering, *J. Pharmacol. Exp. Ther.* **123**, 171 (1958).
1108. M. Matsumoto and H. Terayama, *Gann.* **52**, 239 (1961).
1109. IARC monographs on the evaluation of the carcinogenic risk of chemicals to man, Vol. 8, "Some Aromatic Azo Compounds". International Agency for Research on Cancer, Lyon, France, 1975.
1110. N. I. Golub, T. S. Kolesnichenko and L. M. Shabad, *Bull. Exp. Biol. Med.* **78**, 1402 (1974).
1111. O. M. Bulay and L. W. Wattenberg, *J. Nat. Cancer Inst.* **46**, 397 (1971).
1112. H. Druckrey, S. Ivankovic and R. Preussmann, *Nature*, **210**, 1378 (1966).
1113. S. Ivankovic and H. Druckrey, *Z. Krebsforsch.* **71**, 320 (1968).
1114. L. M. Shabad, T. S. Kolesnichenko and T. G. Nikonova, *Int. J. Cancer*, **11**, 88 (1973).

1115. T. Tanaka, IARC Scientific Publication No. 4, 100 (1973).
1116. A. L. Herbst, H. Ulfelder and D. C. Poskanzer, *New Engl. J. Med.* **284**, 878 (1971).
1117. A. L. Herbst, H. Ulfelder and D. C. Poskanzer, *New Engl. J. Med.* **285**, 407 (1971).
1118. J. R. Bend, G. E. R. Hook, R. E. Easterling, T. E. Gram and J. R. Fouts, *J. Pharmacol. Exp. Ther.* **183**, 206 (1972).
1119. F. J. Wiebel, J. C. Leutz and H. V. Gelboin, *Arch. Biochem. Biophys.* **154**, 292 (1973).
1120. K. Hartiala, *Pharmacol. Rev.* **53**, 496 (1973).
1121. R. M. Welch, J. Cavallito and A. Loh, *Toxicol. Appl. Pharmacol.* **23**, 749 (1972).
1122. L. W. Wattenberg, J. L. Leong and P. J. Strand, *Cancer Res.* **22**, 1120 (1962).
1123. L. W. Wattenberg, *Cancer*, **28**, 99 (1971).
1124. L. W. Wattenberg, *Toxicol. Appl. Pharmacol.* **23**, 741 (1972).
1125. L. W. Wattenberg, *Progr. Exp. Tumor Res.* **14**, 89 (1970).
1126. J. R. Gillette in D. Shugar, Ed., "Biochemical Aspects of Antimetabolites and of Drug Hydroxylation", Academic Press, New York, 1969, pp. 109–124.
1127. P. L. Morselli, H. H. Ong, E. R. Bowman and H. McKennis, *J. Med. Chem.* **10**, 1033 (1967).
1128. P. J. Jose and T. F. Slater, *Xenobiotica*, **3**, 357 (1973).
1129. P. N. Magee, "Alkylating Compounds," Verlag Paul Parey, Hamburg, 1968, p. 79.
1130. P. N. Magee and R. Schoental, *Brit. Med. Bull.* **20**, 102 (1964).
1131. A. Reisenstart and J. L. Rosner, *Genetics*, **49**, 343 (1964).
1132. S. Grilli and G. Prodi, *Gann.* **66**, 473 (1975).
1133. J. A. Wittkop, R. A. Prough and D. J. Reed, *Arch. Biochem. Biophys.* **134**, 308 (1969).
1134. E. Freese, in A. Hollander, Ed., "Chemical Mutagens", Plenum Press, New York, 1971, p. 1.
1135. E. B. Freese and E. Freese, *Proc. Nat. Acad. Sci. U.S.* **52**, 1289 (1964).
1136. H. V. Malling, *Mutation Res.* **4**, 559 (1967).
1137. C. Biancifiori and R. Ribacchi, *Nature*, **194**, 488 (1962).
1138. H. Oswald and F. W. Krüger, *Arzneimittel-Forsch.* **19**, 1891 (1969).
1139. H. Shimizu, D. Nagel and B. Toth, *Int. J. Cancer*, **13**, 500 (1974).
1140. B. Wiebecke, U. Lohrs, J. Gimmy and M. Eder, *Z. Gesamte Exp. Med.* **19**, 277 (1969).
1141. S. Grilli, M. R. Tosi and G. Prodi, *Gann.* **66**, 481 (1975).
1142. W. K. Paik and S. Kim, *Advances in Enzymology*, **42**, 227 (1975).
1143. C. Turberville and V. M. Craddock, *Biochem. J.* **124**, 725 (1971).
1144. V. M. Craddock, *Nature*, **245**, 386 (1973).
1145. C. C. Clayton and C. A. Baumann, *Cancer Res.* **9**, 575 (1949).
1146. L. J. Cole and P. C. Nowell, *Science*, **150**, 1782 (1965).
1147. L. D. Engelse, M. Gebbink and P. Emmelot, *Chem.-Biol. Interactions*, **11**, 535 (1975).
1148. F. Stenback, A. Ferrero, R. Montesano and P. Shubik, *Z. Krebsforsch.* **79**, 31 (1973).
1149. L. Tomatis, P. N. Magee and P. Shubik, *J. Nat. Cancer Inst.* **33**, 341 (1964).
1150. K. M. Herrold, *J. Nat. Cancer Inst.* **39**, 1099 (1967).
1151. W. Dontenwill, *Food Cosmet. Toxicol.* **6**, 571 (1968).
1152. H. Haas, U. Mohr and F. W. Kruger, *J. Nat. Cancer Inst.* **51**, 1295 (1973).
1153. V. J. Feron, P. Emmelot and T. Vossenaar, *Eur. J. Cancer*, **8**, 445 (1972).
1154. P. Pour, F. W. Kruger, A. Cardesa, J. Althoff and U. Mohr, *J. Nat. Cancer Inst.* **51**, 1019 (1973).
1155. J. Althoff, *Z. Krebsforsch.* **82**, 153 (1974).

1156. J. Althoff, R. Wilson, A. Cardesa and P. Pour, *Z. Krebsforsch.* **81**, 251 (1974).
1157. S. F. Yang, *Anal. Biochem.* **32**, 519 (1969).
1158. H. Schreiber, K. Schreiber and D. H. Martin, *J. Nat. Cancer Inst.* **54**, 187 (1975).
1159. D. McGregor, *Mutation Res.* **30**, 305 (1975).
1160. E. C. Miller and J. A. Miller, *J. Nat. Cancer Inst.* **15**, 1571 (1955).
1161. G. Mohn, J. Ellenberger and D. McGregor, *Mutation Res.* **25**, 187 (1974).
1162. F. F. Becker, *Cancer Res.* **35**, 1734 (1975).
1162a A. G. Gilman and A. H. Conney, *Biochem. Pharmacol.* **12**, 591 (1963).
1163. J. M. Machinist, E. W. Dehner and D. M. Ziegler, *Arch. Biochem. Biophys.* **125**, 858 (1968).
1164. J. R. Bend, G. E. R. Hook and T. E. Gram, *Drug Metab. Disposition*, **1**, 358 (1973).
1165. S. S. Nauran and R. T. Louis-Ferdinand, *Res. Commun. Chem. Pathol. Pharmacol.* **12**, 713 (1975).
1166. R. A. Scanlon, S. M. Lohsen, D. D. Bills and L. M. Libbey, *J. Agr. Food Chem.* **22**, 149 (1974).
1167. J. Velisek, J. Davidek, S. Klein, M. Karaskova and I. Vykoukova, *Z. Lebensm. Unters.-Forsch.* **159**, 97 (1975).
1168. W. Lijinsky and S. S. Epstein, *Nature*, **225**, 21 (1970).
1169. W. Lijinsky, L. Keefer and J. Loo, *Tetrahedron*, **26**, 5137 (1970).
1170. P. N. Magee and J. M. Barnes, *Advances Cancer Res.* **10**, 168 (1967).
1171. P. D. Lotlikar, W. J. Baldy, Jr. and E. N. Dwyer, *Biochem. J.* **152**, 705 (1975).
1172. J. LaBar and J. Sander, *Z. Krebsforsch.* **84**, 299 (1975).
1173. S. S. Mirvish, B. Gold, M. Eagan and S. Arnold, *Z. Krebsforsch.* **82**, 259 (1974).
1174. W. Fiddler, J. W. Pensabene, R. C. Doerr and A. E. Wasserman, *Nature*, **236**, 307 (1972).
1175. W. J. Serfontein and L. S. de Villiers, *Res. Commun. Chem. Pathol. Pharmacol.* **12**, 605 (1975).
1176. L. L. Poulsen, F. F. Kadlubor and D. M. Zeigler, *Arch. Biochem. Biophysics*, **164**, 774 (1974).
1177. F. Feigl and E. Silva. *Analyst*, **82**, 582 (1957).
1178. B. B. Brodie and J. R. Gillette, Eds. "Concepts in Biochemical Pharmacology", Vol. 28, Part 2, in "Handbook of Experimental Pharmacology". Berlin, Springer-Verlag, 1971.
1179. R. V. Smith, P. W. Erhardt and S. W. Leslie, *Res. Commun. Chem. Pathol. Pharmacol.* **12**, 181 (1975).
1180. L. A. Goldblatt, Ed., "Aflatoxins", Academic Press, New York, 1969.
1181. G. N. Wogan, *Methods in Cancer Research*, **7**, 309 (1973).
1182. M. E. Alpert, M. S. R. Hutt, G. N. Wogan and C. S. Davidson, *Cancer*, **28**, 253 (1971).
1183. F. Dickens and H. E. H. Jones, *Brit. J. Cancer*, **17**, 691 (1963); **19**, 392 (1965).
1184. P. Newberne and G. N. Wogan, *Toxicol. Appl. Pharmacol.* **11**, 309 (1968).
1185. P. Newberne and W. H. Butler, *Cancer Res.* **29**, 236 (1969).
1186. D. H. Swenson, E. C. Miller and J. A. Miller, *Biochem. Biophys. Res. Commun.* **60**, 1036 (1974).
1187. M. F. Dutton and J. G. Heathcote, *Biochem. J.* **101**, 21P (1966).
1188. M. F. Dutton and J. G. Heathcote, *Chem. Ind.* 418 (March 30, 1968).
1189. P. J. Andrellos and G. R. Reid, *J. Assoc. Offic. Anal. Chem.* **47**, 801 (1964).
1190. J. I. Dalezios, G. N. Wogan and S. M. Weinreb, *Science*, **171**, 584 (1971).
1191. O. Bassir and P. O. Emafo, *Biochem. Pharmacol.* **19**, 1681 (1970).
1192. O. Bassir and G. O. Emerole, *Eur. J. Biochem.* **47**, 321 (1974).

1193. O. Bassir and A. A. Adekunle, *J. Pathol.* **102**, 49 (1970).
1194. O. Bassir and A. A. Adekunle, *FEBS Lett.* **10**, 198 (1970).
1195. J. Axelrod, *Biochem. J.* **63**, 634 (1956).
1196. C. Mitoma, R. L. Dehn and M. Tanabe, *Biochim. Biophys. Acta* **237**, 21 (1971).
1197. R. E. McMahon, H. W. Culp, J. Mills and F. J. Marshall, *J. Med. Chem.* **6**, 343 (1963).
1198. H. G. Bray, V. M. Craddock and W. V. Thorpe, *Biochem. J.* **60**, 225 (1955).
1199. K. J. Netter and G. Seidel, *J. Pharmacol. Exp. Therap.* **146**, 61 (1964).
1200. D. W. Nebert, N. Considine and I. S. Owens, *Arch. Biochem. Biophys.* **157**, 148 (1973).
1201. D. W. Nebert, J. E. Gielen and F. M. Goujon, *Mol. Pharmacol.* **8**, 651 (1972).
1202. M. C. Green, *Biochem. Genet.* **9**, 369 (1973).
1203. N. A. Broadhurst, M. L. Montgomery and V. H. Freed, *J. Agr. Food Chem.* **14**, 585 (1966).
1204. H. Blank, *A.M.A. Arch. Dermatol.* **75**, 184 (1957).
1205. M. J. Barnes and B. Boothroyd, *Biochem. J.* **78**, 41 (1961).
1206. C. Lin, R. Chang, J. Magat and S. Symchowicz, *J. Pharm. Pharmacol.* **24**, 911 (1972).
1207. C. Bedford, D. Busfield, K. J. Child, I. MacGregor, P. Sutherland and E. G. Tomich, *Arch. Dermatol.* **81**, 735 (1960).
1208. S. Symchowicz and K. K. Wong, *Biochem. Pharmacol.* **15**, 1601 (1966).
1209. I. P. Kapoor, R. L. Metcalf, R. F. Nystrom and G. K. Sangha, *J. Agr. Food Chem.* **18**, 1145 (1970).
1210. H. Ohira, H. Kaseda and R. Kido, *Biochem. Pharmacol.* **23**, 1918 (1974).
1211. J. M. Price and L. W. Dodge, *J. Biol. Chem.* **223**, 699 (1956).
1212. J. K. Roy and J. M. Price, *J. Biol. Chem.* **234**, 2759 (1959).
1213. Y. Kotake, N. Hasegawa and M. Yamamoto, *Proc. Japan Acad.* **36**, 445 (1960).
1214. G. T. Bryan, *Cancer Res.* **28**, 183 (1968).
1215. G. T. Bryan, R. R. Brown and J. M. Price, *Cancer Res.* **24**, 596 (1964).
1216. G. T. Bryan and P. P. Springberg, *Cancer Res.* **26**, 105 (1966).
1217. J. L. Gaylor and H. S. Mason, *J. Biol. Chem.* **243**, 4966 (1968).
1218. W. L. Miller, M. E. Kalafer, J. L. Gaylor and C. V. Delwiche, *Biochemistry*, **6**, 2673 (1967).
1219. G. Bartholini, I. Karuma and A. Fletscher, *Nature*, **230**, 533 (1971).
1220. O. Hornykiewicz, *Life Sci.* **15**, 1249 (1974).
1221. K. Schubert, L. D. Phai, G. Kaufmann and R. Knöll, *Acta Biol. Med. Germ.* **34**, 167 (1975).
1222. H. Rosengarten, E. Meller and A. J. Friedhoff, *Trans. Am. Soc. Neurochem.* **5**, 117 (1974).
1223. E. Meller, H. Rosengarten and A. J. Friedhoff, *Life Sci.* **14**, 2167 (1974).
1224. C. Boehme and F. Baer, *Food Cosmet. Toxicol.* **5**, 23 (1967).
1225. J. F. Henderson and P. Mazel, *Biochem. Pharmacol.* **13**, 1471 (1964).
1226. E. C. Shearer, "Relationships among atmospheric formaldehyde, methane and krypton". Arkansas Univ., Fayetteville, Thesis (Ph.D.), 1969, 135 pp., 79 Refs.
1227. J. A. Yoe and L. C. Reid, *Ind. Eng. Chem., Anal. Ed.* **13**, 238 (1941).
1228. Y. Iwasa and T. Imoto, *J. Chem. Soc. Japan, Pure Chem. Sect.* **84**, 29 (1963).
1229. C. J. Kensler and S. P. Battista, *Am. Rev. Resp. Dis.* **93** (Part 2), 93 (1966).
1230. F. Watanabe and S. Sugimoto, *Gann.* **46**, 365 (1955).
1231. K. Edwards, H. Jackson and A. R. Jones, *Biochem. Pharmacol.* **19**, 1791 (1970).
1232. F. B. Erkis and M. M. Ratpan, *Tsitol. Genet.* **7**, 543 (1973).

1233. I. A. Rapoport, *Dokl. Akad. Nauk SSR*, **59**, 1183 (1948).
1234. C. Auerbach, *Science*, **110**, 419 (1949).
1235. T. Alderson, *Nature*, **187**, 485 (1960).
1236. A. M. Poverenny, Y. A. Siomin, A. S. Saenko and B. I. Sinzinis, *Mutation Res.* **27**, 123 (1975).
1237. N. A. Renzetti and R. J. Bryan, *J. Air Poll. Control Assoc.* **11**, 421, 427 (1961).
1238. J. C. Romanowsky, R. M. Ingels and R. J. Gordon, *J. Air Poll. Control Assoc.* **15**, 362 (1965).
1239. E. A. Schuck, E. R. Stephens and J. T. Middleton, *Arch. Env. Health*, **13**, 570 (1966).
1240. V. P. Melekhina, in "Maximum Permissible Concentration of Formaldehyde in Atmospheric Air", in Survey of USSR Literature on Air Pollution and Related Occupational Diseases (translated from Russian by B. S. Levine), **3**, 135 (1960).
1241. R. G. Smith, R. J. Bryan, M. Feldstein, B. Levadie, F. A. Miller, E. R. Stephens and N. G. White, *Health Lab. Sci. Suppl.* **7**, 87 (1970).
1242. V. Masek, *Gesundh. Ingr.* **8**, 234 (1970).
1243. N. V. Avdukova, T. P. Antsiferova and L. D. Loiko, *Gig. Sanit.* **37**, 69 (1972).
1244. E. A. Sidorova and S. A. Proskurina, *Gig. Sanit.* **38**, 77 (1973).
1245. E. Lahmann and K. Jander, *Gesundh. Ingr.* **89**, 18 (1968).
1246. B. S. Levine, Ed., Survey of USSR Literature on Air Pollution and Related Occupational Diseases, **8**, 55 (1963).
1247. N. Yamate, *Sangyo Kogai* (*Ind. Public Nuisance*), **5**, 709 (1969).
1248. N. Yamate, T. Matsumura and M. Tonomura, *Eisei Shikensho Hokoku* (*Bull. Natl. Inst. Hyg. Sci., Tokyo*), **86**, 58 (1968).
1249. N. Yamate and T. Matsumura, *J. Japan Soc. Air Pollution*, **4**, 121 (1969).
1250. G. I. Benzina, *Gig. Sanit.* **33**, 411 (1968).
1251. N. Tajima and Y. Kurosaka, *Bunseki Kagaku*, **23**, 464 (1974).
1252. C. S. Wodkowski and E. E. Weaver, The effects of engine parameters, fuel composition, and control devices on aldehyde exhaust emissions, presented at West Coast Air Pollution Control Association Meeting, San Francisco, California, 8 October, 1970.
1253. E. E. Wiggs, R. J. Campion and W. L. Petersen, The effect of fuel hydrocarbon composition on exhaust and oxygenate emissions, SAE Paper No. 720251, presented at the Society of Automotive Engineers, Automotive Engineering Congress, Detroit, Michigan, 10–14 January, 1972.
1254. P. E. Oberdorfer, *SAE Prog. Tech. Ser.* **14**, 32 (1971).
1255. M. F. Fracchio, F. J. Schuette and P. K. Mueller, *Env. Sci. Technol.* **1**, 915 (1967).
1256. S. N. Mahajan, M. U. Sivaramakrishnan and G. R. Rao, *Indian J. Pharm.* **34**, 98 (1972).
1257. Y. N. Gladchikova and N. I. Shumarina, *Gig. Sanit.* **23**, 83 (1958).
1258. I. Szelejewska, *Chemia Analit.* **20**, 325 (1975).
1259. M. J. Houle, D. E. Long and D. Smette, *Anal. Letters*, **3**, 401 (1970).
1260. D. E. Long, *Anal. Chim. Acta*, **46**, 193 (1969).
1261. F. P. Czech, *JAOAC*, **56**, 1489 (1973).
1262. F. P. Czech, *JAOAC*, **56**, 1496 (1973).
1263. S. Suzuki and H. Nobutoshi, *Environ. Sci. Res. Rep. Chiba Univ.* **1**, 99 (1973).
1264. H. Denffer and V. Heidbrink, *Acta Histochem.* **48**, 62 (1974).
1265. C. C. Musselwhite and K. W. Petts, *J. Inst. Water Poll. Control*, **73**, 443 (1974).
1266. J. M. Harkin, J. R. Obst and W. F. Lehmann, *Forest Prod. J.* **24**, 27 (1974).
1267. Y. Ohkura and K. Zaitsu, *Talanta*, **21**, 547 (1974).

1268. E. F. Ullman, J. H. Osiecki, D. G. B. Boocock and R. Darcy, *J. Am. Chem. Soc.* **94**, 7049 (1972).
1269. J. W. Munson and T. Hodgkins, *Microchem. J.* **20**, 39 (1975).
1270. "The United States Pharmacopeia", 18th rev., Mack Publishing Co., Easton, Pa., 1970, p. 843.
1271. J. W. Munson and T. G. Hodgkins, *J. Pharm. Sci.* **64**, 1043 (1975).
1272. M. J. Cormier, D. M. Hercules and J. Lee, Eds, "Chemiluminescence and Bioluminescence", Plenum Press, New York, N.Y., 1973, pp. 393, 403, 427.
1273. M. Trautz and P. Schorigin, *Z. Wiss. Photogr. Photochem.* **3**, 121 (1905).
1274. D. Slawinska, D. Golebiowska and J. Slawinski, *Chem. Anal. (Warsaw)*, **11**, 1117 (1965).
1275. D. Slawinska and J. Slawinski, *Anal. Chem.* **47**, 2101 (1975).
1276. A. R. Shoaf and R. H. Steele, *Biochem. Biophys. Res. Commun.* **61**, 1363 (1974).
1277. E. McKeown and A. W. Waters, *J. Chem. Soc. B.* 1040 (1966).
1278. E. J. Bowen and R. A. Lloyd, *Proc. Chem. Soc. London*, 305 (1963).
1279. E. J. Bowen, *Pure Appl. Chem.* **9**, 473 (1964).
1280. H. Auterhoff and K. J. Aymanns, *Arch. Pharm.* **307**, 885 (1974).
1281. M. Koivusalo and L. Uotila, *Anal. Biochem.* **59**, 34 (1974).
1282. P. Strittmatter and E. G. Ball, *J. Biol. Chem.* **213**, 445 (1955).
1283. Z. B. Rose and E. Racker, *Methods in Enzymol.* **9**, 357 (1966).
1284. S. Ratner and H. T. Clarke, *J. Am. Chem. Soc.* **59**, 200 (1937).
1285. J. O. McGhee and P. H. von Hippel, *Biochemistry*, **14**, 1281 (1975).
1286. J. O. McGhee and P. H. von Hippel, *Biochemistry*, **14**, 1297 (1975).
1287. W. W. Dean and J. Lebowitz, *Nature, New Biol.* **231**, 5 (1971).
1288. P. H. von Hippel and K. Y. Wong, *J. Mol. Biol.* **61**, 587 (1971).
1289. H. Utiyama and P. Doty, *Biochemistry*, **10**, 1254 (1971).
1290. A. M. Poverennyi, T. L. Aleinikova and A. D. Mar'yasina, *Ukr. Biokhim. Zh.* No. 3, 459 (1965).
1291. V. V. Simonov, N. I. Ryabchenko and A. M. Poverennyi, *Radiobiologiya*, **9**, 663 (1969).
1292. N. I. Ryabchenko, V. V. Simonov and A. M. Poverennyi, *Radiobiologiya*, **9**, 171 (1969).
1293. Y. L. Lyubchenko, E. N. Trifonov, Y. S. Lazurkin and M. D. Frank-Kamenetskii, *Molekul. Biol.* **5**, 772 (1971).
1294. Y. N. Kosaganov, M. I. Zarudnaya, Y. S. Lazurkin and M. D. Frank-Kamenetskii, *Nature New Biol.* **231**, 212 (1971).
1295. V. V. Simonov, Y. A. Semin, S. I. Suminov and A. M. Poverennyi, *Biokhimiya*, **39**, 436 (1974).
1296. D. Brutlag, C. Schlehuber and J. Bonner, *Biochemistry*, **8**, 3214 (1969).
1297. R. P. P. Fuchs and M. P. Daune, *Biochemistry*, **13**, 4435 (1974).
1298. F. Traganos, Z. Darzynkiewicz, T. Sharpless and M. R. Melamed, *J. Histochem. Cytochem.* **23**, 431 (1975).
1299. M. Hitchcock, D. M. Piscitelli and A. Bouhuys, *Arch. Environ. Health*, **26**, 177 (1973).
1300. W. Dankelman and J. M. H. Daemen, *Anal. Chem.* **48**, 401 (1976).
1301. E. J. Kerfoot and T. F. Mooney, *Am. Ind. Hyg. Assoc. J.* **36**, 533 (1975).
1302. Intersociety Committee, Methods of Air Sampling and Analysis, American Public Health Association, Washington D.C., 1972, p. 194.
1303. E. H. Cordes and H. G. Bull, *Chem. Revs.* **74**, 581 (1974).
1304. A. Kankaanpera, *Acta Chem. Scand.* **23**, 1723 (1969).

1305. A. Skrabal, W. Stockmair and H. Schreiner, *Z. Physik. Chem.* **A169,** 177 (1934).
1306. B. W. Fox and H. Jackson, *Brit. J. Pharmacol.* **24**, 24 (1965).
1307. E. Binderup, W. O. Godtfredsen and K. Roholt, *J. Antibiot.* **24**, 767 (1971).
1308. E. Marchi, G. Mascellani and D. Boccali, *J. Pharm. Sci.* **63**, 1299 (1974).
1309. H. Wullen, *Arch. Pharm. Chemi.* **68**, 197 (1961).
1310. A. P. Mathers and M. J. Pro, *Anal. Chem.* **27**, 753 (1955).
1311. British Pharmacopoeia, Pharmaceutical Press, London, 1968, p. 1131.
1312. British Pharmaceutical Codex, Pharmaceutical Press, London, 1968, p. 579.
1313. L. Berrens, E. Young and L. H. Janson, *Brit. J. Dermatol.* **76**, 110 (1964).
1314. A. A. Fisher, N. B. Kanof and E. M. Biondi, *Arch. Dermatol.* **86**, 753 (1962).
1315. W. F. Schorr, E. Keran and E. Plotka, *Arch. Dermatol.* **110**, 73 (1974).
1316. Z. S. Markova and S. A. Vasileva, *Gig. Sanit.* **38**, 21 (1973).
1317. T. Suzuki and S. Sakamoto, *Odor Res. J. Japan*, **1**, 29 (1970).
1318. K. Y. Vengerskaya, G. N. Nazyrov, L. S. Bodrova, S. Y. Dubrovskii and V. P. Dumko, *Gig. Sanit.* **33**, 340 (1968).
1319. T. I. Kravchenko, K. I. Stankevich, E. F. Malygina and T. G. Zakharova, *Gig. Sanit.* **5**, 19 (1974).
1320. H. Neusser and M. Zentner, *Holzforsch. Holzver.* **20**, 101 (1968).
1321. L. Plath, *Holz als Roh-und Werkstoffe*, **26**, 125 (1968).
1322. U.S. Forest Products Laboratory, "Eliminating odor from urea-resin-bonded plywood and particle board. An annotated bibliography", Madison, Wis., 1969, 15 pp.
1323. R. H. Gillespie, "That panel odor—when it may occur—what you can do about it", U.S. Forest Product Laboratory Unnumbered report, Madison, Wis., 1972, 7 pp.
1324. G. Stoger, *Holzforsch. Holzver.* **17**, 93 (1965).
1325. O. Wittmann, *Holz. Roh—Werkstoff.* **20**, 221 (1962).
1326. E. P. Sheppard and C. H. Wilson, *J. Soc. Cosmet. Chem.* **25**, 655 (1974).
1327. J. Blass, C. Verriest, A. Leau, H. Detruit and M. Weiss, *Pathologie-Biologie*, **22**, 593 (1974).
1328. P. Alexander, D. Carter and J. G. Johnson, *Biochem. J.* **48**, 435 (1951).
1329. H. Fraenkel-Conrat and H. Olcott, *J. Am. Chem. Soc.* **70**, 2673 (1948).
1330. H. Fraenkel-Conrat and H. Olcott, *J. Biol. Chem.* **174**, 827 (1948).
1331. J. Blass, *Biol. Med.* (*Paris*), **53**, 202 (1964).
1332. J. Blass, B. Bizzini and M. Raynaud, *Ann. Inst. Pasteur*, **116**, 50 (1969).
1333. J. Blass, *Compt. Rend.* **271**, 94 (1970).
1334. F. A. Quiocho and F. M. Richards, *Biochemistry*, **5**, 4062 (1966).
1335. W. H. Bishop and F. M. Richards, *J. Mol. Biol.* **33**, 415 (1968).
1336. J. H. Bowes and C. W. Cater, *Biochim. Biophys. Acta* (*Amsterdam*), **168**, 341 (1968).
1337. W. F. Happich, M. F. Taylor and S. H. Feairheller, *Text. Res. J.* **40**, 768 (1970).
1338. D. French and J. T. Edsall, *Adv. Prot. Chem.* **2**, 277 (1945).
1339. A. G. E. Pearse, Histochemistry. Theoretical and applied, Vol. 1, London, J. A. Churchill, Ltd., 1968.
1340. P. van Duijn, *J. Histochem. Cytochem.* **9**, 234 (1961).
1341. R. E. Hayes, *Stain Technol.* **24**, 19 (1949).
1342. V. Knight, J. W. Draper, E. A. Brady and C. A. Attmore, *Antibiot. Chemother.* **2**, 615 (1952).
1343. S. Belman, *Anal. Chim. Acta*, **29**, 120 (1963).
1344. S. Murakami and J. Tsurugi, *J. Rubber Ind. Japan*, **34**, 253 (1961); through *Chem. Abstr.* **56**, 2542d (1962).

1345. E. E. Potapov, I. A. Tutorskii, I. D. Khodzhaeva and B. A. Dogadkin, *Sov. Rubber Technol.* **24**, No. 12, 19 (1965).
1346. H. Nitschmann and H. Hadron, *Helv. Chim. Acta*, **26**, 1075 (1943); **27**, 299 (1944).
1347. W. R. Middlebrook, *Biochem. J.* **44**, 17 (1949).
1348. E. C. Miller and J. A. Miller, *Cancer Res.* **7**, 468 (1947).
1349. E. C. Miller, J. A. Miller, R. W. Sapp and G. M. Weber, *Cancer Res.* **9**, 336 (1949).
1350. I. Taki and T. Miyaji, *Gann.* **41**, 194 (1950).
1351. H. Nishioka, *Mutation Res.* **17**, 261 (1973).
1352. D. French and J. Edsall, in M. L. Anson and J. Edsall, Eds, "Advances in Protein Chemistry", Vol. 2, Academic Press, New York, 1945, pp. 278–335.
1353. Y. A. Siomin, E. N. Kolomiytceva and A. M. Poverenny, *Mol. Biol.* (U.S.S.R.), **8**, 276 (1974).
1354. K. Ogura and G. Tsuchihashi, *Tetrahedron Lett.* 3151 (1971).
1355. W. R. Middlebrook and H. Phillips, *Biochem. J.* **36**, 294, 428 (1942).
1356. R. S. Lane, A. Shapley and E. E. Dekker, *Biochemistry*, **10**, 1353 (1971).
1357. V. Maitra and E. E. Dekker, *J. Biol. Chem.* **239**, 1485 (1964).
1358. E. Adams, *Methods Enzymol.* **17B**, 280 (1971).
1359. R. D. Kobes and E. E. Dekker, *J. Biol. Chem.* **244**, 1919 (1969).
1360. H. Nishihara and E. E. Dekker, *J. Biol. Chem.* **247**, 5079 (1972).
1361. B. A. Hansen, R. S. Lane and E. E. Dekker, *J. Biol. Chem.* **249**, 4891 (1974).
1362. M. S. Chen and L. Schirch, *J. Biol. Chem.* **248**, 3631 (1973).
1363. W. Weinberger, *Ind. Eng. Chem., Anal. Ed.* **3**, 365 (1931).
1364. R. D. Stewart *et al.*, *Science*, **176**, 295 (1972).
1365. R. S. Ratney, D. H. Wegman and H. B. Elkins, *Arch. Env. Health*, **28**, 223 (1974).
1366. G. G. Fodor, D. Prajsnar and H. Schlipkoter, *Staub*, **33**, 260 (1973).
1367. R. D. Stewart and C. L. Hake, *J. Am. Med. Assoc.* **235**, 398 (1976).
1368. Anonymous, *Chemical Week*, 65 (Oct. 20, 1971).
1369. L. A. Heppel and V. T. Porterfield, *J. Biol. Chem.* **176**, 763 (1948).
1370. C. N. Frantz and H. V. Malling, *Mutation Res.* **31**, 365 (1975).
1371. B. N. Ames *et al.*, *Proc. Nat. Acad. Sci. USA*, **70**, 782, 2281 (1973).
1372. H. V. Malling and C. N. Frantz, *Env. Health Perspectives*, **6**, 71 (1971).
1373. H. Sutton, "Mutagenic Effects of Environmental Contaminants", Academic Press, London, 1972.
1374. M. Y. Feldman, in J. Y. Davidson and W. E. Cohn, Eds, "Progress in Nucleic Acid Research and Molecular Biology", Vol. 13, Academic Press, pp. 1–49.
1375. T. Alderson, *Nature*, **215**, 1281 (1967).
1376. W. E. Ratnayake, *Mutation Res.* **9**, 71 (1970).
1377. H. Slizynska, *Proc. Roy. Soc.* (*Edinburgh*), **B66**, 288 (1957).
1378. R. Chanet, C. Izard and E. Moustacchi, *Mutation Res.* **33**, 179 (1975).
1379. T. Alderson, *Nature*, **191**, 251 (1961).
1380. H. Nafei and C. Auerbach, *Z. Verebungsl.* **95**, 351 (1964).
1381. R. I. Salganik, *Biol. Zbl.* **91**, 49 (1972).
1382. F. H. Sobels and H. van Steenis, *Nature*, **179**, 29 (1956).
1383. L. L. Hsu and A. J. Mandell, *Life Sci.* **14**, 877 (1974).
1384. L. A. Ordonez and F. Caraballo, *Psychopharmacol. Commun.* **1**, 253 (1975).
1385. L. A. Ordonez and R. J. Wurtman, *J. Neurochem.* **21**, 1447 (1973).
1386. E. Meller, H. Rosengarten, A. J. Friedhoff, R. D. Stebbins and R. Silber, *Science*, **187**, 171 (1975).
1387. E. G. Burton and H. J. Sallach, *Arch. Biochem. Biophys.* **166**, 483 (1975).
1388. A. J. Mandell, S. K. Knapp and L. L. Hsu, *Life Sci.* **14**, 1 (1974).

1389. H. M. Katzen and J. M. Buchanan, *J. Biol. Chem.* **240**, 825 (1965).
1390. L. L. Hsu and A. J. Mandell, *J. Neurochem.* **24**, 631 (1975).
1391. P. Laduron, *Nature New Biol.* **238**, 212 (1972).
1392. S. P. Banerjee and S. H. Snyder, *Science*, **182**, 74 (1973).
1393. L. Hsu and A. J. Mandell, *Life Sci.* **13**, 847 (1973).
1394. J. Leysen and P. Laduron, *FEBS Letters*, **47**, 299 (1974).
1395. L. R. Mandell, A. Rosegay, R. W. Walker, W. J. A. Vanden Heuvel and J. Rokach, *Science*, **186**, 741 (1974).
1396. K. O. Donaldson and J. C. Keresztesy, *Biochem. Biophys. Res. Commun.* **5**, 289 (1961).
1397. G. Cohen, C. Mytilineou and R. E. Barrett, *Science*, **175**, 1269 (1972).
1398. K. M. Taylor and S. H. Snyder, *J. Pharm Exp. Ther.* **173**, 619 (1971).
1399. M. Sandler, S. B. Carter, K. R. Hunter and G. M. Stern, *Nature*, **241**, 439 (1973).
1400. P. Laduron and J. Leysen, *Biochem. Pharmacol.* **24**, 929 (1975).
1401. A. C. Collins, J. L. Cashaw and V. E. Davis, *Biochem. Pharmacol.* **22**, 2337 (1973).
1402. R. Heikkila, G. Cohen and D. Dembee, *J. Pharmacol. Exp. Ther.* **179**, 250 (1971).
1403. R. J. Wyatt, E. Erdelyi, J. R. DoAmaral, G. R. Elliott, J. Renson and J. D. Barchas, *Science*, **187**, 853 (1975).
1404. K. O. Donaldson and J. C. Keresztesy, *J. Biol. Chem.* **234**, 3235 (1959).
1405. J. D. Barchas, G. R. Elliott, J. DoAmaral, E. Erdelyi, S. O'Connor, M. Bowden, H. K. H. Brodie, P. Berger, J. Renson, and R. J. Wyatt, *Arch. Gen. Psychiatry*, **31**, 862 (1974).
1406. L. Z. Meduna, *Ges. Neur. Psych.* **152**, 235 (1935).
1407. E. H. Reynolds, *Lancet*, **i**, 398 (1968).
1408. E. Slater, A. W. Beard and E. Glithero, *Brit. J. Psychiat.* **109**, 95 (1963).
1409. E. H. Reynolds, I. Chanarin, G. Milner and D. M. Mathews, *Epilepsia*, **7**, 261 (1966).
1410. E. H. Reynolds, *Brit. J. Psychiat.* **113**, 911 (1967).
1411. L. T. Ch'ien, C. L. Krumdieck, C. W. Scott and C. E. Butterworth, *Am. J. Clin. Nutrition*, **28**, 51 (1975).
1411a. R. N. Levi and S. Waxman, *Lancet*, **11**, 11 (1975).
1411b. N. C. Rassidakis, M. Kelopouris and S. Fox, *Int. Ment. Hlth. Res. News Lett.* **13**, 6 (1971).
1411c. J. Katz, S. Kunofsky, R. E. Patton and N. C. Attaway, *Cancer*, **20**, 2194 (1967).
1411d. R. Lindelius and D. W. K. Kay, *Acta Psychiat. Scand.* **49**, 315 (1973).
1412. K. Fuxe and G. Jonsson, *J. Histochem. Cytochem.* **21**, 293 (1973).
1413. H. Corrodi and G. Jonsson, *J. Histochem. Cytochem.* **15**, 65 (1967).
1414. A. Bjorklund and U. Stenevi, *J. Histochem. Cytochem.* **18**, 794 (1970).
1415. F. W. D. Rost, *Medical Biol.* **52**, 73 (1974).
1416. H. Corrodi and G. Jonsson, *Helv. Chim. Acta*, **49**, 798 (1966).
1417. W. R. Martin, J. W. Sloan, S. T. Christian and T. H. Clements, *Psychopharmacologia*, **24**, 331 (1972).
1418. G. Jonsson and M. Sandler, *Histochemie*, **17**, 207 (1969).
1419. R. Hakanson and F. Sundler, *Biochem. Pharmacol.* **20**, 3225 (1971).
1420. R. Hakanson, A. K. Sjoberg and F. Sundler, *Histochemie*, **28**, 367 (1971).
1421. W. D. Denckla and H. K. Dewey, *J. Lab. Clin. Med.* **69**, 160 (1967).
1422. B. W. Gay, Jr., R. C. Noonan, P. L. Hanst and J. J. Bufalini, in V. R. Deitz, Ed., "Removal of Trace Contaminants from the Air", ACS Symposium Series, Number 17, 1975, pp. 132–151.
1423. A. P. Altshuller and I. R. Cohen, *Intern. J. Air Water Poll.* **8**, 611 (1964).

1424. S. W. Nicksic, J. Harkins and B. A. Fries, *J. Air Poll. Control Assoc.* **14**, 158 (1964).
1425. A. P. Altshuller, S. L. Kopczynski, W. A. Lonneman, T. L. Becker and R. Slater, *Environ. Sci. Technol.* **1**, 899 (1967).
1426. A. P. Altshuller, S. L. Kopczynski, W. A. Lonneman, F. D. Sutterfield and D. L. Wilson, *Environ. Sci. Technol.* **4**, 44 (1970).
1427. J. Bergerman and J. S. Eliot, *Anal. Chem.* **27**, 1014 (1955).
1428. V. Anger and G. Fischer, *Mikrochim. Acta*, 592 (1960).
1429. T. R. Darling and C. S. Foote, *J. Am. Chem. Soc.* **96**, 1625 (1974).
1430. P. Lechten, A. Yekta and N. J. Turro, *J. Am. Chem. Soc.* **95**, 3027 (1973).
1431. H. Steinmetzer, A. Yekta and N. J. Turro, *J. Am. Chem. Soc.* **96**, 282 (1974).
1432. T. Wilson, M. E. Landis, A. L. Baumstark and P. D. Bartlett, *J. Am. Chem. Soc.* **95**, 4765 (1973).
1433. N. J. Turro and P. Lechtken, *J. Am. Chem. Soc.* **94**, 2886 (1972).
1434. N. J. Turro and P. Lechtken, *Pure Appl. Chem.* **33**, 363 (1973).
1435. N. J. Turro, P. Lechtken, N. E. Schore, G. Schuster, H. Steinmetzer and A. Yekta, *Accounts Chem. Res.* **4**, 97 (1974).
1436. N. C. Yang and R. V. Carr, *Tetrahedron Letters*, 5143 (1974).
1437. R. Criegee, *Angew. Chem., Internat. Edit.* **14**, 745 (1975).
1438. A. Rieche, R. Meister and H. Sauthoff, *Ann.* **553**, 246 (1942).
1439. R. Criegee and P. Gunther, *Chem. Ber.* **96**, 1546 (1963).
1440. R. J. Cvetanovic and L. C. Doyl, *Can. J. Chem.* **35**, 605 (1957).
1441. R. L. Settene, G. L. Parks and G. L. K. Hunter, *J. Org. Chem.* **29**, 616 (1964).
1442. F. W. McLafferty, "Mass Spectrometry of Organic Ions", Academic Press, New York, 1963.
1443. N. I. Shuikin and V. V. An., *Izv. Akad. Nauk. SSSR, Otd. Khim. Nauk.*, 1508 (1960); through *Chem. Abstr.* **55**, 524 (1961).
1444. L. Malaprade, *Bull. Soc. Chim. France*, **43**, 683 (1928).
1445. L. Malaprade, *Compt. Rend.* **186**, 382 (1928).
1446. L. Fishbein, W. G. Flamm and H. L. Falk, "Chemical Mutagens", Academic Press, New York, 1970, p. 215.
1447. R. K. Boutwell, N. H. Colburn and C. C. Muckerman, *Ann. N. Y. Acad. Sci.* **163**, 751 (1969).
1448. F. Dickens, *Brit. Med. Bull.* **20**, 96 (1964).
1449. C. E. Searle, *Brit. J. Cancer*, **15**, 804 (1961).
1450. L. J. Lilly, *Nature*, **207**, 433 (1965).
1451. H. H. Smith and A. M. Srb, *Science*, **114**, 490 (1951).
1452. C. P. Swanson and T. Merz, *Science*, **129**, 1364 (1959).
1453. R. W. Kaplan, *Naturwissenschaften*, **49**, 457 (1962).
1454. F. Mukai, S. Belman, W. Troll and I. Hawryluk, *Proc. Am. Assoc. Cancer Res.* **8**, 49 (1967).
1455. F. J. C. Roe and M. H. Salaman, *Brit. J. Cancer*, **9**, 177 (1955).
1456. A. L. Walpole, D. C. Roberts, F. L. Rose, J. A. Hendry and R. F. Homer, *Brit. J. Pharmacol.* **9**, 306 (1954).
1457. E. D. Palmes, L. Orris and N. Nelson, *Am. Ind. Hyg. Assoc. J.* **23**, 257 (1962).
1458. H. H. Smith and T. A. Lofty, *Am. J. Botany*, **42**, 750 (1955).
1459. N. H. Colburn and R. K. Boutwell, *Cancer Res.* **28**, 642 (1968).
1460. E. Koenigs, K. Kohler and K. Blindow, *Ber.* **58B**, 933 (1925).
1461. J. Epstein, R. W. Rosenthal and R. J. Ess, *Anal. Chem.* **27**, 1435 (1955).
1462. R. Preussmann, H. Schneider and F. Epple, *Arzneimittel-Forsch.* **19**, 1059 (1969).

1463. M. T. Fayet, M. G. Petermann, J. Fontaine, J. Terre and M. Roumiantzeff, *Ann. Inst. Pasteur*, **112**, 65 (1967).
1464. J. Pellerin and J.-F. Letavernier, *Ann. Pharm. Franc.* **32**, 535 (1974).
1465. E. L. Jackson, *Organic Reactions*, **2**, 341 (1944).
1466. C. W. M. Adams, *Advances Lipid Res.* **7**, 1 (1969).
1467. A. J. Cain, *Quart. J. Micr. Sci.* **90**, 411 (1949).
1468. C. Litchfield, "Analysis of Triglycerides", Academic Press, New York, 1972.
1469. R. Nicolaysen and A. P. Nygaard, *Scand. J. Clin. Lab. Invest.* **15**, 79 (1963).
1470. A. Randrup, *Scand. J. Clin. Lab. Invest.* **14**, 578 (1962).
1471. G. H. McLellan, *Clin. Chem.* **17**, 535 (1971).
1472. J. Giegel, F. Soloni, E. Trinidad, *et al., Clin. Chem.* **18**, 693 (1972).
1473. J. Erikson and H. G. Biggs, *J. Chem. Educ.* **50**, 631 (1973).
1474. A. L. Levy and C. Keyloun, *Clin. Chem.* **17**, 640 (1971).
1475. R. P. Noble and F. M. Campbell, *Clin. Chem.* **16**, 166 (1970).
1476. P. Edora and R. A. Rockerbie, *Clin. Biochem.* **8**, 5 (1975).
1477. J. Hoeflmayr and R. Fried, *Arzneimittel-Forsch.* **24**, 904 (1974).
1478. G. Moses, E. Olivereo and T. F. Draisey, *Clin. Chem.* **21**, 428 (1975).
1479. P. J. Martin, *Clin. Chim. Acta*, **62**, 79 (1975).
1480. H. G. Biggs, J. M. Erikson and W. R. Moorehead, *Clin. Chem.* **21**, 437 (1975).
1481. S. V. Isay and V. E. Vaskousky, *Izv Sibirsk. Otd. Akad. Nauk SSSR*, **2**, 142 (1969).
1482. Anonymous, *Lab. Management*, 18 (April, 1975).
1483. L. A. Carlson and L. B. Wadstrom, *Clin. Chim. Acta*, **4**, 197 (1959).
1484. W. D. Block and K. J. Jarrett, *Am. J. Med. Technol.* **35**, 1 (1969).
1485. J. Mendez, B. Franklin and H. Gahagen, *Clin. Chem.* **21**, 768 (1975).
1486. D. S. Fredrickson, R. I. Levy and R. S. Lees, *New Engl. J. Med.* **276**, 148 (1967).
1487. D. Steinberg, *Circulation*, **45**, 247 (1972).
1488. E. W. Rice, in R. P. MacDonald, Ed., "Standard Methods in Clinical Chemistry", Vol. 6, Academic Press, New York, 1970, p. 215.
1489. D. S. Fredrickson, R. I. Levy and P. S. Lees, *New Engl. J. Med.* **276**, 94 (1967).
1490. G. Schlierf, G. Weinans, W. Reinheimer and W. Kahlke, *Deut. Med. Wochenschr.* **97**, 1371 (1972).
1491. H. A. I. Newman, E. A. Gordon, D. W. Heggen and M. D. Keller, *Clin. Chem.* **18**, 290 (1972).
1492. W. D. Block and K. J. Jarrett, *Am. J. Med. Technol.* **35**, 93 (1969).
1493. L. Edwards, C. Falkowski and M. E. Chilcote, *Std. Methods Clin. Chem.* **7**, 69 (1972).
1494. R. L. Dryer, in N. W. Tietz, Ed., "Fundamentals of Clinical Chemistry", W. B. Saunders Co., Philadelphia, Pa., 1970, p. 329.
1495. J. Folch, M. Lees and G. H. S. Stanley, *J. Biol. Chem.* **226**, 497 (1957).
1496. D. P. Mertz, *Klin. Wschr.* **51**, 96 (1973).
1497. N. Kageyama, *Clin. Chim. Acta*, **31**, 421 (1971).
1498. Tariff Commission Reports, "Miscellaneous Chemicals", Washington, D.C. 1974.
1499. B. W. Gay, Jr., W. A. Lonneman, K. Bridboard and J. B. Moran, *Annals N.Y. Acad. Sci.* **246**, 286 (1975).
1500. A. P. Altshuller, D. L. Miller and S. F. Sleva, *Anal. Chem.* **33**, 621 (1961).
1501. H. Bartsch, C. Malveille and R. Montesano, *Int. J. Cancer*, **15**, 429 (1975).
1502. C. Malaveille, H. Bartsch, A. Barbin, A. M. Camus, R. Montesano, A. Croisy and P. Jacquignon, *Biochem. Biophys. Res. Commun.* **63**, 363 (1975).
1503. U. Rannug, A. Johansson, C. Ramel and C. A. Wachtmeister, *Ambio.* **3**, 194 (1974).

1504. R. Göthe, C. J. Calleman, L. Ehrenberg and C. A. Wachtmeister, *Ambio.* **3**, 234 (1974).
1505. R. E. Hefner, Jr., P. G. Watanabe and P. J. Gehring, *Ann. N.Y. Acad. Sci.* **246**, 135 (1975).
1506. B. L. Van Duuren, *Ann. N.Y. Acad. Sci.* **246**, 258 (1975).
1507. C. Maltoni, "WHO International Agency for Research on Cancer", International Technical Report No. 74/005. Report on a Working Group on Vinyl Chloride, Lyon, 24, 25 June, 1975.
1508. T. Green and D. E. Hathaway, *Chem.-Biol. Interactions*, **11**, 545 (1975).
1509. J. McCann, V. Simmon, D. Streitwieser and B. N. Ames, *Proc. Nat. Acad. Sci.* **72**, 3190 (1975).
1510. O. Schaumann, *Med. Chem.* **2**, 139 (1934).
1511. D. E. Hathway, in D. E. Hathway, Ed., "Foreign Compound Metabolism in Mammals", Vol. 2, The Chemical Society, London, 1972, p. 247.
1512. M. Lederer, *Angew. Chem.* **71**, 162 (1959).
1513. G. A. Razuvaev and K. S. Minsker, *J. Gen. Chem.*, (*U.S.S.R.*), **28**, 957 (1958).
1514. I. Grigorescu and G. Tiba, *Rev. Chim.* (*Bucharest*), **17**, 499 (1966).
1515. E. S. Reynolds, M. T. Moslen, S. Szabo and R. J. Jaeger, *Res. Commun. Chem. Pathol. Pharmacol.* **12**, 685 (1975).
1516. S. Orrenius, *J. Cell Biol.* **26**, 713 (1965).
1517. T. E. Gram, D. H. Schroeder, D. C. Davis, R. L. Reagan and A. M. Guarino, *Biochem. Pharmacol.* **20**, 2885 (1971).
1518. I. J. Selikoff and E. C. Hammond, Eds, Toxicity of vinyl chloride–polyvinyl chloride, *Annals N.Y. Acad. Sci.* **246**, 4–337 (1975).
1519. H. Heimann and R. Lilis, *Annals N.Y. Acad. Sci.* **246**, 322 (1975).
1520. R. Spirtas, A. J. McMichael, J. Gamble and M. Van Ert, *Am. Ind. Hyg. Assoc. J.* **36**, 779 (1975).
1521. B. Holmberg and G. Molina, *Work Env. Health*, **11**, 138 (1974).
1522. T. J. Haley, *J. Toxicol. Env. Health*, **1**, 47 (1975).
1523. W. Schotter, *Chem. Tech.* (*Leipzig*), **21**, 708 (1969).
1524. A. Ducatman, K. Hirschhorn and I. J. Selikoff, *Mutation Res.* **31**, 163 (1975).
1525. L. B. Thomas, H. Popper, P. D. Berk, I. Selikoff and H. Falk, *New Engl. J. Med.* 292, 3 (1975).
1526. D. A. Schanche, *Today's Health*, **52**, 16, 70 (1974).
1527. M. G. Ott, R. R. Langner and B. B. Holder, *Arch. Env. Health*, **30**, 333 (1975).
1528. I. R. Tabershaw and W. R. Gaffey, *J. Occup. Med.* **16**, 509 (1974).
1529. C. Maltoni, *Ambio*, **4**, 18 (1975).
1530. Anonymous, *Chem. Eng. News*, 7 (15 March, 1976).
1531. E. S. Gronsberg, *Khim. Prom.* **7**, 30 (1966).
1532. J. Havlicek and O. Samuelson, *J. Chromatog.* **83**, 45 (1973).
1533. B. Carlsson, T. Isaksson and O. Samuelson, *Anal. Chim. Acta*, **43**, 47 (1968).
1534. J. P. Ferris, E. A. Williams, D. E. Nicodem, J. E. Hubbard and G. E. Voecks, *Nature*, **249**, 437 (1974).
1535. J. S. Hubbard, J. P. Hardy and N. H. Horowitz, *Proc. Nat. Acad. Sci. USA*, **68**, 574 (1971).
1536. J. S. Hubbard, J. P. Hardy, G. E. Voecks and E. E. Golub, *J. Mol. Evol.* **2**, 149 (1973).
1537. E. I. Elpiner, "Ultrasound: Physical, Chemical and Biological Effects" (translated by F. L. Sinclair), New York, Consultants Bureau, 371 pp.

SUBJECT INDEX

Since practically every chapter contains data on the absorption spectra of the aldehyde concerned, these spectra are not referenced in the subject index. The Table of Contents and the Subject Index complement each other.

(Throughout this index "aldehyde" is abbreviated to "ald.")

Acetaldehyde, detect. with 2,3-dimethyl-2,3-bis(hydroxylamino)butane, 201
detm. with MBTH, 167
formn from, azoethane, 54
azoxyethane, 54
carbon monoxide, 277
diethylaryltriazenes, 54
diethylnitrosoamine, 78
ethanol, 167
1-(3-methoxy-4-hydroxyphenyl)-propen-3-ol, 133
propylene, 129
tetrahydrofuran, 250
rate of reaction, 203
reaction with, 2,7-dihydroxynapththalene, 100
dopamine, 237, 238
sepn from CH_2O, 167
spectral data, 203
Acetals, hydrolysis, 213
basicity, 214
Acetate, nitrosamine formn, 167
Acetic acid, 93, 94, 96, 97, 100, 101, 107, 108, 137, 197, 220, 245, 261, 266
extractant, 82
formn from cis-ocimene, 29
purifn, 141
solvent, 22, 28, 30, 32, 43, 55, 215
Acetic anhydride, 5
detm. of cholesterol, 61
reaction with, methionine sulfoxide, 140, 141
methyl sulfoxides, 140, 141
Acetoaminonitrile, anticarc., 74
effect on, dimethylnitrosamine, 75
nucleic acid methylation, 75
rats, 74
Acetone, 108, 111, 132, 153, 154, 228, 251, 256, 257, 259
extn of air particles, 28
formn from, isobutene, 25
cis-ocimene, 29
photooxidn, 26, 27
precursor of CH_2O, 26, 27
yield of CH_2O, 26, 27
Acetonitrile, 259
N-Acetoxy-N-2-acetylaminofluorene, effect on DNA-CH_2O reaction, 211
Acetoxymethyl methyl sulfide, precursor of CH_2O, 20
Acetoxymethylsulfides, formn from methyl sulfoxides, 140
structure, 140
Acetylacetone, *see* 2,4-pentanedione
2-Acetylamino-2-deoxy-D-glucose diethyl dithioacetal, oxidation with periodate, 2
precursor of CH_2O, 2
2-Acetylaminofluorene, *see* N-2-fluorenylacetamide
Acetylation, dimethylsulfoxide, 5
methionine hydantoin sulfoxide, 10
methionine sulfoxide, 10
N-Acetyl-2-(2-chloroethyl)cysteine, formn from vinyl chloride, 272
Acetylcholine, precarc., 186
N-Acetyl-D-glucosamine, detm. with MBTH, 34
location, 109
oxidn with periodate, 2
precursor of CH_2O, 2, 34, 109
spectral data, 34
N-Acetylhexosamines, oxidn with periodate, 2
precursor of CH_2O, 2

Acetylmethadol, deacetylation, 156
N-demethylation, 156
effect on N-demethylase, 144
metabolism, 155, 156
eqn, 156
precursor of, dinoracetylmethadol, 156
formald., 144, 155, 156
noracetylmethadol, 156
structure, 156
treatment of opiate dependence, 155
N-Acetylneuraminic acid, 111
detm. with, MBTH, 111
2,4-pentanedione, 34
distribution, 111
oxidn, eqn, 110
with periodate, 2
precursor of, carbohydrate ald., 110
formald., 2, 34, 110
structure, 110
Acid, hydrolysis, 2–14
Acidolysis, guaiacylglycerol-β-(2-methoxyphenyl) ether, 59
Acrolein, collection, 275
detm. in air, 275
detm. with CA, 275
lachrymator, 128, 196
precursor of CH_2O, 275
reaction with Schiff's reagent, 223
Acrylamide, detm. with CA, 31
dimedone, 31, 34
ethyl acetoacetate, 30
J-acid, 31, 34
MBTH, 30, 34
2,4-pentanedione, 31, 34
oxidn, 2, 3, 30, 31
precursor of CH_2O, 2, 30, 31
spectral data, 30
Acrylic acid, detm. with, CA, 31
dimedone, 31, 32, 34
ethyl acetoacetate, 30
J-acid, 34
MBTH, 30, 34
2,4-pentanedione, 31, 32, 34
oxidn. 2, 30–32
precursor of CH_2O, 2, 15, 30–32
spectral data, 30, 34
yield of CH_2O, 15
Acrylonitrile, oxidn., 2
precursor of CH_2O, 2, 15
yield of CH_2O, 15
Actinomycete, growth with methanol, 138
Acyl anhydrides, reaction with CH_2O, 205
structure of product, 205
Adduct formation, 5, 11, 14
Adenine, dimers-methylene bridges, 226
reaction with, formald., 209
5-hydroxymethyl BaP, 209
structure of product, 209
Adenocarcinoma, vaginal, 172
Adenosine, detm, with MBTH, 34
oxidn, 2
precursor of CH_2O, 2
spectral data, 34
S-Adenosylmethionine, effect on schizophrenia, 239
formn from methionine and ATP, 192
methyl donor, 236
precursor of, active methyl, 192
formald., 192, 236–238
methanol, 238
reaction with CH_2O, 182
structure, 192, 236
Adenylic acid, reaction with CH_2O, 233
Adonitol, 253
detm. with MBTH, 34
oxidn, 2
precursor of CH_2O, 2, 34
spectral data, 34
ADP, 171
Adrenalectomy, effect on, aminopyrine N-dimethylase, 144, 167, 168
testosterone, 167, 168
Adrenaline, anal. proc., 48
detm. with MBTH, 48
precursor of CH_2O, 21
spectral data, 242
Aflatoxin B_1, carc., 187
detm. with, MBTH, 188
2,4-pentanedione, 188
effect on rat liver, 187
formn of, 2,3-oxide, 188
hemiacetal, 188
oxidn eqn, 189
metabolism, 188
precursor of CH_2O, 189
structure, 189
Aflatoxin B_1 2,3-oxide, detm. with MBTH, 188
formn by, human liver microsomes, 188
rat liver microsomes, 188
structure, 188
ulticarc., 188
Aflatoxin B_2, carc., 187
Aflatoxin B_2a, detm. with MBTH, 188
formn from aflatoxin B_1, 188
isolation from fungi, 188
precursor of, ald., 180
formald., 188
structure, 188
Aflatoxin G, carc., 187
Aflatoxin G_2, carc., 187
Aflatoxin P_1, formn from aflatoxin B_1, 188
isolation from urine, 188

oxidn. eqn., 189
structure, 189
Aflatoxins, 187–189
carcinogens, 81
carc. to, duck, 187
mouse, 187
rainbow trout, 187
rat, 187
contamination of foods, 187
demethylation, 188
metab., 187, 188
precursors of CH_2O, 187, 188
relative carc. activity, 187
ulticarcinogens, 188
Age, effect on, aminopyrine, 146
N-demethylase, 144, 156
dimethylaniline metab., 147
embryo N-demethylase, 51
Aging, role of, formald., 237
malonald., 237
Agricultural land, nitrate, 65
nitrosamines, 65
Ah locus, genetic region, 189
Air, animal room, 68
BaP concn., 275
bis-chloromethyl ether and cancer, 57
carc. of hempa, 154
detm. of acrolein, 275
acrylic acid, 31
allyl alcohol, 31
1-chloro-2,3-epoxypropane, 278
chloroethanol, 278
diethanolamine, 278
ethanolamine, 81
ethylene, 278
ethylene glycol, 278
ethylene oxide, 80, 81, 278
ethylenimine, 81
formald., 31, 196–198, 213, 217–219
1,2-glycols, 107
methacrylic acid, 31
2-methylaminoethanol, 281
1-octene oxide, 80
propylene oxide, 80
styrene oxide, 80
vinyl butyl ether, 275
vinyl chloride, 274, 281
5-vinyl-2-picoline, 275
2-vinylpyridine, 275
formald., 196–198
formate, 87
formn of, epoxides, 80
hydroperoxides, 80
nitrosamines, 64
peroxides, 80
lachrymators, 128, 196
nitrosamine concns., 64
oxidn. of, 1-alkenes, 2
aminopyrine, 3
azoxymethane, 3
diethyl ether, 5
dimethyl ether, 5
ethylene, 6
hydrocarbons, unsaturated, 9
ketones, dialkyl, 9
ketones, methyl, 9
mesitylene, 10
mesityl oxide, 10
styrenes, 13
triethylamine, 14
m-xylene, 15
permissible dose of Me_2NNO, 77
presence of, alkylating agents, 63
methyl n-butyl-nitrosamine, 68
nitrosamines, 64
reaction of CH_2O and dimethylamine, 225
smog formn, 128
solvent vapors, 166
source of CH_2O, 197
stability of nitrosamines, 64
threshold limit for nitrosamines, 76
toxic syntheses, 32
Airborne particulates, aliph. fraction, 28
determ. of, carbohydrates, 278
1,2-glycols, 107, 108, 280
olefins, 28, 30
extn. with, acetone, 28
benzene, 28
cyclohexane, 28
Airflow test, for CH_2O, 219
monitoring air quality, 219
α-Alanine, effect on d-AMP, 210
β-Alanine, effect on d-AMP, 210
Alar, anal. eqn., 124
anal. proc., 124
detm. in apples, 124, 278
detm. with HBT, 124, 278
hydrol., 2, 124, 281
oxidn., 2
precursor of, dimethylhydrazine, 124
formald., 2, 124, 281
succinic acid, 124
structure, 124
Albumin, oxidn. with periodate, 2
precursor of CH_7O, 2
reaction with β-propiolactone, 251
Alcoholic beverages, detm. of methanol, 136, 280
Alcohols, effect on aminopyrine N-demethylase, 144
precursors of formald., 32–45
Aldehydes, anal. eqn., 160

Aldehydes—*continued*
atm. formn., 28
formn. and genotoxicity, 226
formn. from, amino acids, 49, 96
auto exhaust, 23
nitrosamines, 79
trialkylamines, 160
fuchsin reaction products, 199
Alditols, detm. thro CH_2O, 60
detm. with, orcinol, 117
2,4-pentanedione, 117
precursors of, formald., 60, 117
sepn., 117
Aldoheptitols, sepn., 114
Aldohexoses, dehydration, 2
precursor of CH_2O, 2
Aldoses, oxidn. with periodate, 2
precursors of CH_2O, 2
Aldosterone, detm. with 2,4-pentanedione, 47
oxidn., 2
precursor of CH_2O, 2, 19, 47
Alfalfa, detm. of aramite, 278
Algin, 59
Aliphatic amines, degradation to ald., 165
detm. with ninhydrin, 49
precursors of, ald., 49
nitrosamines, 64–67, 160, 162, 185, 186
reaction with 2,4-dinitrochlorobenzene, 165
Aliphatic fraction, air particulates, 28
Alkali, hydrolysis, 2, 3, 6, 7, 9–14, 33, 47, 91, 92, 94, 124, 256, 259, 261–267
Alkanolamines, oxidn. with periodate, 2
precursor of CH_2O, 2
1-Alkene oxides, cocarcinogens, 81
inhibitors of epoxide hydrolase, 81
precursors of CH_2O, 81
1-Alkenes, 23–29
anal. proc., 29
detm. with CA, 29, 34
J-acid, 29
environmental, 21
identifn. with CA, 15
oxidn, 2
precursor of CH_2O, 2, 22
yield of CH_2O, 15
1-Alkenic compounds, anal. proc., 30–32
detm. with, CA, 31
dimedone, 31, 32
ethyl acetoacetate, 30
MBTH, 30
2,4-pentanedione, 31, 32
Alkoxyacetic acids, 102, 103
detm. with β-naphthol, 102, 103
precursors of, formald., 102
glycolic acid, 102
Alkylamines, anal. eqn., 160
detm. with DCPIP, 160
oxidn. to ald., 160
N-Alkylaromatic amines, oxidn. eqn., 164
precursor of ald., 164
Alkylating agents, 63, 120
cycasin, 54
derived from, nitrosoureas, 73, 74
1-phenyl-3-methyltriazenes, 125
β-propiolactone, 252
formn. in brain, 54
genotoxicity, 51
methoxymethyl methanesulfonate, 142
methylazoxymethanol, 54
methylazoxymethyl acetate, 55
methyl iodide, 139
methyl methanesulfonate, 139
presence in, air, 63
combustion products, 63
Alkylation, nucleic acids, 55, 74
Alkylazomethanes, precursor of CH_2O, 53
tautomerism, 53
Alkyl cation, formn. from, azoalkanes, 52
1,2-dialkylhydrazine, 52
1-aryl-3,3-dialkyltriazene, 52
O-Alkylglycerols, oxidn. with periodate, 2
precursor of CH_2O, 2
Alkyl nitrites, photolysis, 242
Alkylperoxy radicals, 28
Allantoic acid, detm. with phenylhydrazine, 113
precursor of, formald., 20, 113
glyoxylic acid, 113
reaction eqn., 113
structure, 113
Allantoin, detm. with phenylhydrazine, 113
formn. from uric acid, 268
precursor of, allantoic acid, 113
formald., 113
glyoxylic acid, 113
reaction eqn., 113
structure, 113
Allende meteorite, paraformald., 213
Allergens, glycoproteins, 104
house dust, 104
proteins, 104
ragweed pollen, 104
structure, 104
Allitol, sepn., 114
Allium, effect of clastogen, 251
Allopregnane-3,11,17,21-tetrol-20-one, oxidn., 254
precursor of CH_2O, 254
Allopregnane-3,17,20,21-tetrol., oxidn., 254
precursor of CH_2O, 254
Allopregnane-3,17,21-triol-11,20-dione, oxidn., 254

precursor of CH_2O, 254
Allotetrahydrocortisol, oxidn. to CH_2O, 131
structure, 132
Allyl acetate, oxidn., 2
precursor of CH_2O, 2
Allyl alcohol, detm. with, CA, 31
dimedone, 31, 34
ethyl acetoacetate, 30, 34
J-acid, 31, 34
MBTH, 30, 34
2,4-pentanedione, 31, 34
oxidn., 2, 31, 32
precursor of CH_2O, 2, 30, 31
spectral data, 30
Allylamine, detm. with, CA, 31
dimedone, 31
ethyl acetoacetate, 30
J-acid, 31
MBTH, 30, 34
2,4-pentanedione, 31
oxidn., 2, 30, 31
precursor of CH_2O, 2, 30, 31
spectral data, 30, 34
Allylbenzene, detmn. with J-acid, 34
fluor. anal., 34
oxidn., 2
precursor of CH_2O, 2, 34
Allyl bromide, oxidn., 2
precursor of CH_2O, 2
Allyl chloride, oxidn., 2
precursor of CH_2O, 2
Allyl compounds, detm. with chromotropic acid, 34
Allyl ethyl ether, oxidn., 2
precursor of CH_2O, 2
N-Allyl-N-methylaniline, detm. with MBTH, 34
spectral data, 34
1-Allyl-2-thiourea, oxidn., 2
precursor of CH_2O, 2
Alpha-polyoxymethylene, *see* α-polyoxymethylene
Alumina, 260, 261, 264
catalyst, 250
column chrom., 267
phospholipid removal, 264
sepn. of Imidan, 280
Alveolar parenchyma, 69
Amberlite 1R-120, 136
Amides, reaction with CH_2O, 205
structure of products, 205
Amidopyrine, *see* aminopyrine
Amines, aromatic, 163–166
nitrosamine formn., 67
N-oxidn., 155
precursors of nitrosamines, 64
rate of oxidn., 155
reaction accelerators, 210
reaction with CH_2O, 205, 211
secondary, 63, 64, 155
sewage, 65
structure of CH_2O products, 205
tertiary, 155
Aminex A-6, sepn., 115
Aminoacetonitrile, effect on Me_2NNO, carcinogenesis, 73
metabolism, 72
toxicity, 73
inhibitory of N-demethylase, 148
2-Aminoacetophenone, detm. with MBTH, 34
spectral data, 34
α-Amino acids, decarboxylation, 277
detm. with ninhydrin, 49
effect on d-AMP, 210
oxidn. eqn., 99
precursors of ald., 99
reaction with CH_2O, 227
role in crosslinkage, 210
stimulators of N-demethylase, 148
1,2-Aminoalkanols, anal. proc., 48
detm. with, CA, 34
MBTH, 34, 48
oxidn., 3
precursors of CH_2O, 3, 47
4-Aminoantipyrine, effect of, N-demethylase, 144
vinyl chloride, 272
formn. from aminopyrine, 152
precursor of CH_2O, 144
structure, 152
o-Aminoazotoluene, carcinogen, 172
2-Amino-2-deoxy-D-galactose, oxidn. with periodate, 3
precursor of CH_2O, 3
2-Amino-2-deoxy-D-glucose, detm. with MBTH, 35
oxidn., 3
precursor of CH_2O, 3, 15
spectral data, 35
2-Amino-2-deoxy-D-glucose diethyl dithioacetal, 3
precursor of CH_2O, 3
2-Aminoethanol, anal. proc., 48
detm. with, acetoacetate, 34
MBTH, 34, 48
oxidn., 3
precursor of CH_2O, 3
spectral data, 34, 48
3-(2-Aminoethyl)indoles, reaction with CH_2O, 208
structure, 208
of product, 208

α-Aminoglutamic γ-semialdehyde, formn. from hydroxylysine, 131
structure, 131
Aminohexoses, detm. in glycoproteins, 278
detm. with 2,4-pentanedione, 278
4-Amino-3-hydrazino-5-mercapto-1,2,4 triazole, *see* Purpald.
2-Amino-2-hydroxymethyl-1,3-propanediol, detm. with CA, 35
oxidn., 3
precursor of CH_2O, 3, 35
4(5)-Aminoimidazole-5(4)-carboxamide, metabolite of DIC, 125
Aminomethyl ketones, detm. with CA, 50
precursor of CH_2O, 50
Aminomethylol compounds, 233
Aminonitrile formn., 231
Aminopentamide, antispasmodic, 162
precursor of, dimethylnitrosamine, 162
formald., 162
o-Aminophenols, nitrosamine formn., 66
p-Aminophenols, nitrosamine formn., 66
Aminopyrine, analgesic, 162, 185
demethylation, 72, 138
effect of age, 146
inhibitors, 144, 145
stimulators, 144
detm. of N-demethylase, 50, 51
detm. with 2,4-pentanedione, 50, 51
effect of, adrenalectomy, 144, 167, 167
carbon disulfide, 166
castration, 167, 168
decreased selenium, 168
N-demethylase, 144
TCDD, 168
vit. E lack, 168
effect on, ethanol, 167
methanol, 167
phenobarbital, 167
hydrol., 11, 16
metab. eqn., 152
oxidn., 3
oxidative demethylation, 166
precarc., 67, 185
precursor of, 4-aminoantipyrine, 152
dimethylnitrosamine, 67, 162
formald., 3, 11, 16, 50, 51, 144, 152, 162, 166–168
4-methylaminoantipyrine, 152
reaction with, N-demethylase, 3, 72
nitrite, 67
structure, 152
yield of dimethylnitrosamine, 67
Aminopyrine N-demethylase, activity, decrease, 72
increase, 51
detm. with, aminopyrine, 51
2,4-pentanedione, 166
effect of, age, 51
ascorbic acid, 51
carbon monoxide, 51
carbon tetrachloride, 166
DDT, 166
lipid peroxidn., 51
TCDD, 156
embryonal activity, guinea pig, 76, 77
human, 76, 77
rat, 76, 77
formn. of CH_2O, 74
inducers, 50
inhibition by nitrite, 74
mouse liver microsomes, 74
Amino resin molding process, 217, 218
Amino sugars, detm. with 2,4-pentanedione, 34
Amitriptyline, precursor of, dimethylnitrosamine, 162
formald., 162
stimulant, 162
Ammonia, 47, 91, 95, 103, 122, 153, 224, 249
Ammonium, acetate, 30, 32, 47, 91, 94, 96, 97, 107, 137, 153, 197, 220, 262, 262, 267, 275
sulfate, 97
d-Amp, effect of amino acids, 210
DL-Amphetamine, N-demethylation, 145
precursor of CH_2O, 144
Amyl alcohol, extractant, 184
Analginium, *see* Dipyrone
Analyzer, three-channel, 275
Angelic acid, oxidn. with periodate, 3
precursor of CH_2O, 3
Angiosarcoma, induction in rats, 271
Anhydro-D-xylo-benzimidazole, oxidn. with periodate, 3
precursor of CH_2O, 3
Aniline, formn. from, 1-phenyl-3-methyltriazene, 125, 126
N-methylaniline, 166
hydroxylation, 138
Aniline hydroxylase, inhibition by nitrite, 74
Animal room, air contamination, 68
Anion exchange chromatography, 275
Anisaldehyde, precursor of CH_2O, 16
reaction with 2,7-dihydroxynaphthalene, 100
o-Anisidine, detm. of, glycoproteins, 112
polysaccharides, 112
Anisoles, demethylation, 189–192
precursors of CH_2O, 20
Anisyl alcohol, detm. with CA, 35
oxidn., 3

precursor of CH_2O, 3
spectral data, 35
Antibacterial, hexamine, 223
Anticarcinogen, DIC, 125
3-methylcholanthrene, 72
Natulan, 151
1-phenyl-3-methyltriazenes, 125
Anticonvulsants, effect on, epilepsy, 239
serum folate, 239
Antifreeze solutions, detm. of glycolic acid, 280
Antihistamines, N-oxidn., 155
Antihyperlipidaemic drugs, 264
Antimony pentafluoride, 79
(Antipyrinylmethylamino)methanesulfonic acid, *see* Dipyrone
Antitoxicant, 3-methylcholanthrene, 172
Aortic valves, fixation of, crotonald., 221
formald., 221
glutarald., 221
Apiole, oxidn. with H_2SO_4, 3
precursor of CH_2O, 3
Apples, detm. of Alar, 124, 178
Aquarium fish, carc. of diethylnitrosamine, 68
Arabinose, detm. with, CA, 35
MBTH, 35
oxidn., 3
precursor of CH_2O, 3
spectral data, 35
Arabitol, detm. in urine, 278
detm. with, CA, 35, 278
dimedone, 35
J-acid, 35
MBTH, 35
2,4-pentanedione, 35, 41
location, 109
oxidn., 3
precursor of CH_2O, 3, 109, 278
Aramite, anal. eqn., 33
detm. in, alfalfa, 278
clover, 278
corn, 278
detm. with, 2,4-pentanedione, 35
phenylhydrazine, 33, 35, 278
hydrol., 3, 33
oxidn., 3, 33
precursor of, ethylene oxide, 33
formald., 3, 33, 278
structure, 15, 33
Aroclor, 254, 272
N-demethylase, inducer of, 183
inhibitor of, 148
liver tissue proliferation, 183
repressor of dimethylnitrosamine demethylase, 183
Aromatic amines, 163–166
Aromatic hydroxylation, 187
Aromatic olefins, 128
Arsenious oxide, 85, 256
Arsenite, reducant, 29, 84–86
Arterenol, anal. proc., 48
detm. with MBTH, 48
Artifacts, in detm. of formald., 196
Arylalkylamines, N-demethylation, 165
effect of pH, 165
Aryl azomethanes, precursor of CH_2O, 53
tautomerism, 53
1-Aryl-3,3-dialkyltriazenes, carcinogens, 121, 125
mutagens, 125
procarcinogens, 52
Aryl hydrocarbon hydroxylase, induction, 189
N-Arylnitrones, precursors of CH_2O, 187
Aryloxyacetic acids, 99
Asarinin, hydrolysis with H_2SO_4, 3
precursor of CH_2O, 3
Ascorbic acid, detm. with CA, 35
effect on, demethylase, 51
lipid peroxidn., 51
oxidn., 3
precursor of CH_2O, 3
Aspartidol, detm. with CA, 44
Aspergillus niger, lactic dehydrogenase, 114
Asthma, embalmers, 213
Atherosclerosis, levels, of, cholesterol, 255
triglyceride, 255
ATP, 169
formn. of S-adenosylmethionine, 192
Auto Analyzer, 94, 115
detm. of, ethylene glycol, 84
formald., 200
glycerol, 265
triglycerides, 265, 267
Auto exhaust, ald. emissions, 138
detm. of, ethylene, 82
formate, 87
formald., 198
formald., 137, 138
formals, 86
irradiation, 23
Auto fuel, source of CH_2O, 197
Automated analysis, alditols, 60
cholesterol, 139
erythritol, 114, 115
ethylene glycol, 93, 114, 115
formald., 129
glucose, 90, 93
glycerol, 93, 114, 115
glycerophosphate, 93
glycine, 95
glycolald., 93, 114, 115
methenamine, 119
mandelate, 119

Automated analysis—*continued*
propane-1,2-diol., 114, 115
sugars, 60
D-threitol, 114, 115
triglycerides, 264–268
xylitol, 114–116
Automobile, ald. emissions, 138
Azoalkanes, carcinogens, 121
procarc., 52
Azo dyes, chromogens, 56
N-demethylase, 183
N-demethylation, 145
precursors of CH_2O, 144
Azoethane, precursor of acetald., 54
teratogen, 53
Azomethanes, 123
anal. proc., 53
detm. with CA, 34, 53
precursors of CH_2O, 3, 20
tautomerism, 3, 123
Azoxyalkanes, carc., 121
procarc., 52
Azoxyethane, precursor of acetald., 53, 54
teratogen, 52
Azoxymethane, carc., 53
hydroxylation, 123
non-teratogenic, 53
oxidn., 3
precursor of CH_2O, 53, 123
structure, 123
Azulenes, reaction with CH_2O, 207
structure, 207
of products, 207

BaA or Benz(a)anthracene
Baboon, carc. of diethylnitrosamine, 68
Bacitracin, detm. with phloroglucinol, 99
Bacon, nitrosamines, 65
Bacteria, cells, 3, 15
flora in gut, 123
growth with methanol, 138
nitrate reductase, 65
nitrates, 65
Baltimore, dimethylnitrosamine, 64
BaP, *see* benzo(a)pyrene
6-$BaPCH_2^+$, 18
Barium, chloride, 103
hydroxide, 14, 135, 184, 258, 259
Barthrin, detm. with CA, 46
Basal ganglia, conversion of L-dopa to dopamine, 191
Basicity, diacyloxymethanes, 214
dialkoxymethanes, 214
effect on nitrosamine formn., 67
Belle, concn. of dimethylnitrosamine, 64
Benadryl, antihistamine, 162
N-demethylation, 151
effect of BaP, 151
precursor of, dimethylnitrosamine, 162
formald., 151, 162
Benzaldehyde, 134
Benzene, 128, 278
effect on, aminopyrine demethylase, 166
cytochrome P450, 166
eluent, 92
extractant, 278
extraction of air particulates, 28
photooxidn., 26, 27
precursor of CH_2O, 26, 27
solvent, 215
TLC, 55
yield of CH_2O, 26, 27
Benzoic acid, 249
Benzo(a)pyrene, atm. concn., 275
effect of cytochrome P-448, 157
effect on aminopyrine N-demethylase, 144
3-hydroxy-, 18
6-hydroxy-, 18
6-hydroxymethyl-, 18
hydroxymethylation, 18
induction of azo dye N-demethylase, 147
inhibition of N-demethylase, 148, 149
metabolism, 18, 175
methylation, 18
4,5-oxide, ulticarc., 81
1,6-quinone, 18
3,6-quinone, 18
reaction with CH_2O, 207, 209
stimulator of N-demethylase, 145
structure, 207
of products, 207, 209
p-Benzoquinone, formn. from, dialkylanilines, 165
p-hydroxybenzyl alcohol, 130
Benzoyl peroxide, oxidant, 12, 223, 248
structure, 223
Benzphetamine, N-demethylase substrate, 156
N-demethylation, 145
detm. of N-demethylase, 50
detm. with 2,4-pentanedione, 50
effect of, cytochrome P-450, 157
oxidn., 3
precursor of CH_2O, 3, 50, 145, 156
structure, 155
suppression of N-demethylation, 156
Benzyl alcohol, precursor of CH_2O, 33
Benzylhydroxylamine, formn. from N-benzyl N-methylhydroxylamine, 186
structure, 186
N-Benzyl-N-methylhydroxylamine, metab. eqn., 186

precursor of, benzylhydroxylamine, 186
N-benzylnitrone, 186
formald., 186
α-phenyl N-methylnitrone, 186
Benzyl nitrite, photolysis, 242, 243
precursor of benzald., 243
peroxyacyl nitrate, 243
N-Benzylnitrone, formn. from, N-benzyl-N-methylhydroxylamine, 186
precursor of CH_2O, 186
structure, 186
Benzyls, 128
Betaine, oxidn., 3
precarc., 186
precursor of, dimethylnitrosamine, 162
formald., 3, 162
trimethylamine, 162
yield of CH_2O, 15
Beverages, nitrosamines in, 65
Biacetyl, precursor of CH_2O, 21
Bidrin, N-demethylation, 145, 157, 158
insecticide, 157
metab. eqn, 158
metab. in rats, 157, 158
precursor of, Bidrin acid, 158
N-demethyl Bidrin, 158
O-demethyl Bidrin, 158
N,N-dimethylacetoacetamide, 158
dimethyl phosphate, 158
formald., 145, 157, 158
hydroxymethyl Bidrin, 158
methyl phosphate, 158
phosphoric acid, 158
structure, 158
teratogen, 157
Bidrin acid, formn. from N-demethylbidrin, 158
hydroxymethylbidrin, 158
precursor of dimethyl phosphate, 158
structure, 158
Bilamide, hydrolysis, 11, 16
precursor of CH_2O, 11, 16
Bile, 123
Bilirubin, 97, 267
detm. with, ArN_2^+, 55, 56
HBT, 56
MBTH, 3
diazotizn., 3
oxidn., 3
precursor of CH_2O, 3, 15, 55, 56
structure, 56
Biological fluids, detm. of methanol, 280
Bioassay, host-mediated, 73, 74
mutation of, *Neurospora crassa*, 73
Salmonella typhimurium, 73
2,2′-Bipyridyl, inhibitor of N-oxidn., 147
Birth defects, vinyl chloride, 274
1,3-Bis(β-chloroethyl)-1-nitrosourea, anticarc., 73
host-mediated assay, 74
hydrol., 73, 74
mutagenic to *S. typhimurium*, 74
precursor of alkylating agent, 73, 74
Bis-chloromethyl ether, anal. proc., 57
detm. with, CA, 58
MBTH, 35, 57
formn. from CH_2O, 206
hydrol., 3
lung carc., 57
permeation standard, 58
precursor of CH_2O, 3, 21, 57, 58
spectral data, 35, 37
4,4′-Bis(dimethylamino)benzophenone, anal. proc., 164, 165
detm. with MBTH, 164
precursor of CH_2O, 164
spectral data, 164
Bis(dimethylamino)methane, *see* N,N,N′,N′-tetramethylmethylenediamine
1,1-Bis-*p*-ethoxyphenyl-2,2-dimethyloxetane, photodec., 246
precursor of, 1,1-di-*p*-chlorophenyl-2,2-dimethylethylene, 246
formald., 246
structure, 246
1,2-Bis(4-hydroxy-3-methoxyphenyl)-1,3-propanediol, disproportionation, 3
precursor of CH_2O, 3
Bismetan, detm. with CA, 46
Bismuthate, oxidatn., 4, 13, 130, 131
Bisulfite, 187
antioxidant, 152
effect on N-demethylation, 152
see Sulfite, Sodium sulfite
Bladder cancer, 68
Blank, problems in detm. of CH_2O precursors, 177, 178, 184
turbidity, 184
Blood, detm. of, ethylene glycol, 85
glucose, 279
glycerol, 279
glycine, 95–98
methanol, 280
sorbitol, 281
trypsin, 281
dimethylnitrosamine, 75
concn., 70
demethylation, 70
half-life, 70
Natulan metabolites, 151
platelets, methylene reductase in, 234
serum, detm. of glycerophosphate, 94

Board manufacturing plants, monitoring for CH_2O, 219
Body fluids, detm. of, ethylene glycol, 279
formate, 279
Bologna, nitrosamines, 65
Boric acid, 136
Borohydride, reductant, 8, 11
Bovine, liver, prepn. of 2-keto-4-hydroxy-glutarate aldolase, 230
milk PCB, 150
Brachydanio rerio, cancer, 68
Brain, alkylation of nucleic acids, 55
carc. of 1-aryl-3,3-dialkyltriazenes, 125
demethylating activity, 150
detm. of tryptamine, 242
tumor in rat, 54
Brij-35, effect on N-demethylase, 143
2-Bromo-1-butene, detm. with MBTH, 35
spectral data, 35
Bromochloromethane, activation, 232
dehalogenation, 232
hydrol. eqn., 232
precursor of formald., 232
4-Bromo-N,N-dimethylaniline, anal. proc., 164, 165
detm. with MBTH, 164
precursor of CH_2O, 164
spectral data, 164
2-Bromo-2-nitropropan-1,3-diol, detm. with 2,4-pentanedione, 47
hydrol., 3
precursor of CH_2O, 3, 47
yield of CH_2O, 47
3-Bromopropene, oxidn., 3
precursor of CH_2O, 3
Bromothymol blue, indicator, 245
Bronchoconstriction, 212
Bronopol, *see* 2-bromo-2-nitropropan-1,3-diol
B. subtillis, mutation by N-methylhydroxyl-amine, 171
Bufotenin, precursor of, dimethylnitrosamine, 162
formald., 162
psychotomimetic, 162
Building construction, formald. release, 218
disadvantages of, fiberboard, 218
particleboard, 218
Buildings, formald. pollution, 218
1,3-Butadiene, 128
atmospheric, 21
detm. with CA, 29
photooxidn., 18
precursor of, acrolein, 18
aliph. ald., 18
formald., 18, 29
yield of, acrolein, 18
aliph. ald., 18
formald., 18
n-Butanal, detect. with 2,3-dimethyl-2,3-bis-(hydroxylamine)butane, 201
formn. from, methylazobutane, 43
1-methyl-2-butylhydrazine, 43
rate of reaction, 203
spectral data, 203
2,3-Butanedione, detm. with CA, 35
oxidn., 3
precursor of CH_2O, 3
spectral data, 35
1,2,4-Butanetriol, detm. with MBTH, 35
oxidn., 3
precursor of CH_2O, 3
spectral data, 35
n-Butanol, eluent, 92
TLC sepn. with, 55
1-Butene, photoxidn., 18
precursor of, aliph. ald., 18
formald., 18
yield of, aliph. ald., 18
formald., 18
trans-2-Butene, photooxid., 18, 26, 27
precursor of, aliph. ald., 18
formald., 18, 26, 27
yield of, aliph. ald., 18
formald., 18, 26, 27
3-Butenenitrile, detm. with CA, 35
oxidn., 3
precursor of CH_2O, 3, 35
spectral data, 35
yield of CH_2O, 35
2-Butene oxide, photolysis, 250
precursor of 2-methylpropanal, 250
structure, 250
1-Buten-3-ol, detm. with, CA, 31
dimedone, 31, 35
ethyl acetoacetate, 30
J-acid, 31, 35
MBTH, 30, 35
2,4-pentanedione, 31, 35
oxidn., 3, 30, 31
precursor of CH_2O, 3, 30, 31
spectral data, 30, 35
3-Buten-1-ol, detm. with, CA, 31
dimedone, 31, 35
ethyl acetoacetate, 30
J-acid, 31, 35
MBTH, 30, 35
2,4-pentanedione, 31, 35
oxidn., 4, 30, 31
precursor of CH_2O, 4, 30, 31
spectral data, 30, 35
tert.-Butylbenzene, photooxidn., 26, 27

precursor of CH_2O, 26, 27
yield of CH_2O, 26, 27
Butylhydroxytoluene, induction of Me_2NNO demethylase, 73
n-Butyraldehyde, *see* n-butanal

CA, *see* chromotropic acid
Caffeine, precursor of CH_2O, 20, 173
structure, 173
Calciferol, detm. with MBTH, 35
oxidn., 4
precursor of CH_2O, 4
spectral data, 35
Calcium, chloride, 59
hydroxide, 85, 86
phospholipid removal, 264
oxalate, 245
sulfate, 245
Calf-thymus DNA, 211
Camphene, detm. with MBTH, 35
oxidn., 4, 11, 16
precursor of CH_2O, 4, 11, 16
spectral data, 35
Cancer, bladder, 68
dimethylnitrosamine, 74, 75, 79
effect of diet, 74
esophagus, 68
factors, 183
fore-stomach, 68
initiation, 54
kidney, 68
liver, 68
lung, 68
memory, 54
nasal sinus, 68
neurogenic, 54
ovary, 68
pathways, 74, 75, 79
subcarc. doses, 182
susceptibility, 182
transplacental, 54
uterine cervix, 66
Captamine, cutaneous depigmenter, 162
precursor of, dimethylnitrosamine, 162
formald., 162
Carbaryl, N-demethylation, 145
precursor of CH_2O, 145
Carbazole, detm. of, cellobionic acid, 276
cellotrionic acid, 276
xylobionic acid, 276
xylotrionic acid, 276
Carbohydrate aldehydes, 105
formn. from, sophoritol, 252
sophorose, 252
structures, 105, 252
Carbohydrates, detm. in air particles, 278
detm. with, CA, 35
MBTH, 35, 278
2,4-pentanedione, 36
HCG, 121
HFSH, 121
inhibitors of N-demethylase, 148
location on, PC, 35
TLC, 35
oxidn., 4
precursor of CH_2O, 4, 121
spectral data, 35
Carbon, adsorbent, 66
collection of, ethylene, 82
vinyl chloride, 281
Carbon dioxide, 272, 268
^{14}C, 139
formn. from ethylene, 23
precursor of CH_2O, 4
redn., 4
Carbon disulfide, effect on, aminopyrine N-demethylase, 144
metab. of aminopyrine, 166
inhibits demethylase in, man, 51
rat, 51
reduces hexobarbital sleeping time, 51
Carbon monoxide, effect on, demethylase, 51
lipid peroxidn., 51
enz. inhibitor, 189
formn. from, chloroacetates, 60
1,1-dichloroethylene, 62
1,2-dichloroethylene, 63
ethylene, 23, 24
glycolic acid, 60, 101
isobutene, 25
tartronic acid, 101
inhibition of, N-demethylase, 146
N-oxidn., 147
irradiation, 276, 277
precursor of, acetald., 277
formald., 276, 277
formic acid, 276, 277
formamide, 276, 277
glycolic acid, 276
hydrogen cyanide, 276
urea, 275
Carbon tetrachloride, 278
atm. residence time, 193
carc., 193
of subcarc. doses, 183
cocarc. with dimethylnitrosamine, 69
effect of, drugs, 171
LD_{95}, 166
protein, 71
effect on, aminopyrine N-demethylase, 72, 144, 166

Carbon tetrachloride—*continued*
cellular necrosis, 72
cytochrome P_{450}, 166
DMN demethylase, 71, 72
rats, 166
hepatotoxicity, 71, 72
inhibitor of N-demethylase, 148
protection against dimethylnitrosamine, 71, 72
TLC sepn., 55
Carbophenothion, *see* Trithion
Carboxyhemoglobin, reaction with CH_2O, 225
Carboxylic acids, reaction with CH_2O, 205
structure of product, 205
Carboxymethylcellulose, detm. in food products, 58
detm. of glycolic acid, 280
detm. with, CA, 59
2,7-naphthalenediol, 36, 59
oxidn. with sulfuric acid, 4
precursor of CH_2O, 4
Carboxymethyloxysuccinate, anal. proc., 102, 103
detm. in, river water, 102
sewage, effluent, 102
detm. with, CA, 102
β-naphthol, 102, 103
fluor. anal., 102
hydrol., 4, 102, 103
precursor of, formald., 4, 102
glycolic acid, 102, 103
structure, 102
trisodium salt, 102
Carcinogen, aflatoxin B_1, 187
aflatoxin B_2, 187
aflatoxin G_1, 187
aflatoxin G_2, 187
aflatoxins, 81
o-aminoazotoluene, 172
aminopyrine and nitrite, 67
aromatic azo dyes, 171
1-aryl-3,3-dialkyltriazenes, 121, 124–226
azoalkanes, 121
azoxyalkanes, 121
azoxymethane, 53
bis-chloromethyl ether, 37
carbon tetrachloride, 183, 193
cycasin, 53, 54, 121
cyclic nitrosamines, 183
1,2-dialkylhydrazines, 121–123
dialkylnitrosamines, 63, 68, 78, 121
diazomethane, 180
DBahA, 146
dibutylnitrosamine, 183
DIC, 125
3,3-dichlorobenzidine, 172
dl-1,2,3,4-diepoxybutane, 81
meso-1,2,3,4-diepoxybutane, 81
1,2,6,7-diepoxyheptane, 81
1,2,5,6-diepoxyhexane, 81
1,2,7,8-diepoxyoctane, 81
1,2,4,5-diepoxypentane, 81
diethylnitrosamine, 68, 69, 75, 183
diethylstilbestrol, 172
N,N-dimethyl-4-aminostilbene, 169
1,1-dimethylhydrazine, 122, 181
1,2-dimethylhydrazine, 121, 122, 181
dimethylnitrosamine, 67–76, 184, 185
N,N-dimethyl-4-phenylazoaniline, 169, 170
di-n-propylnitrosamine, 183
effect on DNA-CH_2O reaction, 211
epoxides, 81
1,2-epoxyhexadecane, 80, 81
3,4-epoxy-6-methyl-cyclohexylmethyl-3,4-epoxy-6-methylcyclohexanecarboxylate, 81
DL-ethionine, 171
l-ethyleneoxy-3,4-epoxycyclohexane, 81
ethylhydrazine, 181
formald., 1
N-2-fluorenylacetamide, 182, 184
glycidald., 81
Hempa, 154
6-hydroxymethyl BaP, 18
isosafrole, 81
methylazoxymethanol, 55
methylazoxymethanol acetate, 55
methyl n-butylnitrosamine, 68
methylhydrazine, 181
methyl iodide, 193
methyl methanesulfonate, 139
methylnitrosourea, 183
N-methyl-4-phenylazoaniline, 169, 225
N-methyl-N-phenylnitrosamine, 67
Natulan, 151
nitrosamines, 67, 68, 77, 160
N-nitroso-N-methylaniline, 185
N-nitrososarcosine, 185
PAH, 81
1-phenyl-3,3-dimethyltriazene, 124
1-phenyl-3-methyltriazenes, 125
placental transmission, 172
postnatal transmission, 172
proximate, 18
1-(3-pyridyl)-3,3-dimethyltriazene, 124
pyrrolizine alkaloids, 81
radiation, 183
safrole, 81
styrene oxide, 81
o-tolidine, 172
transplacental dose, 54

vinylcyclohexene-3-diepoxide, 81
vinylidene chloride, 62
Carcinogenesis, altered by, choline, 143
ethanolamine, 143
folic acid, 143
methionine, 143
vitamin B_{12}, 143
factors, 183
mechanism, colon, 123
transplacental, 172
Carcinogenicity, nitrosamines and man, 77
organotropic, 52
repression by 3-methyl CHOL., 72
subcarc. doses, 182, 183
transplacental, 52
Carcinoid, treatment with mandelamine, 223
Carcinoma, esophagus, 68, 125
forestomach, 125
local, 125
rat esophagus, 77
rat nasal cavity, 154
Carnitine, precarc., 186
Casein, CH_2O product, mutagenicity, 1
reaction with CH_2O, 225
Castor oil, 249
Castration, effect on, aminopyrine, 167, 168
aminopyrine N-demethylase, 144
testosterone, 167, 168
Catalase, 61, 138, 140, 268
inhibitors, mutation enhancement, 81
Catecholamine uptake, inhibitors, 238
Catecholamines, 238
reaction with CH_2O, 239
CC, *see* column chromatography
Celite, 59, 136
CC sepn., 92
Cell, ana-telophase stage, 195
differentiation, 143
microspectrophotofluorimetry, 242
regulation, 143
Cellobionic acid, detm. with, carbazole, 276
chromic acid, 276
periodate, 276
precursor of CH_2O, 276
sepn., 276
Cellobiose, oxidn. with periodate, 4
precursor of CH_2O, 4
Cellotrionic acid, detm. with, carbazole, 276
chromic acid, 276
periodate, 276
precursor of CH_2O, 276
sepn., 276
Cellulose, 59, 60
anal. proc., 59, 60
automated analysis, 60
detm. of ald. endgroups, 59
hydrol., 4, 59
oxidn., 4
precursor of CH_7O,4
Cellulose acetate formal, detm. with pararosaniline, 36
hydrol., 4
precursor of CH_2O, 4
Cellulose formal, detm. with pararosaniline, 36
hydrol., 4
precursor of CH_2O, 4
Centrilobular necrosis, effect of Me_2NNO, 77
Cephalosporins, acyloxymethyl derivs, 216
Cerebrospinal fluid, detm. of methanol, 136
Cetylpyridinium chloride, pptn. of sulfated hydrocolloids, 59
Cetyl trimethylammonium bromide, effect on N-demethylase, 143
Charcoal, collection of vinyl chloride, 274
Chavicine, detm. with gallic acid, 46
Chemiluminescence, 28
detm. of CH_2O, 203, 204
Cherries, detm. of dimethoate, 278
Chicken, azo dye and liver protein, 226
Chipmunk, azo dye and liver protein, 226
Chloramine T, oxidant, 8, 96, 97
structure, 97
Chloramphenicol, decarboxylation, 4
detm. with chromotropic acid, 36
hydrolysis, 4
oxidn. with periodate, 4
precursor of CH_2O, 4
Chlorocyclizine, N-demethylation, 145
precursor of CH_2O, 145
Chloride, nitrosamine formn., 67
Chloroacetaldehyde, formn. from vinyl chloride, 272
mutagenicity, 273
Chloroacetic acid, 272
detm. with J-acid, 36
fluor. anal., 36
formn. from vinyl chloride, 272
hydrol., 4
precursor of, carbon monoxide, 60
formald., 4, 36, 60
glycolate, 60
Chloroacetyl chloride, formn. from vinylidene chloride, 62
1-Chloro-2,3-epoxypropane, detm. in air, 278
detm. limit, 36
detm. with CA, 36, 278
hydrol., 4
oxidn., 4
precursor of CH_2O, 4, 278
2-Chloroethanol, detm. in air, 278
detm. with, 2,4-pentanedione, 36, 278
phenylhydrazine, 33

2-Chloroethanol—*continued*
hydrol., 4
oxidn., 4
precursor of CH_2O, 4, 21, 33, 36, 278
S-(2-Chloroethyl)cysteine, formn. from vinyl chloride, 272
Chloroform, 94, 220, 224, 256–258, 264, 266
extractant, 66, 249, 278
Chlorogenic acid, anal. proc., 132
coffee constituent, 66
detm. with, chromotropic acid, 31
dimedone, 31
ethyl acetoacetate, 30
J-acid, 31
MBTH, 30, 36, 132
2,4-pentanedione, 31
nitrosamine formn., 66
oxidn., 4, 30, 31, 132
precursor of CH_2O, 4, 30, 31, 132
spectral data, 30, 36, 132
Chlorohydrins, formn. from epoxides, 80
p-Chloromercuribenzoate, effect on aminopyrine N-demethylase, 144
p-Chloro-N-methylaniline, N-demethylation, 145, 169, 185
precursor of CH_2O, 145, 169
Chloromethyl ethers, 56–58
benefit-risk, 57
carcinogens, 56
formn. of exchange resins, 57
interferences in CA detm. of CH_2O, 58
precursor of CH_2O, 57, 58
Chloromethyl methyl ether, anal. proc., 57
detm. in industrial air, 58
detm. with, CA, 58
MBTH, 36, 57
hydrol., 4
precursor of CH_2O, 4, 57, 58
spectral data, 36, 57
2-Chlorophenoxyacetic acid, detm. in water, 278
detm. with, CA, 45, 278
J-acid, 36
phenyl J-acid, 36
hydrol., 4
precursor of CH_2O, 4, 15, 278
spectral data, 36
3-(4-Chlorophenyl)-1-methylurea, enz. demethylation, 60, 61
metabolism eqn., 60
oxidn., 4
precursor of CH_2O, 4, 60
structure, 60
N-Chloroquinone imide, oxidant, 130
Chlorosulfonic acid, 58
(4-Chloro-*o*-tolyl)oxyacetic acid, *see* Methoxone
Chlorphenesin, 1-carbamate, 278
contaminant, 278
detm. with CA, 36, 278
oxidn., 4
precursor of CH_2O, 4, 36, 278
spectral data, 36
Chlorpromazine, antischizophrenic, 159
N-demethylation, 146, 159
metab. eqn., 159
metabolism, 147
neuroleptic, 147
precursor of, demethylated derivs, 147
dimethylnitrosamine, 162
formald., 146, 147, 159
hydroxylated derivs., 147
sulfoxides, 147
structure, 159
tranquilizer, 162
Chlortrimeton, antihistamine, 162
precursor of, dimethylnitrosamine, 162
formald., 162
CHOL, Cholanthrene, *see* 3-methylcholanthrene
Cholangiosarcoma, rat, 187
Cholate, effect on N-demethylase, 143
Δ^4-Cholestenone, formn. from cholesterol, 61
Cholesterol, 264
detm. with, acetic anhydride, 61
ferric chloride, 266
MBTH, 61
2,4-pentanedione, 61, 139
formn. of Δ^4-cholestenone, 61
interferences, 256, 257
levels in atherosclerosis, 255
Cholesterol digitonide, detm. with MBTH, 61
oxidn. with periodate, 61
precursor of CH_2O, 61
Cholesterol esterase, hydrol. of cholesterol esters, 61
Cholesterol esters, anal. proc., 61
enz. detm., 61
hydrol., 61
Cholesterol oxidase, 61
Choline, effect on carc., 143
precarc., 186
Cholinergic receptor, 248
Chromate, oxidant, 10, 165
Chromatography, anion exchange, 275
ion exchange, 35
reversed phase, 259
see also PC, TLC, CC
Chromic acid, detm. of, cellobionic acid, 276
cellotrionic acid, 276
gluconic acid, 276
xylobionic acid, 276

xylonic acid, 276
xylotrionic acid, 276
Chromium trioxide, oxidant, 5, 165
Chromobacter cholinophagum, 3, 15
Chromogen, azo dye, 45, 56
azonium cation, 163
3,5-diacetyl-1,4-dihydrolutidine, 61, 153, 261, 268
dibenzoxanthylium cation, 88, 99, 100, 140, 229
3,5-diethoxycarbonyl-1,4-dihydrolutidine, 30
1,5-diphenylformazan, 89, 90
formazan, 112, 126, 127
anion, 40
cation, 22, 33, 39, 104, 140, 163, 224, 228
free radical zwitterion, 201
Chromosomal aberrations, 81
Allium, 251
Neurospora, 251
Vicia faba, 251
vinyl chloride workers, 273
Chromotropic acid, anal. of industrial air, 31
anal. proc., 199, 213
artifacts in detm. of CH_2O, 196
detm. of, acrolein, 275
acrylic acid, 31
1-alkenes, 29, 34
allyl alcohol, 31
allyl compounds, 34
2-amino-1-alkanols, 34
2-amino-2-hydroymethyl-1,3-propanediol, 34
anisyl alcohol, 35
arabinose, 35
arabitol, 35, 278
ascorbic acid, 35
aspartidol, 45
azomethanes, 35, 53
Barthran, 46
Bismetan, 46
1,3-butadiene, 29
2,3-butanedione, 35
3-butenenitrile, 35
2-butoxyethyl-2,4-D, 278
carbohydrates, 35
carbophenothion, 46
carboxymethylcellulose, 59
chloramphenicol, 36
1-chloro-2,3-epoxypropane, 36, 278
chlorophenoxyacetic acid, 278
chlorphenesin, 36, 278
corticosteroids, 36
cycasin, 55
cyclohexylazomethane, 36
Dazomet, 46
diacyloxymethanes, 214
dialkyloxymethanes, 214
diallyl ether, 37
2,4-dichlorophenoxyacetic acid, 37, 278
diethanolamine, 37, 278
dihydrostreptomycin, 37
dihydroxyacetone, 37
dimethoate, 37, 278
2,2′-dimethyldioxane, 85
1,1-dimethylhydraxine, 37
1,2-dimethylhydrazine, 37
p-dioxane, 85
dipyrone, 198
Dowco, 46, 184
epichlorohydrin, 84, 85
erythritol, 38
ethanolamine, 38, 45, 81
Ethion, 38, 46
ethylene, 38, 82, 278
ethylene glycol, 38, 84–86, 278
ethylene oxide, 38, 80, 81, 85, 279
Ethylguthion, 46
folinic acid, 89
formald., 47, 213, 271
formald. 2,4-DNPH, 39
formald. hydrazone, 39
formald.-tanned collage, 39
formic acid, 39, 87, 279
N-formylaniline, 89
N-formyldiethylaminomalonate, 89
N-formyldiphenylamine, 89
formylhydrazine, 89
N-formylindoline, 89
fructose, 39
galactose, 39
gelsemine, 39
glucosamine, 39
glucose, 39, 139, 279
glycerald., 40
glyceric acid, 40, 279
glycerides, 40, 279
glycerol, 40, 91
glycine, 40, 49, 50, 95, 97, 98, 280
glycolic acid, 40, 100, 280
1,2-glycols, 108
glyoxal, 41
guaiacylglycerol-β-(2-methoxyphenyl) ether, 58
guaran, 112
Guthion, 41, 280
hexamine, 41, 223, 224
hydroxylactone, 42
imidan, 42, 280
3-indoleacetic acid, 42
isoatisine, 42
isoeugenol, 42

Chromotropic acid—*continued*
isoleucinol, 45
isonitrile compounds, 89
isoprene, 29
isosafrole, 42
leucinol, 45
mandelamine, 223, 224
mannitol, 42, 135, 280
methacrylamide, 42
methacrylic acid, 31
methanol, 42, 136–140, 280
methionine hydantoin sulfoxide, 43, 141
methionine sulfoxide, 43, 140
methylalkylhydrazines, 122, 123
2-methylaminoethanol, 43, 281
methylazobutane, 43
methylazoxymethanol, 55
1-methyl-2-butylhydrazine, 43
1-methyl-2-cyclohexylhydrazine, 43
N,N′-methylene-bis-acrylamide, 43
methylene chloride, 43
methylenedioxyphenyl group, 43
1-methyl-2-propylhydrazine, 43
methyl sulfoxides, 140
methyltrithion, 46
Norda Ketal Synergist, 46
oxalate, 244
paraformald., 43, 213
pectin methyl esterase, 136, 282, 283
Phenkapton, 46
phenylalaninol, 45
phenylazomethane, 43
Phorate, 43, 281
pilocarpine, 248, 249
piperettine, 43
piperine, 43, 214, 281
piperonal, 46
piperonylacrylic acid, 46
piperonyl butoxide, 46
piperonylic acid, 46
pivaloyloxymethyl cephaloglycin, 216
pivampicillin, 216
podophyllotoxin, 44, 281
polymethylhydrazines, 122, 123
polyols, 281
propanolamine, 45
β-propiolactone, 252
propylene oxide, 80
quinine, 44
riboflavin, 281
ribose, 44
safrole, 46
Safroxan, 46
serine, 44, 281
Sesamex, 46
sorbitol, 44, 281
styrene, 44
N,N,N′,N′-tetramethylmethylenediamine, 225
tetramethyltetrazene, 122
Thimet, 43
triglycerides, 255–260, 279
1,3,5-trimethylhexahydro-s-triazine, 225
trimethylhydrazine, 45
sym-trioxane, 45
sym-trithiane, 45
tyrosinol, 45
valinol, 45
vincristine sulfate, 89
vinyl butyl ether, 275
vinyl chloride, 45, 281
5-vinyl-2-picoline, 275
2-vinylpyridine, 275
d-xylose, 45
interferences, 29
spectral data, 44, 45
structure, 140, 229
Cider distillate, nitrosamines, 65
Cigarette smoke, nitrosamines, 63
Cinchonidine, detm. with MBTH, 36
precursor of CH_2O, 36
spectral data, 36
Cinnamaldehyde, detm. with MBTH, 36
precursor of CH_2O, 36
spectral data, 36
Cinnamic acid, detm. with MBTH, 36
spectral data, 36
Cirrhosis, effect of dimethylnitrosamine, 77
Citric acid, 119
CL, *see* Chemiluminescence
Clastogen, effect on, *Allium*, 251
Neurospora, 251
Vicia faba, 251
β-propiolactone, 251
vinyl chloride, 273
Clophedianol, antitussive, 162
precursor of, dimethylnitrosamine, 162
formald., 162
Clover, detm. of aramite, 278
CMOS, *see* Carboxymethyloxysuccinate, 102
$^{14}CO_2$, formn. from, diethylnitrosamine, 75
dimethylnitrosamine, 75
CoA, 114
Cobalt, oxidn. catalyst, 177
Cobalt (II), 8
Cobalt (III), oxidant, 68
Cobalt acetate, oxidant, 108
Cobalt (III)ethylenediaminetetraacetate, precursor of CH_2O, 48
Cobaltous salt, oxidant, 106
Cocaine, precursor of CH_2O, 16
Cocarcinogen, 1-alkene oxides, 81

carbon tetrachloride and DMNA, 69
ferric oxide and DENA, 69
styrene oxide, 81
Codeine, N-demethylation, 146
detm. of N-demethylase, 50, 51
detm. with 2,4-pentanedione, 50, 51
oxidn., 4
precursor of CH_2O, 4, 16, 50, 51, 146, 173
Codeine demethylase, activity increase, 51
detm. with codeine, 51
effect of ascorbic acid, 51
carbon monoxide, 51
lipid peroxidn., 51
inhibition, 51
Coffee, catalyst in nitrosamine formn., 66
Cold, effect on aminopyrine N-demethylase, 144
inhibitor of ethylmorphine N-demethylase, 149
Cold trap, collection of nitrosamine, 64
Collagen, fixation of, crotonald., 221
formald., 221
glutarald., 221
Colon, carcinogenesis in man, 123
glucuronidase, 123
rat, 121, 122
Colon cancer, cycasin, 121
1,2-dimethylhydrazine, 121
rats, 121
Column chromatography, alumina, 267
2-butoxyethyl 2,4-D, 278
2,4-dichlorophenoxyacetic acid, 278
DNA hydrolyzate, 77
ethylene oxide, 279
glyceric acid, 279
glycerol, 92
glycine, 96, 98
glycolate, 114
N-glycolylneuraminic acid, 41
imidan, 280
oxalate, 114
Phorate, 281
polyols, 281
serine, 281
sorbitol, 44, 281
triglycerides, 279
Combustion effluents, detm. of methanol, 280
Compound "E" (Reichstein), precursor of CH_2O, 19
Compound *epi* "E", precursor of CH_2O, 19
Compound "S" (Reichstein), precursor of CH_2O, 19
Compound "U" (Reichstein), precursor of CH_2O, 19
Comutagen, catalase inhibitors, 81
hydrogen peroxide, 1, 81
ultraviolet light, 81
Contact dermatitis, 217
Contamination, dimethylamine, 67
Continuous measurement, formald., 197
Coplanarity, effect on enz. inducability, 146
Copper, oxide, 216
sulfate, 85, 86
Copper-lime, lipid removal, 267
Corn, detm. of aramite, 278
Corn oil, solvent, 251
Coronary heart disease, triglycerides, 255
Cortexone, precursor of CH_2O, 19
Corticosteroids, 104
detm. with, CA, 36
ethyl acetoacetate, 36
J-acid, 36
MBTH, 36
2,4-pentanedione, 36
location on, PC, 36
TLC, 36
oxidn., 4
precursor of CH_2O, 4, 36, 45, 131
Corticosterone, 254
precursor of CH_2O, 19
stimulator of N-demethylase, 149
epi-Corticosterone, precursor of CH_2O, 19
Cortisol, oxidn., 4
precursor of CH_2O, 4, 19
epi-Cortisol, precursor of CH_2O, 19
Cortisone, precursor of CH_2O, 19
Cortisone acetate, inhibitor of N-demethylase, 149
Corynebacterium sp., fermentation with 3-O-methylestratrienes, 192
Cosmetic preservatives, anal. proc., 220
detm. of CH_2O, 220, 221
free CH_2O, 220
1-hydroxymethyl-5,5-dimethylhydantoin, 220
methane bis(N,N′(5-ureido-2,4-diketo-tetrahydroimidazole)N,N′-dimethylol), 220
2-nitro-2-bromo-1,3-propanediol, 220
Cosmetics, detm. of CH_2O, 47
Cotinine, N-demethylation, 146
enz. demethylation, 179
formn. from nicotine, 179
precursor of, demethylcotinine, 179
formald., 146
structure, 179
Cotton, formald. content, 217
Cotton leaf, microsomal fraction, 60
Cotton rat, azo dye and liver protein, 226
carc.-mutagen anomaly, 184
Cottonseed, detm. of Guthion, 280
Creatinine, 95

Cross-link, 120, 121
adenine-CH_2O, 226
carc. amine and bipolymer, 226
CH_2O with protein and DNA, 1
histone-DNA, 211
N-methylol groups, 226, 234, 235
nucleotide-amine, 211
protein and DNA, 210
Crotonaldehyde, detm. in, aortic valves, 221
collagen, 221
elastin, 221
detm. with MBTH, 221
Croton oil, promotion, 251
Cumene, photooxidn., 26, 27
precursor CH_2O, 26, 27
yield of CH_2O, 26, 27
Cupric chloride, 216
Curtis Bay, nitrosamines, 65
Cyanide, effect on methylene halides, 232
Cycads, cycasin in, 54, 55
human cancer, 54
presence of macrozamin, 55
Cycasin, 54, 55
alkylating agent, 54
carcinogen, 15, 53, 121
detm. with, CA, 55
MBTH, 55
2,4-pentanedione, 55
extraction, 55
hydrol., 4, 54, 121, 122
oral effect, 54
PC, 55
precursor of, electrophilic cation, 55
formald., 4, 54, 55
presence in cycads, 55
procarc., 54
structure, 15, 54, 123
TLC, 55
transplacental carc., 55
Cyclohexane, 249
extn. of air particulates, 28
photooxidn., 26, 27
precursor of CH_2O, 26, 27
yield of CH_2O, 26, 27
Cyclohexanone, photooxidn., 26, 27
precursor of CH_2O, 26, 27
yield of CH_2O, 26, 27
N′-(2-Cyclohexenylpropionyl)-N-methylurea, formn. from hexobarbital, 178
structure, 178
Cyclohexylazomethane, detm. with CA, 36
hydrol., 4
precursor of CH_2O, 4
spectral data, 36
tautomerism, 4
yield of formald., 36
Cyclopentolate, cyclopegic, 162
mydriatic, 162
precursor of, dimethylnitrosamine, 162
formald., 162
p-Cymene, photooxidn., 26, 27
precursor of CH_2O, 26, 27
yield of CH_2O, 26, 27
Cysteine, 272
detm. with 2,4-pentanedione, 36
effect on methylene halides, 232
interf. in 2,4-pentanedione test, 210
precursor of, CH_2O, 36
thiazolidine-4-carboxylate, 210
reaction with, CH_2O, 206, 210
β-propiolactone, 251
structure of CH_2O product, 206
Cystine, reaction with CH_2O, 229
Cytidine, detm. with MBTH, 36
oxidn., 4
precursor of CH_2O, 4
spectral data, 36
Cytochrome C reductase, 50, 138, 185
Cytochrome P-448, 147, 157
hydroxylation of BaP, 157
Cytochrome P-450, 74, 170, 185, 186
dealkylation, 143, 147, 157, 166, 167
effect of, benzene, 166
carbon tetrachloride, 166
dimethylnitrosamine, 179
methylene chloride, 166
phenobarbital, 166, 167
TCDD, 147, 190
trichloroethylene, 166
vinyl chloride, 272
effect on dimethylnitrosamine, 185
relation to ethylmorphine N-demethylase, 177
stimulation of benzphetamine N-demethylase, 147, 157

2,4-D, *see* 2,4-dichlorophenoxyacetic acid
Dacron, formald. content, 217
Dazomet, detm. with CA, 46
DBacA, stimulation of N-demethylase, 145
DBahA, carcinogen, 146
stimulation of N-demethylase, 145
DDT, 57
effect on aminopyrine N-demethylase, 144, 166
Dealkylation, 190
Deamination, glycine, 7, 8, 96, 280
1,2-Decanediol, oxidn. eqn., 106
precursor of, formald., 106
nonanoic acid, 106
structure, 106

Decarboxylation, amino acids, 277
chloramphenicol, 4
glycine, 7, 8, 48, 49, 96, 277
glyoxylic acid, 8, 277
N-nitrososarcosine, 185
tartronate, 13, 277
1-Decene, detm. with MBTH, 36
oxidn., 5, 11, 16
precursor of CH_2O, 5, 11, 16, 36
spectral data, 36
Decontaminator, β-propiolactone, 251
O-De-ethylase, effect of TCDD, 147
De-ethylation, 53
Dehalogenation, methylenehalides, 231, 232
Dehydration, aldohexoses, 2
11-Dehydrocorticosterone, periodate oxidn., 254
precursor of CH_2O, 19, 254
Demerol, formald. precursor, 173
structure, 173
N-Demethylase, 34, 179
automated detm., 50
effect of TCDD, 147
liver, 50–52, 70
microsomal marker, 50
O-Demethylase, effect of TCDD, 147
N-Demethylation, 143–152, 154–159, 166–170, 232
aminopyrine, 50, 51, 138, 166–168, 272
anal. proc., 70, 171
arylalkylamines, 165
azo dye amines, 168–170
azomethane, 53, 54
azoxymethane, 53, 54
benzphetamine, 50
p-chloro-N-methylaniline, 185
3-(4-chlorophenyl)methylurea, 60, 61
chlorpromazine, 159
codeine, 50, 51
dimethylaminoantipyrine, 272
6-dimethylaminopurine, 147
N,N-dimethylaniline, 165
N,N-dimethyl aromatic amines, 165
dimethylhydrazine, 53
dimethylnitrosamine, 53, 69, 70, 74, 75
effect of vinyl chloride, 272
equation, 53
ethylmorphine, 50, 51, 272
inhibitors, 144–150
methadone, 159
6-methylaminopurine, 147
N-methylaniline, 166
3-phenyl-1-methylureas, 60
puromycin, 147
puromycin aminonucleoside, 147
stimulators, 144–150
Demethylation, 69, 143, 155, 167–171, 175–180, 182, 184–192
aflatoxin B_1, 189
anisoles, 189–192
brain, 150
inhibition, 70
liver, 150
lung, 150
muscle, 150
relation to mutagenicity, 75
spleen, 150
N-Demethylbidrin, formn. from hydroxymethylbidrin, 158
precursor of, bidrin acid, 158
formald., 158
structure, 158
O-Demethyl bidrin, formn. from Bidrin, 158
precursor of methyl phosphate, 158
structure, 158
Demethylcotinine, formn. from, cotinine, 179
nornicotine, 179
structure, 179
Demethylethylmorphine, formn. from ethylmorphine, 176
structure, 176
DENA, *see* diethylnitrosamine
Denaturation, DNA, 211, 212
Densitometry, 46
detm. of, piperine, 43
piperonal, 46
piperonyl butoxide, 46
safrole, 46
Deodorants, vinyl chloride usage, 269
2-Deoxy-D-arabinohexose diethyl dithioacetal, oxidn. with periodate, 5
precursor of CH_2O, 5
2-Deoxy-D-galactose, location, 109
precursor of CH_2O, 109
2-Deoxy-D-glucose, detm. with, dimedone, 36
J-acid, 36
2,4-pentanedione, 36
location, 109
oxidn., 5
precursor of CH_2O, 5, 36, 109
1-Deoxy-2-ketose sugars, 104
2-Deoxy-D-ribose, anal. proc., 61, 62
detm. with, dimedone, 37
J-acid, 37
MBTH, 37, 61, 62
2,4-pentanedione, 37, 61, 62
fluor. anal., 61, 62
location, 109
oxidn., 5
precursor of CH_2O, 5, 37, 61, 62, 109
spectral data, 37
Depolymerization, 1,3,5-trioxane, 215

Deproteinization, 135
  blood, 279–281
  body fluids, 279
  feces, 281
  liver, 279
  plasma, 279, 280
  serum, 279
  tissues, 279
  urine, 280, 281
Desimipramine, N-demethylation, 146
  precursor of CH_2O, 146
Desmethylimipramine, inhibitor of N-oxidn., 147
Desoxycholate, effect on aminopyrine N-demethylase, 144
Detergents, effect on aminopyrine N-demethylase, 144
  stimulator of N-demethylase, 149
Detoxication, alkene oxides, 21
  arene oxides, 21
  carc. epoxides, 81
Dextromethorphan, formald. precursor, 172
  structure, 172
Dextrose, detm. with MBTH, 37
  spectral data, 37
Diacetone alcohol, photooxidn., 26, 27
  precursor of CH_2O, 26, 27
  yield of CH_2O, 26, 27
3,5-Diacetyl-1,4-dihydrolutidine, chromogen, 261, 268
  effect of, rhamnose, 91
    sulfite, 91
    thiosulfate, 91
  fluorogen, 47, 91, 95, 139
Diacyloxymethanes, basicity, 213, 214
  detm. on thin layer plates, 214
  detm. with CA, 214
  precursor of CH_2O, 214
Dialkoxymethanes, basicity, 213, 214
  detm. on thin layer plates, 214
  detm. with CA, 214
  precursor of CH_2O, 214
N,N-Dialkylanilines, formn. from dialkylamines, 5
  oxidn. eqn., 165
  precursor of ald., 165
  reaction with CH_2O, 206
  structure of products, 206
N,N-Dialkyl 2,4-dinitroanilines, formn. from dialkylamines, 165
  oxidn. eqn., 165
  precursors of ald., 165
  structure, 165
1,2-Dialkyhydrazines, carcinogens, 121
  oxidn. eqn., 122
  precursors of aliph. ald., 122
  procarcinogens, 52
Dialkylnitrosamines, cancer of the, esophagus, 68
    liver, 68
  carcinogens, 63, 121
  formn., 63, 64
  hydroxylation, 78
  presence in, cigarette smoke, 63
    liquid smoke, 63
    smoked fish, 63
    wheat flour, 63
  procarcinogens, 52
Diallyl ether, detm. with chromotropic acid, 37
  precursor of CH_2O, 37
  spectral data, 37
1,2-Diamines, precursors, of CH_2O, 47
1,2-Diaminonaphthalene, anal. proc., 201
  detm. of, ald., 201
    formald., 201
1,2-Diaryl-1,3-propanediols, precursor of CH_2O, 58
Diazomethane, carcinogen, 180
  formn. from methylnitrosamine, 180
  Diazotization, bilirubin, 3
  glyoxylic acid, 8
  1-methyl-1-*p*-nitrophenyl-2-methoxydiazonium tetrafluoroborate, 11
  1-quinoline oxide, 11
Dibenamine, inhibitor of N-demethylase, 149
Dibenzalacetone, detm. with MBTH, 37
  spectral data, 37
Dibenzoxanthylium cation, chromogen, 88, 99, 229
  fluorogen, 103
  structure, 88, 99, 100, 103, 140, 229
Dibutylnitrosamine, carc., 183
Di-t-butyloxymethane, anal. proc., 214
  basicity, 214
  detm. with CA, 214
  location on TLC, 214
  precursor of CH_2O, 214
DIC, anticarcinogen, 125
  carc., 125
  metab., 125
  precursor of, 4(5)-aminoimidazole-5(4)-carboximide, 125
    formald., 125
    methyl cation, 125
Dicamba or 3,6-dichlori-*o*-anisic acid, demethylation, 190
3,3′-Dichlorobenzidine, carcinogen, 172
2,2-Dichloro-1,1-difluoroethyl methyl ether, *see* methoxyflurane
Di-(2-chloroethoxy)methane, anal. proc., 214
  basicity, 214
  detm. with CA, 214

location on TLC, 214
precursor of CH_2O, 214
1,1-Dichloroethylene, *see* vinylidene chloride
1,2-Dichloroethylene, form. of ozone, 63
photoxidn., 63
precursor of, carbon monoxide, 63
formic acid, 63
HCl, 63
nitric acid, 63
Dichloromethane, *see* methylene chloride
2,6-Dichlorophenolindophenol, detm. of, primary amines, 160
trimethylamine, 162
2,4-Dichlorophenoxyacetic acid, detm. in, grain, 278
milk, 278
seed, 278
surface waters, 278
water, 278
detm. with, CA, 37, 278
J-acid, 37
phenyl J-acid, 37
hydrol., 5
precursor of CH_2O, 5, 37, 278
spectral data, 37
1,1-Di-p-chlorophenyl-2,2-dimethylethylene, 246
structure, 246
β,β-Dicyano-N,N-dimethyl-4-aminostyrene, anal. proc., 164, 165
detm. with MBTH, 164
precursor of CH_2O, 164
spectral data, 164
Di-N-demethylchlorpromazine, metabolite of chlorpromazine, 147
dl-1,2,3,4-Diepoxybutane, carc., 81
meso-1,2,3,4-Diepoxybutane, carc., 81
1,2,3,4-Diepoxybutane, carcinogen, 81
effect on chromosomes, 81
mutagen, 81
1,2,6,7-Diepoxyheptane, carc., 81
1,2,5,6-Diepoxyhexane, carcinogen, 81
mutagen, 81
1,2,7,8-Diepoxyoctane, carcinogen, 81
mutagen, 81
1,2,4,5-Diepoxypentane, carcinogen, 81
mutagen, 81
Di-(2,3-epoxypropyl) ether, mutagen, 81
Diet, aflatoxin and liver cancer, 187
dose of DENA in rat, 77
effect on carc., 70
rats and liver tumors, 74
Diethanolamine, collection, 278
detm. in air, 278
detm. with, CA, 5, 37, 278
MBTH, 37
oxidn., 5
precursor of CH_2O, 5, 21, 37, 278
spectral data, 37
Diethoxymethane, anal. proc., 214
basicity, 214
detm. with CA, 214
hydrolysis, 5
location on TLC, 214
precursor of CH_2O, 5, 20, 193, 214
purification, 193
standard, 15, 193
Diethylamine, 249
fish products, 65
industrial effluents, 65
photooxidn., 26, 17
precursor of CH_2O, 26, 27
yield of CH_2O, 26, 27
2-Diethylaminoethyl 2,2-diphenyl valerate, *see* SKF 525-A
Diethylaryltriazenes, precursors of acetald., 54
Diethylene glycol, detm. in polyethylene terephthalate, 278
detm. with resorcinol, 106, 107
precursor of CH_2O, 106, 278
Diethyl ether, 256, 257
eluent, 259
photooxidn., 5
precursor of CH_2O, 5
solvent, 215
1,2-Diethylhydrazine, precursor of acetald., 53
teratogen, 53
Diethylnitrosamine, carc., 183
carcinogenic to, aquarium fish, 68
baboon, 68
dog, 68
grass parakeet, 68
guinea pig, 68
hamster, 68
mouse, 68
pig, 68
rabbit, 68
rainbow trout, 68
rat, 68, 75
cocarc. with feric oxide, 69
dose response, 69
effect of, 3-methyl CHOL, 74
phenobarbital, 74
marginal effect dose, 77
non-teratogenic, 63
precursor of acetald., 78
tumors of, esophagus, 69
liver, 69
nasal cavity, 69
Diethylnitrosamine (^{14}C), metab., 76
Diethylstilbestrol, effect on female offspring, 172

Diethylstilbestrol—*continued*
vaginal adenocarcinoma, 172
Diethyl sulfate, carc., 251
mutagen, 251
2,2-(Diethylthio) ethanol, oxidn. with periodate, 5
precursor of CH_2O, 5
Di-(2-fluoroethoxy)methane, anal. proc., 214
basicity, 214
detm. with CA, 214
location on TLC, 214
precursor of CH_2O, 214
Digestive tract, formn. of nitrosamines, 66
tumor induction, 66
Digitonin, pptn of cholesterol, 61
Diglyceride, anal. proc., 91
detm. with 2,4-pentanedione, 91
hydrol., 91
3,4-Dihydro-β-carbolines, 239
detm. with MBTH, 242
formn. from, formald., 242
3-indolylethylamines, 242
20-Dihydro-compound "S", precursor of CH_2O, 19
4,5-Dihydrocorticosterone, periodate oxidn., 254
precursor of CH_2O, 254
5α-Dihydrocortisone, oxidn. to CH_2O, 131
structure, 132
Dihydrodehydrodiconiferyl alcohol, disproportionation, 5
precursor of CH_2O, 5
Dihydrofolate reductase, 1-carbon metab., 152
3,4-Dihydroisoquinolines, 239
formn., 240
structures, 240
Dihydromorphinone, formald. precursor, 173
structure, 173
Dihydronorharman derivative, formn., 241
structure, 241
Dihydrosphingosine, oxidn. with periodate, 5
precursor of CH_2O, 5
Dihydrostreptomycin, detm. with, CA, 37
ethyl acetoacetate, 37
oxidn., 5
precursor of CH_3O, 5, 37
spectral data, 37
Dihydroxyacetone, detm. with, CA, 37, 282
dimedone, 37
J-acid, 37
MBTH, 37
2,4-pentanedione, 37
oxidn., 5, 131, 253
precursor of CH_2O, 5, 37, 253, 282
spectral data, 37
o-Dihydroxybenzenes, nitrosamine formn., 66
p-Dihydroxybenzenes, nitrosamine formn., 66
11β,21-Dihydroxy-18-al-pregn-4-ene,3,20-dione, precursor of CH_2O, 19
7α,14α-Dihydroxy-cortexone, precursor of CH_2O, 19
17α,19-Dihydroxy-cortexone, precursor of CH_2O, 19
Dihydroxydihydrobetulin, oxidn., 254
precursor of CH_2O, 254
Dihydroxydimethyl peroxide, mutagen, 212
structure, 212
17,21-Dihydroxy-20-ketosteroids, detm. with 2,4-pentanedione, 45
oxidn., 131
precursors of CH_2O, 45, 131
structure, 131
2,7-Dihydroxynaphthalene, detm. of, carboxymethylcellulose, 36, 59
glycolic acid, 40, 99, 100, 114, 280
N-glycolylneuraminic acid, 41
glyoxylic acid, 41
inositol, 42
methanol, 42, 280
oxalate, 114
β-propiolactone, 252
RDX, 44
reaction with, acetald., 100
anisald., 100
formald., 100
salicylald., 110
structure, 100
4,5-Dihydroxy-2,7-naphthalenedisulfonate, *see* CA
11α,21-Dihydroxy-pregn-4-ene-3,20-dione, precursor of CH_2O, 19
11β,21-Dihydroxy-pregn-4-ene-3,20-dione, precursor of CH_2O, 19
14α,21-Dihydroxy-pregn-4-ene-3,20-dione, precursor of CH_2O, 19
16α,21-Dihydroxy-pregn-4-ene-3,20-dione, precursor of CH_2O, 19
17α,21-Dihydroxy-pregn-4-ene-3,20-dione, precursor of CH_2O, 19
18,21-Dihydroxy-pregn-4-ene-3,20-dione, precursor of CH_2O, 19
19,21-Dihydroxy-pregn-4-ene-3,20-dione, precursor of CH_2O, 19
17α,21-Dihydroxy-pregn-4-ene-3,11,20-trione, precursor of CH_2O, 19
2,4-Dihydroxypyrimidine 5-aldehyde, formn. from pseudouridine phosphates, 106
structure, 106
2,4-Dihydroxypyrimidine-5-carboxylic acid, 106
structure, 106
6,7-Dihydroxy-1,2,3,4-tetrahydroisoquinoline, formn. from dopamine, 237
structure, 237

Di-isobutyl ketone, photooxidn., 26, 27
precursor of CH_2O, 26, 27
yield of CH_2O, 26, 27
Di-isopropyloxymethane, anal. proc., 214
basicity, 214
detm. with CA, 214
location on TLC, 214
precursor of CH_2O, 214
Dimedone, anal. proc., 31
detm. of, acrylamide, 31, 34
acrylic acid, 31, 34
allyl alcohol, 31, 34
allyl amine, 31
arabitol, 35
1-buten-3-ol, 35
3-buten-1-ol, 31, 35
2-deoxy-D-glucose, 36
2-deoxy-D-ribose, 37
dihydroxyacetone, 37
dulcitol, 38
erythritol, 38
ethylene glycol, 38
eugenol, 31
formald., 193
fructose, 39
glycerol, 40
1,2-glycols, 41, 107
hexitols, 41
mannitol, 42
quinine, 31
safrole, 31
sorbitol, 44
terpenes, 31
vinyl compounds, 31
water soluble olefins, 31
xylitol, 45
fluor. anal., 31, 32, 44, 45
reaction with CH_2O, 205
structure, 205
Dimethoate, detm. in cherries, 278
detm. with CA, 37, 278
hydrolysis, 5
precursor of CH_2O, 5, 37, 278
structure, 15
Dimethoxymethane, anal. proc., 214
basicity, 214
detm. with CA, 214
location on TLC, 214
precursor of CH_2O, 193, 214
purification, 193
standard, 193
N,N-Dimethylacetoacetamide, formn. from Bidrin, 158
structure, 158
N,N-Dimethylalkylamine oxides, adduct formn., 5, 15, 163
detm. with 2,4-pentanedione, 37
precursor of CH_2O, 5, 37, 163
reaction with SO_2, 5, 15
spectral data, 37
Dimethylamine, aniline formn., 5
contamination, 67
fish products, 65
industrial effluents, 65
precursor of, dimethylnitrosamine, 64, 65, 161, 185
formald., 5, 161
methyl cation, 161
methyldiazonium cation, 161
N,N,N′,N′-tetramethylmethylene-diamine, 225
reaction with, 2,4-dinitrochlorobenzene, 5
formald., 225
nitrite, 161
Dimethylamines, oxidn. with H_2O_2, 14
precursor of CH_2O, 14
Dimethylaminoantipyrine, N-demethylation, 272
N,N-Dimethyl-p-aminobenzaldehyde, anal. proc., 164, 165
detm. with MBTH, 37, 164
diazotization, 5
precursor of CH_2O, 5, 20, 164
spectral data, 37, 164
4-Dimethylaminobenzil, anal. proc., 164, 165
detm. with MBTH, 164
precursor of CH_2O, 164
spectral data, 164
N,N-Dimethyl-4-aminobenzylidene 1-amino-pyrene, anal. proc., 164, 165
detm. with MBTH, 37, 164
diazotizn. with MBTH, 5
precursor of CH_2O, 5, 164
spectral data, 37, 164
4-Dimethylaminocinnamaldehyde, anal. proc., 164, 165
detm. with MBTH, 164
precursor of CH_2O, 164
spectral data, 164
2-Dimethylaminoethyl acetate, precarc., 186
6-Dimethylaminopurine, N-demethylation, 146, 147
constitn. of liver RNA, 147
precursor of CH_2O, 146
N,N-Dimethyl-4-aminostilbene, anal. proc., 164, 165
carcinogen, 169
detm. with MBTH, 164
precursor of CH_2O, 164
spectral data, 164
N-Dimethylaminosuccinamic acid, *see* Alar
N,N-Dimethylaniline, anal. eqn., 165
N-demethylation, 146
inhibitors of N-oxidn., 147

N,N,-Dimethylaniline—*continued*
  modes of CH_2O, formn., 147
  monooxygenase (N-oxide forming), 165
  oxidn., 5, 165
  precursor of, *p*-benzoquinone, 165
    N,N-dimethylaniline N-oxide, 165
    formald., 5, 20, 146, 165
  structure, 165
Dimethylaniline N-oxide, aldolase, 166
  N-demethylation, 146
  precursor of CH_2O, 146
Dimethylanilines, diazotizn. of *p*-substitd. derivs., 5
  precursors of CH_2O, 45, 147
7,12-Dimethyl BaA, effect on pregnant mice, 54
  transplacental effect, 54
N,N-Dimethylbenzylamine, N-demethylation, 146
  precursor of CH_2O, 146
2,3-Dimethyl-2,3-bis(hydroxylamine)butane, anal. eqn., 201
  anal. proc., 202
  detm. of, aliph. ald., 201, 202
    formald., 202
  purification, 202
  spot test results, 202
  structure, 201
2,6-Dimethyl-3,5-diacetyl-1,4-dihydropyridine, chromogen, 61, 153
  structure, 153
2,6-Dimethyl-3,5-diethoxycarbonyl-1,4-dihydropyridine, chromogen, 30
  fluorogen, 47
  structure, 30, 47
2,2′-Dimethyl-*p*-dioxane, detm. with CA, 85
  precursor of CH_2O, 85
Dimethyl ether, photooxidation, 5
  precursor of CH_2O, 5
Dimethylethylamine oxide, precursor of CH_2O, 20
Dimethylformamide, 124, 224
2,3-Dimethylglucose, oxidn. with periodate, 17, 253
  precursor of CH_2O, 17, 253
Dimethylglycine, precarc., 186
  precursor of CH_2O., 15
1,1-Dimethylhydrazine, anal. eqn., 122
  carcinogen, 122, 181
  N-demethylation, 181
  detm. with, CA, 37
    2,4-pentanedione, 122
  formn. from, Alar, 124
    dimethylnitrosamine, 64, 78, 180, 181
  precursor of, dimethylnitrosamine, 67
    formald., 5, 37, 122
  rocket fuel, 67
  spectral data, 37
  tautomerism, 5
1,2-Dimethylhydrazine, carcinogen, 121, 122
  detm. with CA, 37
  metabolism, 52, 70
  non-teratogenic, 53
  oxidn., 5, 123
  precursor of CH_2O, 5, 52, 123
  procarcinogen, 52
  spectral data, 37
  structure, 52, 123
  tautomerism, 5, 123
N,N-Dimethylmelamine, anticarc., 120
Dimethyl methylenedisulfate, biol. activity, 195
  precursor of CH_2O, 195
Dimethylnitrosamine, 63–79
  acidic redn., 180
  acute necrosis, 77
  air concns., 64
  analysis, 67
  anal. by, GC-CL,*f*64
    GC-MS-COMP, 64
  anal. through CH_2O, 78, 184
  cancer pathways, 79
  carcinogen, 67, 69, 70, 72–75
  carc. to, guinea pig, 68
    hamster, 68
    mouse, 68, 184
    rabbit, 68
    rainbow trout, 68
    rat, 68, 184
  carc. of subcarc. doses, 182, 184
  centrilobular necrosis, 77
  chemistry of, 67
  cirrhosis, 77
  cocarc. with CCl_4, 69
  concn. in, air, 64
    bologna, 65
    cider distillate, 65
    frankfurters, 65
    herring, 65
    rain puddle, 64
    soil, 64
  contaminant in dimethylamine, 67
  degradation, 181
  N-demethylase, 70, 72
  demethylating activity of, brain, 150
    kidney, 150
    liver, 150
    lung, 150
    muscle, 150
    spleen, 150
  demethylation, inhibitors, 148
    stimulators, 148
  dose-response, 68, 69
  effect of, acetoaminonitrile, 74, 75

aminoacetonitrile, 72
carbon tetrachloride, 71
cofactors on metab., 180
drugs, 71
fat, 74
hepatectomy, 182
3-methyl CHOL., 75
phenobarbital, 75
protein, 71, 74
SKF525A, 73, 74
effect on, cytochrome P-450, 179
hamsters, 183
lipid peroxidn., 179
malonald. formn., 179
Salmonella typhimurium, 46, 73, 75
enz. demethylation, 70–73
formn., 63, 64
formn. from, acetylcholine, 186
aminopyrine, 67, 185
betaine, 186
carnotine, 186
choline, 186
dimethylamine, 65, 162
dimethylaminoethyl acetate, 186
dimethylglycine, 186
1,1-dimethylhydrazine, 67
drugs, 67
neurine, 186
N-nitrososarcosine, 185
oxytetracycline, 67
trimethylamine, 160
trimethylamine oxide, 161–163
genotoxicity, 69
genotoxic pathways, 179–181
half-life in blood, 70, 75
heptatotoxin, 69–72
host-mediated assay, 74
industrial, effluent, 64
leakage, 64
intake from air, 77
kidney tumors, 68, 69
LD_{50}, 71, 72
lethal syntheses, 67
lung tumors, 68, 69
metabolism, 70–79
metab. and kidney tumors, 183
metab. studies, 185
methylation of, DNA, 69
protein, 69, 182
RNA, 69
mutagen, 68–70, 73–75, 233
mutagenesis and, carc., 74–76
formald. formn., 232, 233
mutagenesis to *S. typh.*, 74
non-teratogenic, 53
organotropic effect, 76
oxidn., 5
oxid. demethylation, 180
photoirradiation, 181
physical properties, 67
physiology, 67
precursor of, t-butyl cation, 79
dimethylamine, 79, 181, 182
1,1-dimethylhydrazine, 78, 180
formald., 5, 69–72, 74, 75, 77–79, 148, 180–182, 184, 232, 233
formald. methylamine, 79
formic acid, 180–182
methylamine, 180–182
CH_3^+, 70
$CH_3N_2^+$, 70
methylhydrazine, 180–182
N-methylhydroxylamine, 180–182
precursors, 185, 186
sampling by, cold trap, 64
Tenax GC, 64
sens. to UV, 64
stability, 64
structure, 52
synthesis, 64
toxicity, 70, 71, 73
to man, 77
to mice, 74
Dimethylnitrosamine (^{14}C), effect on rats, 77
metabolism, 75
Dimethylnitrosamine N-demethylase, 74, 75
activity of, C3Hf/A mouse lung, 183
GRS/A mouse lung, 183
hamster bronchus, 183
hamster lung, 183
hamster trachea, 183
human lung, 183
Sprague-Dawley rat lung, 183
amino acid induction, 183
anal. proc., 184
carbohydrate repression, 183
detm. with 2,4-pentanedione, 78, 184
effect of, carbon tetrachloride, 71, 72
3-methyl CHOL, 72, 232, 233
protein, 74
embryonal activity in, guinea pig, 76, 77
human, 76, 77
liver, 76, 77
rat, 76, 77
induction by, butyl hydroxytoluene, 73
3-methyl CHOL, 73
phenobarbital, 73
inhibition by, Aroclor 1254, 183
PAH, 183
phenobarbital, 183
SKF525A, 73
2,2-Dimethyloxetane, photodec., 246
precursor of, acetone, 246

2,2-Dimethyloxetane—*continued*
ethylene, 246
formald., 246
isobutylene, 246
2,3-Dimethyloxirane, *see* 2-butene oxide
N,N-Dimethylphenethylamine, N-demethylation, 148
precursor of CH_2O, 148
N,N-Dimethylphentermane, N-demethyla-ation, 148
precursor of CH_2O, 148
N,N-Dimethyl-4-phenylazoaniline, carc., 169, 170
N-demethylation, 145
effect of, DL-ethionine, 170
phenobarbital, 170
propylene glycol, 171
liver protein binding and carc., 226
precursor of CH_2O, 145, 171
N,2-Dimethyl-4-phenylazoaniline, comparison with ethylmorphine, 169
N-demethylation, 169
detm. with CA, 169
metabolism, 169
precursor of, formald., 169
4-phenylazo-2-methylaniline, 169
N,3-Dimethyl-4-phenylazoaniline, N-demethylation, 145
precursor of CH_2O, 145
N,N-Dimethyl p-phenylenediamine, precursor of CH_2O, 17
N,N-Dimethylphenylpropylamine, N-demethylation, 148
precursor of CH_2O, 148
Dimethyl phosphate, formn. from, Bidrin, 158
bidrin acid, 158
precursor of methyl phosphate, 158
structure, 158
O,O-Dimethyl S-phthalimidomethyl phosphorodithioate, *see* Imidan
2,2-Dimethylpropanoic acid, 216
Dimethyl sulfoxide, 212
acetylation, 5
precursor of CH_2O, 5
Dimethyl terephthalate, detm. in air, 138
detm. with CA, 138
4(5)-3,3-Dimethyl-1-triazenoimidazole-5(4)-carboximide, *see* DIC
N,N-Dimethylurea, metabolism, 159
2,4-Dinitrochlorobenzene, formn. of dialkylanilines, 5
2,4-Dinitrophenylhydrazine, 103, 126
detm. of glycerides, 40
Dinoracetylmethadol, formn. from noracetylmethadol, 156
metab. eqn., 156
structure, 156
Dioxane, 58, 94
detm. with CA, 85
precursor of CH_2O, 85
1,2-Dioxetane, photodec., 247
precursor of CH_2O, 247
structure, 247
2,6-Dioxo-3-carboxyhexanoic acid, formn. from tryptophan, 191
Diphenamide, N-demethylation, 148
precursor of CH_2O, 148
Diphenoxymethane, anal. proc., 214
basicity, 214
detm. with CA, 214
location on TLC, 214
precursor of CH_2O, 214
1,4-Diphenyl-1,3-butadiene, oxidn., 6
precursor of CH_2O, 6
1,1-Diphenylethylene, detm. with MBTH, 37
oxidn., 5
precursor, 5
spectral data, 37
N,N′-Diphenylethylenediamine, detm. with MBTH, 37
spectral data, 37
1,5-Diphenylformazan, chromogen, 89, 90
Diphenylhydantoin, effect on, epilepsy, 239
serum folate, 239
structure, 239
1,9-Diphenyl-1,3,6,8-nonatetraen-5-one, oxidn., 6
precursor of CH_2O, 6
1,8-Diphenyl-1,3,5,7-octatetraene, detm. with MBTH, 37
oxidn., 6
precursor of CH_2O, 6, 37
spectral data, 37
1,5-Diphenyl-3-pentadienone, detm. with MBTH, 37
oxidn., 5
precursor of CH_2O, 5, 37
spectral data, 37
Dipotassium nitrosodisulfonate, oxidant, 130
Di-n-propyl ketone, photooxidn., 26, 27
precursor of CH_2O, 26, 27
yield of CH_2O, 26, 27
Di-n-propylnitrosamine, carc., 183
α,α′-Dipyridyl, stimulator of N-demethylase, 149
Dipyrone, anal. proc., 198
detm. with CA, 198
precursor of formald., 198
Disaccharides, detm. with CA, 46
oxidn., 6
precursor of CH_2O, 6
Dismutation, glyoxylate, 114
Displacement, bilirubin, 3

Disproportionation, 1,2-bis(4-hydroxy-3-methoxyphenyl)-1,3-propanediol, 3
choloracetic acid, 4
chlorophenoxyacetic acids, 4
2,4-dichlorophenoxyacetates, 5, 278
erythrulose, 6
glucose, 7
glycine, 8
glycolald, 8
glycolic acid, 8, 49, 277, 280
glycolic nitrile, 8
glyoxylic acid, 8
guaiacylglycerol-β-(2-anisyl)-ether, 8
4-hydroxy-2-oxobutyrate, 9
3-indoleacetic acid, 9
2-keto-4-hydroxypyruvate, 230
lignin, 10
oxalic acid, 12
oxetanes, 12
penicillin V, 12
phenoxyacetic acids, 12
tartronate, 13
Dissociation equilibrium, 82
Distillation, formald., 49
Distribution coefficients, sugar alcohols, 116
Dithiomethylenes, precursors of CH_2O, 228–230
Dithiothreitol, 187
Di-(2,2,2-trichloroethoxy)methane, anal. proc., 214
basicity, 214
detm. with CA, 214
location on TLC, 214
precursor of CH_2O, 214
Di-(2,2,2-trifluoroethoxy)methane, anal. proc., 214
basicity, 214
detm. with CA, 214
location on TLC, 214
precursor of CH_2O, 214
DMNA, 70, *see* dimethylnitrosamine
DNA, alkylated, 52
binding and tumor initiation, 251
binding with, propylene oxide, 80
β-propiolactone, 251
breaks in *E. coli*, 234
crosslink, to, histone, 211
protein, 1
denaturation, 211
effect of, aminomethylols, 227
formald., 211, 212
N-methylhydrazine, 181
effect on mol. wt., 227
excission repair, 227
hydrolysis, 77
methylation, 69, 70
polymerase I, 227, 233
reaction with CH_2O, 233
single strand breaks, 227
thymine fraction, 77
DNPH, *See* 2,4-dinitrophenylhydrazine
Docosene, detm. with MBTH, 28
Dog, brain, detm. of tryptamine, 242
carc. of diethylnitrosamine, 68
metab. of Natulan, 151
L-Dopa, 237, 238
formn. from 3-O-methyldopa, 191
precursor of dopamine, 191
treatment of Parkinson's disease, 191
Dopamine, 238
formn. of salsolinol, 238
reaction with, acetald., 237
formald., 237
ring formn. eqn., 237
structure, 237
Dose, marginal effect, 77
permissible, 77
Dose response, diethylnitrosamine, 69
dimethylnitrosamine, 68, 69
Dowco 184, detm. with CA, 46
Dowex 1, sepn. of sugar alcohols, 114, 115
Dowex 1x8, 276
Dowex-50, CC of thymine fraction, 77
Dowex-AG-1, 121
Dowex AG2-X8, 114
Dowex AG50-WX8, 180
Dowex 50W-X2, 97, 98
Dowex 50W-X8, chromatography, 60
Drosophila, effect of CH_2O, 234
mutated by, 1,2,3,4-diepoxybutane, 81
formald., 1, 81
hexamethylmelamine, 120
methylazoxymethanol, 55
trimethylolmelamine, 120
replication point, 234
Drosophila melanogaster, effect of CH_2O, 233
Drugs, dimethylamines, 67
nitrosamine precursor, 67
reaction with nitrite, 67
Duck, carc. of aflatoxins, 187
Dulcitol, detm. with, dimedone, 38
J-acid, 38
MBTH, 38
2,4-pentanedione, 38, 41
location, 109
precursor of CH_2O, 38, 41, 109
spectral data, 38
Durable press resins, 217

E. coli, effect of, activated DMNA, 184
N-methylols, 227
mutated by, formald., 1
β-propiolactone, 251

E. coli—continued
vinylidene chloride, 62
see Escherichia coli,
single strand DNA breaks, 227
EDTA, effect on ethylmorphine N-demethylase, 177
increase in demethylase activity, 51
inhibition of, N-demethylase, 149
lipid peroxidn., 50, 51
interference, 317
oxidn., 6
precursor of CH_2O, 6
stimulator of N-demethylase, 149
Elastin, fixation of, crotonald., 221
formald., 221
glutarald., 221
Electrophilic cation, 6-$BaPCH_2^+$, 18
benzenediazonium, 185
methyl, 52, 53, 55, 73, 161
methyldiazonium, 52, 53, 55, 73, 161
Electrophoresis, lipoproteins, 24
Elemicin, oxidn. with H_2SO_4, 6
precursor of CH_2O, 6
Embalmers, asthma, 213
Embalming room air, 213
detm. of CH_2O, 213
Embryo, N-demethylase, 51, 78
effect of age, 51
guinea pig, 76
human, metab. of DMNA, 76
liver, 78
microsomal fractions, 51, 78
Environmental contaminants, contact with small intestine, 175
Enzyme-formaldehyde complexes, 230, 231
Ephedrine, N-demethylation, 148
N-oxidn., 155
precursor of CH_2O, 148
Epichlorohydrin, detm. in industrial air, 84
detm. with CA, 84, 85
hydrol., 6
mutagen, 81
oxidn., 6
precursor of CH_2O, 6, 84, 85
Epidemiology, aflatoxin and liver cancer, 187
Epilepsy, effect of anticonvulsants, 239
relation to schizophrenia, 239
Epinine, metab., eqn., 237
ring formn., 237
structure, 237
Epoxide-S-glutathione transferase, 80
Epoxide hydrolase, 80
detoxication, 81
alkene oxides, 21
arene oxides, 21
effect of, alkene oxides, 81
1,2-epoxydecane, 81
inhibitors, 21, 80
Epoxides, 79–81
conversion to glycols, 80
detection, 94
detoxication, 81
formn. from, aflatoxins, 81
isosafrole, 81
olefins, 80
PAH, 81
pyrrolizine alkaloids, 81
safrole, 81
location, 94
precursors of, ald., 80, 250
formald., 80
prevalence, 80
properties, 79
ring, opening, 250
rearrangement, 250
structures, 80, 250
1,2-Epoxydecane, inhibitor of epoxide hydrolase, 21
1,2-Epoxyhexadecane, carcinogen, 21, 80
formn. from 1-hexadecene, 21, 80
skin carcinoma, 22
3,4-Epoxy-6-methyl-cyclohexylmethyl-3,4-epoxy-6-methylcyclohexanecarboxylate, carc., 81
2,3-Epoxy-1-propanol, oxidn. with periodate, 6
precursor of CH_2O, 6
Ergocristin, reaction with metald., 204
Erythritol, detm. with, CA, 38
dimedone, 38
J-acid, 38
2,4-pentanedione, 38, 114, 115
formn. from, hexitols, 114
polysaccharides, 114
location, 109
oxidn., 6, 253
precursor of CH_2O, 6, 109, 253
sepn., 115
Erythromycin, antibiotic, 162
precursor of, dimethylnitrosamine, 162
formald., 162
Erythrulose, disproportionation, 6
precursor of CH_2O, 6
Escherichia coli, DNA breaks, 234
effect of, formald., 233, 234
UV light, 233
mutated by, 1,2,3,4-diepoxybutane, 81
1,2,5,6-diepoxyhexane, 81
1,2,7,8-diepoxyoctane, 81
1,2,4,5-diepoxypentane, 81
glycidald., 81
vinylcyclohexene-3 diepoxide, 81
prepn. of 2-keto-4-hydroxy-glutarate aldolase, 230

see E. coli
Eserine, TLC sepn., 248
salicylate, 249
Esophagus, cancer, 68
carcinoma, 77, 125
tumors from, diethylnitrosamine, 69
methyl n-butyl nitrosamine, 68
Essential oils, detm. of methanol, 280
Estradiol-3-methylether, precursor of, formald., 192
3-hydroxy compd., 192
Estratrienes, microbial 3-O-demethylation, 192
precursors of CH_2O, 192
Estrone-3-methyl ether, precursor of, formald., 192
3-hydroxy compd., 192
Ethanol, 59, 60, 92, 103, 111, 114, 118, 136, 171, 177, 215–217, 245, 249, 256, 259
effect on aminopyrine N-demethylase, 144
extractant, 66
inhibition of N-demethylase, 145, 146, 149
interference, 136, 137
precursor of acetald., 167
Ethanolamine, detm. with, CA, 38, 44, 81, 97
MBTH, 38
2,4-pentanedione, 38, 97
effect on carcinogenicity, 143
formn. from ethylenimine, 81
oxidn., 6
precursor of CH_2O, 6, 81
spectral data, 38
17α-Ethinyl-estradiol-3-methyl ether, precursor of CH_2O, 192
Ethion, detm. with CA, 11, 16, 38, 46
hydrol., 11
inhibition of N-demethylase, 145
oxidn., 6, 228
precursor of CH_2O, 6, 11, 15, 16, 38, 46, 228, 229
structure, 15, 229
Ethionine, carcinogen, 171
effect on, aminopyrine N-demethylase, 144
carc. of DAB, 170, 171
4-Ethoxybenzoate, monooxygenase, 190
precursor of, formald., 190
4-hydroxybenzoate, 190
7-Ethoxycoumarin O-deethylase, induction, 189
2-Ethoxyethanol, photooxidn., 26, 27
precursor of CH_2O, 26, 27
yield of CH_2O, 26, 27
Ethyl acetate, eluent, 92
TLC sepn. with, 55
Ethylacetoacetate, detm. of,
allyl alcohol, 30, 34
2-aminoethanol, 34
corticosteroids, 36
dihydrostreptomycin, 37
ethylene glycol, 38
formald., 30
1,2-glycols, 41
hexitols, 41
2-nitro-1-alkanols, 47
polar, $R_2C{=}CH_2$, 30
quinine, 30, 44
serine, 44
sorbitol, 44
fluor, anal., 44
reaction with CH_2O, 30
Ethyl acrylate, oxidn., 6
precursor of CH_2O, 6
Ethylamine, detm. with DCPIP, 160
Ethylbenzene, photooxidn., 26, 27
precursor of CH_2O, 26, 27
yield of CH_2O, 26, 27
Ethyl N,N-bis-hydroxymethyl carbamate, hydrolysis, 6
precursor of CH_1O, 6
Ethylene, atmospheric, 21
collection, 278
on carbon, 82
detm. in, air, 278
auto exhaust, 82
detm. with CA, 38, 82, 278
formn. from, oxetane, 250
tetrahydrofuran, 250
oxidn., 6, 23
photooxidn., 6, 18, 23, 24, 26, 27, 82, 243
precursor of, carbon monoxide, 24
formald., 6, 18, 23, 24, 26, 27, 82, 243, 278
formic acid, 24
reaction with NO_2, 23, 24
yield of CH_2O, 18, 26, 27
Ethylenediamine, 82
anal. proc., 48
detm. with MBTH, 48
oxidn., 6
precursor of CH_2O, 6, 21, 48
Ethylenediaminetetracetic acid, *N*-demethylase inducer, 50
Ethylene dichloride, *see* 1,2-dichloroethylene
Ethylene glycol, 84–86, 100
anal. eqn., 103
anal. proc., 84, 85
automated anal., 84
collection, 278
detm. in, air, 278
body fluids, 278
industrial air, 84
surface waters, 84
tissues, 278

Ethylene glycol—*continued*
detm. with, CA, 38, 84, 85, 278
dimedone, 38
ethyl acetoacetate, 38
J-acid, 38, 103
MBTH, 38, 84
2,4-pentanedione, 38, 84, 85, 93, 115, 279
phenylhydrazine, 38
p-rosaniline, 38, 279
diglycid ester, detm. with CA, 85
fluor. anal., 84
formn. from ethylene oxide, 81
interferences, 86
oxidn., 6, 253, 278
reaction rate, 93
precursor of CH_2O, 6, 21, 38, 84, 93, 253, 278
sepn., 115
toxicity, 85
yield of CH_2O, 38
Ethylenimine, 57
hydrolysis, 6
oxidn. with periodate, 6
precursor of CH_2O, 6
Ethylene oxide, adsorption on plastics, 79, 81
cleavage, 94
collection from air, 70
detm. in, air, 81, 279
gases, 279
pharmaceuticals, 279
plastics, 81
spices, 279
detm. with, CA, 38, 80, 81, 85, 279
2,4-pentanedione, 38
phenylhydrazine, 81
hydrol., 6, 80, 81
mutagen, 81
oxidn., 6
persistence, 79, 80
photolysis, 250
precursor of, acetald., 250
ethylene glycol, 81
formald., 6, 81, 85, 279
reaction with, HCl, 79
water, 79
sterilization, 79
structure, 250
1-Ethyleneoxy-3,4-epoxycyclohexane, *see* vinyl-cyclohexene-3-diepoxide
Ethylenimine, detm. with CA, 81
hydrol, 81
Ethyl guthion, detm. with CA, 46
Ethylhydrazine, carcinogen, 181
Ethyl *N*-hydroxymethyl carbamate, hydrolysis, 6
precursor of CH_2O, 6
Ethylmaleimide, inhibitor of *N*-oxidn., 147
Ethylmorphine, anal. proc., 143, 178
demethylation, 6, 143, 148, 156, 169, 170, 176, 178, 272
detm. of *N*-demethylase, 50, 51
detm. with, CA, 176
2,4-pentanedione, 38, 50, 51, 176
effect of, corticosterone, 151
detergents, 143
TCDD, 156
vinyl chloride, 272
metabolism, 175
metab. eqn., 176
precursor of CH_2O, 6, 50, 51, 148, 151, 175
structure, 176
Ethylmorphine *N*-demethylase, activity in, guinea pig, 175
hamster, 175
mouse, 175
rabbit, 175
rat, 175
activity increase, 51
detm. with ethylmorphine, 51
effect of, ascorbic acid, 51
carbon monoxide, 51
corticosterone, 151
EDTA, 176
ferrous ion, 176
lipid peroxidn., 51
SKF-552A, 176
inhibition, 51
intestinal activity in, guinea pig, 175
hamster, 175
mouse, 175
rabbit, 175
rat, 175
liver activity in, guinea pig, 175
hamster, 175
mouse, 175
rabbit, 175
rat, 175
relation to cytochrome P-450, 177
Ethyl nitrite, photolysis, 243
precursor of, acetald., 243
peroxyacyl nitrate, 243
1-Ethylquinaldinium salts, detm. of formic acid, 87
Eucaryotes, mutated by CH_2O, 233, 234
recombinogenesis by CH_2O, 234
Eugenol, detm. with, J-acid, 31
MBTH, 31, 38
pararosaniline, 86
oxidn., 86
oxidn. reaction, 31
precursor of, formald., 6, 31, 86
vanillin, 31
spectral data, 38

structure, 31
Excision repair, DNA, 227
Extractant, acetic acid, 82
alcohol, 66
chloroform, 66
Extraction, triglycerides, 260, 264, 267
Extrapolation, animal to man, 77
safety factor, 77
Eye, response to smog, 27
Eye drops, detm. of pilocarpine, 249
Eye irritation, acrolein, 196
formald., 196
photochemical smog, 196
Eye ointment, detm. of pilocarpine, 249

Fabrics, formald. content, 217
FAD, cofactor, 234–236
oxidn. of methanol, 138
structure, 234
$FADH_2$, 234, 235
Fasting, stimulator of *N*-demethylase, 148
Fat, effect on rat, 74
Fatty acid, interference, 256, 257
Fe(II), inhibitor of *N*-demethylase, 149
Feces, detm. of sorbitol, 281
Feedlots, nitrate, 65
nitrosamines, 65
Fehling solution, 136
Female rat, effect of TCDD, 156
Fermentation products, detm. of methanol, 280
Fermentation solutions, detm. of glycerol, 279
Ferric chloride, detm. of cholesterol, 266
oxidant, 5, 29, 48, 56, 57, 62, 84, 88, 92, 107, 111, 132, 133, 135, 139, 164, 169, 177, 221, 228, 268
Ferric oxide, cocarc. with Et_2NNO, 69
particle size, 69
Ferric salt, oxidant, 106
Ferricyanide, oxidant, 8
Fetus, activation of carcinogens, 53
alkylation of nucleic acids, 55
de-ethylation, 54
demethylation, 54
effect of, azoalkanes, 54
azoxyalkanes, 54
dialkylaryltriazenes, 54
female germ cells, 54
formn. of alkylating agents, 54
Feulgen stain, 108
Fiberboard, construction of, buildings, 218
furniture, 218
mobile homes, 218
detm. of CH_2O in, 200
formald. release, 218
Fish, detm. of 2-butoxyethyl 2,4-D, 278
meal, nitrosamines, 65
products, nitrosamines, 65
Flavin-adenine dinucleotide, *see* FAD
Flavin mononucleotide, *see* FMN
Florisil, 257
phospholipid removal, 264
N-2-Fluorenylacetamide, carcinogen, 182
carc. of subcarc. doses, 182, 184
mutagen, 184
non-carc. to cotton rat, 184
2-(2-Fluorenylamino) ethanol, anal. proc., 48
detm. with MBTH. 39, 48
oxidn., 6
precursor of CH_2O, 6
spectral data, 39
Fluorescein, visualization of triglycerides, 259
Fluorogen, 3,5-diacetyl-1,4-dihydrolutidine, 47, 91, 95, 139
dibenzoxanthylium cation, 103
3,5-diethoxycarbonyl-1,4-dihydrolutidine, 47
Fluorosulfuric acid, 79
FMN, oxidn. of methanol, 138
Folic acid, effect on carc., 143
Folinic acid, detm. with CA, 89
precursor of CH_2O, 89
Food, contamination by aflatoxins, 187
detm. of hexamethylenetetramine, 119
gums, 59
nitrosamines, conc., 65
formn., 64, 65
stability, 64
Forensic materials, detm. of methanol, 280
Forestomach, cancer, 68
carcinoma, 125
Foramaldehyde, absorbant, 197
acetoacetate reaction eqn., 30
aging process, 237
anal. artifacts, 196
anal. eqn., 153
anal. of oxidative precursors, 282
anal. proc., 153, 154, 177, 199, 213
atmospheric, 191–198
sampling, 213
auto exhaust, 198
basic precursor, examples, 20, 21
structures, 20, 21
binding to SH, 254
biosynthesis, 143
carcinogenicity, 1
catalyst in DMNA formn., 185
cellular differentiation, 143
chemiluminescence, 28
collection by, bisulfite, 153
semicarbazide, 153, 157, 283

Foramaldehyde—*continued*
concns. in, atmospheres, 21
industrial air, 21
industrial waters, 21
Tokyo, 197
continuous meas., 197
crosslink of protein to DNA, 1
dehydrogenase, 204
demonstration in keratin, 222
derived from (^{14}C) DMNA, 77
detect. with 2,3-dimethyl-2,3-bis(hydroxyl-amino)butane, 202
detm. in, air, 213
aortic valves, 221
collagen, 221
elastin, 221
funeral homes, 213
maple syrup, 200
sewage, 200
sewage effluents, 200
detm. with, CA, 29, 196, 198, 199, 213, 271
1,2-diaminonanaphthalene, 201
dimedone, 193
2,3-dimethyl-2,3-bis(hydroxylamino)-butane, 201, 202
ethyl acetoacetate, 30
fuchsin, 197
gallic acid, 203, 204
J-acid, 29
MBTH, 197, 198, 221
2-naphthol-6-sulfonate, 196
pararosaniline, 86, 197, 199
2,4-pentanedione, 70, 138, 143, 147, 152, 153, 197, 199, 200
peroxide, 202
phenylhydrazine, 197
purpurald., 200, 218, 219
tryptophan, 177
distillation, 49
effect on, ciliary activity, 195
DNA, 227
vision, 197
enzyme complexes, 230, 231
fixation, 254
formn. of, acetal, 206
adenine dimers, 226
bis-chloromethyl ether, 206
hemiacetal, 206
methylolamides, 205
nitrosamine, 67
ring, 205, 206
Schiff base, 205
semithioacetal, 206
thioacetal, 206
m-trioxane, 206
formn. reactions, 277
genotoxicity, 143, 195, 196, 232–234
histochem. reactions, 239, 242
histone-DNA cross-links, 211
industrial emissions, 217
iodometric standardization, 215
lachrymator, 128, 196
metabolism, 152
mutagen, 1, 73, 81, 150, 196, 226, 232–234
mutagenesis theories, 196, 233
mutagenic for, eucaryotes, 233
procaryotes, 233
viruses, 233
mutagenic pathway, 233
oxidn. to, CO_2, 153
formate, 153
permeation tubes, 194, 195
physiol. activity, 195, 196
precursor of bis-chloromethyl ether, 57, 58
presence in, fabrics, 217
fiberboard, 200, 218
formalin, 212
prevention of loss, 153
probe of DNA structure, 210
rate of reaction, 203
reaction with, acetoacetate, 30
acids, 205
active methylenes, 205
acyl anhydrides, 205
adenine, 209, 210
adenosine, 233
amides, 205
amino acids, 226, 227
3-(2-aminoethyl)indoles, 208
asparagine, 225
azulenes, 207
BaP, 207, 209
bisulfite, 152, 153
carboxyhemoglobin, 225
casein, 1, 225
cysteine, 206
cystine, 229
cytosine, 210
dialkylanilines, 206
2,7-dihydroxynaphthalene, 100
dimedone, 205
dimethylamine, 225
DNA, 77, 78, 210, 226
epinine, 237
gelatine, 225
glycine, 227
guanine, 210
hydrochloric acid, 206
hydroxyl compds, 206
imines, 205
indoles, 208
keratin, 229

ketene, 251
lysine, 210, 225
mercaptans, 206
mercaptoamines, 205
4-methoxy-3-hydroxyphenylethylamine, 237
N-methyltryptamine, 237
β-naphthol, 101
nitromethane, 205
nucleic acids, 1, 233
nucleotides, 1, 210, 211
PAH, 207
peptides, 205
phenethylamines, 236
phenols, 206
β-phenylethanolamine, 236
polyamines, 205
poly(inosinic acid), 211
polysaccharides, 210
poly(uridylic acid), 211
protein, 1, 210, 221, 222, 226, 233
pyrrols, 208
RNA, 210
sulfate, 209
sulfite, 205
thymidine, 211
thymine, 211
tryptamine, 237
tryptophanyl-dipeptide, 241
uracil, 211
wool, 225
recombinogenesis, 234
relation to carc., 143
release from, amino resin molding, 218
fabrics, 217
fibreboard, 218
microporous rubber, 218
particle board, 218
SH groups, 222
role in aging, 237
simplex optimization, 199
source, 197
standardization, 231
standards, 193–195
stimulates mucous membrane, 197
sulfite interference, 152, 153
thiol-bound, 254
trapping by semicarbazide, 283
unwinding of DNA, 211
Formaldehyde dimethylacetal, detm. with phenylhydrazine, 39
hydrol., 6
precursor of CH_2O, 6, 39
release parameters, 218
Formaldehyde 2,4-dinitrophenylhydrazone, detm. with, CA, 39
J-acid, 39
MBTH, 126, 127
phenyl J-acid, 39
formn. of precursors, 204–212
hydrol., 7
precursor of CH_2O, 7, 39, 126, 127
spectral data, 39
Formaldehyde hyrazone, 126–128
anal. eqn., 126
detm. with, CA, 39
J-acid, 39
MBTH, 126
NAD^+, 204
phenyl J-acid, 39
hydrol., 7
precursor of CH_2O, 7, 39, 126
spectral data, 39
Formaldehyde methylhydrazone, formn. from azomethane, 123
precursor of CH_2O, 123
structure, 123
Formaldehyde-tanned collagen, hydrolysis, 6
detm. with CA, 39
Formalin, contains, acetals, 212
formald., 212
methylal, 212
methylene glycol, 212
methyl formate, 212
polyoxymethylene glycols, 212
trioxane, 212
stabilization, 212
Formals, auto exhaust, 86
Formamide, formn. from CO, 276
Formate, air, 87
auto exhaust, 87
redn. to CH_2O, 87
Formazan, chromogen, 112, 113
structure, 112, 113
Formazan anion, chromogen, 40
Formazan cation, 39
chromogen, 22, 33, 104, 113, 163, 224
spectral data, 127, 228
structures, 22, 104, 113, 127, 140, 163, 228
Formic acid, 93
detect. with, 1-ethylquinaldinium salts, 87
HBT, 87
J-acid, 87
MBTH, 87
1-methylquinaldinium salts, 87
detm. in, body fluids, 279
fruit juice, 279
natural waters, 279
wine, 279
detm. with, CA, 39, 87
HBTH, 87
indole, 244

Formic acid—*continued*
J-acid, 87
MBTH, 87
formn. from, carbon monoxide, 276, 277
1,1-dichloroethylene, 62
1,2-dichloroethylene, 62
dimethylnitrosamine, 180, 181
ethylene, 23, 24
glucose phenylosazone, 90
methyl nitrite, 243
cis-ocimene, 29
oligosaccharides, 134, 135
polyols, 251
sophoritol, 252
sophorose, 252
precursor of CH_2O, 7, 39, 87, 279
redn., 7, 87
spectral data, 39
yield of CH_2O, 39
Formiminoglutamic acid, 1-carbon metab., 152
N-Formylaniline, detm. with CA, 89
precursor of CH_2O, 89
Formylase, 1-carbon metab., 152
N-Formyldiethylaminomalonate, detm. with CA, 89
precursor of CH_2O, 89
N-Formyldiphenylamine, detm. with CA, 89
precursor of CH_2O, 89
Formyl group, precursor of CH_2O, 7
redn. with Mg, 7
Formylhydrazine, detm. with CA, 89
precursor of CH_2O, 89
N-Formylindoline, detm. with CA, 89
precursor of CH_2O, 89
L-Formylkynurenine, formn. from tryptophan, 190
Formylmethylstearate, formn. from glycidyl stearate, 94
Frankfurters, nitrosamines, 65
Free radical zwitterion, chromogen, 201
Freon, residence time, 193
Friedel–Crafts chloride salt, carc. contaminant, 58
Fructose, detm. with, CA, 39
dimedone, 39
J-acid, 39
MBTH, 39
2,4-pentanedione, 39
oxidn., 7, 253
precursor of CH_2O, 39, 253
spectral data, 39
Fruit, detm. of aramite, 278
Fruit flies, mutagenicity of hempa, 154
Fruit juices, detm. of formate, 279
Fuchsin, 42
detm. of atm. CH_2O, 197
Funeral home, detm. of CH_2O, 213
Fungus, aflatoxins, 187
growth with methanol, 138
Furfural, detm. with MBTH, 39
oxidn., 7
precursor of CH_2O, 7
spectral data, 39
Furfurals, form. from, ketoses, 116
sugar alcohol glycosides, 116
Furniture, formald. pollution, 218
Furniture polishes, vinyl chloride usage, 269
β-2-Furylacrylophenone, detm. with MBTH, 39
oxidn., 7
precursor of CH_2O, 7
spectral data, 39

Galactonic acid, detm. with MBTH, 39
location, 109
oxidn., 7
precursor of CH_2O, 7, 109
spectral data, 39
Galactosamine, location, 109
precursor of CH_2O, 109
Galactose, detm. with, CA, 39
MBTH, 39
oxidn., 7, 17, 253
precursor of CH_2O, 7, 17, 39, 253
spectral data, 39
α-D-Galactose diethyl dithioacetal, oxidn. with periodate, 7
precursor of CH_2O, 7
Gallic acid, detm. of, chavicine, 46
formald., 203, 204
glyceric acid, 46
glycolic acid, 46
glyoxal, 203
glyoxylic acid, 46
methylenedioxyphenyl group, 43
piperettine, 46
piperic acid, 46
piperine, 43, 214
pyruvald., 203
tartaric acid, 46
3-O-β-Galp-D-arabinitol, distribution coeff., 116
precursor of, formald., 116, 117
furfurals, 116, 117
sepn., 116, 117
4-O-β-D-Galp-D-glucitol, distribution coeff., 116
precursor of, formald., 116
furfurals, 116, 117
sepn., 116, 117

6-O-α-D-Galp-D-glucitol, distribution coeff., 116
precursors of, formald., 116, 117
furfurals, 116, 117
sepn., 116, 117
4-O-β-D-Galp-D-mannitol, distribution coeff., 116
precursors of, formald., 116
furfurals, 116
sepn., 116
Gases, detm. of ethylene oxide, 279
Gasoline, content of, aromatics, 129
lead, 129
formald. yield, 129
methanol as extender, 137
Gastric mucosa, effect of PCB, 150
pH, 67
GC, anal. of, dimethylnitrosamine, 64
GC–CL, anal. of nitrosamines, 66
GC–MS–COMP, anal. of dimethylnitrosamine, 64
Geigy 28029, *see* Phenkapton
Gelatin, reaction with CH_2O, 225
Gelsemine, detm. with CA, 39
oxidn., 7
precursor of CH_2O, 7, 39
spectral data, 39
Genetics, tumor etiology, 76
Genotoxicity, ald. formn. and., 226
alkylating agents, 51
demethylase activity, 74
pathway, 69
Genotoxic pathways, dimethylnitrosamine, 179–181
Geraniol, 33
oxidn., 11, 16
oxidn. eqn., 32
precursor of, acetone, 32
β-acetylpropionald., 32
formald., 11, 16, 32
formic acid, 32
structure, 32
Germall 115, detm. with 2,4-pentanedione, 47
precursors of CH_2O, 47
Germany, nitrosamines, 64
Germ cells, damage by carcinogens, 54
Glaucoma, 248
3-O-α-D-Glcp-D-glucitol, distribution coeff., 116
precursors of, formald., 116, 117
furfurals, 116, 117
sepn., 116, 117
4-O-α-D-Glcp-D-glucitol, distribution coeff., 116
precursor of, formald., 116, 117
furfurals, 116, 117
sepn., 116, 117
4-O-β-D-Glcp-D-glucitol, distribution coeff., 116
precursors of, formald., 116, 117
furfurals, 116, 117
sepn., 116, 117
6-O-α-D-Glcp-D-glucitol, distribution coeff., 116
precursors of, formald., 116
furfurals, 116
sepn., 116
6-O-β-D-Glcp-D-glucitol, distribution coeff., 116
precursors of, formald., 116
sepn., 116
3-O-α-D-Glcp-D-mannitol, dist. coeff., 116
precursors of, formald., 116
furfurals, 116
sepn., 116
6-O-α-D-Glcp-D-mannitol dist. coeff., 116
precursors of, formald., 116
furfurals, 116
sepn., 116
Glucaric acid, 89, 90
detm. with phenylhydrazine, 89, 90
oxidn., 7
precursors of, CH_2O, 7, 89
glyoxylic acid, 89
sepn. from urine, 89
Glucitol, detm. with 2,4-pentanedione, 117
distribution coeff., 116
glucopyranosides, 118
precursor of CH_2O, 116, 117
prepn. from D-glucose, 136
sepn., 116, 117
Glucoheptose, 253
Gluconic acid, detm. with, chromic acid, 276
MBTH, 39
periodate, 276
oxidn., 7, 253, 276
precursor of CH_2O, 7, 253, 276
sepn., 276
spectral data, 39
Gluconolactone, detm. with MBTH, 39
spectral data, 39
D-Glucosamine, anal. proc., 48
detm. with, CA, 39
MBTH, 48
oxidn., 7, 48
precursor of CH_2O, 7, 39, 48
spectral data, 48
Glucose, 59, 139, 266, 267
anal. interference, 244, 245
anal. proc., 139
automated anal., 90
decomp. by H_2SO_4, 90

Glucose—*continued*
detm. in, blood, 279
plasma, 279
serum, 279
detm. with, CA, 39, 139, 279
MBTH, 39, 139, 279
orcinol, 117
2,4-pentanedione, 90, 93
disproportionation, 7
fluor. anal., 90, 93
formn. of 1,2-phenylosazone, 90
location, 109
oxidn., 7, 90, 93, 140, 253
precursor of, formald., 7, 90, 93, 109, 253, 279
furfural, 117
5-hydroxymethyl-2-furaldehyde, 60
levulinic acid, 60
redn., 136
removal from cycasin, 54, 55
sepn., 117
D-Glucose-D-glucitol, α-(1 → 6)-linked, 118
β-(1→3) linked, 118
β-(1→4) linked, 118
Glucose oxidase, 139, 140
Glucose phenylosazone, oxidn., 7, 90, 253
precursor of, formald., 7, 90, 253
formic acid, 90
structure, 90
Glucose-6-phosphate, 154, 171, 178
β-Glucosidase, bacterial, 121
microbial, 54, 55
Glucosone, oxidn. with periodate, 7
precursor of CH_2O, 7
Glucuronic acid, detm. with MBTH, 39
oxidn., 7
precursor of CH_2O, 7
spectral data, 39
β-Glucuronidase, colon, 123
Glutamic acid, 272
γ-carboxymethyl ester, 100
Glutaraldehyde, anal. proc., 221
detm. in, aortic valves, 221
collagen, 221
elastin, 221
tanned heterografts, 221
detm. with MBTH, 221
reaction with proteins, 221
Schiff reaction, 223
Glutathione, 204
effect on methylene halides, 232
Glyceraldehyde, detm. with CA, 40
oxidn., 7
precursor of CH_2O, 7
Glyceric acid, chromatography, 279
detm. in urine, 279
detm. with, CA, 40, 279
gallic acid, 46
MBTH, 40
oxidn., 7
precursor of CH_2O, 7, 279
spectral data, 40
Glycerides, 91–93
anal. proc., 91, 92
detm. in, lipids, 279
plasma, 279
rat adrenal glands, 279
serum, 279
detm. with, CA, 40, 279
DNPH, 40
MBTH, 40, 92
2,4-pentanedione, 40, 91, 93
phenylhydrazine, 40, 92
oxidn., 7
precursor of CH_2O, 7, 40, 279
saponification, 92, 279
D-Glycero-D-galacto-heptitol, dist. coeff., 116
precursor of, formald., 116, 117
furfurals, 116, 117
sepn., 116
D-Glycero-D-gulo-heptitol, dist. coeff., 116
precursors of, formald., 116, 117
furfurals, 116, 117
sepn., 116
Glycerol, 114, 257–259, 266, 267
anal. proc., 92
automated detm., 265
detm. in, blood, 279
fermentation solns., 279
liquors, 279
phosphatides, 279
phospholipid hydrolyzate, 279
rat plasma, 279
wines, 279
detm. with, CA, 40, 92, 279
dimedone, 40
MBTH, 40
2,4-pentanedione, 40, 93, 115, 279
phenylhydrazine, 40
formn. from, glycerides, 91
triglycerides, 260, 264
oxidn., 7, 93, 253, 260, 264
oxidn., eqn., 93
oxidn. reaction rate, 93
precursor of, formald., 7, 40, 92, 93, 253, 260, 264, 279
formic acid, 264
glycolald., 260
sepn., 92, 115
spectral data, 40
structure, 264
D-Glycero-D-manno-heptitol, dist. coeff., 116
precursor of, formald., 116, 117

furfurals, 116, 117
sepn., 116
α-Glycerophosphates, detm. with 2,4-pentanedione, 93
oxidn., 7
oxidn. eqn., 93
precursor of CH_2O, 7, 93
glycolald. phosphate, 93
structure, 93
Glycerophosphatides, 93, 94
anal. proc., 94
detm. in blood serum, 94
detm. with 2,4-pentanedione, 94
hydrol., 7, 94
oxidn., 7
precursor of CH_2O, 7, 94
Glycidaldehyde, carcinogen, 81
mutagen, 81
Glycidol, mutagen, 81
Glycidyl stearate, anal. proc., 94
oxidn., 7
precursor of, formald., 7, 94
formylmethylstearate, 94
Glycine, 94–99, 268
anal. proc., 49, 50, 96–98
automated anal., 95
col. chrom., 96, 98
deamination, 7, 8, 96, 97
decarboxylation, 7, 8, 96, 98, 277
detm. in, biol. fluids, 95, 96
blood, 95, 96, 280
spinal cord, 280
urine, 95, 96, 280
detm. with, CA, 40, 49, 50, 95, 97, 98, 280, 283
2,4-pentanedione, 96–98
resorcinol, 99
disproportionation, 8, 49
effect on d-AMP, 210
oxidn., 7, 8, 95, 283
oxidative deamination, 49
oxidn., eqn., 97
precursor of CH_2O, 7, 8, 48–50, 95–99, 280, 283
reaction with CH_2O, 227
structure, 49, 97
Glycinosis, hyperglycinemia, 95
Glycoladehyde, anal. proc., 113
decomposition, 93
detm. with, MBTH, 260
2,4-pentanedione, 93, 114, 115
pyrogallol, 40, 113
disproportionation, 8
formn. from, glycerol, 93, 260
triglycerides, 260
oxidn., 8
oxidn. reaction rate, 93
precursor of CH_2O, 8, 93, 260
formic acid, 260
sepn., 115
structure, 93, 260
Glycolaldehyde phosphate, formn. from α-glycerophosphates, 93
Glycolic acids, 99–103
anal. interf., 244
anal. proc., 99, 100
detm. in, antifreeze soln., 280
carboxymethylcellulose, 280
irradiated oxalates, 280
oxidn. products, 280
resin products, 280
detm. with, CA, 40, 100, 101, 244, 280
2,7-dihydroxynaphthalene, 40, 99, 100, 114, 280
gallic acid, 46
J-acid, 40
β-naphthol, 101
disproportionation, 8, 49, 277
formn. from, CO, 276
chloroacetates, 60
glyoxylate, 114
oxalate, 114, 244
tartronate, 100
precursor of CH_2O, 8, 99–101, 114, 280
spectral data, 40
structure, 49
Glycolic nitrile, disproportionation, 8
precursor of CH_2O, 8
Glycolipids, detm. with pararosaniline, 40
oxidn., 8
precursor of CH_2O, 8
TLC, 40
1,2-Glycols, 103–111
anal. proc., 106–108
collection, 280
detm. in, air, 107
air particles, 280
tissue, 280
detm. with, acetoacetate, 41
CA, 108
dimedone, 41, 107
J-acid, 41, 103, 104, 107, 108, 111
MBTH, 41, 104, 107, 280
pararosaniline, 41, 280
2,4-pentanedione, 41, 107
resorcinol, 106
formn. from olefins, 22
location, 108
oxidn., 8, 15, 80, 103, 108
precursor of, aliph. ald., 22, 80
formald., 8, 22, 80, 108
spectral data, 41
structure, 80

N-Glycolylneuraminic acid, detm. with, 2,7-dihydroxynaphthalene, 41
hydrol., 8
oxidn., 8
precursor of CH_2O, 8
α-N-Glycolyl-D-arginine, γ-carboxymethyl ester, 100
Glycoproteins, 111, 112
allergens, 104
detm. of aminohexoses in, 278
detm. with, 2,4-pentanedione, 278
phenylhydrazine, 111, 112
electrophoresis, 111
NANA, 111
oxidn., 8
precursor of CH_2O, 8, 111, 278
Glycosides, precursors of, formald., 116–118
furfurals, 116–118
sepn., 116–118
Glyoxal, 204
chemiluminescence, 28
detm. with, CA, 41
gallic acid, 203
oxidn., 8
precursor of CH_2O, 8, 21
spectral data, 41
Glyoxalase, 204
Glyoxylate, 113, 114
acid hydrol., 282
anal. proc., 113
characterization, 41
decarboxylation, 8, 277
detm. with, CA, 282
2,7-dihydroxynaphthalene, 41
gallic, acid, 46
J-acid, 41
phenylhydrazine, 41, 89, 113
pyrogallol, 113, 114
disproportionation, 8
formn. from, allantoic acid, 113
allantoin, 113
glucaric acid, 89
ureidoglycolate, 113
PC, 41
precursor of, formald., 8, 113, 114, 277, 282
glycolate, 114
oxalate, 114
reaction with phenylhydrazine, 8
redn., 8
TLC, 41
Golden hamster, teratogenicity of methylazoxymethanol, 55
GP, *see* Genotoxic pathway
Grafts, sterilization, 251
Grain, detm. of 2,4-dichlorophenoxyacetic acid, 278
Griseofulvin, carc., 190
demethylation, 190
effect on, griseofulvin, 190
3-methylcholanthrene, 190
mouse, 147
phenobarbital, 190
fungicide, 190
metab. in, animals, 190
man, 190
rat tissue, 190
precursor of, desmethylgriseofulvins, 190
formald., 190
stimulator of N-demethylase, 145
GSH, 180
Guaiacylglycerol-β-(2-anisyl)-ether, disproportionation, 8
precursor of CH_2O, 8
Guaiacylglycerol-β-(2-methoxyphenyl) ether, detm. with CA, 58
hydrol. eqn., 59
precursor of, CH_2O, 58
homovanillin, 58
structure, 59
Guanosine, methylation by 1-phenyl-3-methyltriazene, 125
Guaran, detm. with CA, 112
precursor of, formald., 8, 112
D-galactose, 112
D-mannose, 112
redn., 8
Guinea pig, azo dye and liver protein, 226
cytochrome P-450, 177
N-demethylase activity, 51, 177
embryo, 76, 77
tumors from nitrosamines, 68
Gut, 123
bacterial flora, 123
Guthion, detm. in, cotton seed, 280
cotton seed oil, 280
detm. with CA, 41, 280
hydrol., 8
precursor of CH_2O, 8, 229, 280
structure, 16, 229

Hair spray, detm. of methanol, 136
vinyl chloride usage, 269
Hair tonic, detm. of pilocarpine, 249
Hallucinogen, harmaline, 238
harmine, 238
Halogen, oxidant, 12
Hamster, azo dye and liver protein, 226
effect of ferric oxide, 69
tumors from, 1,1-dimethylhydrazine, 122
nitrosamides, 183
nitrosamines, 68, 69, 183

vinyl chloride, 274
resistance to DMNA, 183
Hamster bronchus, Me_2NNO demethylase, 183
Hamster lung, DMNA N-demethylase, 183
effect of, cyclic nitrosamines, 183
dibutylnitrosamine, 183
diethylnitrosamine, 183
dimethylnitrosamine, 183
di-n-propylnitrosamine, 183
methylnitrosourea, 183
Hamster lung, Me_2NNO demethylase, 183
Hamster trachea, DMNA N-demethylase, 183
Hantzsch condensation, 261
Harmaline, hallucinogen, 238
structure, 238
Harman, formn. from tryptophan, 242
structure, 242
Harmine, hallucinogen, 238
structure, 238
HBT, *see* 2-hydrazinobenzothiazole
HCG, 121
Hemel, *see* Hexamethylmelamine, 120
Hemiacetals, formn. eqn., 213
formn. from CH_2O, 212
Hempa, carcinogen, 154
formn. from thiohempa, 155
metabolism, 155
mutagen, 154
structure, 154
Hen eggs, teratogenicity of Bidrin, 157
Heparin, 85, 96
Hepatectomy, effect on carcinogenesis, 182
Hepatic angiosarcomas, vinyl chloride workers, 273, 274
Hepatic microsomes, 21, 138
Hepatic tumors, effect of acetoaminonitrile, 74
Hepatocarcinogenesis, 143
repression, 72
Hepatocellular, carcinoma, rat, 187
necrosis, 72
Hepatotoxin, carbon tetrachloride, 71
dimethylnitrosamine, 69, 70, 71, 77
vinylidene chloride, 62
Heptane, extractant, 260, 263, 264
1-Heptene, detm. with MBTH, 41
oxidn., 8
precursor of CH_2O, 8
spectral data, 41
Heroin, formald. precursor, 172
structure, 172
Herring, nitrosamines, 65
Heterografts, detm. of ald., 221
detm. with MBTH, 221
fixation of CH_2O, 221
tanning, 221
1-Hexadecene, atmospheric, 21
biotransformation, 21
carcinogen, 21
epoxidation, 21
formn. of 1,2-epoxyhexadecane, 80
precarcinogen, 21
Hexaethylmelamine, anticarc., 120
Hexahydro-1,3,5-trinitro-*s*-triazine, hydrolysis, 8
precursors of CH_2O, 8, 118
Hexamethylenetetramine, 119, 120
anal. proc., 119, 120, 224, 225
antibacterial, 223
automated anal., 119, 223
detm. in food, 119
tablets, 119, 223, 224
vulcanizates, 119, 120
detm. with, CA, 41, 119, 120, 224, 225, 282
HBT, 224
J-acid, 41
2,4-pentanedione, 224
phenyl J-acid, 41
extn. with chloroform, 224
formn. from, mandelamine, 223
resotropin, 224
hydrol., 8, 282
mandelate, 119
methylene donor, 224
precursor of, bis-chloromethyl ether, 58
formald., 8, 119, 193, 282
sarcoma, 195
spectral data, 41
standard, 43, 193
structure, 193
TLC, 224, 225
use in tires, 224
Hexamethylmelamine, 120, 121
anticarcinogen, 120
metabolism, 121
mutagenic to, *Drosophila*, 120
Musea domestica, 120
precursor to formald., 121
structure, 120
Hexamethylphosphoramide, formn., 155
metabolism, 155
mutagen, 154
nasal carcinoma, 154
precursor of CH_2O, 155
structure, 154
Hexamine, *see* Hexamethylenetetramine
Hexane, solvent, 40
TLC sepn. with, 55
1-Hexene, anal. parameters, 28
atmospheric, 21
detm. with, J-acid, 41
MBTH, 28, 29, 41

1-Hexene—*continued*
oxidn., 8
precursor of CH_2O, 8, 41
cis-2-Hexene, photooxidn., 18
precursor of aliph. ald., 18
yield of aliph. ald., 18
5-Hexen-1-ol, oxidn. with ozone, 8
precursor of CH_2O, 8
5-Hexen-2-one, oxidn. with ozone, 8
precursor of CH_2O, 8
Hexitols, detm. with, acetoacetate, 41
dimedone, 41
J-acid, 41
MBTH, 41
2,4-pentanedione, 41
oxidn., 8
precursors of, erythritol, 114
formald., 8, 41
threitol, 114
spectral data, 41
Hexobarbital, N-demethylation, 149
effect of, carbon disulfide, 51
nitrogen dioxide, 51
ozone, 51
phenobarbital, 166, 167
inhibits N-demethylase, 147
metab., 178
precursor of, N′-(2-cyclohexenylpropionyl)-N-methylurea, 178
formald., 149, 178
3′-hydroxyhexobarbital, 178
3′-ketohexobarbital, 178
3′-ketonorhexobarbital, 178
norhexobarbital, 178
sleeping time effect, 178, 179
structure, 178
Hexosamines, anal. proc., 48
detm. with MBTH, 48
oxidn., 8
precursor of CH_2O, 8, 47
Hexoses, detm. with CA, 39
Hexuronic acids, oxidn. with periodate, 9
precursors of CH_2O, 9
HFSH, 121
Hg^{+2}, stimulation of N-demethylase, 146
$HgCl_2$, inhibition of N-demethylase, 146
High risk group, 67
Histidase, 1-carbon metab., 152
Histamine release, 212
Histidine, reaction with $HOCH_2NR_2$, 226
Histone, crosslink to DNA, 211
effect on nucleotide reaction, 211
Hodgkin's disease, effect of Natulan, 151
Homocysteine, 182
L-Homoserine, degradation eqn., 230
precursor of, formald., 230
2-keto-4-hydroxybutyrate, 230
pyruvic acid, 230
structure, 230
Homovanillin, formn. from guaiacylglycerol-β-(2-methoxyphenyl) ether, 59
HPLC-CL, anal. of nitrosamines, 66
Human, chorionic gonadotrophin, 121
follicle-stimulating hormone, 121
intrauterine microsomes, 78
see Man
Human adipose tissue, detm. of triglycerides, 267
PCB, 150
Human blood, formald. from S-adenosylmethionine, 192
Human brain, detm. of tryptamine, 242
enz. conversions, 238
Human cancer, bis-chloromethyl ether, 57
nitrosamines, 77
Human embryo, carc. of, DMNA, 78
vinyl chloride, 273, 274
N-demethylase, 5, 78
effect of age, 51
metab. of nitrosamines, 76–78
microsomal fractions, 51, 76, 77
prenatal carcinogenicity, 77
Human liver, cytochrome P-450, 157
detm. of tryptophan, 242
effect of DMNA, 77
effect on vinyl chloride, 273
formn. of aflatoxin B_1 2,3-oxide, 188
metab. of DMNA, 76
microsomes, 147, 157, 273
Human lung, DMNA demethylase, 183
effect of methyl piperonylate, 212
histamine release, 212
Human milk, PCB in, 150
Human plasma, chlorpromazine metabolites, 147
detm. of tryptophan, 242
Human platelets, formn. of tryptolines, 238
Human scalp, Schiff test, 254
thiol-bound CH_2O, 254
Human urine, acetylmethadol, 155
chlorpromazine, 147
detm. of, glyceric acid, 279
tryptophan, 242
metabolites, 147, 155, 191
8-methoxykynurenic acid, 191
H. vulgare, 162
Hydrastine, detm. with CA, 214
location on TLC, 214
precursor of CH_2O, 214
Hydrazine, formn. from methylnitrosamine, 180
carc., 251

mutagen, 251
Hydrazines, 121–124
carc., 121–123, 251
detm., 122–123
metabolism, 122, 123
2-Hydrazinobenzothiazole, detm. of, Alar, 124, 278
formate, 87
hexamine, 224
17-hydroxy-17-ketosteroids, 42
mandelamine, 224
spot test for formate, 88
Hydrindantin, formn. from ninhydrin, 95
Hydriodic acid, reductant, 11
Hydrocarbons, non-methane, 25
oxidn., 9
precursors of, formald., 9, 28
PAN, 128
product yields, 26–28
rate measurements, 26
reactivity, 25, 26
unsaturated, 9, 128, 129
Hydrochloric acid, 7, 9, 12, 47, 61, 88, 89, 91, 97, 98, 102, 103, 107, 111, 124, 135, 177, 180, 215, 216, 220, 224, 245, 249, 266, 268
hydrolysant, 5, 8, 11, 14
precursor of bis-chloromethyl ether, 57, 58
protein pptn., 117
reaction with CH_2O, 206
structure of products, 206
Hydrochlorothiazide, detm. with pararosaniline, 41
hydrolysis, 9
precursor of CH_2O, 9, 41
ring opening, 41
Hydrocortisone, detm. with, J-acid, 41
MBTH, 41
oxidn., 9
precursor of CH_2O, 9
spectral data, 41
Hydrogen cyanide, formn. from CO, 276
Hydrogen peroxide, comutagen, 1
formn., 140
mutation enhancement, 81
oxidant, 9, 10, 14, 61, 106, 139, 140
Hydrolysis, acetals, 213
alar, 2, 277
aramite with alkali, 3
asarin with H_2SO_4, 3
bis-chloromethyl ether, 3
bromochloromethane, 232
2-bromo-2-nitropropan-1,3-diol, 3
2-butoxyethyl 2,4-D, 278
carboxymethylcellulose, 4
cellulose, 4
cellulose acetate formal, 4
cellulose formal with H_2SO_4, 4
chloracetic acid with H_2SO_4, 4
chloramphenicol with acid, 4
1-chloro-2,3-epoxypropane, 4
chloromethyl methyl ether, 4
chlorophenoxyacetic acids, 4
cosmetic preservatives, 220
cycasine with acid, 4
cyclohexylazomethane, 4
2,4-dichlorophenoxyacetates, 5
diethoxymethane, 5
dimethoate with HCl, 5
N,N-dimethylaniline, 5
N,N-dimethylaniline N-oxide, 5
1,1-dimethylhydrazine, 5
1,2-dimethylhydrazine, 5
dimethylsulfoxide, 5
DNA, 77
epichlorohydrin, 6
epoxides, 81
Ethion, 6
ethyl N,N-bis-hydroxymethyl carbamate, 6
ethylene oxide, 6, 81
ethylenimine, 6, 81
ethyl N-hydroxymethylcarbamate, 6
formald. dimethylacetal, 6
formald.-tanned collagen, 6
formald. 2,4-dinitrophenylhydrazone, 7
formald., hydrazone, 7
formyl group, 7, 89
glucaric acid, 7
glycerides, 91
glycerophosphatides, 7
N-glycolyneuraminate, 8, 41
glyoxylic acid, 277
Guthion with HCl, 8
hexahydro-1,3,5-trinitro-5-triazine, 8
hexamethylenetetramine, 8, 282
hydrochlorothiazide, 9
1-hydroxymethyl-5-,5-dimethylhydantoin, 9
N-hydroxymethylurea, 9
imidan, 9
isonitrile group, 9, 89
methionine hydantoin sulfoxide, 10
methionine sulfoxide, 10
methoxone, 141, 142
methoxy compounds, 10
methylal, 10
methylazobutane, 10
methylazoxymethanol acetate, 10
1-methyl-2-butylhydrazine, 11
1-methyl-1-cyclohexylhydrazine, 11
1-methyl-2-cyclohexylhydrazine, 11
N,N-methylene-bis-acrylamide, 11
methylene chloride, 11, 277

Hydrolysis—*continued*
  methylenedioxo compounds, 11
  methylenedioxyphenol compounds, 11
  methylenedithio compounds, 11
  methyl esters, 11
  methyl methacrylate, 11, 277
  methyl methanesulfonate, 139
  methyl methylthiomethyl sulfoxide, 228
    1-methyl-1-p-nitrophenyl-2-methoxy azonium tetrafluoroborate, 11
  methylpropylhydrazine, 11
  methylsulfoxymethyl methyl sulfide, 229
  morphine-N-oxide, 11
  myosalvarson, 11
  narceine, 11
  narcotine, 11
  nialate, 11
  2-nitro-1-alkanols, 11, 47, 231
  nitromethanol, 11, 277
  2-oxapentoate, 12
  paraformald., 12
  penicillin V, 12
  phenoxyacetic acids, 12
  phenylazomethane, 12
  phenylephrine, 277
  phorate, 12, 229
  phosphatides, 279
  phospholipids, 279
  S-piperidinomethylthiobenzoate, 161
  piperine, 12, 282
  piperonal, 282
  piperonyl butoxide, 12, 282
  piperonylic acids, 12, 282
  pivampicillin, 216
  podophyllotoxin, 12
  α-polyoxymethylene, 12
  1,2-propanediol phosphate, 12
  protoporphyrin monomethyl ester, 13
  protein-CH_2O, 225
  raceophenidol, 13
  safrole, 13, 282
  sesamin, 13
  N,N-α,α-tetramethylheptylamine oxide, 14
  thiazolidine-4-carboxylic acid, 14
  thiol-bound formald., 14
  p-tosyl arginine methyl ester, 14
  triglycerides, 14
  trimethylamine oxide, 14
  trimethylhydrazine, 14
  sym-trioxane, 14
  sym-trithiane, 14, 282
  trithion, 14
  trioxane, 215
  tropine N-oxide, 14
  urea-CH_1O, 282
Hydroperoxides, formn. from olefins, 80
Hydroquinone, formn. from *p*-hydroxybenzyl alcohol, 130
Hydroxyacetone, detm. with MBTH, 42
  oxidn., 9, 131
  precursor of CH_2O, 9, 131
α-Hydroxyalkyl-5-glutathiones, formn., 80
3-Hydroxyanthranilic acid, formn. from tryptophan, 191
p-Hydroxyarylglycerol-β-aryl ethers, 134
p-Hydroxybenzaldehyde, 134
3-Hydroxy BaP, metabolite of BaP, 18
6-Hydroxy BaP, metabolite of BaP, 18
4-Hydroxybenzoate, formn. from, 4-ethoxybenzoate, 190
    4-methoxybenzoate, 190
p-Hydroxybenzyl alcohol, oxidn. eqn., 130
  precursor of, p-benzoquinone, 130
    formald., 130
    hydroquinone, 130
    1-oxaspiro(2.5) octa-4,7-dien-6-one, 130
  structure, 130
p-Hydroxybenzyl alcohols, 129, 130
  oxidn., 129
  precursors of, benzoquinones, 129, 130
    formald., 129, 130
p-Hydroxycinnamyl alcohol, formn. from lignin, 134
16α-Hydroxy-compound "S", precursor of CH_2O, 19
14α-Hydroxy-cortexone, precursor of CH_2O, 19
16α-Hydroxy-cortexone, precursor of CH_2O, 19
18-Hydroxy-cortexone, precursor of CH_2O, 19
19-Hydroxy-cortexone, precursor of CH_2O, 19
6β-Hydroxy-corticosterone, precursor of CH_2O, 19
2α-Hydroxy-cortisol, precursor of CH_2O, 19
2β-Hydroxy-cortisol, precursor of CH_2O, 19
16α-Hydroxy-cortisol, precursor of CH_2O, 19
6β-Hydroxy-cortisone, precursor of CH_2O, 19
16α-Hydroxy-cortisone, precursor of CH_2O, 19
3-Hydroxy-N,N-dimethyl-cis-crotonamide, *see* Bidrin
3-Hydroxyestratrienes, O-methylation, 192
3′-Hydroxyhexobarbital, formn. from hexobarbital, 178
  precursor of, formald., 178
    3′-ketohexobarbital, 178
    3′-ketonorhexobarbital, 178
  structure, 178
5-Hydroxy-3-indolylacetic acid, precursor of CH_2O, 9

reaction with CH_2O, 223
17-Hydroxy-17-ketolsteroids, detm. with HBT, 42
oxidn., 9
precursors of CH_2O, 9
spectral data, 42
3-Hydroxykynurenine, formn. from tryptophan, 191
Hydroxylactone, detm. with CA, 42
oxidn., 9
precursor of CH_2O, 9, 16
spectral data, 42
yield of CH_2O, 16
Hydroxylamine, collection of ald., 81
formn. from methylnitrosamine, 180
Hydroxylamines, 186, 187
Hydroxylase, effect of TCDD, 147
Hydroxylation, aniline, 138
dialkylnitrosamines, 78
dimethylnitrosamine, 70
fetal, 53
methyl BaP, 18
methyl groups, 18
Hydroxyl compounds, reaction with CH_2O, 206
structure of products, 206
5-Hydroxylysine, detm. with, MBTH, 42
2,4-pentanedione, 42
phenylhydrazine, 42
oxidn., 9, 131, 253
oxidn. eqn., 131
precursor of, α-aminoglutaric γ-semiald., 131
formald., 9, 42, 131, 253
reaction, 221
spectral data, 42
structure, 131
4-Hydroxy-3-methoxyacetophenone, formn., 133
structure, 133
Hydroxymethylamine formn., 221, 222
Hydroxymethylation, benzene, 18
benzo(a)pyrene, BaP, 18
6-Hydroxymethyl BaP, carcinogen, 18
formn. from, BaP, 18
6-methyl BaP, 18
proximate carcinogen, 18
TLC sepn., 18
UV abs. spectra, 18
Hydroxymethylbidrin, formn. from Bidrin, 158
precursor of, N-demethylbidrin, 158
formald., 158
structure, 158
1-Hydroxymethyl-5,5-dimethylhydantoin, anal. proc., 220, 221
cosmetic preservative, 220
detm. with 2,4-pentanedione, 47, 220
hydrol., 9
precursor of CH_2O, 9, 47, 193, 220
standard, 193
structure, 193, 220
yield of CH_2O, 47
5-Hydroxymethyl-2-furaldehyde, formn. from glucose, 60, 90
interference, 137
precursor of CH_2O, 90
N-Hydroxy methyl groups, 226
crosslinkage, 226
reaction with, histidine, 226
trytophan, 226
tyrosine, 226
Hydroxymethyl methylnitrosamine, 79
H-Hydroxymethylpiperidine, nitrosation, 161
precursor of, formald., 161
N-nitrosopiperidine, 161
structure, 161
N-Hydroxymethylurea, hydrolysis, 9
precursor of CH_2O, 9
4-Hydroxy-2-oxobutyrate, disproportionation, 9
precursor of CH_2O, 9
Hydroxyoxobutyrate aldolase, (EC4.1.2.1), 9
3-Hydroxyphenylethylamines, reaction with CH_2O, 239, 242
21-Hydroxy-pregn-4-ene-3,20-dione, precursor of CH_2O, 19
21-Hydroxy-pregn-4-ene-3,11,20-trione, precursor of CH_2O, 19
Hydroxyproline, oxidn. with periodate, 9
precursor of CH_2O, 9
yield of CH_2O, 16
4-Hydroxystyrenes, 132–134
anal. proc., 132, 133
detm. with MBTH, 132–134
precursor of CH_2O, 132, 133
5-Hydroxytryptamine, precursor of 5-hydroxytryptoline, 238
reaction with CH_2O, 238
5-Hydroxytryptoline, formn. from 5-hydroxytryptamine, 238
5-Hydroxytryptophan, Fexc/em, 242
Hyflo Super-Cel, 59
Hyperglycinemia, 94, 97
kerotic, 95
nonkerotic, 95
Hyperlipidemia, 255, 264, 265
Hyperoxaluria, 245

Imidan, detm. in milk, 280
detm. with CA, 42, 280

Imidan—*continued*
extraction, 280
hydrol., 9
precursor of CH_2O, 9, 42, 227, 280
structure, 16, 227
Imines, reaction with CH_2O, 205
structure of products, 205
Indole, detm. of, ald., 244
formate, 244
oxalate, 244
detm. with MBTH, 242
3-Indoleacetic acid, detm. with CA, 42
disproportionation, 9
precursor of CH_2O, 9, 42
spectral data, 42
Indole compounds, reaction with metaldehyde, 204
Indoleethylamine N-methyltransferase, 159
Indoles, reaction with CH_2O, 208
structure, 208
of product, 208
3-Indolylethylamines, reaction with CH_2O, 239, 242
Industrial air, detm. of, acrolein, 275
1-alkenes, 29
2,4-butadiene, 29
2,2′-dimethyldioxane, 85
dimethyl terephthalate, 138
p-dioxane, 85
epichlorohydrin, 85
ethylene glycol, 85
ethylene oxide, 85
isoprene, 29
methacrylate, 138
methanol, 137
methoxone, 141
2-methoxyethanol, 85
vinyl butyl ether, 275
vinyl compounds, 274
5-vinyl-2-picoline, 275
2-vinylpyridine, 275
formn. of nitrosamines, 74
solvent vapors, 166
Industrial effluents, DMNA, 64
formald., 217–219
nitrosamines, 64, 65
Industrial exposure, DMNA, 77
vinyl chloride, 273
Industry, vinyl chloride and hepatic angiosarcoma, 273, 274
Ingestion, env. contaminants, 175
Inhalants, benzene, 166
carbon disulfide, 51
carbon tetrachloride, 166
cotton flax, 212
effect on metab., 166
env. contaminants, 175
ferric oxide, 68
hempa, 154
hemp dust, 212
methyl n-butyl nitrosamine, 68
methylene chloride, 166
trichloroethylene, 166
Inheritance, cancer memory, 54
effect of, 7,12-dimethyl BaA, 54
nitrosomethylurethan, 54
intergenerational, 172
short term, 54
Inhibitors, effect on N-demethylases, 144–150
Initiation, 251
Inositol, detm. in pharm. prepns., 280
detm. with 2,7-naphthalenediol, 42, 280
oxidn., 9
precursor of CH_2O, 280
Injection, intraperitoneal, 55
subcutaneous, 55
Insecticide, Bidrin, 157
Interference removal, bisulfite in CH_2O detm., 153
detm. of ethylene glycol, 85
Interferences, solvent partition, 256
sulfite in detm. of CH_2O, 152
Intestines, ethylmorphine N-demethylase, 175
metab. of xenobiotics, 1
tumors, 55
Iodine, 108, 267
oxidant, 14, 153, 154, 228
3-Iodopropene, detm. with MBTH, 42
oxidn., 9
precursor of CH_2O, 9, 42
spectral data, 42
Ion exchange CC, 89
Iron, oxidn. catalyst, 177
Irradiation, 23
ethylene and NO_2, 24
isobutene, NO and air, 25
solvents, 100, 101
Isoatisine, detm. with CA, 42
oxidn., 9
precursor of CH_2O, 9, 42
spectral data, 42
Isobutene, photooxidn., 18
precursor of CH_2O, 18
yield of CH_2O, 18
Isobutyl acetate, photooxidn., 26, 27
precursor of CH_2O, 26, 27
yield of CH_2O, 26, 27
Isobutyraldehyde, detection with 2,3-dimethyl-2,3-bis(hydroxylamino)butane, 202
rate of reaction, 203
spectral data, 203
Isocitrate, 171, 184

dehydrogenase, 184
Isoeugenol, detm. with MBTH, 42
oxidn., 9
precursor of CH_2O, 9, 42
spectral data, 42
Isoleucinol, detm. with CA, 84
Isomaltitol, oxidn. with periodate, 9
precursor of CH_2O, 9
Isonitrile group, anal. eqn., 89
detm. with CA, 89
hydrol., 9, 89
precursor of, formald., 8, 89
formic acid, 89
redn., 9
Isophorone, photooxidn., 26, 27
precursor of CH_2O, 26, 27
yield of CH_2O, 26, 27
Isoprene, detm. with CA, 29
precursor of CH_2O, 29
Isopropanol, 91, 94
extractant, 200
photooxidn., 26, 27
precursor of CH_2O, 26, 27
yield of CH_2O, 26, 27
Isopropyl ether, 92, 256
Isosafrole, carc., 81
detm. with CA, 42
oxidn., 9
precursor of CH_2O, 9, 42
spectral data, 42
Isotenulin, detm. with CA, 42
oxidn., 9
precursor of CH_2O, 9, 42
Itaconic acid, detm. with CA, 42
oxidn., 9
precursor of CH_2O, 9
spectral data, 42

J-Acid, 107
detect. of 1,2-glycols, 109, 111
detm. of, acrylamide, 31, 34
acrylic acid, 31, 34
1-alkenes, 29
allyl alcohol, 31, 34
allyl amine, 31
allylbenzene, 34
arabitol, 35
1-buten-3-ol, 35
3-buten-1-ol, 31, 35
chloroacetate, 36
2-chlorophenoxyacetate, 36
corticosteroids, 36
2-deoxy-D-glucose, 36
2-deoxy-D-ribose, 37
2,4-dichlorophenoxyacetates, 37
dihydroxyacetone, 37
dulcitol, 38
erythritol, 38
ethylene glycol, 38, 103
eugenol, 31
formald., 199
formald. 2,4-DNPH, 39
formald. hydrazone, 39
formic acid, 87
fructose, 39
glycerol, 40
glycolic acids, 40
1,2-glycols, 41,
glyoxylic acid, 41
hexamethylenetetramine, 41
1-hexene, 41
hexitols, 41
hydrocortisone, 41
mannitol, 42
N,N′-methylene-bis-acrylamide, 43
phenoxyacetic acids, 101
piperonal, 46
β-piperonylacrylic acid, 46
piperonylic acid, 44
quinine, 31
ribose, 44
safrole, 31, 44
sorbitol, 44
terpenes, 31
sym-trioxane, 45
sym-trithiane, 45
vinyl compounds, 31
water-sol. olefins, 29
xylitol, 45
eqn. for formate detm., 88
fluor. anal., 31, 32, 44, 45
spectral data, 44, 45
spot test for formate, 87, 88
structure, 88, 103

Keratin, reaction with CH_2O, 229
SH-bound CH_2O, 222
Ketene, reaction with CH_2O, 251
structure, 251
3′-Ketohexobarbital, formn. from, hexobarbital, 178
3′-hydroxyhexobarbital, 178
precursor of, formald., 178
3′-ketonorhexobarbital, 178
structure, 178
2-Keto-4-hydroxyglutarate aldolase, binding with CH_2O, 230
inactivation by cyanide, 231
prepn., 230

2-Keto-4-hydroxypyruvate, disproportionation, 230
precursor of, formald., 230
pyruvate, 230
structure, 230
20,21-Ketolic steroids, detm. with 2,4-pentanedione, 45
20,21α-Ketols, oxidn., 9
precursor of CH_2O, 9
Ketones, dialkyl, photooxidn., 9
formn. from auto exhaust, 23
methyl, photooxidn., 9
3′-Ketonorhexobarbital, formn. from hexobarbital, 178
3′-hydroxyhexobarbital, 178
3′-ketohexobarbital, 178
structure, 178
Ketoses, 116
precursors of, formald, 116
furfurals, 116
Ketotetrose aldolase, 6
Kidney, activation of DMNA, 76
cancer, 68
demethylating activity, 150
detm. of demethylase, 191
DNA methylation, 70
effect of acetoaminonitrile, 75
effect on mutagenicity, 273
metab. of DMNA, 76
microsome fractions, 76
tumors, 76
tumors from, 1-aryl-3,3-dialkyltriazenes, 125
DMNA, 68, 69
Kieselgel, 266
Kieselguhr G, 259
phospholipid removal, 264
Kinetics, DMNA, demethylation, 73
mutagenesis, 73
Kynurenic acid, formn. from tryptophan, 190
Kynurenine, formn. from tryptophan, 190

Lachrymators, acrolein, 28
atmospheric precursors, 128
formald., 128
PAN, 128
Lactic acid, contamination, 280
detm. of methanol, 280
Lactic dehydrogenase, dismutation of glyoxylate, 114
in, *Aspergillus niger*, 114
pig heart, 114
rabbit muscle, 114
Lactoflavin, see Riboflavin, 254
Lactose, oxidn., 9, 134
precursor of CH_2O, 9, 134
Lactulose, oxidn., 9
oxidn. eqn., 105
precursor of, carbohydrate ald., 105
formald., 9, 105
formic acid, 105
glycolic acid, 105
structure, 105
Lake water, nitrosamines, 65
Laminarin, oxidn., 10
precursor of CH_2O, 10
LD_{50}, DMNA, 7, 68, 172
Lead dioxide, oxidant, 201
Lead tetraacetate, oxidant, 10, 164, 278
Lecithin, detm. in serum, 93
Lethal synthesis, aminopyrine, 67
N-methyl N-phenyl nitrosamine, 67
nitrite, 67
nitrosamines, 63–65
Lethal syntheses, BaP→6BaPCH_2^+(?), 18
ultimate carcinogens, 52
Leucine, effect on d-AMP, 210
Leucinol, detm. with CA, 44
Leucodrin methyl ether, 254
Leukemia, effect of Natulan, 151
Levulinic acid, formn. from glucose, 60
Levulose, oxidn., 10
precursor of CH_2O, 10
Light petroleum, 136
Lignin, 130
degradation, 134
disproportionation, 10
precursor of CH_2O, 10, 58
Limonene, ald. formn., 250
oxidn., 10
precursor of CH_2O, 10
ring rearrangement, 250
structure, 250
Linalool, oxidn., 11, 16
precursor of CH_2O, 11, 16
Linalyl acetate, oxidn., 11, 16
precursor of CH_2O, 11, 16
Lincomycin, precursor of CH_2O, 173
structure, 173
Lipid peroxidation, drug metabolism, 50
effect of, dimethylnitrosamine, 179
EDTA, 176
ferrous ion, 176
effect on, demethylation, 176
microsomal membranes, 51
formation of malonald., 176
inhibition, 50, 51
microsomal, 138
relation to N-demethylation, 176
stimulation with, ascorbic acid, 51
carbon monoxide, 51
Lipids, detm. of glycerides, 279

Lipoprotein metabolism, abnormalities, 255
Lipoproteins, electrophoresis, 264
Liquid smoke, nitrosamines, 63
Liquors, detm. of glycerol, 279
Lithium hydroxide, 97, 98
Lithium sulfate, 118
Liver, activation of Me_2NNO, 76
  activity of ethylmorphine N-demethylase, 175
  angiosarcoma from vinyl chloride, 274
  cancer, 68
  carc. with, dimethylhydrazines, 181
    ethylhydrazine, 181
    methylhydrazine, 181
  cytochrome P-450, 143
  N-demethylase, 70
  demethylating activity, 150, 168, 178
  detm. of, demethylase, 191
    ethylene glycol, 85
    ethylmorphine, 279
    tryptophan, 242
  DNA, effect of $(^{14}C)Me_2NNO$, 77
  effect of, acetoaminonitrile, 75
    aminopyrine and nitrite, 67
    pollutants, 166
  embryo N-demethylase, 78
  endoplasmic reticulum, 123
  formn. of epoxides, 80
  guinea pig embryo, 76, 77
  homogenate, 85
  human embryo, 76, 77
  male rat N-demethylase, 156
  metabolism of, Me_2NNO, 76
    xenobiotics, 175
  microsomal enzymes, 50, 51, 74, 75, 272
    effect of vinyl chloride, 272
  microsome fraction, 76, 122, 125, 126, 153, 165, 168, 178, 232, 233
    prepn., 154
  nucleic acid methylation, 74
  rat embryo, 76, 77
  tumor formn., 67
  tumors, 55, 76
  tumors from DMNA, 74, 75
Liver fraction, effect on mutagenicity, 273
Liver microsomal fractions, 273
Liver microsomal hydroxylation, activity factors, 174
  inducers, 174
  nomenclature, 174
  reactions catalyzed, 174
  substrates, 174
Liver microsomes aminopyrine demethylase, 166
  N-demethylation activity, 70, 147
  enzyme induction, 50
  hydroxylation, 74
  metab. of Me_2NNO, 78
Liver N-oxidase, 147
Liver RNA, 6-dimethylaminopurine, 147
  6-methylaminopurine, 147
Liver tissue proliferation, 183
    induction by, Aroclor 1254, 183
    phenobarbital, 183
Liver tumors, 68, 69
  lack of protein dye, 226
  DMNA, 75
Lloyd's reagent, phospholipid removal, 264, 267
Los Angeles, smog, 128
L. plantarum, redn. of glyoxylate, 114
Lubricating oil, source of CH_2O, 197
Lung, activation of DMNA, 76
  carc. of bis-chloromethyl ether, 57
  demethylating activity, 150
  DNA methylation, 70
  metabolism of, DMNA, 76
    xenobiotics, 175
  squamous carcinoma, 57
  tumors, 76
  tumors from dimethylnitrosamine, 68, 69
Lung adenoma, lethal synth., 185
Lung cancer, 68
Lung fraction, effect on mutagenicity, 273
Lung N-oxidase, 147
Lymphocyte cultures, vinyl chloride workers, 273
Lymphosarcoma, effect of Natulan, 151
Lysine, 104
  reaction of amine group, 221
Lysine-methylene tyrosine formn., 221
Lysyl residue, binding with CH_2O, 230, 231

Macrozamin, precursor of CH_2O, 55
  presence in cycad seeds, 55
Magnesium, 169
  chloride, 70, 154, 184
  reductant, 4, 7–9, 12, 87–89
  stimulation of N-demethylase, 146
  sulfate, 119, 178
Magic acid, effect on DMNA, 79
Male rat, effect of, age, 156
    TCDD, 156
Maleic acid hydrazide, detm. with MBTH, 42
  oxidn., 10
  precursor of CH_2O, 10, 42
  spectral data, 42
Maleylpyruvic acid isomerase, 204
Malonaldehyde, 179
  detm. with TBA, 138
  effect on demethylase, 51

Malonaldehyde—*continued*
formn. from cis-ocimene, 29
lipid peroxidn., 176
role in aging, 237
Malonaldehydic acid, formn. from cis-ocimene, 29
Malonic acid, formn. from cis-ocimene, 29
Maltitol, detm. with, orcinol, 117
2,4-pentanedione, 117
oxidn., 10
oxidn. eqn., 112
precursor of, carbohydrate polyald., 112
formald., 10, 112, 117
formic acid, 112
furfurals, 117
sepn., 117
Maltose, oxidn., 10, 135
oxidn. eqn., 112
precursor of, carbohydrate polyald., 112
formald., 10, 112, 135
formic acid, 112, 135
maltitol, 112
furfurals, 117
reduction, 112, 135
structure, 112
sepn., 117
Maltotetraose, detm. with orcinol, 117
precursor of furfurals, 117
sepn., 117
Maltotetraitol, detm. with, orcinol, 117
2,4-pentanedione, 117
precursor of, formald., 117
furfurals, 117
sepn., 117
Maltotriitol, detm. with, orcinol, 117
2,4-pentanedione, 117
precursor of, formald., 117
furfurals, 117
sepn., 117
Maltotriose, detm. with ordinol, 117
precursor of furfurals, 117
sepn., 117
Mammalian liver enzymes, cleavage of, 6-methythiopurine, 192
prometryne, 192
effect of PAH, 192
Man, acute necrosis, 77
anticarc. of, hemel, 120
trimethylolmelamine, 121
carc. of nitrosamines, 76, 77
chronic leukemia, 151
cirrhosis of liver, 77
effect of, carbon disulfide, 51, 166
phenylbutazone, 144, 146
Hodgkin's disease, 151
lymphosarcoma, 151
metab. of, DIC, 125
Natulan, 151
urinary metabolites, 125
Mandelamine, anal. proc., 224
precursor of CH_2O, 223
reaction with hydroxyindoleacetate, 223
treatment of carcinoid, 223
Manganese (III), ethylenediaminetetraacetate, precursor of CH_2O, 48
oxidant, 6
Manganous sulfate, removal of SO_2, 198
Mannitol, 135, 136
anal. proc., 135
detm. in, plasma, 280
serum, 280
urine, 280
detm. with, CA, 42, 135, 280
dimedone, 42
J-acid, 42
MBTH, 42, 135
2,4-pentanedione, 42
distribution coefficient, 116
formald. standard, 135
location, 109
oxidn., 10, 17, 253
precursor of CH_2O, 10, 17, 21, 109, 116, 135, 253
preparation, 135, 136
primary standard, 42
sepn., 116
spectral data, 42
yield of CH_2O, 16
Mannonic acid, location, 109
precursor of CH_2O, 109
Mannose, oxidn., 253
precursor of CH_2O, 15, 253
redn., 135, 136
Mannose phenylhydrazone, oxidn., 253
precursor of CH_2O, 253
Marginal effect dose, DMNA, 77
MBTH, *see* 3-methyl-2-benzothiazolinone hydrazone
Meat, effect on, mouse liver, 169
rat liver, 169
Melamine, 217
Membranes, effect of lipid peroxidn., 51
Memory, 7,12-dimethyl BaA in mice, 54
nitrosomethylurethan, 54
transplacental carc., 54
Menadione, 138
activation effect, 236
effect in CH_2O formn., 236
structure, 236
Menthol, effect on aminopyrine N-demethylase, 144
Meperidine, N-demethylation, 149

precursor of CH_2O, 149
Mercaptans, reaction with CH_2O, 206
structures of products, 206
Mercaptoamines, reaction with CH_2O, 206
structure of product, 206
Mercaptoethanol, inhibitor of N-oxidn., 147
Mercuric chloride, hydrolysant, 14
release of CH_2O, 222
Schiff test, 222
Mercuric sulfate, oxidant, 14, 122, 123
tautomerism, 5, 11, 14
Mesitylene, photooxidn., 10, 18, 26, 27
precursor of, aliph. ald., 18
formald., 10, 18, 26, 27
yield of aliph. ald., 18
formald., 18, 26, 27
Mesityl oxide, photooxidn., 10, 26, 27
precursor of CH_2O, 10, 26, 27
yield of CH_2O, 26, 27
Mesoxalaldehyde osazone, formn. from glucose phenylosazone, 90
structure, 90
Metabolism, chloropromazine, 159
dealkylation, 74, 143
dimethylnitrosamine, 68, 70–77
N,N-dimethylurea, 159
embryonal, 77
ethylmorphine, 143
kidney, 185
link to carc. and mutag., 76
liver, 185
lung, 185
methadone, 159
methyl methanesulfonate, 139
Natulan, 151
one-carbon, 152
pesticides, 143
vinyl chloride, 271, 272
Metaldehyde, reaction with, ergocrostin, 204
indole compds., 204
Metapyrone, inhibitor of N-oxidn., 147
Metaraminol, anal. eqn., 240
histochem. detm., 240
spectral data, 242
structure, 240
Metasaccharinic acids, oxidn. with periodate, 10
precursor of CH_2O, 10
Methacrylamide, detm. with CA, 42
oxidn., 10
precursor of CH_2O, 10, 42
spectral data, 42
Methadol, demethylation, 156
formn. from acetylmethadol, 156
precursor of, formald., 156
normethadol, 156
structure, 156
Methadone, 155
N-demethylation, 149, 159
metab. eqn., 159
precursor of CH_2O, 149
structure, 159
TLC of metabolites, 159
treatment of opium habit, 159
dl-Methamphetamine, N-demethylation, 149
precursor of CH_2O, 149
Methane, atm. residence time, 193
oxidn., 10
precursor of CH_2O, 10, 193
Methane bis (N,N′-(5-ureido-2,4-diketo-tetrahydroimidazole)N,N′-dimethylol), anal. proc., 220, 221
cosmetic preservative, 220
detm. with 2,4-pentanedione, 220
precursor of CH_2O, 220
structure, 220
Methanol, 94, 109, 133, 136–140, 164, 168, 169, 220, 268
anal. proc., 136, 137
detm. in, alc. beverages, 136, 280
biol. fluids, 280
blood, 280
cerebrospinal fluid, 136
combustion effluents, 280
essential oils, 280
fermentation products, 280
forensic materials, 280
impure lactic acid, 280
industrial air, 137
serum, 136
soil, 280
wines, 280
detm. of, cholesterol, 139
glucose, 139, 140
uric acid, 139
detm. with, CA, 42, 136, 137, 280
2,7-dihydroxynaphthalene, 42, 280
pararosaniline, 42, 280
2,4-pentanedione, 42, 137
phenylhydrazine, 43
effect on aminopyrine N-demethylase, 144
enz. oxidn., 138
formn. from S-adenosylmethionine, 238
gasoline extender, 137
inhibition of N-demethylase, 146
irradiation, 100
oxidn., 7, 10, 61, 238, 282
precursor of CH_2O, 7, 61, 100, 136–140, 167, 268, 280, 282
source of energy, 138
stabilizer of CH_2O, 212
triglyceride extn., 264

Methanol peroxidase, detm. with CA, 139
Methenamine, *see* Hexamethylenetetramine
Methenamine mandelate, *see* Mandelamine
Methionine, effect on carcinogenesis, 143
  formn. from vinyl chloride, 272
  formn. of S-adenosylmethionine, 192
Methionine hydantoin sulfoxide, acetylation, 10
  anal. proc., 141
  detm. with CA, 43, 141
  precursor of CH_2O, 10, 141
    spectral data, 43
Methionine sulfoxide, acetylation, 10
  anal. eqn., 140
  anal. proc., 141
  detm. in protein, 141
  detm. with CA, 43, 140, 141
  precursor of CH_2O, 10, 16, 43, 140, 141
  spectral data, 43
Methoxone, 101
  detm. in industrial air, 141
  detm. with CA, 141, 142
    hydrol., 141
  oxidn., 10
  precursor of CH_2O, 10, 141
  structure, 141
Methoxyarenes, metab. to CH_2O, 187
  oxidn. to CH_2O, 187
  precursors of phenols, 187
p-Methoxybenzenediazonium tetrafluoroborate, detm. of urobilinogen, 56
4-Methoxybenzoate, metab. eqn., 190
  monooxygenase, 190
  precursor of, formald, 190
    4-hydroxybenzoate, 190
Methoxychlor, precursor of formald., 190
Methoxy compounds, hydrolysis, 10
  oxidn., 10
  precursors of CH_2O, 10, 16
4-Methoxybenzoate, oxidn., 10
  precursor of CH_2O, 10
Methoxy compounds, detm., 139, 282
2-Methoxyethanol, detm. with CA, 85
  precursor of CH_2O, 85
Methoxyflurane, human metabolism, 142
  precursor of CH_2O, 142
  structure, 142
4-Methoxy-3-hydroxyphenylethylamine, reaction with CH_2O, 237
1-(3-Methoxy-4-hydroxyphenyl)propen-3-ol, oxidn., 133
  precursor of, acetald., 133
    formald., 133
    4-hydroxy-3-methoxyacetophenone, 133
    vanillin, 133
  structure, 133
8-Methoxykynurenic acid, carcinogen, 191
  demethylation, 191
  detm. with 2,4-pentanedione, 191
  formn. from tryptophan, 190, 191
  metab. eqn., 191
  precursor of, formald., 190, 191
    xanthurenic acid, 191
  structure, 191
  urinary metabolite in, humans, 191
    monkeys, 191
    swine, 191
Methoxymethyl methanesulfonate, alkylating agent, 142
  detm., 142
  precursor of CH_2O, 142
  structure, 142
4-(*p*-Methoxyphenyl)-3-buten-2-one, detm. with MBTH, 32
  oxidn., 10
  precursor of CH_2O, 10
  spectral data, 43
Methyl acrylate, detm. in air, 138
  detm. with CA, 138
  precursor of CH_2O, 138
5-Methoxytryptamine, fluor. anal., 242
Methylal, hydrolysis, 10
  interf. in CH_2O detm., 213
  precursor of CH_2O, 10
  presence in formalin, 212
  structure, 213
Methylene bridge, and mutagenicity, 226
Methylene chloride, alk. hydrol., 277
  precursor of CH_2O, 277
Methyl methacrylate, alk. hydrol., 277
  precursor of CH_2O, 277
N-Methyl groups, crosslinkage, 226
  mutagenicity of CH_2O, 226
Methylakylamines, aniline formn., 5
  precursor of CH_2O, 5
  reaction with 2,4-dinitrochlorobenzene, 5
N-Methyl-N-alkylarylamines, oxidn., 10
  precursor of CH_2O, 10
Methylalkylhydrazines, anal. proc., 123
  detm. with CA, 123
  precursor of CH_2O, 123
Methylamine, detm. with, DCPIP, 160
    ninhydrin, 49
  formn. from, DMNA, 180, 181
  methylnitrosamine, 180, 181
  1-phenyl-3,3-dimethyltriazene, 126
  oxidn., 160
  oxidn. eqn., 160
  precursor of, formald., 47, 49
    methyl cation
    methyl diazonium cation, 161
  reaction with nitrite, 161

Methylamines, lake water, 65
precursors of nitrosamines, 65
sewage, 65
4-Methylaminoantipyrine, formn. from aminopyrine, 152
precursor of CH_2O, 152
structure, 152
4-Methylaminoazobenzene, *see* N-methyl-4-phenylazoaniline
N-Methylaminoazo dyes, detm. of N-demethylation, 171
N-Methyl 4-aminobenzoate, monooxygenase, 190
precursor of, 4-aminobenzoate, 190
formald., 190
2-Methylaminoethanol, collection, 281
detm. in air, 281
detm. with, CA, 43, 281
MBTH, 43
oxidn., 10
precursor of CH_2O, 10, 43, 281
spectral data, 43
6-Methylaminopurine, constitn. of liver RNA, 147
N-demethylation, 147
N-Methylaniline, detm. with 2,4-pentanedione, 167
enz. demethylation, 10, 149, 166
oxidn. eqn., 166
precarcinogen, 67, 185
precursor of, aniline, 166
formald., 16, 149, 166, 167
phenylhydroxylamine, 166
Methylanilines, detm. with MBTH, 163–166
precursors of CH_2O, 163–166
N-Methylarylamines, oxidn., 187
oxidn. eqn., 164
precursors of CH_2O, 164, 187
Methylation, BaP, 18
DMNA, 70
DNA, 69, 70, 74
nucleic acids, 75
protein, 69, 182
RNA, 69
Methylazobutane, detm. with CA, 43
hydrol., 10
precursor of, n-butanal, 43
formald., 10
spectral data, 43
tautomerism, 10
yield of n-butanal, 43
Methylazoxymethanol, alkylating agent, 54
carcinogen, 54
formn. from, azoxymethane, 53
cycasin, 55
mutagen, 55
precursor of CH_2O, 53, 54
electrophilic cation, 55
proxicarcinogen, 53, 55
structure, 52, 54
teratogen, 53, 55
TLC sepn., 55
Methylazoxymethanol acetate, alkylation of nucleic acids, 55
carcinogen, 55
hydrol., 10
precursor of, formald., 10, 55
electrophilic cation, 55
Methylazoxymethyl-B-D-glucopyranoside, *see* cycasin
6-Methyl BaP, formn., 18
metabolism, 18
Methylbenzenes, precursors of CH_2O, 23
Methyl benzoate, precursor of CH_2O, 26, 27
yield of CH_2O, 26, 27
3-Methyl-2-benzothiazolinone hydrazone, detm. of, acetals, 107
N-acetylglucosamine, 34
acrylamide, 34
acrylic acid, 30, 34
adenosine, 34
adonitol, 34
adrenaline, 48
aflatoxins, 188, 189
ald. 2,4-DNPH, 126
allyl alcohol, 34
allyl amine, 34
N-allyl N-methylaniline, 34
2-aminoacetophenone, 34
2-amino-1-alkanols, 34, 48
2-amino-2-deoxy-D-glucose, 35
2-aminoethanol, 34, 38, 48
arabinose, 35
arabitol, 35
arterenol, 48
bis-chloromethyl ether, 35
2-bromo-1-butene, 35
1,2,4-butanetriol, 35
1-buten-3-ol, 35
3-buten-1-ol, 35
calciferol, 35
camphene, 35
carbohydrates, 35, 278
chlorogenic acid, 36, 132
2-chlorophenoxyacetic acid, 36
cholesterol, 61
corticosteroids, 36
cycasin, 55
2-deoxy-D-glucose, 36
2-deoxy-D-ribose, 37, 61, 62
dextrose, 37
dibenzalacetone, 37

3-Methyl-2-benzothiazolinone hydrazone—*continued*
diethanolamine, 37
dihydro-β-carbolines, 242
dihydroxyacetone, 37
N,N-dimethyl-p-aminobenzaldehyde, 5, 37
N,N-dimethyl-4-aminobenzylidene aniline, 5, 37
N,N-dimethylanilines, p-substituted, 5
1,1-diphenylethylene, 37
N,N'-diphenylethylenediamine, 37
1,8-diphenyl-1,3,5,7-octatetraene, 37
1,5-diphenyl-3-pentadienone, 37
dulcitol, 38
erythritol, 38
ethylenediamines, 48
ethyleneglycol, 38, 84
eugenol, 31, 38
2-(2-fluorenylamine)ethanol, 39, 48
formald., 111, 197, 198
formic acid, 87
fructose, 39
furfural, 39
β-2-furylacrylophenone, 39
galactonic acid, 39
galactose, 39
gluconic acid, 39
gluconolactone, 39
glucosamine, 39, 48
glucose, 39, 139, 179
glucuronic acid, 39
glyceric acid, 40
glycerides, 40, 92
glycerol, 40
1,2-glycols, 41, 107, 280
glyoxylic acid, 111
1-heptene, 41
1-hexene, 41
hexitols, 41
hexosamines, 48
hydrocortisone, 41
hydroxyacetone, 42
4-hydroxybenzylideneacetophenone, 133
hydroxylysine, 42
4-hydroxystilbene, 133
4-hydroxystyrenes, 132, 133
indole, 242
3-iodopropene, 42
isoeugenol, 42, 133
isosafrole, 133
maleic acid, hydrazide, 42
mannitol, 42
3-methoxy-4-hydroxypropenylbenzene, 133
4-(p-methoxyphenyl)-3-buten-2-one, 43
4-methoxystilbene, 133
4-methoxystyrene, 133
N-methyl 4-aminobenzylide anilines, 168
2-methylaminoethanol, 43
N-methyl 4-aminostilbenes, 168
N-methylbenzylidene p-phenylenediamine, 168
9-methylcarbazole, 242
N-methyl 4-phenylazoanilines, 168
α-methylstyrene, 43
methylvinylketone, 43
N-(1-naphthyl)ethylenediamine, 43
1-nonadecene, 43
1-octene, 43
olefins, 28, 29
phenylephrine, 48
piperonylidene acetone, 44
4-propenylveratrole, 44, 133
propylene glycol, 44
quinine, 31
raffinose, 44
$R_2C{=}CH_2$, 22
ribose, 44
safrole, 31, 44
serine, 44, 48
sialoglycoprotein, 111
sorbitol, 44, 135
terpenes, 31
1-tetradecene, 45
1,2,3,4-tetrahydro-β-carbolines, 242
thiazolidine-4-carboxylic acid, 45
2,4,4-trimethyl-1-pentene, 45
vinyl chloride, 31
xylitol, 45
interference of SO_2, 197
spectral data, 44, 45
spot test for formate, 88
structure, 22
N-Methylbenzylamine, N-demethylation, 149
precursor of CH_2O, 149
Methyl borate, 136
2-Methyl-2-butene, photooxidn., 18
precursor of, aliph. ald., 18
formald., 18
yield of, aliph. ald., 18
formald., 18
1-Methyl-2-butylhydrazine, detm. with CA, 43
hydrolysis, 11
oxidn., 11
precursor of, n-butanal, 43
formald., 11, 43
tautomerism, 11
yield of n-butanal, 43
Methyl isobutyl ketone, precursor of CH_2O, 26, 27
yield of CH_2O, 26, 27
Methyl n-butyl ketone, precursor of CH_2O, 26, 27

yield of CH_2O, 26, 27
Methyl tert-butyl ketone, precursor of CH_2O, 26, 27
yield of CH_2O, 26, 27
Methyl n-butyl nitrosamine, carcinogen, 68
carcinoma of the esophagus, 77
inhalation, 77
permissible dose, 77
9-Methylcarbazole, detm. with MBTH, 242
4-Methylcatechol, nitrosamine formn, 66
Methyl cation, 79, 125, 126
formn. from DMNA, 70, 73, 232
1-phenyl-3-methyltriazene, 125
Methyl cellosolve, precursor of CH_2O, 16
Methyl cellulose, precursor of CH_2O, 16
N-Methyl *m*-chlorobenzylamine, N-demethylation, 149
precursor of CH_2O, 149
3-Methylcholanthrene, anticarc., 72
antitoxicant, 72
N-demethylase inducer, 50, 73, 144, 151, 170, 181
effect on, aminopyrine N-demethylase, 144, 146
cytochrome P-450, 170
DMNA N-demethylase, 75, 148
DMNA mutagenicity, 75
mice, 74
rats, 74
enzyme, depressant, 72
induction, 189, 232, 233
inhibitor of N-demethylase, 148, 149
pretreatment of, mice, 232, 233
rats, 69
repression of DMNA demethylase, 72
N-Methyl compounds, 142–187
demethylation, 143–152, 155–173, 175–187
enzymic substrates, 143–152, 155–162, 166–176, 178
metabolism, 142–152, 155–162, 166–176, 178–186
precursor of CH_2O, 143–152, 155–173, 175–187
O-Methyl compounds, 187–192
S-Methyl compounds, precursors of CH_2O, 192
1-Methyl-1-cyclohexylhydrazine, hydrolysis, 11
oxidn., 11
precursor of CH_2O, 11
tautomerism, 11
1-Methyl-2-cyclohexylhydrazine, detm. with CA, 43
hydrolysis, 11
oxidn., 11
precursor of CH_2O, 11, 43
spectral data, 43
tautomerism, 11
Methyldialkylamines, oxidn. with nitrite, 11
precursor of CH_2O, 11
Methyldiazonium cation, 70, 73, 123
formn. from, azomethane, 52
cycasin, 123
1,2-dimethylhydrazine, 52
1-phenyl-3,3-dimethyltriazene, 52
ulticarcinogen, 52
Methyldiethylamine oxide, precursor of CH_2O, 20
2-Methyl-6,7-dihydroxy-1,2,3,4-tetrahydroisoquinoline, formn. from epinine, 237
ident. by TLC, 237
metab. eqn., 237
structure, 237
3′-Methyl-4-dimethylaminoazobenzene, carc., 182
carc. of subcarc. dose, 182
effect of hepatectomy, 182
3-O-Methyldopa, precursor of, dopa, 191
dopamine, 191
formald., 191
structure, 191
S-Methylene amines, 227, 228
N,N′-Methylene-bis-acrylamide, detm. with, CA, 43
J-acid, 43
phenyl J-acid, 43
hydrol., 11
precursor of CH_2O, 11, 43
spectral data, 43
Methylene chloride, 92, 111
detm. with CA, 43
effect on, aminopyrine N-demethylase, 166
carboxyhemoglobin, 231
cytochrome, P-450, 166
hydrolysis, 11
metab. to, carbon monoxide, 231
formald., 231
precursor of CH_2O, 11, 43, 231
prodn., 231
use, 231
Methylene compds., hydrolysis, 193–234
precursors of CH_2O, 193–234
reaction with CH_2O, 205
Methylenediamines, 223–227
precursors of CH_2O, 223–227
Methylene dimethanesulfonate, half-life, 216
hydrol. eqn., 216
precursor of, formald., 216
methanesulfonic acid, 216
structure, 216
o-Methylenedioxybenzenes, detm. with, CA, 43

o-Methylenedioxybenzenes—*continued*
gallic acid, 43
tannic acid, 43
distribution, 212
environmental aspects, 16
examples, 212
hydrolysis, 11
pesticides, 212
precursors of CH_2O, 11
utilization, 212
Methylenedioxy compounds, 212–217
detm. with gallic acid, 46
hydrolysis, 11
precursors of CH_2O, 11, 212–217
Methylenedithio compounds, detm. by CA, 16
hydrolysis, 11
precursors of CH_2O, 11, 228–230
presence in polycaprolactam fibers, 16
Methylene glycol, 153
presence of formalin, 212
Methylene reductase, in brain regions, 234
5,10-Methylene tetrahydrofolate, effect on schizophrenia, 239
metab. eqn., 235
precursor of CH_2O, 234, 235
structure, 235
transfer of CH_2O, 231
Methylene tetrahydrofolate dehydrogenase, 1-carbon metab., 152
Methyl esters, alkylating agents, 138, 139
detm. with CA, 137, 138
enz. hydrol., 11
oxidn., 11
precursors of CH_2O, 11, 137–139
Methyl ethyl ketone, precursor of CH_2O, 26, 27
yield of CH_2O, 26, 27
Methyl formate, present in formalin, 212
3-Methylglucosamic acid, periodate oxidn., 253
precursor of CH_2O, 253
2-Methylglucose, periodate oxidn., 253
precursor of CH_2O, 253
3-Methylglucose, 253
7-Methylguanine, formn. from, guanosine, 125
nucleic acids, 125
Methylhydrazine, 123
carcinogen, 181
clastogen, 181
effect on DNA, 181
formn. from DMNA, 180
precursor of CH_2O, 181
3-Methyl-6-hydroxyisoquinoline, formn., 240
structure, 240
N-Methylhydroxylamine, formn. from DMNA, 180, 181
mutagen, 181
precursor of CH_2O, 181
reaction with cytosine, 181
Methylhydroxylamines, precursors of CH_2O, 186
N-Methylimidazoles, oxidn. eqn., 223
precursors of CH_2O, 223
structure, 223
Methyl iodide, atm. residence time, 193
carc., 193
Methylmalonic aciduria, hyperglycinemia, 95
N-Methylmelamine, anticarc., 120
Methyl methacrylate, detm. with pararosaniline, 43
hydrol., 11
oxidn., 11
precursor of CH_2O, 11, 43
Methyl methanesulfonate, alkylating agent, 139
carcinogen, 139
hydrol., 139
metabolism, 139
mutagen, 138
precursor of, formald., 139
methanol, 139
teratogen, 139
3-Methyl-4-methylaminoazobenzene, comparison with ethylmorphine, 169
N-demethylation, 169
detm. with CA, 169
metabolism, 169
precursor of, formald., 169
3-methyl-4-aminoazobenzene, 169
3-Methyl-4-methylaminoazobenzene N-demethylase, induction, 189
Methyl methylthiomethyl sulfoxide, hydrolysis, 228
precursor of CH_2O, 228
structure, 228
N-Methylmorpholine, precursor of, formald., 160
N-nitrosomorpholine, 160
reaction with nitrite, 160
N-Methylmorpholine N-oxide, detm. with 2,4-pentanedione, 16
formation of adduct, 11, 163
hydrolysis, 11
precursor of CH_2O, 11, 16, 163
Methyl nitrite, photolysis, 243
precursor of, carbon monoxide, 243
formald., 243
formic acid, 243
nitric acid, 243
nitrogen dioxide, 243
nitrogen pentoxide, 243
1-Methyl-1-*p*-nitrophenyl-2-methoxyazonium tetrafluoroborate,

diazotization, 11
hydrolysis, 11
precursor of CH_2O, 11, 222
reaction with quinoline N-oxide, 222
structure, 222
Methylnitrosamine, 79
formn. from DMNA, 180
precursor of, diazomethane, 180
hydrazine, 180
hydroxylamine, 180
methylamine, 180
nitrous acid, 180
N-Methyl N-nitrosourea, carc., 183
hydrol., 73
precursor of alkylating agent, 73, 74
Methylol groups, precursors of CH_2O, 218
N-Methyloctylamine, N-demethylation, 149
precursor of CH_2O, 149
N-Methylols, amides, 217
carbamates, 217
crosslinkage, 121, 233, 234
effect on DNA, 227
mutagenicity, 234
precursors of CH_2O, 217–222
reaction with, amides, 227
guanidines, 227
imidazoles, 227
indoles, 227
nucleic acids, 227
nucleotides, 227, 234
phenols, 227
ureas, 217
4-Methyl-3-pentene-2-one, *see* mesityl oxide
N-Methyl-4-phenylazoaniline, anal. eqn., 163
anal. proc., 164, 165, 168
carcinogen, 169
N-demethylation, 145, 163, 164
detm. with MBTH, 163, 164, 168
precursor of CH_2O, 145, 163, 171
spectral data, 165
structure, 163
N-Methyl 4-phenylazoanilines, anal. proc., 171
binding to proteins, 226
carc., 225, 226
detm. of N-demethylation, 171
metab. eqn., 225
precursor of CH_2O, 225
structure, 225
N-Methyl-*p*-phenylenediamine, precursor of CH_2O, 171
N-Methyl N-phenyl nitrosamine, carcinogen, 67
formn. from N-methylaniline, 67
Methyl phosphate, formn. from, O-demethylbidrin, 158
dimethylphosphate, 158
precursor of phosphoric acid, 158
structure, 158
Methyl piperonylate, effect on human lungs, 212
1-Methyl-2-propylhydrazine, detm. with CA, 43
hydrolysis, 11
oxidn., 11
precursor of, formald., 11, 43
n-propanal, 43
tautomerism, 11
spectral data, 43
yield of n-propanal, 43
Methyl n-propyl ketone, precursor of CH_2O, 26, 27
yield of CH_2O, 26, 27
N-Methyl pyrrolidone, photooxidn., 26, 27
precursor of CH_2O, 26, 27
1-Methylquinaldinium salts, detm. of formic acid, 87
α-Methylstyrene, detm. with MBTH, 43
oxidn., 11
precursor of CH_2O, 11, 26, 17
spectral data, 43
yield of CH_2O, 26, 27
Methyl sulfoxides, anal. eqn., 140
detm. with CA, 140, 141
precursors of CH_2O, 140
Methylsulfoxymethyl methyl sulfide, hydrol. eqn., 229
precursor of, ald., 229
dimethyl disulfide, 229
5-Methyltetrahydrofolate-N-methyltransferase, 236
5-Methyltetrahydrofolic acid, 234–239
metab. eqn., 235
precursor of, formald., 234–237
5,10-methylene tetrahydrofolate, 236
tetrahydrofolate, 236
reaction with tryptamines, 238
role in schizophrenia, 239
structure, 235
6-Methylthiopurine, metabolism, 192
enhanced by PAH, 192
precursor of CH_2O, 192
structure, 192
Methyl trithion, detm. with CA, 46
precursor of CH_2O, 46, 229
structure, 230
N-Methyltryptamine, precursor of 1-methyltryptoline, 238
reaction with CH_2O, 237, 238
1-Methyltryptoline, formn. from N-methyltryptamine, 238
Methylvinyl ketone, detm. with MBTH, 43
oxidn., 11

Methylvinyl ketone—*continued*
precursor of CH_2O, 11
spectral data, 43
Methysergide, formald. precursor, 173
structure, 173
Mice, *see* Mouse
Microbial, nitrosamine formn., 65
Microporous rubber, release of, carbon dioxide, 218
carbon monoxide, 218
formald., 218
Microsomes, enzymes, 50, 51, 71–73
epoxidase, 80
fractions from, kidney, 76
liver, 50, 51, 76
lung, 76
pork liver, 155
spleen, 76
stomach, 76
testes, 76
liver peroxidn., 51
membranes, 51
Microspectrophotofluorimetry, organelles, 242
single cells, 242
tissues, 242
Milk, carcinogens, 172
detm. of Imidan, 280
Miscarriages, vinyl chloride and wives, 274
Mixed function oxidases, 142
Mobile homes, formald. pollution, 218
Molybdate, 114
Monkey, effect of PCB, 150
effect of gastric mucosa, 150
formn. of 8-methoxykynurenate, 191
urine metabolite, 191
Monoamine oxidase, effect on primary amines, 160
Mono-demethylchlorpromazine, metabolite of chlorpromazine, 147
Monoglycerides, anal. proc. 91
detm. with 2,4-pentanedione, 91
precursors of CH_2O, 91
(1-Mono)glycerides, detm. in ice cream, 279
detm. with CA, 279
precursors of CH_2O, 279
Monomethylolglycine, equil. constant, 227
formn. from glycine and CH_2O, 227
precursor of CH_2O, 227
structure, 227
Monosaccharides, in allergens, 104
detm. with CA, 106
precursor of CH_2O, 106
Monuron, N-demethylation, 149
precursor of CH_2O, 149
Morphine, formald. precursor, 173
structure, 173
Morphine N-oxide, adduct formation, 11
hydrolysis, 11
precursor of CH_2O, 11
Morpholine, effect of nucleotide reaction, 211
Mosquito, detm. of triglycerides, 279
Mouse, carc. of aflatoxins, 187
DENA, 68
1,1-dimethylhydrazine, 122
DMNA, 68, 184
8-methoxykynurenic acid, 191
methyl n-butylnitrosamine, 68
β-propiolactone, 251
vinyl chloride, 274
carcinogenesis, mutag. anomaly, 184
subcutaneous injection, 81
cytochrome P-450, 177
demethylation of griseofulvin, 190
DMNA and CH_2O, 150
effect of Aroclor 1254, 148, 150
DENA, 74
7,12-dimethyl BaA, 54
DMNA, 74, 76
DMNA demethylase, 75
3-methyl CHOL, 74, 75
phenobarbital, 74, 75
starvation, 150
TCDD, 147
ethylmorphine N-demethylase, 177
host-mediated bioassay, 73
metab. of, hempa, 155
thiohempa, 155
organotropic effect, 76
pretreatment with DMNA, 70
second generation effects, 54
strains, 75, 76, 172, 183, 232, 233
Mouse bladder, implantation of 8-methoxykynurenic acid, 191
Mouse kidney, effect on vinyl chloride, 273
tumours, 69
Mouse liver microsomes, 73, 74, 169, 273
N-demethylation of ethylmorphine, 176
effect of, Aroclor 1254, 183
meat products, 169
organic peroxides, 169
PAH, 183
phenobarbital, 183
effect on griseofulvin, 190
lipid peroxidn., 176
toxicity of DMNA, 74
Mouse lung, adenomas, 185
DMNA demethylase, 183
effect on vinyl chloride, 273
tumors, 68, 69
Mouse lymphoreticular system, carc. of 8-methoxykynurenic acid, 191

Mouse skin, carcinogens, 81
carcinoma, 22
DNA and β-propiolactone, 251
tumors, 251
Musca domestica, mutated by, hexamethylmelamine, 120
trimethylolmelamine, 120
Muscle, demethylating activity, 150
trimethylamine oxide, 163
Mutagen, aryldialkyltriazenes, 125
chloroacetald., 273
1,2,3,4-diepoxybutane, 81
1,2,5,6-diepoxyhexane, 81
1,2,7,8-diepoxyoctane, 81
1,2,4,5-diepoxypentane, 81
di-(2,3-epoxypropyl)ether, 81
dihydroxydimethyl peroxide, 212
1,1-dimethylhydrazine, 122
1,2-dimethylhydrazine, 122
DMNA, 69, 70, 73–76, 233
enhancers, 81
epichlorohydrin, 81
epoxides, 80, 81
ethylene oxide, 81
formald., 1, 81, 150, 196, 226, 232–234
formald.-containing resins, 195
formald.-treated, casein, 1
foods, 1
N-2-fluorenylacetamide, 184
glycidald., 81
glycidol, 81
hempa, 154
methylazoxymethanol, 55
N-methylhydroxylamine, 181
methyl methanesulfonate, 138
nitrous acid, 180
β-propiolactone, 251
propylene oxide, 81
2,4,6-tri(hydroxymethylamino)triazine, 222
vinyl chloride, 271, 273
vinylcyclohexene-3 diepoxide, 81
vinylidene chloride, 62
Mutagenesis, *Neurospora crassa*, 73
organotropic effect, 76
relation to CH_2O formn., 75
Salmonella typhimurium, 73, 76
Mutation pathways, 74
plate tests, shortcomings, 184
Myanesin, anal. with CA, 16
oxidn. with periodate, 11
precursor of CH_2O, 11, 16
Mycotoxin, prodn., 123
Mylone, detm. with CA, 46
precursor of CH_2O, 46
Myosalvarson, hydrolysis, 11
precursor of CH_2O, 11
Myrcene, oxidn., 11
precursor of CH_2O, 11
Myristicin, oxidn. with H_2SO_4, 11
precursor of CH_2O, 11

NAD^+, 114, 169
detm. of CH_2O, 204
NADH, 60 61, 114, 171
N-demethylase inducer, 50
$NADP^+$, 154, 165, 169, 171, 184
NADPH, 50, 52, 60, 61, 70, 73, 75, 138, 155, 165, 166, 180, 185, 186, 273
NANA, *see* N-Acetylneuraminic acid
2,7-Naphthalenediol, *see* 2,7-dihydroxynaphthalene
β-Naphthoflavone, inducer, 189
inhibitor of N-demethylase, 148
stimulator of N-demethylase, 145
β-Naphthol, detm. of, acetald., 102
CMOS, 102, 103
formald., 102
glycolic acid, 101, 102
malic acid, 102
2-Naphthol-6-sulfonic acid, detm. of formald., 196
N-1-Naphthylethylenediamine, detm. with MBTH, 43
oxidn. with periodate, 11
precursor of CH_2O, 11
spectral data, 43
Narceine, anal. with CA, 11, 16
hydrolysis, 11
precursor of CH_2O, 11, 16
Narcotics, N-oxidn., 155
Narcotine, detm. with CA, 11, 16, 214
hydrolysis, 11
location on TLC, 214
precursor of CH_2O, 11, 16, 214
Nasal cavity, carcinoma, 154
tumours from DENA, 69
Nasal sinus, cancer, 68
Nasal tumours, bis-chloromethyl ether, 57
Nasal turbinate bones, carcinoma, 154
Natulan, anticarc., 151
carc., 151
N-demethylation, 149
metabolism, 151
precursor of, formald., 149, 151
N-isopropylterephthalamic acid, 151
structure, 151
teratogen, 151
Natural waters, detm. of formate, 279
Necrosis, effect of, CCl_4, 72
DMNA, 72

Neoarsphenamine, hydrolysis, 11, 16
precursor of CH_2O, 11, 16
Nerol, oxidn., 11, 16
precursor of CH_2O, 11, 16
Nervous system, carc. of 1-aryl-3,3-dialkyltriazenes, 125
Neuramine metabolism, inhibitors, 238
Neurine, precarc., 186
Neurogen, 1-phenyl-3-methyltriazene, 125
Neurogenic cancer, 54
Neuroleptic, phenothiazines, 147
Neurospora, effect of clastogen, 251
mutated by, di-(2,3-epoxypropyl)ether, 81
epichlorohydrin, 81
ethylene oxide, 81
glycidol, 81
N-methylhydroxylamine, 181
propylene oxide, 81
Neurospora cassida, mutagenicity of CH_2O, 1
Neurospora crassa, indicator of mutagenesis, 73
Nialate, see Ethion
Nickel, oxidn. catalyst, 177
Nicotinamide, 171
effect on aminopyrine N-demethylase, 144
Nicotine, enz. demethylation, 150, 179
enz. oxidn., 179
metabolism, 179
precursor of, cotinine, 179
demethylcotinine, 179
formald., 150, 170
nornicotine, 179
structure, 179
Ninhydrin, detm. of, α-amino acids, 49
glycine, 48, 49, 283
oxidant, 5, 7, 283
oxidn. of glycine, 95
structure, 49
Nitrate, bacterial, 65
formn, from auto exhaust, 23
plants, 65
precursor of, nitrite, 65
nitrosamine, 65
runoff water, 65
well waters, 65
Nitrate reductase, bacteria, 65
Nitric acid, 245
form. from, 1,1-dichloroethylene, 62
1,2-dichloroethylene, 63
methyl nitrite, 243
oxidant, 10
Nitric oxide, 9, 10, 13, 15, 24, 129
oxidn. to NO_2, 23
Nitrite, deamination, 8
oxidation, 11
photolysis, 242
precursor of nitrosamines, 65, 67, 160
reaction with, aminopyrine, 67
drugs, 67
N-methylmorpholine, 160
oxytetracycline, 67
Nitroalkanes, formn. from 2-nitro-1-alkanols, 33
2-Nitro-1-alkanols, anal. proc., 47
detm. with, CA, 11, 16
ethyl acetoacetate, 11, 16, 47
disproportionation, 33
hydrol., 11
precursors of, formald., 11, 16, 33
nitroalkanes, 33
p-Nitroanisole, demethylation, 189
metab. eqn., 189
precursor of, formald., 189
glucuronide, 189
p-nitrophenol, 189
sulfate, 189
structure, 189
urinary products, 189
p-Nitroanisole O-demethylase, induction, 189
4-(4′-Nitrobenzyl)pyridine, detm. of β-propiolactone, 252
2-Nitro-2-bromo-1,3-propanediol, anal. proc., 220, 221
cosmetic preservative, 220
detm. with 2,4-pentanedione, 220, 231
precursor of CH_2O, 220, 231
structure, 231
4-Nitro-N,N-dimethylaniline, anal, proc., 164, 165
detm. with MBTH, 164
precursor of CH_2O, 164
spectral data, 164
2-Nitroethanol, formn. from CH_2O, 205
precursor of CH_2O, 21
Nitrogen dioxide, depletion, 26
effect of hexobarbital sleep time, 51
formn, 25, 26, 129
from methyl nitrite, 243
interferent, 86
oxidn. of vinyl chloride, 270, 271
photooxidant, 23–25, 63
precursor of ozone, 24, 25
product yields, 26
reaction with, ethylene, 24
isobutene, 25
vinylidene chloride, 62, 63
Nitrogen mustard, carc., 251
mutagen, 251
Nitrogen oxides, smog formn, 128, 129
Nitrogen pentoxide, formn., from methyl nitrite, 243
Nitromethane, reaction with CH_2O, 205
structure of product, 205

Nitromethanol, alk. hydrol., 277
detm. with CA, 11, 16, 277
hydrolysis, 11
precursor of CH_2O, 11, 16, 277
tautomerism, 11, 16
p-Nitrophenol, formn. from *p*-nitroanisole, 189
structure, 189
2-(4-Nitrophenylazo)quinoline, formn., 222
structure, 222
2-Nitropropane, precursor of CH_2O, 26, 27
yield of CH_2O, 26, 27
Nitrosamides, carc. to man, 76
Nitrosamines, air, 64
adsorption from water, 66
agricultural land, 65
ald. formn., 79
anal. by, GC-CL, 66
HPLC–CL, 66
anal. eqn., 160, 161
bacon, 65
Baltimore, 64
Belle, 64
beverages, 65
bologna, 65
carcinogenic, 63, 67, 68, 160
to man, 76, 77
cider distillate, 65
cigarette smoke, 65
concns. in, air, 64
bacon, 65
bologna, 65
cider distillate, 65
fish meal, 65
frankfurters, 65
herring, 65
ocean, 65
tobacco, 65
vagina, 66
Curtus Bay, 65
decomposition, 79
effect of UV light, 79
effect on, guinea pig, 68
rat, 68
feedlots, 65
fishmeal, 65
fish products, 65
formn., 63, 64
formn, catalysts, 66
formn, from, aliph. amines, 64–67
nitrite, 64–67
sec. amines, 162
tert. amines, 162
formn. in, digestive tract, 66
respiratory tract, 66
urinary tract, 66
industrial effluents, 64, 65
isolation, 66
lake water, 64, 65
lethal syntheses, 63–66
liquid smoke, 3
metab. by, human liver, 76
rat liver, 76
ocean, 65
plants, 65
presence in, air, 64
cigarette smoke, 63
liquid smoke, 63
smoked fish, 63
wheat flour, 63
rain puddles, 64
range in concns., 65
sewage, 64–66
soil, 64, 65
stability in, food, 64
sewage, 64
soil, 64
water, 64
threshold limit in air, 76
tobacco, 65
tobacco smoke, 63
well waters, 65
Nitrosation, effect of, acetate, 66
amine bascity, 66
chloride anion, 66
formaldehyde, 66
phenols, 66
pH, 66
substrate concn., 66
thiocyanate, 66, 67
N-methylaniline, 185
Nitrosative dealkylation, tertiary amines, 161
4-Nitroso-N,N-dimethylaniline, anal. proc., 164, 165
detm. with MBTH, 164
precursor of CH_2O, 164
spectral data, 164
N-Nitrosodiphenylamine, reaction with sec. amine, 67
transnitrosator, 67
precarcinogen, 67
N-Nitrosoethylurea, carcinogen, 68
mutagen, 68
structure, 68
teratogen, 68
N-Nitroso-N-methylaniline, carc., 185
formn, from N-methylaniline, 185
precursor of, benzenediazonium cation, 185
formald., 185
N-nitrosomethylurea, carcinogen, 68
mutagen, 68
structure, 68
teratogen, 68

Nitrosomethylurethan, effect on pregnant rats, 54
  transplacental effect, 54
N-Nitrosomorpholine, formn from N-methylmorpholine, 160
Nitrosonaphthol, 223
N′-Nitrosonornicotine, concn. in tobacco, 65
N-Nitrosopiperidine, formn. from, N-hydroxymethylpiperidine, 161
    S-piperidinomethylthiobenzoate, 161
  structure, 161
Nitrosopyrrolidine, concn. in bacon, 65
N-Nitrososarcosine, carc., 185
  decarboxylation, 185
  precursor of DMA, 185
Nitrosoureas, 73
Nitrous acid, formn, from methylnitrosamine, 180
  oxidative deamination, 49
  reaction with nucleic acids, 210
NO_2, see nitrogen dioxide
NO_x, oxidant, 23
  precursors of, dimethylnitrosamine, 64
    nitrosamines, 63, 64, 185
1-Nonadecene, detm. with MBTH, 43
  spectral data, 43
Nonane, extractant, 262, 264
Nonanoic acid, formn. from 1,2-decanediol, 106
  structure, 106
Noracetylmethadol, formn. from acetylmethadol, 156
  metab. eqn., 156
  precursor of, dinoracetylmethadol, 156
    formald., 156
  structure, 156
Norda Ketal Synergist, detm. with CA, 46
Norharman, formn, from tryptophan, 242
  structure, 242
Norhexobarbital, formn, from, hexobarbital, 178
  structure, 178
Norit, 136
Normethadol, formn. from, methadol, 156
    noracetylmethadol, 156
  metab., eqn., 156
  structure, 156
Nornicotine, enz. oxidn., 179
  formn, from nicotine, 179
  precursor of demethylcotinine, 179
  structure, 179
Nortriptyline, N-demethylation, 150
  precursor of CH_2O, 150
Noscapine, see Narcotine
Nucleic acids, alkylation. 55
  methylation, 74, 125
  reaction with, formald., 233
    ionizing radiation, 210
    nitrous acid, 210
    UV radiation, 210
Nucleotides, cross-link to amines, 211
  reaction with CH_2O, 1, 210

Ocean, nitrosamines, 65
cis-Ocimene, oxidn., 11, 16
  oxidn. eqn., 29
  precursor of, formald., 11, 16, 29
    formic acid, 29
    glyoxylic acid, 29
    malonaldehyde, 29
    malonaldehydic acid, 29
    malonic acid, 29
  structure, 29
1-Octadecene, oxidn., 11, 16
  precursor of CH_2O, 11, 16
n-Octane, precursor of CH_2O, 26–28
  yield of CH_2O, 26–28
1-Octene, detm. with MBTH, 43
  spectral data, 43
1-Octene oxide, atmospheric, 80
Oil of lavender, 249
Olefinic compounds, anal. reaction sequence, 22
  detm. with MBTH, 22
  oxidation, 22
  precursor of, aliph. ald., 22
    formald., 22
Olefins, anal. procedures, 28, 29
  atmospheric, 80
  detm. with, J-acid, 29
    MBTH, 28, 29
  formn, of epoxides, 80
  in airborne particulates, 30
  internal, 128
  metabolic oxidn., 80
  source, 80
  terminal, 128
  water-soluble, detm., 29
Oligosaccharides, oxidn., 11, 34, 135
  precursor of, formald., 11, 134, 135
    formic acid, 134, 135
  reduction, 11, 134
Orcinol, automated proc., 60
  detm. of, furfural precursors, 115–117
    sugars, 60
Organelles, microspectrophotofluorimetry, 242
Organic peroxides, effect on, mouse liver, 169
    rat liver, 169
Organotropic carcinogenicity, 51, 52
Organotropy, azoalkanes, 52

azoxyalkanes, 52
cancers of, bladder, 68
esophagus, 68
fore-stomach, 68
kidney, 68
liver, 68
lung, 68
nasal sinus, 68
ovary, 68
1,2-dialkylhydrazines, 52
Orlon, formald, content, 217
Orosomucoid, anal. with 2,4-pentanedione, 11, 16
fluor. anal., 11, 16
oxidn., 11
precursor of CH_2O, 11, 16
Ovary, cancer, 68
Oxalic acid, 244–246
anal. interf., 244
anal. mech., 244
anal. proc., 245, 246
anal. with, CA, 12, 16
2,7-dihydroxynaphthalene, 12, 16, 114
phenylhydrazine, 12, 16
calcium, 245
detm. in urine, 244, 245, 281
detm. with, CA, 244, 245, 281
indole, 244
disproportionation, 12
effect of radiation, 280
formn. from glyoxylate, 114
potassium, 245
precursor of, CH_2O, 12, 16, 114, 281
glycolate, 114
redn., 244, 245
redn. with, Mg, 12
zinc, 12
standard, 245
yield of CH_2O, 16
Oxaloacetic acid, anal. interference, 244, 245
2-Oxapentoate, enz. redn., 12
precursor of CH_2O, 12
1-Oxaspiro(2.5)octa-4,7-dien-6-one, formn. from p-hydroxybenzyl alcohol, 130
precursor of, formald., 130
hydroquinone, 130
Oxetane, photodec., 246
photorearrangement, 250
precursor of, ethylene, 246, 250
formald., 246, 250
structure, 246, 250
Oxetanes, disproportionation, 12
photodecomposition, 12
precursors, of CH_2O, 12
Oxidant, air, 2, 3, 5, 6, 9, 10, 13–15
benzoyl peroxide, 12, 223, 248
bismuthate, 4, 13, 130, 131
chloramine T, 8, 96
N-chloroquinone imide, 130
chromate, 5, 10, 165
chromium trioxide, 5, 165
cobalt acetate, 108
Co^{+2} +O_2, 106
dipot. nitrosodisulfonate, 130
ferric chloride, 3, 5, 28, 29, 48, 56, 57, 62, 84, 88, 92, 107, 111, 132, 133, 135, 139, 164, 169, 177, 228, 268
ferric salt, 106
halogen, 12
hydrogen peroxide, 9, 10, 14, 61, 106
iodine, 14, 153, 228
lead dioxide, 201
lead tetraacetate, 10, 164, 278
manganese (III), 6
mercuric chloride, 14
mercuric sulfate, 5, 11, 14, 122
nitric acid, 10
nitrite, 11, 14
NO_x, 23
nitrogen dioxide, 14, 63, 270
oxygen, 5, 8, 10, 61
ozone, 2, 6, 8–10, 23, 270
paraperiodic acid, 61, 94, 99
perbenzoic acid, 6, 12
perborate, 16
periodate, 2–15, 22, 29, 80–82, 84, 89–94, 130, 131, 134, 135, 252–254, 256–261, 266, 267
periodic acid, 82, 92, 111, 275
permanganate, 2–7, 9–15, 22, 28–30
persulfate, 6, 10, 82
potassium ferricyanide, 92, 124, 197
potassium metaperiodate, 92
potassium permanganate, 82, 136, 137, 275
ruthenium dioxide, 4
selenium dioxide, 2, 124
sodium metaperiodate, 48
sodium periodate, 94
sulfuric acid, 3, 7, 12, 141, 252
tetranitromethane, 160
1,2,3-trioxolanes, 247
vanadate, 12
Oxidants, list of, 282
Oxidase, 179
N-Oxidase activity, effect of pH, 147
Oxidation, amines, 155
antihistamines, 155
ephedrines, 155
narcotics, 155
tranquilizers, 155
tropine alkaloids, 155

Oxidative dealkylation, 51–52, 142–187
  phosphoramides, 155
Oxidative deamination, 97, 99
  glycine, 48, 49
  primary amines, 160
Oxidative N-demethylation, aminopyrine, 166
  dimethylnitrosamine, 180, 185
  N-methylimidazoles, 223
Oxirane, anal. proc., 94
  cleavage, 94
  location, 94
Oxiranes, *see* epoxides
Δ^4-3-Oxo-C_{21}-steroids, detm. with 2,4-pentanedione, 45
  precursors of CH_2O, 45
  TLC sepn., 45
Oxygen, oxidant, 5, 8, 10, 61
N-Oxymethylene amines, 217–222
  hydrol., 217–222
  precursors of CH_2O, 217–222
Oxymethylene sulfides, 222, 223
  precursors of CH_2O, 222
Ozone, detm. with pararosaniline, 31, 86
  effect on hexobarbital sleeping time, 51
  formn., 129
  formn, from NO_x, 24, 25
    NO_2, 62, 63
  oxidant, 2, 6, 8–10, 23, 28
  oxidn. of vinyl chloride, 270
  product yield, 26
  reaction with, olefins, 26, 27
    propene, 28
Ozonides, 247, 248
  formn. eqn., 247
  formn. from alkenes, 247
  precursors of ald., 247, 248
  reaction with tetracyanoethylene, 247, 248
  structure, 247, 248
Ozonization, olefins, 247

PAH, carc., 81
  enhancement of, 6-methylthiopurine → CH_2O, 192
    prometyrne → CH_2O, 192
  inducers of, azo dye N-demethylase, 183
    liver tissue proliferation, 183
  induction of N-demethylase, 146
  inhibitor of N-demethylase, 148
  metab. enhancement of $O_2NC_6H_4OCH_3$ → CH_2O, 189
  reaction with CH_2O, 207
  repressor of Me_2NNO demethylase, 183
  stimulation of N-demethylase, 145
Palatinose, 104
  oxidn. eqn., 105
  precursor of, carbohydrate ald., 105
    formald., 105
    formic acid, 105
    glycolic acid, 105
    glyoxylic acid, 104
  structure, 105
Palm-kernel oil, detm. of triglycerides, 259, 260
Palmotoxin Bo, metabolism, induced by phenobarbitone, 189
    inhibited by CO, 189
  precursor of CH_2O, 188, 189
Palmotoxin Go, metab., induced by phenobarbitone, 189
    inhibited by CO, 189
  precursor of CH_2O, 188, 189
Palmitoxins, induce liver lesions, 188
  metabolism, 188, 189
  precursors of CH_2O, 188
PAN, see Peroxyacetyl nitrate
Panthenol, 249
Papain, digestion of proteins, 58
Papaverine, anal. with Ca, 12, 16
    oxidn., 12
  precursor of CH_2O, 12, 16
Paper chromatography, *see* PC
Paraffin, sepn. of triglycerides, 259
Paraffins, 128
Paraformaldehyde, detm. in funeral homes, 213.
  detm. with CA, 43, 213
  hydrolysis, acid, 12
  thermal, 12
  particle size, 213
  precursor of, bis-chloromethyl ether, 58
    formald., 12, 193, 213
  spectral data, 43
  standard, 193
  structure, 213
Parakeet, carc. of diethylnitrosamine, 68
Paraperiodic acid, oxidant, 61, 62, 94
Pararosaniline, detm. of, atm. CH_2O, 197
    cellulose acetate formal, 36
    cellulose formal, 36
    ethylene glycol, 38, 279
    eugenol, 31
    glycolipids, 40
    1,2-glycols, 41, 280
    hydrochlorothiazide, 41
    methanol, 42, 280
    methyl methacrylate, 43
    ozone, 86
  disadvantages of anal. use, 199
  impurities in, 199
Parkinsonian patients, 238

Parkinson's disease, treatment with L-dopa, 191
Particleboard, construction of, buildings, 218
furniture, 218
mobile homes, 218
formald. release, 218
Partition, effect on, N-demethylase, 145, 146, 148
effect on enz. reactions, 165
Partition chromatography, aldoheptitols, 114
sugar alcohols, 114, 116–118
sugars, 114, 117
PAS test, 222
Pathways, carc., 74
mutagenic, 74
toxic, 74
PC, characterizn. of glyoxylic acid, 41
cycasin, 55
2,4-D, 278
glyceric acid, 279
glycerol, 279
identifn. of ethanolamine, 38
location of, carbohydrates, 36
chloroacetic acid, 36
corticosteroids, 36
1,2-glycols, 108, 109
RDX, 281
sepn. of chlorophenoxyacetic acids, 278
triglycerides, 279
PCB, effect on gastric mucosa, 150
human adipose tissue, 150
human milk, 150
in bovine milk, 150
Peanut meal, aflatoxins, 187
Pectic acid, 59
Pectin, precursor of CH_2O, 16
Pectin ester, anal. proc., 137
detm. with 2,4-pentanedione, 137
hydrol., 137
Pectin methyl esterase, detm. with, CA, 137, 282, 283
phenylhydrazine, 282, 283
Penicillin V, disproportionation, 12
hydrolysis, 12
precursor of CH_2O, 12
Pentamethylphosphoramide, formn, from hempa, 155
precursor of CH_2O, 155
2,4-Pentanedione, *see* acetylacetone, 94
anal. eqn., 153
anal. proc., 31, 32, 171
automated anal., 260
cosmetic preservatives, 220
detm. of, N-acetylneuraminic acid, 34
acrylamide, 31, 34
acrylic acid, 31, 34
allyl alcohol, 31, 34
allyl amine, 31
aflatoxin B_1 metab., 188, 189
aminohexoses, 278
aminopyrine, 50, 167
aminopyrine demethylase, 166
arabitol, 34, 41
aramite, 34
benzphethamine, 50
2-bromo-2-nitropropan-1,3-diol, 47
1-buten-3-ol, 35
3-buten-1-ol, 31, 35
carbohydrates, 36
2-chlorethanol, 36
cholesterol, 61
codeine, 50
corticosteroids, 36
cycasin, 55
cysteine, 36
demethylase, 191
N-demethylation of azo dyes, 171
2-deoxy-D-glucose, 36
2-deoxy-D-ribose, 37, 61, 62
dihydroxyacetone, 37
dimethylalkylamine oxides, 37
dimethylhydrazine, 122
dimethylnitrosamine, 70
DMNA N-demethylase, 78
dulcitol, 38, 41
erythritol, 38
ethanolamine, 38
ethylene glycol, 38, 84, 93, 279
ethylene oxide, 38
ethylmorphine, 38, 50, 279
formald., 47, 50, 70, 143, 147, 153, 154, 166, 167, 191, 197, 199, 200
fructose, 39
Germall, 47
glucose, 90, 93
glycerides, 40, 91
glycerol, 40, 93, 279
glycerophosphate, 93
glycerophosphatides, 94
glycolald., 93
1,2-glycols, 41, 107–109
hexamine, 223, 224
hexitols, 41
hydroxylysine, 42
1-hydroxymethyl-5,5-dimethylhydantoin, 47, 220
mandelamine, 223, 224
mannitol, 42
methane bis(N,N'(5-ureido-2,4-diketo-tetrahydroimidazole)N,N'-dimethylol), 220
methanol, 42, 137, 138

2,4-Pentanedione—*continued*
N-methylaniline, 167
methylazoxymethanol, 55
2-nitro-2-bromo-1,3-propanediol, 220, 231
serine, 44
sorbitol, 41, 45
sugar alcohols, 115–118
sugars, 115–117
triglycerides, 94, 255, 260–267, 279
xylitol, 45
fluor. anal., 31, 32
interferences, 47
location of 1,2-glycols, 108
manual methods, 260
purifn., 220
reagent stabilization, 261
structure, 122
sulfite interference, 152
1-Pentene, photooxidn., 18
precursor of, aliph. ald., 18
formald., 18
yield of, aliph. ald., 18
formald., 18
Pentose, oxidn. with periodate, 12
precursors of CH_2O, 12
Pentoses, furanose configuration, 12
oxidn., 12
precursors, of CH_2O, 12
Pepper, detm. of piperine, 214, 281
Peptides, reaction with CH_2O, 206
structure of product, 206
Perbenzoate, oxidant, 6, 12
Perbenzoic acid, oxidant, 228
Perchloric acid, hydrol. of DNA, 77
protein pptn., 177
Perfume ingredients, precursors of CH_2O, 47
Periodate, 94, 111, 112, 114, 115, 121
detm. of, cellobionic acid, 276
cellotrionic acid, 276
gluconic acid, 276
xylonic acid, 276
oxidant, 2–15, 22, 35, 80–82, 84–86, 89–93,
oxidn of, 2-acetylamino-2-deoxy-D-glucose diethyl dithioacetal, 2
N-acetylglucosamines, 2
N-acetylhexosamines, 2
N-acetylneuraminic acid, 2
acrylamide, 2
acrylic acid, 2
acrylonitrile, 2
adenosone, 2
adonitol, 2
albumin, 2
alditol, 60
aldoses, 2
alkanolamines, 2
1-alkenes, 2
O-alkylglycerols, 2
allopregnane-3,11,17,21-tetrol-20-one, 254
allopregnane-3,17,20,21-tetrol, 254
allopregnane-3,17,21-triol-11,20-dione, 254
allyl acetate, 2
allyl alcohol, 2
allylamine, 2
allylbenzene, 2
allyl bromide, 2
allyl chloride, 2
allyl ethyl ether, 2
1-allyl-2-thiourea, 2
2-amino-1-alkanols, 3
2-amino-D-glucose, diethyl dithioacetal, 3
2-amino-1-ethanol, 3
2-amino-2-hydroxymethyl-1,3-propanediol, 3
2-amino-2-deoxy-D-galactose, 3
2-amino-2-deoxy-D-glucose, 3
angelic acid, 3
anhydro-D-xylo-benzimidazole, 3
arabinose, 3
arabitol, 3
aramite, 3
ascorbic acid, 3
3-bromopropene, 3
1,2,4-butanetriol, 3
3-butenenitrile, 3
1-buten-3-ol, 3
3-buten-1-ol, 4
calciferol, 4

carbohydrates, 4
cellobiose, 4
chloroamphenicol, 4
2-chloroethanol, 4
1-chloro-2,3-epoxypropane, 4
chlorogenic acid, 4
chlorphenesin, 4
corticosteroids, 4, 131
cortisol, 4
cytidine, 4
1-decene, 5, 16
dehydrocorticosterone, 254
2-deoxy-D-arabinohexose diethyl dithioacetal, 5
2-deoxy-D-glucose, 5
2-deoxy-D-ribose, 5
diethanolamine, 5
2,2-(diethylthio)ethanol, 5
4,5-dihydrocorticosterone, 254
dihydrosphingosine, 5

dihydroxyacetone, 5, 131, 253
dihydroxydihydrobetulin, 254
2,3-dimethylglucose, 253
1,1-diphenylethylene, 5
1,5-diphenyl-3-pentadienone, 5
1,8-diphenyl-1,3,5,7-octatetraene, 6
1,9-diphenyl-1,3,6,8-nonatetraene-5-one, 6
1,4-diphenyl-1,3-butadiene, 6
disaccharides, 6
epichlorohydrin, 6
2,3-epoxy-1-propanol, 6
erythritol, 6, 253
ethanolamine, 6
ethyl acrylate, 6
ethylene, 6
ethylenediamine, 6
ethylene glycol, 6, 84–86, 253
ethylene oxide, 6
ethyleneimine, 6
2-(2-fluorenylamino)ethanol, 6
fructose, 7, 253
furfural, 7
β-2-furylacrylophenone, 7
galactonic acid, 7
galactose, 7, 253
α-D-galactose diethyl dithioacetal, 7
gelsemine, 7
geraniol, 16
glucaric acid, 7
gluconic acid, 7, 253
glucosamine, 7
glucose, 253
glucose phenylosazone, 7, 253
glucosone, 7
glucuronic acid, 7
glyceraldehyde, 7
glyceric acid, 7
glyceride glycerol, 7
glycerides, 7
glycerol, 7, 253
α-glycerophosphates, 7
glycerophosphatides, 7
glycidyl stearate, 7
glycolaldehyde, 7
glycolipids, 7
1,2-glycols, 8
glycoproteins, 8
guaran, 8
1-heptene, 8
1-hexene, 8
hexitols, 8
hexosamines, 8
hexuronic acids, 9
hydrocortisone, 9
hydroxyacetone, 9, 131
p-hydroxybenzyl alcohol, 130
hydroxylactone, 9
hydroxylysine, 9, 253
hydroxyproline, 9
inositol, 9
3-iodopropene, 9
isoatisine, 9
isoeugenol, 9
isomaltitol, 9
isoafrole, 9
itaconic acid, 9
20,21α-ketols, 9
lactose, 9
lactulose, 9
laminarin, 10
levulose, 10
linalool, 16
linalyl acetate, 16
maleic acid hydrazide, 10
maltitol, 10
maltose, 10
mannitol, 10, 253
mannose, 253
mannose phenylhydrazone, 253
metasaccharinic acids, 10
methacrylamide, 10
4-(*p*-methoxyphenyl)-3-buten-2-one, 10
2-methylaminoethanol, 10
3-methylglucosamic acid, 253
2-methylglucose, 253
α-methylstyrene, 11
methyl vinyl ketone, 11
myanesin, 11
myrcene, 11
N-1-naphthylethylenediamine, 11
Nerol, 16
cis-ocimene, 16
octadecene, 16
olefins, 28
oligosaccharides, 11, 134
pentose, 12
phosphatidylglycerol, 12
β-pinene, 16
polyols, 12, 251
Δ^4-pregnene-17,21-diol-3,11-20-trione, 254
Δ^4-pregnene-11, 17, 21-triol-3,20-dione, 254
1,2-propanediol phosphate, 12
4-propenylveratrole, 12
β-propiolactone, 252
propylene oxide, 13
pseudouridine phosphates, 13
quinine, 13
raceophenidol, 13
riboflavin, 13, 254

Periodate—*continued*
ribose, 13
ribose-3-phosphoric acid, 253
sabinene, 13, 16
safrole, 13
serine, 13, 253
sialic acids, 13
sialoglycoprotein, 13
sophoritol, 13
sophorose, 13, 252, 253
sorbitol, 13, 253
sorbose, 253
steroids, 13
streptomycin, 13
styrene, 13
styrolene, 13
sugar sulfates, 13
terpenes, 13
1-tetradecene, 13
tetritols, 14
triglycerides, 14
2,3,4-trimethylglucose, 253
2,4,4-trimethyl-1-pentene, 14
10-undecanoic acid, 15
vinyl acetate, 15
vinyl butyl ether, 15
vinyl chloride, 15
5-vinyl-2-picoline, 15
2-vinylpyridine, 15
xylitol, 15
d-xylose, 15, 253
Periodic acid, 99, 107, 108
oxidant, 82, 92, 135, 275
Periodic acid—Schiff, 108
Permanganate, oxidant, 2–7, 9–15, 22
oxidn. of, acrylamide, 2
acrylic acid, 2
acrylonitrile, 2
1-alkenes, 2
allyl acetate, 2
allyl alcohol, 2
allylamine, 2
allylbenzene, 2
allyl bromide, 2
allyl chloride, 2
allyl ethyl ether, 2
1-allyl-2-thiourea, 2
3-bromopropene, 3
3-butenenitrile, 3
1-buten-3-ol, 3
3-buten-1-ol, 4
chlorogenic acid, 4
corticosteroids, 4
1-decene, 5
1,1-diphenylethylene, 5
1,5-diphenyl-3-pentadienone, 5
1,8-diphenyl-1,3,5,7-octatetraene, 6
1,9-diphenyl-1,3,6,8-nonatetraene-5-one, 6
1,4-diphenyl-1,3-butadiene, 6
ethyl acrylate, 6
ethylene, 6
furfural, 7
β-2-furylacrylophenone, 7
gelsemine, 7
hydroxylactone, 9
3-iodopropene, 9
isoatisine, 9
isoeugenol, 9
itaconic acid, 9
maleic acid hydrazide, 10
methacrylamide, 10
methanol, 10
methoxone, 10
methoxy compounds, 10
4-(*p*-methoxyphenyl)-3-buten-2-one, 10
methyl esters, 11
methyl methacrylate, 11
α-methylstyrene, 11
methyl vinyl ketone, 11
myrcene, 11
olefins, 28
orosomucoid, 11
4-propenylveratrole, 12
protoporphyrin monomethyl ester, 13
quinine, 13
sabinene, 13
safrole, 13
styrolene, 13
terpenes, 13
1-tetradecene, 13
p-tosyl argenine methyl ester, 14
2,4,4-trimethyl-1-pentene, 14
10-undecanoic acid, 15
vinyl acetate, 15
vinyl butyl ether, 15
5-vinyl-2-picoline, 15
2-vinylpyridine, 15
Permeation tube, bis-chloromethyl ether, 58
formald., 194
Permissible dose, Me_2NNO, 77
Peroxides, effect on, mouse liver, 169
rat liver, 169
formn. from olefins, 80
Peroxyacetyl nitrate, PAN, formn, from,
auto exhaust, 23
isobutene, 25
propylene, 129
Peroxyacyl nitrates, formn. from, benzyl nitrite, 243
ethyl nitrite, 243

propyl nitrite, 243
Perseitol, location, 109
precursor of CH_2O, 109
Persulfate, oxidant, 6, 10, 82
Pesticide, Alar, 124
Pesticides, env. fate, 143
metabolism, 143
precursors of ald., 143
use, 143
vinyl chloride usage, 269
Petroleum ether, 94, 256
extractant, 267
pH, effect on N-demethylase, 145, 146, 148
effect on N-demethylation, 165
gastric mucosa, 67
nitrosamine, formn., 66
Pharmaceuticals, detm. of ethylene oxide, 279
Pharmaceutical preparations, detm. of inositol, 280
Phenazine methosulfate, 160, 162
o-Phenanthroline, stimulator of N-demethylase, 149
Phenethanol, oxidn., 12
precursor of CH_2O, 12
Phenyethylamines, reaction with CH_2O, 236
Phenkapton, detm. with CA, 46
precursor of CH_2O, 229
structure, 230
Phenobarbital, 272
blockage of induction, 170
effect on, aminopyrine, 167
aminopyrine N-demethylase, 144
carc. of DAB, 170, 171
cytochrome P_{450}, 166, 167
Me_2NNO N-demethylase, 75
Me_2NNO mutagenicity, 75
hexobarbital oxidn., 167
mice, 74
phospholipids, 167
rats, 74
inducer of, azo dye N-demethylase, 183
N-demethylase, 50, 73, 143
liver tissue proliferation, 183
NADPH-cytochrome c oxidase, 50
metab. enhancement $4\text{-}O_2NC_6H_4OCH_3 \rightarrow CH_2O$, 189
precursor of CH_2O, 150
repressor of Me_2NO demethylase, 183
stimulation of N-demethylase, 145–150, 170, 171
Phenobarbitone, enzyme inducer, 189, 273
Phenol, 217
Phenol-formaldehyde resin, 218
Phenol-H_2SO_4, reagent, 59
Phenols, diazotizn, 130
nitrosamine formn., 66
reaction with CH_2O, 206
structure of products, 206
Phenolphthalein, 97
Phenothiazine derivatives, antipsychotogens, 159
Phenothiazines, neutroleptics, 147
N-Phenothiazinyl ethylenediamines, oxidn., 82
precursor of, acetald., 83
formald., 82
structure, 82
Phenoxyacetic acids, anal. with, CA, 12, 16
J-acid, 12, 17
Phenyl J-acid, 12, 17
detm. with, J-acid, 101
phenyl J-acid, 101
disproportionation, 12
hydrolysis, 12
precursors of CH_2O, 12, 101
Phenyl acetate, precursor of CH_2O, 26, 28
yield of CH_2O, 26, 28
Phenylalaninol, detm. with CA, 44
Phenylazomethane, detm. with CA, 43
hydrolysis, 12
precursor of CH_2O, 12
spectral data, 43
tautomerism, 12
yield of CH_2O, 43
2-Phenylazo-2-methylaniline, formn. from 4-phenylazo-N,2-dimethylaniline, 169
Phenylbutazone, effect on aminopyrine N-demethylase, 144
inhibitor of N-demethylase, 149
1-Phenyl-2-butene, formn. from 2-phenyltetrahydropyran, 250
structure, 250
Phenylcoumarin, 58
Phenyldiazonium cation, ulticarc., 126
1-Phenyl-3,3-diethyltriazene, 125
structure, 125
1-Phenyl-3,3-dimethyltriazene, anal. eqn., 126
anal. with CA, 12, 16, 126
anticarc., 125
carc., 124
oxidn., 12, 126
precursor of CH_2O, 12, 125, 126
structure, 126
teratogen, 125
1-Phenyl-3,3-di-n-propyltriazene, 125
structure, 125
Phenylephrine, 249
alk. hydrol., 277
anal. proc., 48
N-demethylation, 150
detm. with MBTH, 48

Phenylephrine—*continued*
oxidn., 12, 17
precursor of CH_2O, 12, 17, 150, 277
β-Phenylethanolamine, reaction with CH_2O, 236
Phenyl ether, 94
Phenyl hydrazine, 103
detm. of, allantoic acid, 113
allantoin, 113
aramite, 33, 35, 278
ethylene glycol, 38
formald., 197
formald. dimethylacetal, 39
glycerides, 40, 92
glycerol, 40
glycoproteins, 111
glyoxylate, 8
glyoxylic acid, 41, 113
hexamethylenetetramine, 41
hydroxylysine, 42
methanol, 43
pectin methyl esterase, 282, 283
polysaccharides, 111
triglycerides, 255, 279
ureidoglycolate, 113
oxidn., 113
purifn. of acetic acid, 141
reaction with glucose, 90
structure, 113
Phenylhydroxylamine, formn. from N-methylaniline, 166
Phenyl J-acid, detm. of, 2-chlorophenoxyacetic acid, 36
2,4-dichlorophenoxyacetates, 37
formald. 2,4-DNPH, 39
formald. hydrazone, 39
hexamethylenetetramine, 41
N,N′-methylene-bis-acrylamide, 43
phenoxyacetic acids, 101
piperonal, 46
piperonylacrylic acid, 46
piperonylic acid, 43
sym-trioxane, 45
sym-trithiane, 45
spectral data. 45
1-Phenyl-3-methyl-3-n-butyltriazene, anticarc., 125
structure, 125
1-Phenyl-3-methyl-3-t-butyltriazene, 125
structure, 125
1-Phenyl-3-methyl-3-ethyltriazene, anticarc., 125
structure, 125
1-Phenyl-3-methyl-3-isopropyltriazene, anticarc., 125
structure, 125
α-Phenyl N-methylnitrone, formn. from N-benzyl N-methylhydroxylamine, 186
structure, 186
1-Phenyl-3-methyltriazene, precursor of, aniline, 125, 126
guanosine, 125
CH_3^+, 125, 126
methylates, 125
nucleic acids, 125
phenyldiaz. cation, 126
rat. carc., 125
structure, 125, 126
3-Phenyl-1-methylureas precursors of CH_2O, 60
2-Phenyltetrahydropyran, precursor of, formald., 250
1-phenyl-2-butene, 250
ring-opening, 250
structure, 250
Phildelphia, concn. of dimethylnitrosamine, 64
Phloroglucinol, detm. of bacitracin, 99
structure, 99
Phorate (O,O-Diethyl-S-(ethylthio)-methyl phosphorodithioate, 12
anal. eqn., 229
detm. in plant tissue, 281
detm. with CA, 43, 229, 281
hydrolysis, 12
oxidn., 12
precursor of, diethyl dithiophosphate, 229
ethyl mercaptan, 229
formald., 12, 20, 228, 229, 281
structure, 229, 230
TLC sepn., 281
Phorate sulfone, precursor of CH_2O, 20
Phosgene, formn. from vinylidene chloride, 62
Phosphate buffer, 70, 139, 143
Phosphates, detm. of glycerol, 279
Phosphatidylglycerol, detect. with *p*-rosaniline-SO_2, 12, 17
oxidn., 12
precursor of CH_2O, 12
TLC sepn., 12, 17
Phospholipids, 260, 264, 266
detm. of glycerol, 279
detm. with CA, 38
hydrol., 38, 279
interference removal, 256
removal by, alumina, 264
calcium hydroxide, 264
florisil, 264
Kieselguhr, 264
Lloyd's reagent, 264
silicic acid, 264
Zeolite, 264

Phosphoric acid, solvent, 43, 92, 136, 158, 216
Phosphorous trichloride, 58
Photochemical smog, 128
  eye irritation, 196
Photodecomposition, 1,1-bis-*p*-ethoxyphenyl-2,2-dimethyloxetane, 246
  2,2-dimethyloxetane, 246
  oxetane, 246
Photolysis, alkyl nitrites, 242
  benzyl nitrites, 242, 243
  carbon monoxide, 276, 277
  dimethylnitrosamine, 182
  ethyl nitrite, 243
  methyl nitrite, 243
  propyl nitrite, 243
Photooxidation, 23
  1-alkenes, 2
  dialkyl ketones, 9
  mannitol, 10
  mesitylene, 10
  methyl ketones, 9
  styrenes, 13
  toluene, 14
  triethylamine, 14
  vinyl chloride, 269
  vinylidene chloride, 62
  m-xylene, 15
Photorearrangement, oxetane, 250
  tetrahydrofuran, 250
Phthalyl chloride, 58
Pictet-Spengler condensation, 237
Pig, carc. of diethylnitrosamine, 68
  metab. of nicotine, 179
Pig aortic valves, fixation of, crotonald., 221
    formald., 221
    glutarald., 221
Pig brain enzyme, 237
Pig heart, lactic dehydrogenase, 114
Pig heart muscle, pyruvic oxidase, 114
Pilocarpine, anal. proc., 248, 249
  detm. in, eye drops, 249
    eye ointments, 249
    hair tonics, 249
    standards, 248
  detm. with CA, 248, 249
  glaucoma treatment, 248
  interferences in detm., 248
  oxidn., 12
  precursor of CH_2O, 12
  stimulation, 248
  structure, 248
  TLC sepn., 248, 249
β-Pinene, oxidn., 11, 16
  precursor of CH_2O, 11, 16
Piperettine, detm. with gallic acid, 46
Piperic acid, detm. with gallic acid, 46
S-Piperidinomethylthiobenzoate, hydrol. eqn., 161
  nitrosation eqn., 161
  precursor of, formald., 161
    N-nitrosopiperidine, 161
  structure, 161
Piperine, anal. proc., 214
  densitometry, 46
  detm. in pepper, 43, 214, 281
  detm. with, CA, 43, 214, 281
    gallic acid, 43, 214, 281
  extn., 281
  hydrolysis, 12, 282
  precursor of CH_2O, 12, 43, 214, 281, 282
Piperonal, acid hydrol., 282
  densitometry, 46
  detm. with, CA, 46, 282
    J-acid, 46
    phenyl J-acid, 46
  precursor of CH_2O, 46, 282
  TLC sepn., 46
Piperonyl butoxide, densitometry, 46
  detm. with CA, 46, 282
  hydrolysis, 12, 282
  inhibitor of N-oxidn., 147
  precursor of CH_2O, 12, 282
  synergist, 212
  TLC sepn., 46
Piperonylic acid, acid hydrol., 282
  detm. with, CA, 43, 282
    J-acid, 44
    phenyl J-acid, 43
  precursor of CH_2O, 28, 43, 282
  spectral data, 43, 44
Piperonylic acids, hydrolysis, 12
  precursor of CH_2O, 12
Piperonylidene acid, detm. with MBTH, 44
  spectral data, 44
Pituitary mammotropic tumour, 146
Pivaloyloxymethyl cephaloglycin, precursor of CH_2O, 216
Pivampcillin, anal. proc., 216, 217
  detm. with CA, 216, 217
  hydrol. eqn., 216
  precursor of CH_2O, 216
Plant, amine oxidase, 160
  oxidn. of methylamines to CH_2O, 160
Plants, nitrosamines, 65
Plant tissue, detm. of Phorate, 281
Plasma, detm. of, glucose, 279
    glycerides, 279
    glycine, 98
    triglycerides, 255–257, 260, 264, 265, 267, 279
    trypsin, 281

Plasma—*continued*
tryptophan, 242
sterilization, 251
Plasma lipids, extn., 267
Plasmal reaction, 222, 254
interference, 223
Plastics, adsorption of ethylene oxide, 81
detm. of ethylene oxide, 81
Podophyllotoxin, contaminant, 281
detm. in proresid, 281
detm. with CA, 44, 281
hydrol., 12
precursor of CH_2O, 12, 44, 281
TLC sepn., 281
Pollution, mobile sources, 80
stationary sources, 80
Polyacrylonitrile, formald. content, 217
Polyalkylbenzenes, 128
Polyamines, reaction with CH_2O, 205
Polyester knit, CH_2O content, 217
Polyethylene terephthalate, detm. of diethylene glycol, 278
Polyhydroxy compounds, oxidn., 253, 254
precursors of CH_2O, 253, 254
Poly(inosinic acid), reaction with CH_2O, 211
Polymethylene oxides, 249–250
Polymethylphosphoramides, formn. from hempa, 155
precursor of CH_2O, 155
Polyols, anal. eqn., 104
detm. in urine, 281
detm. with, CA, 12, 16, 281
MBTH, 104
oxidn., eqn., 251
with periodate, 12, 104, 251
precursors of CH_2O, 12, 16, 104, 251, 281
formic acid, 251
sepn. by CC, 281
structure, 104
α-Polyoxymethylene, permeation tube prepn, 194, 195
precursor of CH_2O. 12, 194, 195
prepn., 194
thermal hydrolysis, 12
Polyoxymethylene glycols, in aq. CH_2O, 212
present in formalin, 212
structure, 212
Polysaccharides, 111, 112
detm. with phenylhydrazine, 111, 112
electrophoresis, 111
precursors of, erythritol, 114
formald., 111, 112
threitol, 114
Poly(uridylic acid), reaction with CH_2O, 211
Polyvinyl chloride, 273, 274
Porapack Q, collection of methylal, 213
Potassium arsenite, 30
Potassium bicarbonate, 92, 106
Potassium bisulfate, 139, 275
Potassium dihydrogen phosphate, 49, 184
Potassium ethoxide, 14
Potassium ferricyanide, oxidant, 88, 92, 124, 197, 224
Potassium hydroxide, 14, 91–94, 124, 154, 184, 256–259
saponification, 259–261, 266, 267
Potassium iodide, 91, 267
detm. of ozone, 86, 228
Potassium metaperiodate, oxidant, 92
Potassium oxalate, standard, 245
Potassium periodate, oxidant, 84, 135, 256
Potassium permanganate, 28, 30, 275
oxidant, 82, 136, 137
Potassium phosphate, 49
buffer, 171
Prebiotic synthesis, 276, 277
Precarcinogen, acetylcholine, 186
aliph. amines, 186
amidopyrine, 186
carnotine, 186
choline, 186
dimethylamine, 186
dimethylaminoethyl actate, 186
dimethylglycine, 186
1-hexadecene, 80
N-methylaniline, 186
neurine, 186
nitrite, 160, 161, 186
NO_x, 186
tertiary amines, 160
trimethylamine, 186
Precursor structures, 276
Prednisolone, effect on aminopyrine N-demethylase, 144
Pregnancy, effect of carcinogens, 54
effect on, aminopyrine N-demethylase, 144
demethylation of aminopyrine, 168
metab. redn. of $4\text{-}O_2NC_6H_4OCH_3 \rightarrow CH_2O$, 189
Pregnant women, effect of diethylstilbestrol, 172
Pregnane-3,21-diol-11, 20-dione, 254
Δ^4-Pregnene-17,21-diol-3,11,20-trione, oxidn., 254
Pregn-4-ene-3,20-dione, precursor of CH_2O, 19
Δ^4-Pregnene-11,17,21-triol-3,20-dione, oxidn., 254
precursor of CH_2O, 254
Pregnenolone-16α-carbonitrile, inhibitor of N-demethylase, 148
stimulation of N-demethylase, 145, 149

Prenatal carcinogenesis, 77
Procarcinogens, 51
 1-aryl-3,3-dialkyltriazenes, 52
 azoalkanes, 52
 azoxyalkanes, 52
 definition, 52
 1,2-dialkylhydrazines, 52
 dialkylnitrosamines, 52
Procaryotes, mutated by CH_2O, 233
Progesterone, precursor of CH_2O, 19
Promethazine, precursor of CH_2O, 84
Prometryne, metabolism, 192
  enhancement by PAH, 192
 precursor of CH_2O, 192
Promotion, 251
n-Propanal, 136
 detect. with 2,3-dimethyl-2,3-bis(hydroxylamino)butane, 201
 formn, from methylpropylhydrazine, 43
 reaction rate, 203
 spectral data, 213
Propane-1,2-diol, 114
 detm. with 2,4-pentanedione, 115
 sepn., 115
1,2-Propanediol phosphate, hydrolysis, 12
 oxidn., 12
 precursor of CH_2O, 12
n-Propanol, effect on aminopyrine N-demethylase, 144
2-Propanol, 202
 extractant, 260–267
 stimulation of N-demethylase, 146
Propanolamine, detm. with CA, 44
4-Propenylveratrole, detm. with MBTH, 44
 oxidn., 12
 precursor of CH_2O, 12
 spectral data, 44
Propene, reaction with ozone, 28
 precursor of CH_2O, 28
5-(2-Propenyl)-1,3-benzodioxole, *see* Safrole
Propionaldehyde, *see* n-Propanal
β-Propiolactone, alkylating agent, 251, 252
 anal. proc., 252
 binding to, DNA, 251
  protein, 251
  RNA, 251
 cancer initiator, 251
 carcinogen, 251
 clastogen, 251
 decontaminant, 251
 detm. with, CA, 252
  2,7-dihydroxynaphthalene, 252
  4-(4′-nitrobenzyl)pyridine, 252
  4-pyridinealdehyde 4-nitrophenylhydrazone, 252
 DNA binding and carc., 251
 half-life, 251
 mutagen, 251
 precursor of CH_2O, 252
 preparation, 251
 reacts with, albumin, 251
  cysteine, 251
 sterilizing agent, 251
 structure, 251
Propoxyphene, analgesic, 162
 narcotic, 162
 precursor of, dimethylnitrosamine, 162
  formald., 162
n-Propylamine, detm. with DCPIP, 160
Propylene, atmospheric, 21
 photooxidn., 18, 129, 243
 precursor of, acetald., 129
  aliph. ald., 18
  formald., 18, 129, 243
  PAN, 129
 yield of, aliph. ald., 18
  formald., 18
Propylene glycol, 268
 detm. of contamination, 106
 detm. with MBTH, 44
 effect on, aminopyrine N-demethylase, 144
  N-demethylation, 171
 spectral data, 44
Propylene oxide, absorbance into plastics, 79
 binding to DNA, 81
 collection from air, 80
 detm. with CA, 80
 hydrol., 80
 mutagen, 81
 oxidn., 13
 persistence, 79
 precursor of CH_2O, 13
 reaction with, HCl, 79
  water, 79
 sterilization agents, 79
Propylhexedrine, N-demethylation, 150
 precursor of CH_2O, 150
Propyl nitrite, photolysis, 243
 precursor of, peroxyacyl nitrate, 243
  propionald., 243
Proresid, contamination, 281
 detm. of podophyllotoxin, 281
Protein, binding by β-propiolactone, 251
 crosslink with DNA, 1
 effect on, N-demethylase, 148
  rat, 74
 methylation, 69
 microsomal, 70
Protein precipitation, 85, 135, 154, 184, 256
 detm. of methionine sulfoxide, 141
 hydrochloric acid, 177

Protein precipitation—*continued*
  perchloric acid, 177
  trichloroacetic acid, 85, 154, 177
Proteins, allergens, 104
  digestion with papain, 58
  reaction with, formald., 221, 233
    glutarald., 221
Protoporphyrin monomethyl ester, anal. with CA, 13, 16
  oxidn., 13
  precursor of CH_2O, 13, 16
Proximate carcinogen, methylazoxymethanol, 53, 55
Pseudoephedrine, N-demethylation, 150
  precursor of CH_2O, 150
Pseudouridine phosphates, 104
  formn. of, dihydroxypyrimidines, 106
    formald., 106
  oxidn., 13
  oxidn. eqn., 106
  precursors of CH_2O, 13
  structure, 106
Psilocybin, precursor of, dimethylnitrosamine, 162
    formald., 162
  psychotomimetic, 162
Psychotogens, interferences of formn., 159
Puromycin, antibiotic, 162
  N-demethylation, 147
  precursor of, dimethylnitrosamine, 162
    formald., 162
Puromycin aminonucleoside, N-demethylation, 147
Purpald, absorption of CH_2O, 200, 218, 219
  anal. eqn., 219
  detm. of CH_2O, 200, 218, 219
  spot test for wood CH_2O, 200
  structure, 200, 219
Purpurald, *see* Purpald
PVC, 269
Pyridine, 108, 111, 202
  solvent, 30
4-Pyridinealdehyde 4-nitrophenylhydrazone, detm. of β-propiolactone, 252
1-(3-Pyridyl)-3,3-dimethyltriazene, carc., 124
  precursor of CH_2O, 124
Pyrogallol, detm. of, glycolald., 40, 113
  glyoxylic acid, 113
Δ′-Pyroline S-carboxylic acid, formn. from hydroxylysine, 131
  structure, 131
Pyrrolizine alkaloids, carc., 81
Pyrrols, reaction with CH_2O, 208
  structure, 208
    of product, 208
Pyruvaldehyde, 204
  chemiluminescence, 28
  detm. with gallic acid, 203
Pyruvate, formn. from 2-keto-4-hydroxybutyrate, 230
  structure, 230
Pyruvic oxidase, cleavage of glyoxylate, 114
  formn. of CH_2O, 114
  pig heart muscle, 14

Quinine, detm. with, CA, 44
    ethyl acetoacetate, 30, 44
    J-acid, 31
    MBTH, 31
  fluor. anal., 44
  oxidn., 13
  precursor of CH_2O, 13
  spectral data, 44
Quinoline N-oxide, 11, 222
  structure, 222

Rabbit, azo dye and liver protein, 226
  carc. of, diethylnitrosamine, 68
    dimethylnitrosamine, 68
  chlorpromazine metabolites, 147
  cytochrome P-450, 177
  ethylmorphine N-demethylase, 177
  liver N-demethylase, 147, 177
  lung N-demethylase, 147
Rabbit muscle, lactic dehydrogenase, 114
Raceophenidol, alk. hydrol., 277
  hydrolysis, 13
  oxidn., 13
  precursor of CH_2O, 13, 277
Radiation, carc. of subcarc. doses, 183
  effect on nucleic acids, 210
Raffinose, detm. with MBTH, 44
  spectral data, 44
Ragweed pollen, allergens, 104
Rainbow trout, carc. of, diethylnitrosamine, 68
    dimethylnitrosamine, 68
Rat, alkylation of nucleic acids, 55
  brain tumour, 54
  cancer with vinylidene chloride, 62
  carc. of, aflatoxin B_1, 187
    aflatoxin G_1, 187
    1-aryl-3,3-dialkytriazenes, 125
    bis-chloromethyl ether, 57
    cycasin, 55
    diethylnitrosamine, 68, 69
    dimethylnitrosamine, 68, 69..
    methylazoxymethanol acetate, 55
    1-phenyl-3-methyltriazene, 125
    β-propiolactone, 251
  carc-mutagenic anomalies, 184

carc. nitrosamines, 68
chlorpromazine metabolites, 147
colon, 122
cytochrome P-450, 177
N-demethylase activity, 151
demethylating activity of, brain, 150
  kidney, 150
  liver, 150
  lung, 150
  muscle, 150
  spleen, 150
demethylation of griseofulvin, 190
DMNA, 68
Donryu strain, 191
diet and Me_2NNO, 74
effect of, acetoaminonitrile, 75
  Aroclor 1254, 148, 150, 183
  CCl_4, 166
  CS_2, 51
  DMNA demethylase, 75
  fat, 74
  hepatectomy, 182
  methylbutylnitrosamine, 77
  3-methylcholanthrene, 75, 170
  nitrosomethylurethan, 54
  PAH, 183
  phenobarbital, 75, 183
  phenylbutazone, 144, 146
  starvation, 150
  TCDD, 156
  thioacetamide, 170
  vinyl chloride, 271, 272
elimination of vinyl chloride, 272
ethylmorphine N-demethylase, 177
kidney O-demethylase, 191
kidney tumors, 68
lipid perioxidation and N-demethylation, 176
liver, 77
liver carcinoma, 187
liver homogenates, 18
liver tumours, 67, 69
marginal effect dose, Me_2NNO, 77
metab. of, aflatoxins, 188
  BaP, 175
  Bidrin, 157, 158
  cotinine, 179
  hempa, 155
  palmitoxins, 188
  thiohempa, 155
  vinyl chloride, 271
Natulan, carc., 151
  metabolism, 151
  teratogenicity, 151
neonatal period and N-demethylase, 146
Nitrosoethylurea, 68
Nitrosomethylurea, 68
phenobarbital-treated, 143, 147
rectum, 122
sarcoma from hexamethylenetetramine, 195
S.C. injection, 18
second generation effect, 54
Sprague-Dawley, 71
teratogenicity, 52
teratogenic phenyldimethyltriazene, 125
transplacental carcinogenicity, 52
treatment with (^{14}C)Me_2NNO, 77
tumor and demethylation, 146
tumors from, Et_2NNO, 75
  Me_2NNO, 75
tumors of, esophagus, 69
  intestines, 55
  liver, 55, 69
  nasal cavity, 69
Rat adrenal glands, detm. of glycerides, 279
Rat biofluids, acetylmethadol metabolites, 155
Rat brain, metab. of 3-O-methyldopa to dopa, 191
  formald., 191
Rat brain tissue, methylene reductase in, 234
Rat cells, chromosome aberrations, 195
  effect of formald-contng. resins, 195
Rat embryo, 76, 77
  carc. of nitrosamines, 78
  metab. of nitrosamines, 78
Rat kidney, N-demethylation, 169
  effect on vinyl chloride, 273
Rat liver, cytochrome c oxides, 50
  N-demethylase, 50
  $Et_2NNO \rightarrow CH_3CHO$, 78
  $Me_2NNO \rightarrow CH_2O$, 78
  effect of aflatoxin B_1, 187
  effect on vinyl chloride, 273
  ethylmorphine metab., 177
  formn. of aflatoxin B_1 2,3-oxide, 188
  microsomal fraction, 12, 180
  prepn. of 2-keto-4-hydroxyglutarate aldolase, 230
Rat liver homogenate, 50
Rat liver homogenates, 169
  effect of, meat products, 169
    organic peroxides, 169
Rat liver microsomal fraction, 273
Rat liver microsomes, effect of age, 146
  effect on griseofulvin, 190
  malonald. and formald., 176
  metab. of, aflatoxins, 188
    demethylation of ethylmorphine, 143, 176
    Me_2NNO, 78
    palmotoxins, 188
Rat lung, effect on vinyl chloride, 273
Rat plasma, detm. of glycerol, 279

Rats, BD, 121
  carc. of hempa, 154
  colon cancer, 121
  effect of, adrenalectomy, 167
    castration, 144, 146, 167
    prednisolone, 168
    pregnancy, 168
    selenium, 168
    TCDD, 168
    testosterone propionate, 168
    vit. E, 168
  s.c. injection carcinogens, 81
    Sprague-Dawley, 154
  teratogenicity of, azoxyethane, 53
    1,2-diethylhydrazine, 53
    methylazoxymethanol, 53
  tumors from, 1-aryl-3,3-dialkytriazenes, 52
    azoalkanes, 52
    azoxyalkanes, 52
    1,2-dialkylhydrazines, 52
    dialkylnitrosamones, 52
    vinyl chloride, 274
Rat tongue, Schiff test, 254
  thiol-bound CH_2O, 254
Rayon, formald. content, 217
RDX or Hexahydro-1,3,5-trinitro-s-triazine, 44, 46
  contaminant 281
  detm. in tetranitrotetrazine, 281
  detm. with 2,7-dihydroxynaphthalene, 44, 281
  hydrol., 13
  precursor of CH_2O, 13, 44, 281
  sepn. by PC, 281
Recombinogenesis, formald., 234
Rectum, rat, 122
Reductant, borohydride, 8, 11
  hydrogen iodide, 11
  magnesium, 4, 7–9, 12, 87–89
  rhamnose, 91
  sodium arsenite, 84–86, 107, 111, 135, 137, 257, 259
  sodium bisulfite, 260
  sodium borohydride, 112, 134, 135
  sodium metaarsenite, 61
  sodium sulfite, 81, 82, 136
  sodium thiosulfate, 91, 267
  zinc, 12, 114, 245
Replication point, mutation at, 234
Resin-bonded boards, anal. proc., 218, 219
  detm. of CH_2O, 218
Resorcinol, 249
  detm. of diethylene glycol, 106, 107
Resotropin, anal. proc., 224, 225
  bonding agent, 224
  extn., 224
  formn. from hexamine and resorcinol, 224
  precursor of, formald., 224
    hexamine, 224
  Respiratory tract, form of nitrosamines, 66
    tumors, 69
Rhamnose, reductant, 91
Rhesus monkeys, effect of PCB, 150
  isolation of aflatoxin P_1, 188
Ribitol, location, 109
  precursor of CH_2O, 109
  sepn., 114
Riboflavin, detm. in tablets, 281
  detm. with CA, 13, 16, 281
  oxidn., 13, 254, 281
  precursor of CH_2O, 13, 16, 254, 281
D-Ribose, detm. with, CA, 44
    J-acid, 44
    MBTH, 44
  fluor. anal., 44
  location, 109
  oxidn., 13
  precursor of CH_2O, 13, 109
  spectral data, 44
Ribose-3-phosphoric acid, oxidn., 253
  precursor of CH_2O, 253
Ring closure, purpald-CH_2O, 219
Ring formation, 205, 206
  adrenalin, 242
  aflatoxin B_1, 188
  CA and CH_2O, 229
  2,3-dimethyl-2,3-bis(hydroxylamine)-butane, 20
  dopamine, 237, 238, 242
  epinine, 237
  5-hydroxytryptamine, 238
  5-hydroxytryptophan, 242
  3-indolylethylamines, 238
  metaraminol, 240
  5-methoxytryptamine, 242
  methylamine and CH_2O, 225
  5-methyltetrahydrofolic acid, 225
  N-methyltryptamine, 238
  purpald, 219
  quinoline N-oxide, 222
  sulfate, 209
  tryptamine, 238, 242
  tryptophan, 208, 242
  tryptophanyl-dipeptide, 241
  L-tryptophanyl-L-glycine, 242
  m-tyramine, 240
Ring opening, 222, 235
  2-butene oxide, 250
  2,2-dimethyloxetane, 246
  1,2-dioxetane, 247
  ethylene oxide, 250
  hexamethylenetetramine, 193

hexobarbital, 178
limonene, 250
oxetane, 246
2-phenyltetrahydropyran, 250
thiazolidines, 228
1,3,5-trimethylhexahydro-s-triazine, 225
1,3,5-trioxane, 215
tryptophan, 190
Ring rearrangement, limonene, 250
River water, detm. of CMOS, 102
RNA, alkylated, 52
binding by β-propiolactone, 251
methylation, 69
presence of, 6-dimethylaminopurine, 147
6-methylaminopurine, 147
RNA ase—polymerase, 210
Rodents, hepatotoxicity with vinylidene chloride, 62
urine, DIC metabolite, 125
Rogor, see dimethoate
p-Rosaniline, see pararosaniline
Ruthenium dioxide, oxidant, 4

Sabinene, oxidn., 11, 13, 16
precursor of CH_2O, 11, 13, 16
Saccharomyces cerevisiae, effect of CH_2O, 233
Safrole, acid hydrol., 282
anal. with gallic acid, 13
carc., 81
detm. with, CA, 46, 282
J-acid, 31, 44
MBTH, 31, 44
fluor. anal., 44
hydrolysis, 13
oxidn., 13
precursor of CH_2O, 13, 195, 282
spectral data, 44
TLC sepn., 46
Safroxan, detm. with CA, 46
Salicylaldehyde, reaction with 2,7-dihydroxynaphthalene, 100
Salicyclic acid, 249
Salmonella typhimurium, bioassay anomalies, 184
bioassay for Me_2NNO, 76
dimethylnitrosamine, 232
effect of N-2-fluorenylacetamide, 184
effect of microsomal fractions, 76
G-46, 74, 75
effect of DMNA, 75
host-mediated, 74
indicator of mutagenesis, 73
mutagenicity of methylazoxymethanol, 55
mutated by, 1,2,3,4-diepoxybutane, 81
diethyl sulfate, 251
hydrazine, 251
nitrogen mustard, 251
β-propiolactone, 251
vinyl chloride, 273
strain, G-46, 273
TA 1530, 273
TA 1535, 273
Salsolinol, formn, from dopamine, 237, 238
structure, 238
urinary, 238
Sampling, cold trap, 64
Tenax GC, 64
Sampling, formald., 213
Sawdust, 218
Saponification, glycerides, 92, 279
triglycerides, 259, 261, 263, 267
Sarcoma, rats, 195
Sarcosine, detm. with 2,4-pentanedione, 97
precursor of CH_2O, 15
Schiff-mercuric chloride reaction, 254
Schiff tests, bound CH_2O, 254
Schiff's reagent, detm. of, ethylene glycol, 85
N,N,N',N'-tetramethylmethylenediamine, 225
1,3,5-trimethylhexahydro-s-triazine, 225
reaction with, acrolein, 223
glutaraldehyde, 223
test for OCH_2S, 222
Schizophrenia, aldehyde reactions, 238
labile methyl groups, 239
relation to, cancer, 239
epilepsy, 239
treatment with chloropromazine, 147
Screening tests, mutagenesis, 73
Secondary amines, N-oxidn., 155
precursor of, ald., 155
hydroxylamine, 155
primary amine, 155
Second generation effect, 171, 172
Sedoheptulose anhydride, location, 109
precursor of CH_2O, 109
Seed, detm. of 2,4-dichlorophenoxyacetic acid, 278
Seed fat, detm. of triglycerides, 279
Selenium, effect on, aminopyrine N-demethylase, 145
demethylation, 168
Selenium dioxide, oxidant, 2, 124
Semicarbazide, stabilization of CH_2O, 78
trapping of CH_2O, 153, 157, 167, 169, 178, 184, 283
Serine, 272
anal. proc., 48
automated assay, 44
detm. in soft drinks, 281
detm. with, CA, 281

Serine—*continued*
ethyl acetoacetate, 44
MBTH, 44, 48
2,4-pentanedione, 44
fluor. anal., 44
formn, from vinyl chloride, 272
oxidn., 13, 253
precursor of CH_2O, 13, 253, 281
sepn. by CC, 281
spectral data, 48
Serine hydroxymethylase, 1-carbon metab., 152
Serine transhydroxymethylase, binding with CH_2O, 231
Serratia marcescens, mutated by β-propiolactone, 251
Serum, detm. of, bilirubin, 55
glucose, 140, 279
glycerides, 91, 92, 279
glycine, 97, 98
lecithin, 93, 94
mannitol, 135
methanol, 136
sorbitol, 135
triglycerides, 255–258, 260–267, 279
triglyceride levels, 255
Serum folate, effect of, anticonvulsants, 239
diphenylhydantoin, 239
Serum transaminase, 272
Sesamex, detm. with CA, 46
Sesamin, detm. with CA, 214
hydrolysis, 13
location on TLC, 214
precursor of CH_2O, 13, 214
Sewage, amines, 65
nitrate, 65
nitrosamines, formn., 64
microbial contribn., 65
stability, 64
concn., 66
Sewage effluent, detm. of CMOS, 102
Shampoos, detm. of CH_2O, 47
Shellfish, detm. of 2-butoxyethyl 2,4-D, 278
Sialic acids, detect. with 2,4-pentanedione, 13, 17
oxidn., 13
PC sepn., 13, 17
precursors of CH_2O, 13, 17
Sialoglycoprotein, anal. proc., 111
detm. with MBTH, 111
oxidn., 13
precursor of CH_2O, 13
Silica gel, 94
Silica gel G, 249
TLC sepn., 120
Silicic acid, 92
interference removal, 256, 258
phospholipid removal, 264, 265
Silver nitrate, sepn. of triglycerides, 259
Simplex optimization, formald, detm. with, CA, 199
J-acid, 199
2,4-pentanedione, 199
Sin, of the parents, 172
Singlet oxygen, 272
Sinus problems, 213
SKF-525-A, β-diethylaminoethyldiphenylpropyl acetate, 151
inhibitor of, bacterial killing, 74
N-demethylase, 145, 146, 176
O-demethylation, 176
dimethylnitrosamine, 73
ethylmorphine, 149
formn. of morphine glucuronide, 176
hydrol. of procaine, 176
microsomal hydroxylation, 74
mutagenesis, 74
N-oxidn., 147
oxidn. of barbiturates, 176
structure, 176
Skin, metabolism of xenobiotics, 175
Skin tumors, initiation by β-propiolactone, 251
mouse and β-propiolactone, 251
Small intestine, env. contaminants, 175
Smog, formation, 128
hydrocarbon rankings, 26
eye response, 26
formald. prodn., 26–28
NO_2 t-max, 26
oxidant prodn., 26, 28
lachrymators, 128
photochemical, 23, 25
role of, hydrocarbons, 23, 25
NO_x, 23
Smog chamber experiments, 23, 128, 129
Smoked fish, nitrosamines, 63
Sodium acetate, 137, 276
Sodium arsenite, reductant, 29, 84, 85, 92, 107, 108, 111, 135, 137, 256, 257, 258, 262
Sodium azide, blank improvement, 178
Sodium bicarbonate, 108, 111
Sodium bisulfite, 153, 154
reductant, 260
trapping of CH_2O, 153
Sodium borohydride, reductant, 59, 112, 134 135
Sodium chloride, 59, 136, 197, 248
Sodium cyanide, inhibitor of CH_2O formn., 178

Sodium dichlorosulfitomercurate, 86
Sodium ethoxide, 14, 262
Sodium fluoride, enhancer of CH_2O formn., 178
Sodium hydroxide, 9, 11–14, 70, 81, 85, 89, 98, 101, 124, 137, 201, 204, 245, 249, 256, 266
  hydrolysis, 47
Sodium isocitrate, 184
Sodium metaarsenite, 48
  reductant, 61
Sodium metaperiodate, 261–263
  oxidant, 30, 48
Sodium methoxide, 14, 261, 263–265, 267
Sodium nitrite, 50
  anticarc., 74
  antimutagen, 74
  antitoxicant, 74
  inhibitor of, aminopyrine demethylase, 74
    aniline hydroxylase, 74
  precursor of Me_2NNO, 185
Sodium periodate, oxidant, 81, 91, 94, 201, 202, 257–260, 265–267
Sodium sulfate, 59
Sodium sulfite, reductant, 81, 82, 136
Sodium thiosulfate, reductant, 267
Sodium tungstate, protein precipitant, 96–98
Soft drinks, detm. of serine, 281
Soil, detm. of methanol, 280
  nitrosamines, formn., 64
    microbial contribn., 65
    stability, 64
Solvents, irradiation, 101
Sophoritol, formn. from sophorose, 252
  oxidn., 13, 252
  precursor of, carbohydrate ald., 252
    formald., 13, 252
    formic acid, 252
  structure, 252
Sophorose, oxidn., 13, 253
  precursor of, formald., 13, 252
    formic acid, 252
    sophoritol, 252
  reduction, 252, 253
  structure, 252
Sorbitol, anal, proc., 135
  detm. in blood, 281
    feces, 281
    urine, 281
  detm. with, CA, 44, 281
    dimedone, 44
    ethyl acetoacetate, 44
    J-acid, 44
    MBTH, 44, 135
    2,4-pentanedione, 41, 44
  location, 109
  oxidn. with periodate, 13, 253, 281
  precursor of CH_2O, 13, 109, 135, 281
  sepn. by CC, 281
  spectral data, 44
Sorbose, oxidn., 253
  precursor of CH_2O, 253
Species differences, 184
  cell membrane permeability, 184
  damage repair, 184
  excretion rates, 184
  process fidelity, 184
  transport mechanisms, 184
Spices, detm. of ethylene oxide, 279
Spironolactone, stimulator of N-demethylase, 149
Spleen, demethylating activity, 150
  metab. of Me_2NNO, 76
  microsome fractions, 76
Sprague–Dawley rat lung, Me_2NNO demethylase, 183
Sprague–Dawley rats, 154
Squamous carcinoma, bis-chloromethyl ether and lungs, 57
Stain, Feulgen, 108
  periodic acid—Schiff, 108
Standard, aqueous CH_2O, 193
  bis-choromethyl ether, 58
  diethoxymethane, 193
  dimethoxymethane, 193
  formald. permeation tube, 194, 195
  hexamethylenetetramine, 193
  1-hydroxymethyl-5,5-dimethylhydantoin, 193
  mannitol, 135
  paraformaldehyde, 193, 194
  α-polyoxymethylene, 194, 195
  potassium oxalate, 245
  triolein, 262
  tripalmitin, 256
  tristearin, 258
Standardization, formald., 215, 216, 231
Stannous chloride, 50
Starch, 59
Starvation, induction of Me_2NNO N-demethylase, 150
  inhibitor of N-demethylase, 148
  stimulator of N-demethylase, 148
Steer brain, detm. of tryptamine, 242
Sterilization, β-propiolactone, 251
Steroids (*see* Corticosteroids), anal. by CA, 13, 17
  enhancement of, aliph. hydroxylation, 151
    N-demethylation, 151
  oxidn., 13

Steroids—*continued*
  precursors of CH_2O, 13, Table 3, 19
  structures, 19
Stillbirths, vinyl chloride and wives, 274
Stimulators, effect on N-demethylases, 144–150
Stomach, lethal synth, 185
  metab. of Me_2NNO, 76
  microsome fractions, 76
Streptomycin, oxidn., 13
  precursor of CH_2O, 13
Stress, stimulator of N-demethylase, 149
Styrene, atmospheric, 21
  detm. with CA, 45
  oxidn., 13
  photooxidn., 26, 28
  precursor of CH_2O, 13, 26, 28
  yield of CH_2O, 26, 28
Styrene oxide, atmospheric, 80
  carc., 81
  cocarc., 81
  structure, 81
Styrenes, eye irritation, 196
  photooxidn., 13
  precursors of CH_2O, 13
Styrolene, anal. with ethyl acetoacetate, 13, 16
  oxidn., 13
  precursor of CH_2O, 13, 16
Subcarcinogenic doses, 182
Sublimate, 254
Succinic acid, formn. from Alar, 124
Sulfur dioxide, 5, 11, 14
  detm. with, orcinol, 60, 115–117
    2,4-pentanedione, 115–117
  sepn., 114–117
Sugar alcohols, 114–118
  detm. with 2,4-pentanedione, 115–118
  sepn., 114–117
Sugar sulfates, oxidn., 13
  precursor of CH_2O, 13
Sulfamic acid, 84, 86
Sulfarsphenamine, hydrolysis, 11, 16
  precursor of CH_2O, 11, 16
Sulfate, reaction with, formald., 209
    6-hydroxymethyl BaP, 18
  structure of products, 209
Sulfated hydrocolloids, 59
Sulfite, effect on fluor., 91
  interference in anal. for CH_2O, 152, 153
  mechanism of interference, 153
  reaction with CH_2O, 205
  structure of products, 205
Sulfomethylation, 153
Sulfotransferase, 18
Sulfoxone sodium, hydrolysis, 11, 16
  precursor of CH_2O, 11, 16
Sulphur, 133, 134

  adducts with N,N-dimethylalkylamines, 163
    N-methylmorpholine-N-oxide, 163
    trimethylamine oxide, 163
    tropine N-oxide, 163
  interf. in MBTH method, 197, 198
  release of bound CH_2O, 222
Sulfuric acid, 29, 35, 37, 49, 50, 53, 58–62, 81, 82, 84, 88, 92, 94, 96–101, 103, 106, 108, 113, 118, 123, 124, 135–137, 140, 141, 177, 198, 201, 204, 213, 215, 217, 248, 252, 256, 259, 260, 263, 266, 278
  decarboxylation, 8, 13, 277
  disproportionation, 8, 9, 12, 277
  hydrolysis, 3–8, 10–14
  oxidant, 3, 4, 6–9, 12, 13, 141, 252, 257, 258, 282
  tautomerism, 3, 4
Sulpyrine, oxidn., 13
  precursor of CH_2O, 13
Supracide, precursor of CH_2O, 227
  structure, 227
Surface waters, detm. of 2,4-D, 278
  ethylene glycol, 84
Surveyor III, 67
Swine urinary metabolite, 8-methoxykynurenic acid, 191
Synergism, CCl_4 and radiation, 183
  DMNA and N-2-fluorenylacetamide, 182, 184
  DMNA and hepatectomy, 182
  3′-methyl-4-dimethylaminoazobenzene, 182

Tablets, detm. of hexamine, 223, 224
    mandelamine, 223, 224
    methenamine, 119
    methenamine mandelate, 119
    riboflavin, 281
Tannic acid, detm. of, methylenedioxyphenyl group, 43
Tartaric acid, detm. with gallic acid, 46
Tartronate, anal. with 2,7-dihydroxynaphthalene, 13, 16
  decarboxylation, 13, 277
  disproportionation, 13
  precursor of CH_2O, 12, 277
Tartronic acid, 99
  oxidn., 100
  precursor of, carbon monoxide, 100
    formaldehyde, 100
  structure, 100
Tautomerism, azomethane, 3
  cyclohexylazomethane, 4
  1,1-dimethylhydrazine, 5

1,2-dimethylhydrazine, 5
methylazobutane, 10
1-methyl-2-butylhydrazine, 11
1-methyl-1-cyclohexylhydrazine, 11
1-methyl-2-cyclohexylhydrazine, 11
methylpropylhydrazine, 11
nitromethanol, 11
phenylazomethane, 12
trimethylhydrazine, 14
TBA, see 2-Thiobarbituric acid
TCDD, effect on, aminopyrine N-demethylase, 50, 145
benzphetamine N-demethylase, 147
N-demethylation, 156, 168
'non-responsive' mice, 146
enz. stimulation of 'non-responsive' mice, 189
inhibition of N-demethylase, 145, 149
stimulation of N-demethylase, 145
structure, 157
teratogen in rat, 146
Technicon T5C, 114
Tenax GC, collection of nitrosamine, 64
Teratogen, azoxyethane, 53
bidrin, 157
1,2-diethylhydrazine, 53
methylazoxymethanol, 53, 55
methyl methanesulfonate, 139
Natulan, 151
TCDD, 146
vinyl chloride, 274
Tetratogenicity, humans, 274
rats, 52
Terpenes, anal. with, CA, 13, 16
J-acid, 31
MBTH, 31
oxidn., 13
precursor of CH_2O, 13, 16
Terpin hydrate, effect on aminopyrine N-demethylase, 145
Tertiary amines, N-oxidn., 155
precursors of, amine oxides, 155
nitrosamines, 160–163
reaction with nitrite, 160–163
Testes, metab. of Me_2NNO, 76
microsome fractions, 76
Testosterone, effect on aminopyrine N-demethylase, 145
Testosterone hydroxylation, 167
Testosterone propionate, effect on aminopyrine N-demethylase, 145
Tetracaine, anesthetic, 162
precursor of, dimethylnitrosamine, 162
formald., 162
2,3,7,8-Tetrachlorodibenzo-*p*-dioxin, see TCDD
Tetrachloroethylene, prodn., 269
Tetracyanoethylene, reaction with 1,2,4-trioxolanes, 248
Tetracyanoethylene oxide, formn, from 1,2,4-trioxolane, 248
structure, 248
Tetracycline, antibiotic, 162
precursor of, dimethylnitrosamine, 162
formald., 162
1-Tetradecene, detm. with MBTH, 45
oxidn., 13
precursor of CH_2O, 13
spectral data, 45
1,2,3,4-Tetrahedro-β-carbolines, detm. with MBTH, 242
formn. from, formald., 242
3-indolylethylamines, 242
Tetrahydrofolate, 182
Tetrahydrofuran, photooxidn., 26, 28
photorearrangement, 250
precursor of, acetald., 250
ethylene, 250
formald., 26, 28
structure, 250
yield of CH_2O, 26, 28
Tetrahydroisoquinoline, 237
1,2,3,4-Tetrahydroisoquinolines, 240
structures, 240
Tetrahydroisoquinolines, derived from dopamine, 238
effect on schizophrenia, 238
Tetrahydronorharman derivative, formn., 241
structure, 241
Tetrahydropapaveroline, 237
Tetrahydropyrans, structure, 250
2α,11β,17α,21-Tetrahydroxy-pregn-4-ene-3,20-dione, precursor of CH_2O, 19
2β,11β,17α,21-Tetrahydroxy-pregn-4-ene-3,20-dione, precursor of CH_2O, 19
11β,16α,17α,21-Tetrahydroxy-pregn-4-ene-3,20-dione, precursor of CH_2O, 19
11α,17α,20β,20-Tetrahydroxy-pregn-4-ene-3-one, precursor of CH_2O, 19
11β,17α,20β,21-Tetrahydroxy-pregn-4-ene-3-one, precursor of CH_2O, 10
Tetramethylammonium hydroxide, 216
N,N-α,α-Tetramethylheptylamine oxide, adduct form, 14
hydrolysis, 14
precursor of CH_2O, 14
Tetramethylhydrazine, oxidn. with $HgSO_4$, 14
precursor of CH_2O, 14
N,N,N′,N″-Tetramethylmelamine, anticarc., 120
N,N,N′,N′-Tetramethylmethylenediamine, detm. with, CA, 225

N,N,N′,N′-Tetramethylmethylenediamine—*continued*
Schiff's reagent, 225
formn. from CH_2O, 225
precursor of CH_2O, 225
N,N-β,β-Tetramethylphenethylamine oxide, adduct form, 14
hydrolysis, 14
precursor of CH_2O, 14
Tetramethylphosphoramide, formn. from hempa, 155
precursor of CH_2O, 155
Tetramethyltetrazene, anal. with CA, 14, 16
oxidn., 14
precursor of CH_2O, 14, 16
Tetranitromethane, oxidant, 160
Tetranitrotetrazine, contamination, 281
detm. of RDX, 281
Tetrazolium blue, 216
Tetritols, oxidn., 14
precursors of CH_2O, 14
Textiles, treated with CH_2O resins, 217
Thermal decomposition, 1,2-dioxetane, 247
Thiazolidine-4-carboxylic acid, anal. eqn., 228
anal. proc., 228
detm. with MBTH, 45, 228
form from cysteine, 210
hydrol., 14
oxidn., 14
precursor of, cystine, 228
formald., 14, 20, 228
structure, 228
Thimet, *see* Phorate
Thin layer chromatography, *see* TLC
Thioacetamide, effect on, cytochrome P_{450}, 170
N-demethylation, 170
blocks phenobarbital induction, 170
2-Thiobarbituric acid, detm. of malonald., 138
use in lipid peroxidn., 51
Thiocyanate, nitrosamine formn., 66, 67
4-Thiocyano-N,N-dimethylaniline, anal. proc., 164, 165
detm. with MBTH, 164
precursor of CH_2O, 164
spectral data, 164
Thiodiglycollic acid, 272
Thiohempa, metabolism, 155
precursor of, formald., 155
hempa, 155
Threitol, detm. with 2,4-pentanedione, 114, 115
formn. from hexitols, 114
polysaccharides, 114
sepn., 115
Thioformaldehyde, 36
Thiol-bound formaldehyde, hydrolysis, 14
precursor of CH_2O, 14
Thiomethylenamines, 227, 228
precursors of CH_2O, 227, 228
Thiosulfate, effect on fluor., 91
reductant, 91
Threshold limit, nitrosamines in air, 76
Thymidine, reaction with CH_2O, 211
structure, 211
Thymine fraction, DNA, 77
Tissue, microspectrophotofluorimetry, 242
Tissues, detm. of, ethylene glycol, 279
1,2-glycols, 280
Titanium dioxide, catalyst, 250
TLC, aldehyde-fuchsin products, 199
BaP metabolites, 18
characterizn. of glyoxylic acid, 41
cycasin, 55
detectn. of O—CH_2—O, 43
detm. of, hexamine, 225
methylenedioxy compds., 214
pilocarpine, 249
triglycerides, 259, 260
ethylene glycol., 38
glycerides, 279
glycerol, 279
glycolipids, 40
ident. of 2-methyl-6,7-dihydroxy-1,2,3,4-tetrahydroisoquinoline, 237
location by CA of CH_2O precursors, 214
location of 1,2-glycols, 108, 109
methadone metabolites, 159
methylazoxymethanol, 55
phorate, 281
piperonal, 46
piperonyl butoxide, 46
Podophyllotoxin, 281
safrole, 46
sepn. of hexamine, 120
Δ^4-3-oxo-C_{21}-steroids, 45
triglycerides, 279
Tobacco, concn. of N′-nitrosonornicotine, 65
nitrosamines, 65
Tokyo, traffic volume, 197
o-Tolidine, carcinogen, 172
Toluene, anal. with CA, 14, 16
photooxidn., 18, 243
photooxidn. with NO_2, 14, 26, 28
precursor of, aliph. ald., 18
formald., 14, 16, 18, 26, 243
solvent, 215
yield of, aliph, ald., 18
formald., 26, 28
p-Toluenesulfonic acid, 215
p-Tosyl arginine methyl ester, anal. with 2,4-pentanedione, 14, 17

enz. hydrolysis, 14
oxidn., 14
precursor of CH_2O, 14, 17
yield of CH_2O, 17
Toxicity, dimethylnitrosamine, 70, 71, 74
dimethylnitrosamine to man, 77
pathways, 74
redn., 74
vinyl chloride, 273
TPN, 178
Tranquilizers, N-oxidn., 155
Transaminase, liver injury, 272
Transamination, L-homoserine, 230
Transesterification, eqn., 264
triglycerides, 264, 267
Transformation, Me_2NNO, 74
interf. by nitrite, 74
Transnitrosation, 67
Transplacental cancer, 54
Transplacental carcinogenesis, 172
o-amineoazotoluene, 172
3,3′-dichlorobenzidine, 172
diethylstilbestrol, 172
N,N-dimethyl-4-phenylazoaniline, 172
o-tolidine, 172
Transplacental exposure, 54, 55
Trialkylamines, nitrosation mechanism, 161
precarcinogen, 160, 161
precursors of, dialkylnitrosamines, 161
aldehydes, 161
Triazine N-demethylase, 125
Trichloroacetic acid, 171, 228
pptn of protein, 85, 154, 177
Trichloroethylene, effect on aminopyrine demethylase, 166
cytochrome P_{450}, 166
prodn., 269
solvent, 215
Triethanolamine, 81
Triethylamine, photooxidn., 14, 26, 28
precursor of CH_2O, 14, 26, 28
yield of CH_2O, 26, 28
Triglycerides, 255–268
anal. eqn., 260
anal. proc., 91, 94, 256–268
clinical lab methods, 255
concns., 263
coronary heart disease, 255
detm. by TLC, 259–260
detm. in, mosquitos, 279
palm-kernel oil, 259, 260
plasma, 255–257, 260, 264, 265, 267, 279
seed fat, 279
serum, 255–258, 260–267, 279
detm. with, acetylacetone, 94
automated methods, 255, 260, 264–267
CA, 255–260, 279
colorimetric methods, 255–266
enzymatic methods, 255
ferric chloride, 255
fluorimetric methods, 255, 260, 263, 266, 267
manual methods, 255–260, 263, 264
MBTH, 255, 260
2,4-pentanedione, 91, 255, 260–267, 279
phenylhydrazine, 255, 279
UV methods, 255
extn. with, heptane-isopropanol, 264
nonane-isopropanol, 264
hydrol., 91, 256
hydrol. eqn., 264
hydrol. with, HCl, 14
pot. ethoxide, 14
pot. hydroxide, 14
sod. ethoxide, 14
sod. hydroxide, 14
sod. methoxide, 14
interf. removal, 256, 264, 268
levels in atherosclerosis, 255
oxidn., 14
precursors of, fatty acids, 260
formald., 14, 255, 260
glycerol, 260
glycolald, 260
serum concns., 255
structure, 264
transesterification, 264
2,4,6-Tri(hydroxymethylamino)-1,3,5-triazine, anticarc., 222
mutagen, 222
precursor of CH_2O, 222
3β,11β,21-Trihydroxy-5α-pregnane-20-one, oxidn. to CH_2O, 131
structure, 132
6β,11β,21-Trihydroxy-pregn-4-ene-3,20-dione, precursor of CH_2O, 19
7α,14α,21-Trihydroxy-pregn-4-ene-3,20-dione, precursor of CH_2O, 19
11α, 17α, 21-Trihydroxy-pregn-4-ene-3,20-dione, precursor of CH_2O, 19
11β,17α,21-Trihydroxy-pregn-4-ene-3,20-precursor of CH_2O, 19
16α,17α,21-Trihydroxy-pregn-4-ene-3,20-dione, precursor of CH_2O, 19
17α,20α,21-Trihydroxy-pregn-4-ene-3,11-dione, precursor of CH_2O, 19
17α,20β,21-Trihydroxy-pregn-4-ene-3,11-dione, precursor of CH_2O, 19
17α,19,21-Trihydroxy-pregn-4-ene-3,20-dione, precursor of CH_2O, 19
17α-20β,21-Trihydroxy-pregn-4-en-3-one, precursor of CH_2O, 19

6β,17α,21-Trihydroxy-pregn-4-ene-3,11,20-trione, precursor of CH_2O, 19
16α,17α,21-Trihydroxy-pregn-4-ene-3,11,20-trione, precursor of CH_2O, 19
17,20,21-Trihydroxysteroids, detm. with 2,4-pentanedione, 45
  precursors of CH_2O, 45
Trimethylamine, anal. eqn., 160
  dec. product of betaine, 162
  detm. eqn., 162
  detm. with trimethylamine dehydrogenase, 162
  formn. from betaine, 162
  in refrigerated seafood, 185
  oxidn., 14
  precursor of, dimethylnitrosamine, 160–162, 185
    formald., 14, 160–162
    methyl cation, 161
    methyldiazonium cation, 161
  present in, lichens, 162
    seaweed, 162
  reaction with nitrite, 161, 162
Trimethylamine dehydrogenase, 162
Trimethylamine oxide, adduct formn, 14
  adduct, with SO_2, 163
    hydrol. to CH_2O, 163
  enz. hydrolysis, 14
  oxidn., 14
  precursor of, dimethylamine, 163
    dimethylnitrosamine, 161–163
    formald., 14, 161–163
    methyl cation, 161
    methyldiazonium cation, 161
  present in, gadoid muscles, 162
    sea fish, 162
  reaction with nitrite, 163
N,N,4-Trimethylaniline, anal. proc., 164, 165
  detm. with MBTH, 164
  precursor of CH_2O, 164
  spectral data, 164
1,3,5-Trimethylbenzene, *see* mesitylene
1,3,5-Trimethylcyclotrimethylenetriamine, *see* 1,3,5-trimethylhexahydro-s-triazine
Trimethylene oxide, *see* Oxetane
2,3,4-Trimethylglucose, oxidn., 253
  precursor of CH_2O, 253
1,3,5-Trimethylhexahydro-s-triazine, detm. with, CA, 225
    Schiff's reagent, 225
  formn. from CH_2O, 225
  precursor of CH_2O, 225
  structure, 225
Trimethylhydrazine, detm. with CA, 45
  hydrolysis, 14
  oxidn., 14
  precursor of CH_2O, 14
  spectral data, 45
  tautomerism, 14
  yield of CH_2O, 45
N,N′,N″-Trimethylmelamine, anticarc., 120
Trimethylolmelamine, anticarc., 121
  mutagenic to, *Musca domestica*, 120
    Drosophila, 120
  precursor of CH_2O, 121
  structure, 120
2,4,4-Trimethyl-1-pentene, detm. with MBTH, 45
  oxidn., 14
  precursor of CH_2O, 14
  spectral data, 45
N,N,3′-Trimethyl 4-phenylazoaniline, carc. of subcarc. doses, 182
  effect of hepatectomy, 182
sym-Trimethylphosphoramide, formn. from hempa, 155
Trinitromethane, 160
Triolein, standard, 262, 263
sym-Trioxane, depolymerization, 215, 216
  detm. with, CA, 45
    J-acid, 45
    phenyl J-acid, 45
  formald. std., 215
  formn. from CH_2O, 206
  hydrol., 14
  physical properties, 215
  precursor of CH_2O, 14, 215
  presence in formalin, 212
  structure, 215
1,2,3-Trioxolanes, formn., from alkenes, 247
  oxidants, 247
  precursor of, aldehydes, 247
    1,2,4-trioxolanes, 247
1,2,4-Trioxolanes, *see* ozonides
Tripalmitin, standard, 92, 256, 257
TRIS, 180
Tris buffer, 154
Tris HCl, 268
2,4,6-Trisdimethylamino-s-triazine, *see* Hexamethylmelamine
Tristearin, standard, 258
sym-Trithiane, detm, with, CA, 45, 282
    J-acid, 45
    phenyl J-acid, 45
  hydrolysis, 14, 282
  precursor of CH_2O, 14, 20, 282
  spectral data, 45
Trithion, detm. with CA, 46
  hydrolysis, 14
  precursor of CH_2O, 14, 229
  structure, 17, 230

Triton X-100, effect on N-demethylase, 143, 146
Triton B-1956, effect on N-demethylase, 143
Triton CF-10, effect on N-demethylase, 143
Triton CF-32, effect on N-demethylase, 143
Triton DF-12, effect on N-demethylase, 143
Triton N-57, effect on N-demethylase, 143
Triton N-101, effect on N-demethylase, 143
Triton QS-15, effect on N-demethylase, 143
Triton X-67, effect on N-demethylase, 143
Tropine alkaloids, N-oxidn., 155
Tropine N-oxide, adduct formn., 14
  adduct with SO_2, 163
    hydrol. to CH_2O, 163
  anal. with 2,4-pentanedione, 14, 17
  hydrolysis, 14
  precursor of CH_2O, 14, 17, 20
  yield of CH_2O, 14, 17
Trout, carc. of Et_2NNO, 68
    Me_2NNO, 68
Trypsin, 11, 14, 138
  detm. in blood plasma, 281
  detm. with 2,4-pentanedione, 281
Tryptamine, conversion to tryptoline, 238
  fluor, anal., 242
  found in brains of, dog, 242
    man, 242
    steer, 242
  reaction with CH_2O, 237, 242
Tryptamines, reaction with, formald., 239
  5-methyltetrahydrofolic acid, 238
Tryptoline, formn. from tryptamine, 238
  structure, 238
Tryptophan, anal, eqn., 242
  catabolism, 190
  detm. in, human liver, 242
    plasma, 242
    urine, 242
  detm. of CH_2O, 177
  2,3-dioxygenase, 190
  formn. of, harman, 242
    norharman, 242
  metab. to, 2,6-dioxo-3-carboxyhexanoic acid, 191
    L-formylkynurenine, 190
    3-hydroxyanthranilic acid, 191
    3-hydroxykynurenine, 191
    kynurenic acid, 190
    kynurenine, 190
    methoxykynurenic acid, 191
    nicotinic acid, 191
    quinolinic acid, 191
  reaction with, aldehydes, 242
    $HOCH_2NR_2$, 226
  sensitivity, 177
  structure, 242
Tryptophanyl-dipeptide, reaction with CH_2O, 241
  ring-closure eqn., 241
  structure, 241
L-Tryptophanyl-L-glycine, Fexc/em, 242
Tryptophanyl peptides, detm. in, liver, 242
    plasma, 242
  fluor. dctm., 242
  reaction with CH_2O, 242
Tumor, rat brain, 54
Tumor etiology, genetic factors, 76
Tumor incidence, mice, 74
  rats, 74
Tumor initiation, β-propiolactone, 251
Tumor promotion, β-propiolactone, 251
Tumors, bladder, 68
  esophagus, 68
  fore-stomach, 68
  intestinal, 55
  kidney, 55, 68
  liver, 55, 68
  lung, 68
  nasal sinus, 68
  ovary, 68
Tungstate, pptn of protein, 85
m-Tyramine, anal. eqn., 240
  histochem. detm., 240
  structure, 240
Tyrosine, reaction with $HOCH_2NR_2$, 226
Tyrosinol, detm. with CA, 44

Ulticarcinogen, acetoxymethyl methyl nitrosamine, 73, 79
  benzenediazonium cation, 185
  benzo(a)pyrene 4,5-oxide, 81
  epoxides, 81
  N-formyl N-methylnitrosamine, 73, 79
  malonaldehyde, 179
  methylacyloxynitrosamine, 179
  methyl cation, 73, 79, 125, 161, 232
  methylformylnitrosamine, 179
  monomethylnitrosamine, 179
  phenyldiazonium cation, 126
  vinyl chloride, 271
Ultrasonic syntheses, 277
Ultraviolet light, dec. of nitrosamines, 79
  mutation enhancement, 81
10-Undecanoic acid, anal. with chromotropic acid, 15, 17
  oxidn., 15
  precursor of CH_2O, 15, 17
  yield of CH_2O, 15, 17
Uracils, reaction with CH_2O, 211
Urate oxidase, 268

Urea, 97, 217, 272
  detm. with CA, 282
  formald. condensate, acid hydrol., 282
  formald. resins, 217
  form. from, CO, 275
    vinyl chloride, 272
  precursor of CH_2O, 282
Ureidoglycolate, detm. with phenylhydrazine, 113
  precursor of, formald., 113
    glyoxylic acid, 113
  reaction eqn., 113
  structure, 113
Uric acid, anal. eqn., 268
  anal. interf., 244
  anal. proc., 268
  detm. with, 2,4-pentanedione, 268
    MBTH, 268
    methanol, 139
  precursor of allantoin, 268
Uricase, 268
Urinary pathogens, 223
Urinary tract, formn. of nitrosamines, 66
Urine, acetylmethadol metabolites, 155
  detm. of, arabitol, 278
    glucaric acid, 89
    glyceric acid, 279
    glycine, 49, 95–98
    glycolate, 114
    oxalate, 114, 244–246
    oxalic acid, 281
    polyols, 281
    sorbitol, 281
    tryptophan, 242
  DIC metabolites, 125
  dimethylnitrosamine, 179
  hydroxyindoleacetic acid, 223
  isolation of aflatoxin P, 188
Urobilinogen, detm. with p-$MeOC_6H_4N_2BF_4$, 56
  oxidn, 15
  precursor of CH_2O, 15, 56
  reaction with ArN_2^+, 15
  structure, 56
Urocanase, 1-carbon metab., 152
Uterine cervix, cancer, 66
  dimethylnitrosamine, 66
UV abs. spectra, 6-hydroxymethyl BaP, 18
UV-Schiff reaction, 222, 254
  interference, 223
UV irradiation, 23

Vaccines, sterilization, 251
Vagina, adenocarcinoma, 172
  dimethylnitrosamine, 66
n-Valeraldehyde, detection with 2,3-dimethyl-2,3-bis(hydroxylamino)butane, 202
  rate of reaction, 203
  spectral data, 203
Valine, effect on d-AMP, 210
Valinol, detm. with chromotropic acid, 44
Vanadate, oxidant, 12
Vanillin, form. from eugenol, 31
    1-(3-methoxy-4-hydroxyphenyl)propen-3-ol, 133
  precursor of CH_2O, 16
  structure, 31, 133
Vicia faba, effect of clastinogen, 251
  mutated by β-propiolactone, 251
Vincristine, detm. with CA, 89
  precursor of CH_2O, 89
Vinyl acetate, oxidn., 15
  precursor of CH_2O, 15
Vinyl butyl ether, collection, 275
  detm. in air, 275
  detm. with CA, 275
  oxidn., 15
  precursor of CH_2O, 15, 275
Vinyl chloride, 269–274
  atm. oxidn., 269, 270
  biotransformation, 272
  carcinogen, 273, 274
  clastogen, 273
  collection, 274, 281
  detm. in air, 281
  detm. with, CA, 45, 274, 281
    GC, 274
    GC-MS-COMP, 274
  effect on N-demethylation, 272
  human exposure, 269
  metabolism, 271, 272
  mutagen, 272, 273
  oxidn. with periodate, 15
  photooxidn., 269, 270
  physical properties, 269
  precursor of, carbon monoxide, 270, 271
    chloroacetald, 272
    chloroacetic acid, 272
    formald., 15, 269–272, 274, 281
    formic acid, 270, 271
    formyl chloride, 270
  prodn., 269
  reaction, with NO_2, 270
    ozone, 270
  relative, mutagenic activity of, human liver, 273
    mouse kidney, 273
    mouse lung, 273
    rat kidney, 273
    rat liver, 273
    rat lung, 273

teratogen, 274
toxicity, 273
transplacental effects, 274
tumors in, hamsters, 274
humans, 274
mice, 274
rats, 274
usage, 269
UV irradiation, 269, 270
workers, 272
(^{14}C)Vinyl chloride, eliminative route, 272
Vinyl compounds, 274, 275
detm. with, CA, 31
dimedone, 31
ethyl acetoacetate, 30
J-acid, 31
MBTH, 31
2,4-pentanedione, 31
oxidn., 30–32
presence in industrial air, 274
Vinylcyclohexene-3 diepoxide, carcinogen, 81
mutagen, 81
Vinylidene chloride, carc., 62
hepatotoxic, 62
mutagenic, 62
photooxidn., 62, 63
precursor of, carbon monoxide, 62, 63
chloroacetyl chloride, 62, 63
formald., 62, 63
formic acid., 62, 63
phosgene, 62, 63
nitric acid, 62, 63
5-Vinyl-2-picoline, collection, 275
detm. in air, 275
detm. with CA, 275
oxidn., 15
precursor of CH_2O, 15, 275
2-Vinylpyridine, collection, 275
detm. in air, 275
detm. with CA, 275
oxidation, 15
precursor of CH_2O, 15, 275
Virus, mutated by CH_2O, 233
Vitamin B_{12}, 182
effect on carc., 143
Vitamin E, effect on, aminopyrine N-demethylase, 144
demethylation, 168
Vulcanizate, detect. of hexamine, 119
Vulcanizates, detect. of hexamine, 224
detm. of hexamine, 225

Water, detm. of chlorophenoxyacetic acids, 278
nitrosamines, formn., 64
stability, 64
Well waters, nitrosamines, 65
Wheat flour, nitrosamines, 63
Wines, detm. of, formic acid, 279
glycerol, 279
methanol, 280
Wistar rat, effect of pregnancy on metab., 189
Wood-particle board, detm. of CH_2O in, 200
see particle board
Wood shavings, 218
Wool, reaction with CH_2O, 225

Xenobiotics, effect of enzymes, 142
genotoxic pathways, 175
metabolism, 175
Xylene, photooxidn., 18
precursor of, aliph. ald., 18
formald., 18
yield of, aliph. ald., 18
formald., 18
m-Xylene, photooxidin., 15, 26, 28, 243
precursor of CH_2O, 15, 26, 28, 243
yield of CH_2O, 26, 28
D-Xylitol, detm. with, dimedone, 45
J-acid, 45
MBTH, 45
2,4-pentanedione, 45, 115, 116
fluor. anal., 45
location, 109
oxidn., 15
precursor of CH_2O, 15, 109, 115, 116
sepn., 115, 116, 118
spectral data, 45
Xylobionic acid, 275
detm. with, carbazole, 276
chromic acid, 276
sepn., 276
Xylonic acid, detm. with, chromic acid, 276
periodate, 276
precursor of CH_2O, 275, 276
sepn., 275, 276
structure, 275
D-Xylopyranosides, sepn., 116
D-Xylose-D-xylitol, β-(1 → 4)-linked, 118
d-Xylose, detm. with CA, 45
oxidn., 15, 253
precursor of CH_2O, 15, 253
spectral data, 45
Xylotrionic acid, 275
detm. with, carbazole, 276
chromic acid, 276
sepn., 276
4-O-β-Xylp-D-xylitol, distribution coeff., 116
precursor of, formald., 116, 117

4-O-β-Xylp-D-xylitol—*continued*
furfurals, 116, 117
sepn., 116, 117

Yeast, growth with methanol, 138

Zectran, N-demethylation, 150
precursor of CH_2O, 150
Zeolite, phospholipid removal, 264, 267
Zinc, reductant, 12, 245
Zinc bromide, 250
Zinc chloride, 58, 70
Zinc dust, reductant, 114
Zinc sulfate, 135, 184

THE